METHANE FROM BIOMASS:
A SYSTEMS APPROACH

METHANE FROM BIOMASS: A SYSTEMS APPROACH

Edited by

WAYNE H. SMITH
*Center for Biomass Energy Systems,
Institute of Food and Agricultural Sciences,
University of Florida,
Gainesville, Florida 32611, USA*

and

JAMES R. FRANK
*Physical Sciences Department—Biology,
Gas Research Institute,
8600 West Bryn Mawr Ave,
Chicago, Illinois 60631, USA*

With Foreword by

PHILIP H. ABELSON
*Former Editor, Science;
now Deputy Editor for
Engineering and Applied Sciences
and Scholar in Residence at
Resources For The Future*

ELSEVIER APPLIED SCIENCE
LONDON and NEW YORK

ELSEVIER APPLIED SCIENCE PUBLISHERS LTD
Crown House, Linton Road, Barking, Essex IG11 8JU, England

Sole Distributor in the USA and Canada
ELSEVIER SCIENCE PUBLISHING CO., INC.
52 Vanderbilt Avenue, New York, NY 10017, USA

WITH 98 TABLES AND 95 ILLUSTRATIONS

© ELSEVIER APPLIED SCIENCE PUBLISHERS LTD 1988

British Library Cataloguing in Publication Data

Methane from biomass: a systems approach.
  1. Methane   2. Biomass energy
  I. Smith, Wayne H.   II. Frank, James R.
  665.7'7   TP761.M4

Library of Congress Cataloging-in-Publication Data

Methane from biomass.

  Bibliography: p.
  Includes index.
  1. Biogas. I. Smith, Wayne H. II. Frank, James R.
TP359.B48M48   1987   665.8'9   87-9120
ISBN 1-85166-102-6

No responsibility is assumed by the Publisher for any injury and/or damage to persons or property as a matter of products liability, negligence or otherwise, or from any use or operation of any methods, products, instructions or ideas contained in the material herein.

Special regulations for readers in the USA
This publication has been registered with the Copyright Clearance Center Inc. (CCC), Salem, Massachusetts. Information can be obtained from the CCC about conditions under which photocopies of parts of this publication may be made in the USA. All other copyright questions, including photocopying outside of the USA, should be referred to the publisher.

All rights reserved. No parts of this publication may be reproduced, stored in a retrieval system, or transmitted in any form or by any means, electronic, mechanical, photocopying, recording, or otherwise, without the prior written permission of the publisher.

Phototypesetting by Keyset Composition, Colchester, Essex
Printed in Great Britain at the University Press, Cambridge

Dedication

Two persons played significant roles in bringing together our respective institutions and establishing the framework for the editors to use in designing this program of research. They counseled and encouraged us while serving as the first Program Directors for the jointly planned, funded, and managed effort between an industry research sponsor and a university research performer. For their signal performance this book is dedicated to those two significant individuals.

1. Ab Flowers

Dr Ab Flowers has been a major 'force' behind the development of biomass energy technology by the natural gas industry over the past decade. Though trained as a metallurgist, Ab Flowers was quick to understand the potential applications of biotechnology to the natural gas industry. Before his career in the natural gas industry, he was educated in Chemical Engineering and Metallurgy at Auburn University and the Universities of Alabama and Tennessee where he taught chemical engineering and metallurgy for 5 years. He then spent 21 years in the metals field with Crucible Steel Company of America, Combustion Engineering, Inc., and Republic Steel Corporation. His work involved carbon, steels and alloys, stainless steel technology, and developing special fabrication methods for power boilers. He holds several patents in the area of titanium production and finishing technology.

Upon joining the American Gas Association (AGA) in 1966, Ab Flowers was responsible for residential, commercial, and industrial utilization research and coal gasification research. From 1971 to 1978 Dr Flowers was Director, Gas Supply Research, and was instrumental in developing and implementing the $360 million Joint Coal Gasification Research

Program with the US Government. He was responsible for initiating and running the Synthetic Pipeline-Gas Symposium from its inception in 1966. He also developed and implemented the gas industry's materials of construction program dedicated to screening, testing and developing metallic and ceramic material to withstand the hostile environments of high temperature, high pressure, and high-sulfur-content, second generation coal gasification technology. He also had staff responsibility for environmental concerns in coal gasification and interfacing between the gas industry and government. Dr Flowers was also instrumental in developing and implementing the Biomass Research Program at AGA. Despite his responsibilities and his involvement with diverse technologies, the biomass program was one of his most consuming interests and one to which he was most dedicated. His intense interest in this technology led to the biomass program becoming a major research activity in the gas supply area at the Gas Research Institute (GRI) when he assumed the role of Director, Gas Supply, in 1978 following transfer of gas research from AGA to the newly formed GRI. The GRI program evolved into a program which included research on marine, land, and freshwater biomass resources as well as wastes and conversion of these resources to fuel gas.

Ab Flowers pioneered the concept of using biomass as a major substitute for natural gas in the early 1970s at a time when few in the gas pipeline industry had any inkling of the future promise of biotechnology. Characteristically, Dr Flowers promoted the highly ambitious and often controversial marine biomass concept for methane production. Despite many difficulties and false starts, much of the work in this initial program laid the foundations for the development of the programs and systems described in this book. Dr Flowers also played a crucial part in the landmark project currently located at Walt Disney World. His vision and drive were crucial ingredients in the successes which have been achieved and it is to his credit that the gas industry has persevered and made significant progress in the production of methane from biomass despite the considerable uncertainty generated by the energy shortage and surplus cycles experienced in the past 15 years.

Currently, Dr Flowers is an Executive Scientist at GRI, a position he has held since 1981, where his assignment is developing natural gas, LNG, coal and oil shale gasification and renewable sources of fuels. One of his major responsibilities is serving on GRI's Senior Research Council which evaluates the complete GRI R&D program. He is a registered Professional Engineer in the State of Tennessee, Life Member and Fellow of the American Society for Metals, and a member of numerous boards and associations.

The editors feel that without Ab Flowers' strength of character, support, and guidance during difficult periods much of the progress made in this program would not have been possible.

2. F. Aloysius Wood

The late F. Aloysius 'Al' Wood (17 November 1932–22 August 1985) was an innovator throughout his career as an academician. Following his BS in forestry and MS in botany at the University of Missouri, Al Wood earned his PhD from the University of Minnesota. Dr Wood began pioneering new programs as an assistant professor at the Pennsylvania State University where he made major contributions that established the foundations for epidemiological research. He successfully applied these 'systems concepts' to the complex effects of air pollutants on plant systems. Later, when he became head of the Department of Plant Pathology at the University of Minnesota, he encouraged the concept of the 'Doctor of Plant Medicine', a plant health technology program that was attractive to students and of great benefit to commercial plant groups and the homeowners who prize an environment comprised of healthy plants. All new programs developed or encouraged by Dr Wood recognized the need for interdepartmental cooperation and multidisciplinary participation in order to solve complex problems. Dr Wood was an early advocate of the benefits of biotechnology to agriculture. When he became Dean for Research at the University of Florida's Institute of Food and Agricultural Sciences (IFAS) in 1977, he chose biotechnology in agricultural sciences as the 'centerpiece' of his program development initiatives. Applications of this emerging opportunity were central to the 'Low Energy Technology Program' conceived by IFAS and funded by the State of Florida in 1979. In addition to encouraging the advances in molecular biology and genetic engineering to improve crop yield and organismal tolerances to stresses, he also recognized the importance of rhizospheric microbial associations of plants and the opportunity to use conventional and emerging biotechnologies to develop new crops with new uses such as biomass for energy. Regionally and nationally he was similarly aggressive in advancing biotechnology. He was the force that led to the Southern regional project 'Molecular and Cellular Genetics for Crop Improvement'. Nationally, he chaired the Committee on Biotechnology for the National Association of State Universities and Land-Grant Colleges that developed a national biotechnology plan. He then took leave from the University of Florida to work with others in obtaining substantial competitive funding for the continuing program of biotechnology research. Subsequently, Dr Wood became Director of Biotechnology in IFAS. He gave positive support and strong encouragement to the GRI/IFAS Methane from Biomass Program. He saw biomass as a new research dimension in agriculture and as an opportunity to involve the new biotechnologies at the outset in programming a new initiative. We are thankful for that support and that he encouraged a program structure that would accommodate a long term science-driven strategy making use of modern instrumentation (mass and nuclear magnetic spectroscopy, electron miscroscopy, etc.) and contemporary biotechnologies. Because of Dean Al Wood, we believe the GRI/IFAS program is positioned to make continued progress.

Acknowledgements

The editors, who served as program managers, wish to acknowledge many of the people who have worked to make this joint industry/university program successful. We would particularly like to thank Professor Paul Smith who sparked the idea for this program, Stephen Ban (President and CEO), GRI Senior Vice-Presidents William Burnett and Robert Rosenberg, and IFAS Vice-President Kenneth Tefertiller, who provided senior management oversight and sound guidance for launching and sustaining the program. We were also greatly aided by Professors John Gander, Gerald Isaacs, and Paul Smith of IFAS and Dr Thomas Hayes of GRI, who as members of our Technical Management Team provided scientific advice and managerial assistance. Thanks also go to Drs Kermit Woodcock, Peter Benson, Ronald Isaacson, and Kimon Bird, GRI program staff, for their advice and support during the development of this program. Special appreciation also goes to William Boyland (Chair) and John Duda (Co-Chair), of the Florida Gas and Agricultural Industry Review Panel. Other industry members of the panel who provided both good advice and in-state program assistance over the years included Kenneth Bailey, William Boardman, Donald Napier, Frank Ray, Harry Stout, Earl Talbot, and Harry Vaughn, along with *ex officio* members Drs Jay Hakes and Katie Tucker of the Florida Governor's Energy Office. There were also many other advisors, too numerous to mention, from the gas industry and academic institutions who have helped to keep us on track. We would particularly like to thank Robert Christopher, who was a strong supporter and biomass innovator in the natural gas industry. Without the strong technical support and active participation of major scientists in a variety of fields this program would not be at its present exciting stage. The chapters in this volume are testimony to the multidisciplinary cooperation and the diligence of participating scientists in this program. Thomas Hayes and Joseph Cinereski have provided continuous, valuable program support and coordination assistance.

Finally, we, the editors, would like to thank our wives, Midge Smith and Kathleen Frank, for their interest, tolerance, and encouragement during these several years. Our office staffs who have helped us keep up with the considerable work generated by this program are due our thanks, especially B. Stanek and C. McRae who have coordinated work flow between our offices. Special appreciation goes to Wendy Shuler, who played a major role in every step in the development of this book. She worked tirelessly with the several authors, the many peer reviewers and the editors in developing the final typed, reviewed, revised and edited manuscripts. Proofreading, typing, and other assistance provided by Julia Andrew, Julie Barickman and Caprice McRae is gratefully acknowledged. Without the able assistance of everyone named here, and many others, this program would not have accrued the progress and scientific accomplishment reflected in this book.

WAYNE H. SMITH
JAMES R. FRANK

Foreword

It has long been known that methane can be produced by anaerobic fermentation, but until recent years, little effort was made to create the highly efficient, dependable technology required for the commercial application of this knowledge. This volume describes substantial progress toward the goal of large-scale conversion of biomass to inexpensive methane.

In 1981, the Gas Research Institute (GRI) of Chicago and the Institute of Food and Agricultural Sciences (IFAS) in Florida began a collaborative study, reported on herein, of the various aspects of converting biomass to methane. At first, the program consisted mainly of efforts to identify crops that would produce the largest yields of biomass. Later, however, it became apparent that even if excellent yields were achieved, more fundamental knowledge of the fermentation process would be necessary to reach cost goals. Further analysis led to the creation of a computer-oriented model that permits studying the effects of the many variables involved in the total production system.

The computer calculations show that using the cropping and technology capabilities of 1981, the cost of 1000 cubic feet of methane—roughly 1 GJ energy—was about $9.20. As a result of efforts from 1981 to 1985, this figure was reduced to $6.50. The research group has set for itself a goal of $3.00 by the year 2000. On the basis of what has been already achieved and in view of new technology to be exploited, this goal seems possible.

As of December 1986, natural gas was selling in the United States for about $1.50 mcf^{-1} at the wellhead. This price reflected a depressed market influenced by the low cost of crude oil. Prospects are that in a few years, the US gas bubble will disappear and that oil prices also will move upward. As early as 1990, the wellhead price for gas could be above $3.00 mcf^{-1}. At places distant from producing wells the cost of natural gas would be substantially higher.

In the longer term, it is inevitable that the world will make expanded use of renewable energy sources as reserves of oil and natural gas are depleted. Coal will be important as an energy source, but in areas where there is no coal and where biomass is available, the latter will be important. In addition, there continues to be the specter of the CO_2 greenhouse effect. We have seen adverse public opinion lead to a severe curtailment of the use of nuclear energy. An event in which deleterious consequences of the greenhouse effect were strikingly demonstrated could result in a public demand for sharply-expanded use of bio-energy. A factor favoring methane is research already carried out for GRI that has resulted in highly efficient water heaters, heat pumps, and other appliances. Fuel cells for production of electricity are proving increasingly practical and competitive.

Until recently little effort was devoted to crops for energy. However, the successful breeding and selection techniques that have led to increased yields of agricultural crops for food, feed, and fiber can be expected to lead to comparable improvements in energy crops. Moreover, the new biotechnology is already facilitating advances in this direction.

Early in the program, the managers chose to emphasize work on Napiergrass, sorghum, sugarcane, and water hyacinth. The first three are grasses. Among the various species there is considerable variation, and it is feasible to select superior specimens for propagation. During the period 1981–86, biomass yields were nearly doubled to 37–50 Mg ha^{-1} yr^{-1} of dry solids. As much as 112 dry Mg ha^{-1} yr^{-1} were obtained with water hyacinth. At the same time, improvements were made in propagation of the grasses by means of tissue culture. Napiergrass does not ordinarily produce fertile seeds. The usual method of propagation is to plant individual pieces of stalk. A tiny leaf explant of Napiergrass in tissue culture gave rise to 93 plants after 8 weeks, and to 24 600 after 26 weeks. This represents a fast and practical way to achieve rapid propagation of a superior plant. In the field, plants derived from tissue culture performed better than those produced by the older technique. Tissue culture propagation also permits selection of mutants that have such desired properties as tolerance to saline conditions or resistance to low temperatures. Feasibility of widespread intergeneric somatic hybridization among the grasses has also been looked into. The efforts have not been fully successful, but through use of protoplast fusion, an encouraging beginning has been made.

The optimum cultivar for a given area will depend in part on future advances in plant biotechnology and in part on the basic characteristics of the various plants. For example, sorghum is an annual that can be grown over a large fraction of the surface of the earth, including, of course, the temperate zone. The roots of Napiergrass, a perennial, cannot tolerate freezing temperatures, and it thrives best in semitropical climates. For the tropics, sugarcane appears to be superior to other grasses in achieving large total biomass and high yields of sugar. Water hyacinth produces high yields but requires fresh water and does not grow at temperatures below 15°C.

Energy from biomass can be obtained by simple combustion, by fermentation, or by thermal gasification. The editors of this volume cite a number of advantages of biogasification. One is the ability to handle wet or dry feedstocks. A second is the high quality and purity of the gas produced. A third is the flexibility in the size of gasification facilities.

As of December 1986, the principal obstacles to achieving low-cost biomethane lay in chemical conversions taking place in the biogasification process. The grasses yielding the best biomass contain a substantial fraction of the complex polymers cellulose and lignocellulose. Lignocellulose is poorly degraded into simple sugars by ruminant animals and both polymers are difficult to convert to simple sugars *in vitro*. Dealing with this problem is usually the rate-limiting process. Conversion of sugars to volatile fatty acids and from thence to methane is comparatively straightforward, but the process requires a mixed culture of microflora with conflicting pH optima. That is, some acid formers prefer a low pH, while the methanogens thrive best at about pH 7.

In the past, the usual methods in attaining biogasification have been to provide anaerobic conditions and to permit chance and nature to establish a mixed culture. When inocula were used, the practice often was to obtain anaerobes from sewage or other anaerobic environments such as the rumen. Efforts to identify the components of the anaerobic microbial populations have disclosed as many as 100 species. Thus improving on nature is a complex undertaking. Research described in this volume provides a solid beginning to this task.

It was noted that in some typical mixed cultures only 0·1% of the organisms were capable of hydrolyzing cellulose. Thus the slow degradation of cellulose might in principle be avoided by using inocula enriched in cellulose-attacking organisms or by supplementing existing populations. Efforts to identify and isolate bacteria capable of superior performance in conducting other steps leading to methane are described in this volume. It seems likely that ultimately a superior collection of anaerobes will be identified and cultured. These may in turn be improved by recombinant DNA techniques. The various anaerobes differ in their phospholipids. Spectrophotometric determinations can detect the populations and provide an estimate of their components. Fourier Transform Infrared Spectroscopy is also a useful approach.

One hard-won lesson was recognition of the need to insure adequate supplies of certain trace elements. When Napiergrass was fermented, it was noted that there was pile-up of volatile fatty acids and very little production of methane. After tiny amounts of molybdenum, cobalt, nickel and selenium ions were added, volumetric productivity of methane improved fivefold.

Because the various steps in biogasification take place at different rates and have different pH optima, current practice is to utilize at least two sets of reactors—leaching bed reactors in which the initial degradation of biomass occurs, and packed bed reactors in which organic acids are converted to methane. The computer modeling described in this book analyzes production on the basis of the two types of reactors.

A goal of the collaborative GRI/IFAS program is to produce pipeline quality gas. This means attaining a product with at least 98% methane and thus no more than 2% CO_2. The usual gasification of biomass yields 60% methane and 40% CO_2. For many purposes such as nearby use as an energy source this product is satisfactory. However, if the gas is to be introduced into a large pipeline system, the CO_2 must be removed. This can be accomplished in large measure by proper engineering of the flows of liquids between the leaching bed and the packed bed reactors. Organic acids from the leaching bed can be used to lower the pH of effluents from the packed beds to release CO_2 while allowing only small amounts of methane to escape.

There is much work to be done to arrive at optimum feedstocks, microbial populations, and conditions in the medium. As this research progresses, it is likely to produce results helpful to many others, both in the United States and in other countries where biogasification is being employed. Already large numbers of biogas generators are in use in China, India and elsewhere. Better understanding of biogasification should lead to more effective production. In addition, the potential for rapid growth of biomass in the moist tropics might well be harnessed to provide energy needs for large populations.

PHILIP H. ABELSON

Contents

Dedication	v
Acknowledgements	ix
Foreword	xi
List of Contributors	xix

1. Introduction to Methane from Biomass: A Systems Approach 1
   J. R. FRANK and W. H. SMITH

2. Information Management 15
   K. M. PORTIER and J. E. CINERESKI

3. Model and Analysis of Biomass to Methane 21
   J. W. MISHOE, C. F. KIKER, R. C. FLUCK, W. G. BOGGESS, S. L. CURRY and M. E. WAELDER

4. Conceptual Design of a Commercial Biomass-to-Methane System 49
   T. D. HAYES, C. S. WARREN and S. W. HINTON

5. Energy Crop Development 79
   W. H. SMITH and J. R. FRANK

6. Model Crop Systems: Sorghum, Napiergrass 83
   G. M. PRINE, L. S. DUNAVIN, B. J. BRECKE, R. L. STANLEY, P. MISLEVY, R. S. KALMBACHER and D. R. HENSEL

7. Water Hyacinth (*Eichhornia crassipes* (Mart) Solms) Biomass Cropping Systems: I. Production 103
   K. R. REDDY

8. Water Hyacinth (*Eichhornia crassipes* (Mart) Solms) Biomass Cropping Systems: II. Harvesting and Handling . . . 141
   L. O. BAGNALL

9. Tissue Culture of Gramineous Biomass Species . . . 155
   K. RAJASEKARAN, I. K. VASIL and S. C. SCHANK

10. Breeding Grasses for Improved Biomass Properties for Energy 169
    S. C. SCHANK and L. S. DUNAVIN

11. Development of Artificial Seeds of Sweet Potato for Clonal Propagation through Somatic Embryogenesis . . . 183
    D. J. CANTLIFFE, J. R. LIU and J. R. SCHULTHEIS

12. Nitrogen Fixation in *Sargassum* Biomass Production Systems 197
    K. T. SHANMUGAM, H. SPILLER and E. J. PHLIPS

13. Manipulation of *Sargassum* for Biomass Production . . 211
    J. F. PRESTON III, T. ROMEO, A. GIBOR and M. POLNE-FULLER

14. Alternative Production Systems: Root and Other Herbacous Crops 235
    S. K. O'HAIR, G. H. SNYDER, J. M. WHITE, S. M. OLSON and L. S. DUNAVIN

15. Alternative Production Systems: Marine Crops . . . 249
    K. BIRD, B. LAPOINTE, D. HANISAK, J. RYTHER and C. DAWES

16. Alternative Production Systems: Nonconventional Herbaceous Species 261
    P. MISLEVY, J. P. GILREATH, G. M. PRINE and L. S. DUNAVIN

17. Alternative Production Systems: Woody Crops D. L. ROCKWOOD and G. M. PRINE	277
18. Biological Production of Methane from Biomass P. H. SMITH, F. M. BORDEAUX, M. GOTO, A. SHIRALIPOUR, A. WILKIE, J. F. ANDREWS, S. IDE and M. W. BARNETT	291
19. Microbial Aspects of Biogas Production P. H. SMITH, F. M. BORDEAUX, A. WILKIE, J. YANG, D. BOONE, R. A. MAH, D. CHYNOWETH and D. JERGER	335
20. Chemical Characteristics and their Relation to Fermentability of Potential Biomass Feedstocks K. A. BJORNDAL and J. E. MOORE	355
21. Cellulase Enzymes for Enhancement of Methane Production from Biomass M. GRITZALI, A. SHIRALIPOUR and R. D. BROWN Jr	367
22. Ultrastructural Analyses of Methanogens H. C. ALDRICH, D. S. WILLIAMS and R. W. ROBINSON	385
23. Properties that Affect Aggregation and Dispersion of Methanogens A. S. BLEIWEIS, H. C. ALDRICH and R. A. MAH	397
24. Phospholipids to Monitor Microbial Ecology in Anaerobic Digesters A. T. MIKELL Jr, T. J. PHELPS and D. C. WHITE	413
25. Fixed Film Reactors: Packing Media, Mathematical Models and Bulk Flow in Packed Beds R. A. NORDSTEDT, M. V. THOMAS and C. Y. CHOU	431
26. Anaerobic Digester Residues: Treatment and Utilization D. A. GRAETZ and K. R. REDDY	445
27. Perspectives on Biomass Research J. R. FRANK and W. H. SMITH	455

Appendix A. List of GRI/IFAS Collaborative Program Publications, 1981–1986 465

Appendix B. Conversion Table: SI to Imperial Units . . . 484

Index 485

List of Contributors

H. C. ALDRICH
  Microbiology and Cell Science Department, University of Florida—IFAS, Gainesville, Florida 32611, USA

J. F. ANDREWS
  Environmental Science and Engineering Department, Rice University, Houston, Texas 77251, USA

L. O. BAGNALL
  Agricultural Engineering Department, University of Florida—IFAS, Gainesville, Florida 32611, USA

M. W. BARNETT
  Environmental Science and Engineering Department, Rice University, Houston, Texas 77251, USA

K. BIRD
  Harbor Branch Oceanographic Institution, Fort Pierce, Florida 33450, USA

K. A. BJORNDAL
  Animal Science Department, University of Florida—IFAS, Gainesville, Florida 32611, USA

A. S. BLEIWEIS
  Microbiology and Cell Science Department, University of Florida—IFAS, Gainesville, Florida 32611, USA

W. G. BOGGESS
Agricultural Engineering Department, University of Florida—IFAS, Gainesville, Florida 32611, USA

D. BOONE
Division of Environmental and Occupational Health Sciences, School of Public Health, University of California at Los Angeles, Los Angeles, California 90024, USA

F. M. BORDEAUX
Microbiology and Cell Science Department, University of Florida—IFAS, Gainesville, Florida 32611, USA

B. J. BRECKE
Agricultural Research and Education Center, University of Florida—IFAS, Jay, Florida 32565, USA

R. D. BROWN Jr
Food Science and Human Nutrition Department, University of Florida—IFAS, Gainesville, Florida 32611, USA

D. J. CANTLIFFE
Vegetable Crops Department, University of Florida—IFAS, Gainesville, Florida 32611, USA

C. Y. CHOU
Agricultural Engineering Department, University of Florida—IFAS, Gainesville, Florida 32611, USA

D. CHYNOWETH
Agricultural Engineering Department, University of Florida—IFAS, Gainesville, Florida 32611, USA

J. E. CINERESKI
Center for Biomass Energy Systems, University of Florida—IFAS, Gainesville, Florida 32611, USA

S. L. CURRY
Agricultural Engineering Department, University of Florida—IFAS, Gainesville, Florida 32611, USA

C. DAWES
  Biology Department, University of South Florida, Tampa, Florida 33620, USA

L. S. DUNAVIN
  Agricultural Research and Education Center, University of Florida—IFAS, Jay, Florida 32565, USA

R. C. FLUCK
  Agricultural Engineering Department, University of Florida—IFAS, Gainesville, Florida 32611, USA

J. R. FRANK
  Gas Research Institute, 8600 W. Bryn Mawr Ave, Chicago, Illinois 60631, USA

A. GIBOR
  Biology Department, University of California, Santa Barbara, California 93106, USA

J. P. GILREATH
  Gulf Coast Research and Education Center, University of Florida—IFAS, Bradenton, Florida 33508, USA

M. GOTO
  Microbiology and Cell Science Department, University of Florida—IFAS, Gainesville, Florida 32611, USA

D. A. GRAETZ
  Soil Science Department, University of Florida—IFAS, Gainesville, Florida 32611, USA

M. GRITZALI
  Food Science and Human Nutrition Department, University of Florida—IFAS, Gainesville, Florida 32611, USA

D. HANISAK
  Harbor Branch Oceanographic Institution, Fort Pierce, Florida 33450, USA

T. D. HAYES
: Gas Research Institute, 8600 W. Bryn Mawr Ave, Chicago, Illinois 60631, USA

D. R. HENSEL
: Agricultural Research and Education Center, University of Florida—IFAS, Hastings, Florida 32045, USA

S. W. HINTON
: Reynolds, Smith and Hills Architects-Engineers-Planners Inc., Jacksonville, Florida 32201, USA

S. IDE
: Environmental Science and Engineering Department, Rice University, Houston, Texas 77251, USA

D. JERGER
: Agricultural Engineering Department, University of Florida—IFAS, Gainesville, Florida 32611, USA

R. S. KALMBACHER
: Agricultural Research and Education Center, University of Florida—IFAS, Ona, Florida 33865, USA

C. F. KIKER
: Agricultural Engineering Department, University of Florida—IFAS, Gainesville, Florida 32611, USA

B. LAPOINTE
: Harbor Branch Oceanographic Institution, Fort Pierce, Florida 33450, USA

J. R. LIU
: Vegetable Crops Department, University of Florida—IFAS, Gainesville, Florida 32611, USA

R. A. MAH
: Division of Environmental and Occupational Health Sciences, School of Public Health, University of California at Los Angeles, California 90024, USA

A. T. MIKELL Jr
Institute for Applied Microbiology, Microbiology/Ecology Department, University of Tennessee, Knoxville, Tennessee 37996-0845, USA

J. W. MISHOE
Agricultural Engineering Department, University of Florida—IFAS, Gainesville, Florida 32611, USA

P. MISLEVY
Agricultural Research and Education Center, University of Florida—IFAS, Ona, Florida 33865, USA

J. E. MOORE
Animal Science Department, University of Florida—IFAS, Gainesville, Florida 32611, USA

R. A. NORDSTEDT
Agricultural Engineering Department, University of Florida—IFAS, Gainesville, Florida 32611, USA

S. K. O'HAIR
Tropical Research and Education Center, University of Florida—IFAS, Homestead, Florida 33031, USA

S. M. OLSON
North Florida Research and Education Center, University of Florida—IFAS, Quincy, Florida 32351, USA

T. J. PHELPS
Institute for Applied Microbiology, Microbiology/Ecology Department, University of Tennessee, Knoxville, Tennessee 37996-0845, USA

E. J. PHLIPS
Fisheries and Aquaculture Department, University of Florida—IFAS, Gainesville, Florida 32611, USA

M. POLNE-FULLER
Biology Department, University of California, Santa Barbara, California 93106, USA

K. M. PORTIER
Statistics Department, University of Florida—IFAS, Gainesville, Florida 32611, USA

J. F. PRESTON III
Microbiology and Cell Science Department, University of Florida—IFAS, Gainesville, Florida 32611, USA

G. M. PRINE
Agronomy Department, University of Florida—IFAS, Gainesville, Florida 32611, USA

K. RAJASEKARAN
Botany Department, University of Florida—IFAS, Gainesville, Florida 32611, USA

K. R. REDDY
Central Florida Research and Education Center, University of Florida—IFAS, Sanford, Florida 32771, USA

R. W. ROBINSON
Microbiology and Cell Science Department, University of Florida—IFAS, Gainesville, Florida 32611, USA

D. L. ROCKWOOD
Forestry Department, University of Florida—IFAS, Gainesville, Florida 32611, USA

T. ROMEO
Microbiology and Cell Science Department, University of Florida—IFAS, Gainesville, Florida 32611, USA

J. RYTHER
Harbor Branch Oceanographic Institution, Fort Pierce, Florida 33450, USA

S. C. SCHANK
Agronomy Department, University of Florida—IFAS, Gainesville, Florida 32611, USA

J. R. Schultheis
  Vegetable Crops Department, University of Florida—IFAS, Gainesville, Florida 32611, USA

K. T. Shanmugan
  Microbiology and Cell Science Department, University of Florida—IFAS, Gainesville, Florida 32611, USA

A. Shiralipour
  Microbiology and Cell Science Department, University of Florida—IFAS, Gainesville, Florida 32611, USA

P. H. Smith
  Microbiology and Cell Science Department, University of Florida—IFAS, Gainesville, Florida 32611, USA

W. H. Smith
  Center for Biomass Energy Systems, University of Florida—IFAS, Gainesville, Florida 32611, USA

G. H. Snyder
  Everglades Research and Education Center, University of Florida—IFAS, Belle Glade, Florida 33430, USA

H. Spiller
  Microbiology and Cell Science Department, University of Florida—IFAS, Gainesville, Florida 32611, USA

R. L. Stanley
  North Florida Research and Education Center, University of Florida—IFAS, Quincy, Florida 32351, USA

M. V. Thomas
  Agricultural Engineering Department, University of Florida—IFAS, Gainesville, Florida 32611, USA

I. K. Vasil
  Botany Department, University of Florida—IFAS, Gainesville, Florida 32611, USA

M. E. WAELDER
> Agricultural Engineering Department, University of Florida—IFAS, Gainesville, Florida 32611, USA

C. S. WARREN
> Reynolds, Smith, and Hills, Architects-Engineers-Planners Inc., Jacksonville, Florida 32201, USA

D. C. WHITE
> Institute for Applied Microbiology, Microbiology/Ecology Department, University of Tennessee, Knoxville, Tennessee 37996-0845, USA

J. M. WHITE
> Central Florida Research and Education Center, University of Florida—IFAS, Sanford, Florida 32771, USA

A. WILKIE
> Microbiology and Cell Science Department, University of Florida—IFAS, Gainesville, Florida 32611, USA

D. S. WILLIAMS
> Microbiology and Cell Science Department, University of Florida—IFAS, Gainesville, Florida 32611, USA

J. YANG
> Microbiology and Cell Science Department, University of Florida—IFAS, Gainesville, Florida 32611, USA

1
Introduction to Methane from Biomass: A Systems Approach

J. R. FRANK
Gas Research Institute, Chicago, Illinois, USA
and
W. H. SMITH
Center for Biomass Energy Systems, University of Florida—IFAS, Gainesville, Florida, USA

INTRODUCTION

In the future, new affordable sources of energy will be needed just as farmers now need new crops that can be grown at a profit. Land for this development appears adequate. Research advances in the past five years have consistently shown that the cost of producing energy from biomass can be significantly decreased when cropping systems are designed for energy rather than food, feed and fiber production and conversion systems are adapted to an energy industry and not to waste treatments and industrial chemicals processing. Furthermore, there is a myriad of further improvements that are possible with the new techniques now at our disposal. This book summarizes some of the research performed which supports these assertions and the research needed to arrive at cost competitive methane from biomass.

The volatility and uncertainties associated with both the energy and agricultural industries in the United States led to the creation of a unique industry/university/state research partnership for developing the technology needed to produce large quantities of pipeline quality gas. In 1980, energy prices were rising precipitously, sufficient gas supply for the short-term was a major concern, and agricultural profit margins were threatened by the rapidly rising energy prices and instability of markets for conventional domestic crops. In response to these trends, the Gas Research Institute (GRI) and the University of Florida's Institute of Food and

Agricultural Sciences (IFAS) agreed to join forces and jointly plan, fund and manage a comprehensive research effort to utilize the strengths of both institutions to develop energy crops and the technology to produce significant quantities of pipeline quality methane. Today even though there is currently an abundance of crude oil, the total is being depleted worldwide (including the US) resulting in increased dependence on imports (Kerr, 1984).

Biomass gasification and energy crop development research programs were initiated by the gas industry and the State of Florida at IFAS in the 1970s. Both organizations shared the perception that the use of biomass feedstocks to produce pipeline quality methane was in their long-term interests. For the natural gas industry, methane from biomass would provide a flexible supply option; for the state of Florida, which is almost totally dependent on imported energy and well suited for biomass resources development, biomass crops would provide the farmer alternative crops. This latter option for farms is especially attractive because present markets appear to be unable to assimilate current and future crop production levels under current market limitations for food, feed and fiber.

This program was initiated and operated with several important premises and definitions relating to biomass feedstocks and conversion processes. We defined biomass as contemporary plant matter formed by photosynthetic capture of solar energy and stored as chemical energy. Biomass may be comprised of energy crops, harvesting and crop processing waste and residues, and the biodegradable fraction of municipal solid waste. Operating premises for the program include:

(1) Food, feed and fiber crop varieties generally are unsuitable for energy production for economic and energetic reasons; thus, biomass energy crops should be developed and grown specifically as energy crops.

(2) Compared to various thermal gasification processes, biogasification using anaerobic digestion processes is the most promising conversion option for the production of pipeline quality gas. Favorable attributes identified for biogasification included: (a) the relatively small facility sizes required, (b) the relatively high quality and purity of the biogas produced, (c) the ability to handle wet or dry feedstocks, (d) the relatively low technology and materials needed in facility construction and operation, and (e) the advent and potential applicability of new biotechnologies which will allow significant improvement in the biologically mediated steps in anaerobic digestion. Furthermore, limited fiscal resources available to this program

precluded comprehensive, in-depth investigations of other options. Existing programs of the Department of Energy (DOE) have focused on thermochemical conversion allowing monitoring of advances in that area.
(3) Conventional anaerobic digestion processes developed solely to meet waste management criteria are unsuitable for energy production. Redesign is necessary for improved energy efficiency, lower capital cost, and the ability to handle diverse, particulate biomass feedstocks for commercial methane productions.
(4) A systems approach is necessary as an integrative process for synthesizing the diverse data required for such a comprehensive program. Systems analysis can aid program management and research prioritization. This approach encourages standardized, compatible data development, and interdisciplinary research leading to interfaced models that translate results mechanistically as well as into economic terms. In this way, program focus can be on those research and development sensitivities and issues with the most significant overall impacts on reducing costs or containing major uncertainties.

A flexible management structure was created which made use of advisory and peer input that allowed access to a wide array of scientific competencies. This process included:

(1) A technical management team (TMT), consisting of the GRI and IFAS program managers (the editors), empowered to make all technical, programmatic, and budgetary decisions once the overall program plans and budgets were approved by the Program Directors. This approach placed management in close contact with researchers and allowed flexibility when program adjustments were needed.
(2) An advisory panel of key Florida agricultural and gas industry leaders and a representative of the Florida Governor's Energy Office was constituted to provide guidance on program objectives and to review potential approaches for technology implementation. This group supplemented the peer reviews already widely used by both partners and program reviews by the existing GRI national team of Project Advisors.
(3) Third party contracts outside the University of Florida were used, where needed, to tap scientific competencies critical to accomplishing program objectives. Also included among these

contracts was one with a firm capable of engineering and economic evaluations of various developmental strategies.

This program was formalized in early 1981 by appropriate agreements between GRI and IFAS and research was initiated in mid-1981. Areas selected for major research included: identification and genetic improvement of promising biomass energy crops; development of appropriate crop management practices and harvesting/handling strategies for energy crops; improvement of anaerobic digester designs; advancement in knowledge of the organisms and biochemistry of anaerobic digestion; application of tissue culture propagation to promising biomass species; development of methods for planting 'artificial seeds' (somatic embryos); measuring microbial activity in bacterial digestion processes; and determination of plant morphology and chemical properties relating to biomass fermentability. Before introducing the summary chapters reporting research results from this program, the external forces and analyses that influenced the program are reviewed.

ENERGY SITUATION

The changing energy perceptions caused continual adjustments in the program. Initially the perception was of serious near-term shortfalls and an almost desperate need for new supply. Today the situation is different; instead of shortfalls, surplus supply and plummeting prices characterize the situation. To illustrate the change, consider the following. The GRI Baseline Projection published in November, 1985 forecast '. . . steady economic growth with stable energy prices' (Holtberg *et al.*, 1985). Three months later the rapid fall of oil prices (allegedly brought on by deliberate overproduction of oil by Saudi Arabia's efforts to regain control of the energy market) led quickly to the observation that '. . . the price of natural gas—the main source of income for independents—is dropping below the finding costs of even the most efficient US explorers' (*Business Week*, February 10th, 1986, p. 84). This problem with prediction has been endemic in previous assessments of future needs for both energy supplies and agricultural commodities (Smith, 1986). Lieber (1986), of Georgetown University, argues that despite today's oil glut, the economics and politics of energy are no more predictable over the long run than in 1972–74 and 1979–80 when specialists projected enduring scarcity and relentless price increases. A Congressional Research Service report (Kerr, 1984) projected a 17% decline in US oil production by the year 2000. This was his optimistic prediction; a more likely 29% decline forecast was offered because of doubts of certain new finds. That decline still assumes an Alaskan discovery rate equal to the one that included the Prudhoe Bay field.

Some have argued that diversification of supplies, products and new approaches could improve stability in the face of the considerable uncertainty and volatility of the energy (Smartt, 1983) and agricultural (Avery, 1985) markets. Several studies suggest that the use of energy sources which are decentralized, smaller-scale, and have lower capital costs per plant than conventional utility plants could provide important contributions to regional and local markets where demand uncertainties may exist because of industrial fuel-switching or industry relocation (Kelly, 1982; Houseman, 1983). Since biomass systems may be economically competitive at relatively small facility sizes (about 1·0–5·3 million GJ), planning and construction periods are expected to be shorter than for very large systems (Ford and Youngblood, 1982, 1983; Sterman, 1983). This strategy of diversifying energy supply is currently being considered by some elements of the electric power industry as a response to demand uncertainties (OTA, 1982; Smartt, 1983; Myers, 1983).

Similar opportunities may be applicable to the gas industry especially in situations where incremental sources of gas are needed, or when more dependence on supplemental sources of gas become necessary. According to the 1985 GRI Baseline Projection, a significant amount of energy (about 6·6 billion GJ) will be required from supplemental sources in the year 2000 as conventional sources decline. Much of the anticipated supplemental gas is expected to come from unconventional sources in the year 2000 (about 2·6 billion GJ) and imports from Canada and Mexico including liquefied natural gas (about 4 billion GJ). By the year 2010, this dependence on supplemental sources could increase to over 10·5 billion GJ yr^{-1} (Holtberg et al., 1985). Clearly, the availability of supplemental sources such as biomass, if they are low cost, could be advantageous if these forecasts are accurate.

BIOMASS ENERGY OPPORTUNITIES

The significance of the potential impact of energy from biomass on overall energy supplies is an important question to consider in planning a long-term research program to support the natural gas and agriculture industries. Biomass already annually contributes about 3·2 billion GJ to the US energy mix which is roughly the amount expected to be supplied from nuclear or hydroelectric power plants (Holtberg et al., 1985). This biomass energy is used primarily through wood burning (Haggin and Krieger, 1983) and, on a limited basis, for producing low and medium energy gas for process energy requirements (Lipinski et al., 1983). For electricity generation in 1983, about 3000 MW (approximately 94·9 million GJ) of biomass-fueled net generating capacity was produced or expected to be produced from 196 facilities which were either operating, under construction, or planned (Easterly and Saris, 1984). Roughly half of this capacity was being produced from currently operating facilities. These facilities ranged in size from 5 to 50 MW (0·16 million GJ to 1·6 million GJ) and used a variety of feedstocks such as wood waste (46·6%), municipal solid waste (41·9%), agricultural residues (8·1%), landfill methane (3%), sewage methane (0·3%), and animal manure (less than 0·1%). Electric utilities account for roughly 10% of the biomass-fueled electric capacity with small power production, cogeneration, and municipal facilities accounting for the remainder. Since 1978, the number of biomass facilities filing for qualifying status under PURPA has approximately doubled each year (Easterly and Saris, 1984).

Locally or regionally the biomass share of the energy mix is even more impressive. In Florida, total energy use has increased 38·6% since 1972 while renewable energy has increased 91·6% (GEO, 1985). Biomass has rapidly become the dominant (97%) renewable energy source comprising over 5·8% of the total energy consumed in this sixth most populous state in America. The significant contribution locally and nationally of biomass to the current energy mix has emerged without a large research and development program characteristic of other energy sources. Also, long-standing, well organized, and strongly supported industry associations or other advocacy groups have not existed to aid bioenergy development. Importantly, the growth in biomass energy applications has not been significantly encouraged by subsidies or other incentives. The potential size and sophistication of the biomass to energy industry, had there existed a strong research base and the policy to encourage development, is interesting to contemplate.

Biomass Energy Potential

In the year 2010, the 1985 GRI Baseline Projection estimate is that 9·9 billion GJ of energy could be supplied by renewables consisting primarily of wood and nonwoody biomass from crop and wastes sources with a small amount of geothermal and solar energy. These estimates are derived from DOE sources and do not include the potential from producing energy crops on prime cropland as an alternative market for the agriculture industry. The energy provided from these limited sources, however, is almost equal to the combined amount of energy anticipated to be generated from both nuclear and hydroelectric power. According to this projection, in the absence of the development of new technology for producing cost competitive gas from renewables, some of this energy will be used to produce electricity or steam and will reduce the overall demand for natural gas (Holtberg et al., 1985). It is likely, therefore, that the development of technologies for capturing these resources for the gas industry could have long-term benefits to the gas industry and its ratepayers.

Land Availability

Land availability is often cited as a constraint to bioenergy development. The Battelle study (Lipinski et al., 1983), concluded that sufficient land would be available in the US to produce up to 5·3 billion GJ of biomass energy crops without adversely affecting the production of food, feed and fiber crops (assuming an average biomass yield of only $13\cdot4$–$22\cdot4$ Mg^{-1} ha^{-1} yr^{-1}). In an average year, about 12·2 million ha of prime farmland, approximately 8% of the total US cropland of about 150 million ha are kept out of production by government programs such as the Payment in Kind (PIK) program that idled 33 million ha in 1983. There is an additional 100 million ha of cropland that is either idle or being used only for pasture (USDA, 1985). These estimates include agricultural cropland but do not consider the use of nonagricultural lands such as the 239 million ha of grassland pasture and the 285 million ha of forest lands not protected by the federal government (USDA, 1985).

Between 1930 and 1976, cropland used for crops declined from 153 to 148 million ha. During the same period productivity increased 2·1 times (Zeimetz, 1980). The USDA Soil Conservation Service (Dideriksen et al., 1977) reported that 14 million ha were available for conversion simply by beginning tillage since they had no development problems. If only moderate success is realized from the applications envisioned for biotechnology in production and high technology in processing (NAS, 1983), large gains in food, feed and fiber supplies will be realized from smaller land areas. For example, injections of the recombinantly derived bovine somatotrophine hormone, now awaiting FDA approval, could increase dairy production by 25+ % while improving feed intake efficiency (Milligan et al., 1985). Since dairy products are already in surplus, this means that adequate supplies could be produced on less land than is now used to produce dairy products. Hall, in several analyses (e.g. Hall and deGroat, 1985), has calculated a worldwide surplus of food even though hunger still exists in the world (even in some neighborhoods of affluent countries). Hunger is not a problem of food supply; rather it is attributable to economic, cultural, political and food distribution problems. Solving hunger problems in both developed and developing countries appears to be as complex as solving the overproduction of present domestic crops in developed countries. Another consideration is the changing perceptions regarding diet (Hall and deGroat, 1985). Should the trend of reduced consumption of red meat and eggs (USDA, 1985) continue, even more land should be available for new cropping opportunities since about 90% of US grain production is consumed by animals.

According to Avery (1985), subsidized commodity prices in the developed countries have provided the incentives for developing countries to expand their crop production by applying conventional technology (new varieties, fertilizer, etc.) or by bringing more lands into cultivation. Already such developments reveal a 33% increase in production in developing countries while developed countries increased output 18% over a recent 10 year interval. As a result of these factors the net trade surplus for agricultural crops is declining rapidly from the high of 27 billion dollars in 1980 to a net deficit in May, 1986. All of these developments suggest that ample land will be available to produce the needed domestic crops and to accommodate new biomass energy crops should support be sustained to develop the new agricultural enterprise.

Energy Crop Yields

Potential contributions of biomass to the future energy mix have been estimated by several groups. Biomass yields of $13 \cdot 4$–$22 \cdot 4$ Mg ha^{-1} yr^{-1} on $16 \cdot 6$–$19 \cdot 8$ million ha of land would produce the biomass resource sufficient for producing $3 \cdot 4$–$4 \cdot 5$ billion GJ yr^{-1} according to estimates by Lipinski *et al.* (1983). The agricultural waste biomass increment, added to the total energy potential the equivalent of $3 \cdot 7$–$5 \cdot 2$ Mg ha^{-1} yr^{-1}. These average yields of $13 \cdot 4$–$22 \cdot 4$ Mg ha^{-1} yr^{-1} are characteristic of fairly productive agricultural crops, but are considerably lower than many of the energy crops currently being considered in this program (Smith and Frank, 1985). Energy crops grown in the southeastern United States in field plots have produced yields up to 112 Mg ha^{-1} yr^{-1} with average yields in the $33 \cdot 6$–

67 Mg ha^{-1} yr^{-1} range for certain terrestrial crops, and up to 67·2 Mg ha^{-1} yr^{-1} for selected aquatic plants.

Somewhat lower plant growth in more temperate climates due to shorter growing seasons is partially offset by longer daylengths. Yields of almost 26·9 Mg ha^{-1} yr^{-1} have been achieved as far north as Minnesota with corn and rye in double-cropping (Crookston *et al.*, 1978). In contrast to these agronomic crops, cattails (*Typha*) grown in Minnesota can produce yields of over 40·3 Mg ha^{-1} yr^{-1} in natural stands (Pratt and Andrews, 1980). If an average dry biomass yield of 33·6 Mg ha^{-1} yr^{-1} is obtainable from suitable energy crops, the Battelle estimates of the probable potential of biomass to methane would be roughly doubled to 6·4–8·4 billion GJ yr^{-1} with a potential upper limit of 16·9–21·1 billion GJ. Estimates of biomass potentials arrived at by Battelle were in agreement with similar estimates carried out by governmental organizations and academic institutions. These estimates indicate that a large proportion of the 16–18 billion GJ yr^{-1} US gas market could potentially be provided from biomass if competitively priced.

Several other studies support the significant potential of biomass energy. For example, a study conducted by the Office of Technology Assessment (OTA, 1981) concluded that biomass could provide from 4·2 to 17·9 billion GJ yr^{-1} of gross energy for the US. A more recent study by the US Department of Energy's Research Advisory Board (ERAB, 1982) indicated a net energy potential of 5·3 billion GJ yr^{-1} by the year 2000 from biomass systems. Pimental *et al.* (1984) calculated that a harvest without adversely impacting food, feed and fiber crop production of 22% of the crop residues could provide 11% of the US energy consumption; but when discounted for energy inputs to harvest and transport, a net of 5% of the national energy consumption (about 84·3 billion GJ) resulted. If this point estimate is accurate, this is a rather impressive net from this one source of biomass. Regardless of the exact contribution, it is clear that biomass can make a significant contribution so long as one accepts the premise that the future energy supply will come not from a single source but from an array of appropriate energy options.

Systems Analysis
The production of energy (methane) from biomass can be better understood through systems analysis methodology that integrates the availability of physical sources (land, climate, etc.) with biomass production functions, conversion options and energy storage, transportation and demand

factors. From the outset this program chose to develop an analysis framework where all of these considerations could be linked by a family of models. This approach differs significantly from analysis procedures reported in the literature because the system is designed to simulate how changes in physical and biological processes are reflected in gas costs.

Assessments of the kind developed here have demonstrated the need for innovative systems to accommodate biomass feedstocks that can be produced with the physical resources available and delivered in significant quantities to conversion sites. Biomass supplies envisioned for these systems will not resemble sludges and the other wastes often subjected to biogasification, nor will the digesters be designed like those at waste treatment facilities. The systems analysis framework chosen for this program was designed to accommodate the production and delivery of both high and low solid biomass supplies, and to array into a continuum the various phases of anaerobic digestion. This approach facilitates the analysis of each of the important steps for assessing improvements. Some of the important steps include crop yield in relation to environment (land quality, rainfall and sunlight); production inputs; seasonal supply; harvesting/handling requirements; plant quality; biomass pretreatment and/or hydrolysis; organic acid production; rate and yield of methane production; and the associated microbial populations and environmental factors relating to digester stability, among other considerations.

Several models are necessarily linked in this approach for describing the unit processes by using the data collected from the several contributing projects. A computer-mediated information management system was developed to facilitate communication among the over 50 investigators and for reporting progress in reports that include over 1000 data tables and about 300 publications. Analysis using program data helped assess progress toward program cost-goals, characterize cost sensitivity to various factors, and identify informational gaps restraining the analytical process. Such information proved useful to both researchers and program managers in making mid-course adjustments for focusing research emphasis on critical elements. The systems integration structure evolved for this computer model (BIOMET—biomass from methane) is in version two and it will continue to evolve as new data are accumulated and the individual models are validated (Mishoe et al., 1985). Ultimately, we expect to evolve a set of generic models that can be properly linked by responding to menus that adapt the BIOMET model to site-specific conditions.

The development of a 'systems engineering' approach has been an integral factor in the development of multidisciplinary research combining both fundamental and applied technologies on cost-sensitive research issues. The need for this 'product-oriented' rather than 'technology-oriented' research and development is essential since any energy-producing system derived from biomass is an interactive one in which some of the tradeoffs may not be obvious. The case for integrative approaches has been made previously by Bell Labs (Morton, 1971) as well as others (Dumbleton, 1986) in the literature on innovation and research management.

SUMMARY

This book describes some of the results of the research efforts of the joint GRI/IFAS Methane from Biomass program. It is a compilation of studies during the past five years which has led from a collection of projects with a broad focus to a more targeted program with clearly defined goals and closely coordinated interdisciplinary research projects. Although this program is the major component of GRI's Methane from Biomass research effort, it is dependent on the activities of other GRI-sponsored projects which have contributed significantly to advances in this field. This includes the development of 'solid-state' fermentation by W. Jewell at Cornell University, research on sorghum genetics and production systems at Texas A&M University, advancement of 'vertical flow reactor' technology and its evaluation with the test unit at Walt Disney World, and others. Similarly, several cooperative projects are going forward in IFAS on other biomass topics and on fundamental biological processes, environmental sciences, and advanced technology development involving molecular biology, genetic engineering, robotics, and microprocessor controls, among others. These projects are not described in this book but are part of the overall program thrust described in the last chapter of this book.

Notably, much of the current work on energy cropping 'solid-state' digesters and vertical flow reactors had its early genesis in the biomass programs sponsored by the natural gas industry and the Department of Energy in the 1970s and early 1980s and biomass studies in IFAS, done mainly in ecological and/or agricultural production contexts. These early days taught us much about the problems, pitfalls and opportunities of developing biomass systems, managing research and development, and introduced the editors of this book to the 'systems approach' of program management. The authors recommend that if readers wish to have a more complete understanding of the framework in which the following research is discussed, they next read Chapter 27.

REFERENCES

Avery, D. (1985). U.S. farm dilemma: The global bad news is wrong. *Science* **230**, 408–11.

Crookston, R., Kent, C., Fox, D., Hill, S. and Moss, D. M. (1978). Agronomic cropping for maximum biomass production. *Agronomy Journal* **70**, 899–902.

Dideriksen, R. I., Hidlebaugh, A. R. and Schmude, K. U. (1977). Potential Cropland Study. USDA Statistical Bulletin No. 578, 105 pp.

Dumbleton, J. H. (1986). *Management of High-Technology Research and Development*. Elsevier Science Publishers, Amsterdam.

Easterly, J. L. and Saris, E. C. (1984). A survey of the use of biomass as a fuel to produce electric energy in the United States. *Public Utilities Fortnightly* **113**(9), 71–7.

Energy Research Advisory Board (ERAB). (1982). Biomass energy. In: *Report of the Energy Research and Advisory Board Panel on Biomass*, US Department of Energy, Washington, D.C., 32 pp.

Ford, A. and Youngblood, A. (1982). Simulating the planning advantages of shorter lead time generating technologies. *Energy Systems and Policy* **6**(4), 341.

Ford, A. and Youngblood, A. (1983). Simulating the spiral of impossibility in the electric utility industry. *Energy Policy*, March, 19.

Frank, J. R. (1980). Studies improve biomass to SNG conversion. *Hydrocarbon Processing*, April, 117–21.

Frank, J. R., Hayes, T. D. and Smith, P. H. (1985). The production of methane from biomass in the United States: economics, tradeoffs, and prospects. In: *Energy from Biomass*, Palz, W., Coombs, J. and Hall, D. O. (eds.), pp. 484–8. Elsevier Applied Science, London.

Governor's Energy Office (GEO). (1985). Energy Data 1970–1984. Florida Governor's Energy Office. Tallahassee, Florida, 106 pp.

Haggin, J. and Krieger, J. H. (1983). Biomass becoming more important in U.S. energy mix. *Chemical and Engineering News*, **61**(11), 28–30.

Hall, D. O. and deGroat, P. J. (1985). Biomass for food or fuel: a world problem? In: *Energy Applications of Biomass*, Lowenstein, M. Z. (ed.), pp. 21–37. Elsevier Applied Science, London.

Holtberg, P. D., Woods, T. J., Ashby, A. B., Dreyfus, D. A. and Hilt, R. C. (1985). 1985 GRI Baseline Projection of U.S. Energy Supply and Demand to 2010. Gas Research Institute, Chicago, Illinois.

Houseman, J. (1983). An assessment of changes in industrial composition and location in the United States and their implications for gas usage. *Gas Research Insights*, January.

Isaacson, H. R., Hayes, T. D. and Chynoweth, D. P. (1984). Pipeline quality methane from biomass and wastes. In: *Proceedings of the Energy Technology Conference XI*, Hill, R. F. (ed.), pp. 1294–305. Government Institutes, Rockville, Maryland.

Kelly, R. W. (1982). Industrial gas demand: The future of existing markets. *Gas Research Insights*, August.

Kerr, R. A. (1984). How fast is oil running out? *Science* **229**, 426.

Klass, D. L. (1983). Energy and synthetic fuels from biomass and wastes. In: *Handbook of Energy Technology and Economics*, Meyers, R. A. (ed.), pp.

712–86, John Wiley, New York.
Lieber, R. J. (1986). The oil glut offers gains, dark clouds. *Gainesville Sun*, February 22.
Lipinski, E. S., Jenkins, D. M., Young, B. A. and Shepard, W. J. (1983). Review of the potential for biomass resources and conversion technology. Final Report, Battelle-Columbus Laboratories, to Gas Research Institute. GRI-83/007 May. Chicago, Illinois.
Milligan, R., Kalfer, R. and Lesser, W. (1985). The economic implications of bovine somatotrophine. In: *Proceedings of Biotech 85, US*, pp. 529–37, October 1985. Washington, D.C., On-Line, London.
Mishoe, J. W., Lorber, M. N., Peart, R. M., Fluck, R. C. and Jones, J. W. (1985). Modeling and analysis of biomass production systems. In: *Energy from Biomass*, Palz, W., Coombs, J. and Hall, D. O. (eds.), pp. 119–30. Elsevier Applied Science, London.
Morton, J. A. (1971). *Organizing for Innovation: A Systems Approach to Technical Management*. McGraw-Hill, New York.
Myers, R. (1983). Small reactors: is the nuclear industry on the verge of a large step backward? *The Energy Daily*, December 22.
National Academy of Science (NAS). (1983). Report of the Briefing on Agricultural Research. National Academy of Science, Washington, D.C., 21 pp.
Office of Technology Assessment (OTA). (1981). Energy from biological processes. Government Printing Office, Washington, D.C., p. 195.
Office of Technology Assessment (OTA). (1982). Economies of scale in the electric-utility industry. A Review. Government Printing Office, Washington, D.C.
Pimental, D., Fried, C., Olson, L., Schmidt, S., Wagner-Johnson, K. and Westmen, A. (1984). Environmental and social cost of biomass energy. *Bioscience* **34**, 89–93.
Pratt, D. C. and Andrews, N. J. (1980). Cattails (*Typhus* spp.) as an energy source. In: *Energy from Biomass and Wastes IV*, Klass, D. (ed.), pp. 43–63. Institute of Gas Technology, Chicago, Illinois.
Smartt, L. E. (1983). Electric utilities a decade from now: another attempt to discern dim outlines on the horizon. *Public Utilities Fortnightly*, April 28, 4.
Smith, W. H. (1986). Biomass energy crops—what, where, when, and who. In: *Proceedings of U.S. Department of Agriculture Solar and Biomass Workshop*, pp. 51–6. Tifton, Georgia.
Smith, W. H. and Frank, J. R. (1985). Comparative biomass yields of energy crops. In: *Energy from Biomass*, Palz, W., Coombs, J. and Hall, D. O. (eds.), pp. 323–9. Elsevier Applied Science, London.
Sterman, J. D. (1983). Economic vulnerability and the energy transition. *Energy Systems and Policy* **7**(4), 259.
US Department of Agriculture (USDA). (1985). 1985 Agricultural Chartbook. In: *Agriculture Handbook No. 652*. Economic Research Service, p. 92. US Department of Agriculture, Washington, D.C.
Zeimetz, K. A. (1980). Growing energy: land for biomass. USDA Agricultural Economics Report No. 425, p. 35.

2
Information Management

K. M. PORTIER[a] and J. E. CINERESKI[b]

[a]*Statistics Department,* [b]*Center for Biomass Energy Systems, University of Florida—IFAS, Gainesville, Florida, USA*

INTRODUCTION

A process for managing information from a complex, interdisciplinary program was a part of the initial planning. The information management system developed was an attempt to satisfy the need for exchange of technical information produced by as many as 125 research projects involving over 80 faculty, graduate students and research staff located at 12 out-state research centers and 11 academic departments on the University of Florida campus in Gainesville (Fig. 1). The system was designed for information acquisition, storage and retrieval. The information archived in the system was for use in systems integration, and for researchers and program managers.

Initial planning recognized the importance of having a logical structure for information collected by program participants. An evaluation of the needs of the individual projects was undertaken, as well as an inventory of existing data sets which could be of use to the program. Special consideration was given to the perceived needs of the systems integration group. A comprehensive list of potential feedstock characteristics was constructed and examined, and an inventory of potential yields of biomass feedstocks was compiled according to highest, lowest and average expected yields. Another document compiled methane production from screening bioassay factors affecting bioconversion, as well as other basic information for documenting experimental data derived from the projects. These lists were used to structure the information management system and as benchmarks for measuring progress.

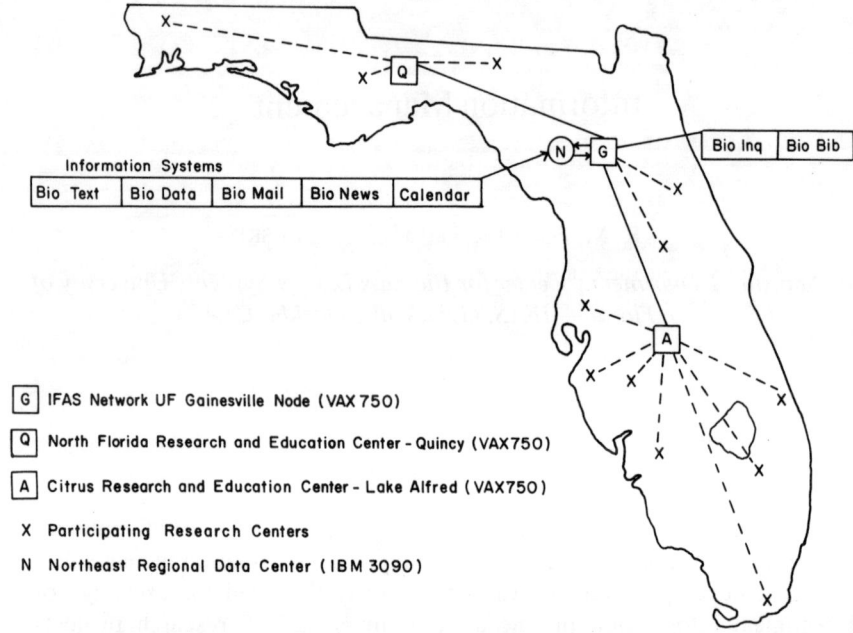

FIG. 1. Network for information management system involving IFAS faculty at several research centers.

SYSTEM STRUCTURE

The information system as initially designed and implemented consisted of a series of computer modules accessed by a simple command which allowed users to input and retrieve programmatic information. The collection of modules which made up the system was given the name CSIMS for Comprehensive Scientific Information Management System (Portier and Hetrick, 1982; Hetrick and Portier, 1982). All modules, unless otherwise indicated, were implemented under IBM's VM/SP operating system on an IBM mainframe computer at the Northeast Regional Data Center in Gainesville, Florida.

Information Acquisition

Program information was divided into four types, text, numerical data, graphic files and bibliographic citations. Text files were perceived to consist of project reports and technical publications. The BIOTEXT module in CSIMS was designed to permit text entry with limited text processing capability. BIOTEXT was comprised of the components, QUARTERLY and ANNUAL. QUARTERLY allowed the user to enter quarterly reports by task, structured for direct printing. The ANNUAL component was for entry of annual reports in a flexible format. The BIOTEXT module has been used extensively for project reporting over the life of the Program.

The power of the mainframe computer for text formatting of technical publications was initially planned for program participants. But, the increased availability of relatively inexpensive microcomputers and word processors made efforts on the mainframe computer seem less efficient, so further development was deferred.

The most important research information is tables of numeric data and graphs of these data. A BIODATA module was designed and implemented into CSIMS as one way of helping researchers to enter and document their data. The ease with which microcomputer systems could be modified for data entry precluded use of this module.

Specially formatted text files for entry of bibliographic information into the database were developed. All bibliographic information was entered by program staff trained to create files in the appropriate format. The files contained in this database are mainly of program generated publications and some of the key citations for each of the research areas.

Information Retrieval

Work on the actual design and implementation of information management software did not begin until after much of the preliminary information acquisition modules were in place. This allowed time to evaluate the quantity and structure of the information being generated within the program. From the outset, more information was being generated than was useful in the first phases of systems analysis. Rather than attempt to determine which information would be useful in future systems integration work, it was decided to build a structure which could hold as much of the program information as possible. The information management system required flexibility, in order to hold information files of different formats and to allow easy access to the information.

Computer software to allow initial exploration of different structures for the information system was obtained in 1982. The system, as currently implemented on a Digital Equipment Corporation VAX 11/750 minicomputer owned by IFAS, is a combination of specially written Fortran 77 code and components of the RIM (1982) database management system. RIM was chosen because of its ease of use, interface to Fortran 77 and low cost due to its being considered public domain software.

The initial information base contained only bibliographic files and was accessed by the BIOBIB module in the CSIMS system. As more types of information were introduced into the database, BIOBIB became part of the larger BIOINQ module. In the BIOINQ module, information retrieval is accomplished through a combination of menu selection and keyword search. The menus allowed selection of the appropriate file type. Conceptually there are six file types defined for the system:

— Bibliographic citations with abstracts of publications.
— Numerical summary files consisting of summary data tables of experimental results.
— Project descriptor files giving the objectives and tasks for each project.
— Data description files consisting of the FORMATTER information.
— Abstract files giving descriptions of experiments run in each project.
— Raw data files consisting of the useful experimental data from research projects.

This particular system had as its basic objective the enabling of rational retrieval of research information and associated documentation. The bibliographic files and project descriptors are made available to individuals new or unfamiliar to the program who wish to examine current literature and program composition. The summary files contain the relevant summary information of all research done within the program. The last three files were to be used to index the summary files and to provide detailed experimental data in one place for future systems analysis. Information on experiments is currently available in the quarterly and annual reports. Data description files and the actual experimental data have remained with the researcher, and hence are not readily accessible from the information system.

Experience with free keywording by the researchers produced inconsistencies in keywords used for the same topic. This resulted in inefficient and slow retrieval of information files. A thesaurus was created of consistent keywords and all files have been keyed to this list. All retrieval of information in BIOINQ is now available through the use of keywords in the biomass thesaurus, which is available electronically and as a printed copy.

Other Activities
In any multidisciplinary research program, communication among participants is important to meeting program objectives. A number of computer modules were developed to aid in that communication. A CALENDAR module was developed to display program related activities and reporting deadlines. A BIOMAIL module was developed to increase the ease of use of electronic mail capabilities on the IBM mainframe. A BIONEWS module was developed to inform users of CSIMS of events and items of immediate importance. A number of modules (BIOANALYSIS, BIOGRAF and FINDGRAF) never got past the conceptual stage.

SUMMARY

At present, the information system has stored much of the bibliographic and statistical summary information from the quarterly reports in a retrievable format. These summary tables, now numbering over 900, can be accessed in a logical manner using the thesaurus of keywords. Similarly, over 300 publication citations and their abstracts are stored and retrievable using keyword search. The list of research publications presented in the Appendix to this volume was generated using the BIOINQ information base.

The information system described in this chapter was designed and implemented during a time of very rapid change in computer technology. Many researchers and their staff, though somewhat familiar with computers, were not accustomed to computer documentation of research projects. The centralized mainframe computing facility was not very user friendly; therefore, computing emphasis within the program quickly transferred to the much more friendly minicomputer (VAX) and personal microcomputers.

The mainframe oriented CSIMS system has since been underutilized because the microcomputers were more flexible and accommodating for research computing. Unfortunately, tasks such as centralized information storage and retrieval and electronic communications are not easily accomplished within a microcomputer-based distributed processing system. Clearly, the microcomputer has become an integral tool of the researcher. Future research in information management will need to incorporate the microcomputer as an integral part of the information acquisition and management process.

Development of microcomputer-based documentation systems, which are user friendly and allow information consolidation into a centralized information base, mainframe or microcomputer based, for use by other researchers, is the suggested next step.

REFERENCES

Hetrick, V. R. and Portier, K. M. (1982). CSIMS—Comprehensive Scientific Information Management System: a tool for managing information in biomass programs. In: *Second European Community Conference on Biomass Research*, Strub, A., Chartier, P. and Scheleser, G. (eds.), pp. 666–70. Elsevier Applied Science, New York.

Portier, K. M. and Hetrick, V. R. (1982). CSIMS—Comprehensive Scientific Information Management System. *Proceedings of COMSTAT 1982*. International Association of Statistical Computing, Phsyica-Verlag, Vienna.

RIM 5.0 (1982). Relational Information Management System, Version 5, User's Guide. Boeing Commercial Airplane Company, Seattle, Washington.

3
Model and Analysis of Biomass to Methane

J. W. MISHOE, C. F. KIKER, R. C. FLUCK, W. G. BOGGESS, S. L. CURRY
and M. E. WAELDER

*Agricultural Engineering Department, University of Florida—IFAS,
Gainesville, Florida, USA*

INTRODUCTION

Biomass is a source of energy that can potentially provide an important contribution to the energy supply of regions having a favorable climate for crop production. The biomass feedstocks can be derived from many sources including wastes, crops grown specifically for energy production and crops produced in combination with other operations such as water treatment. An example of water treatment is the use of water hyacinths to remove nutrients from municipal waste water or from eutrophic lake water (Hayes *et al.*, 1986). For methane production, these various sources of biomass are collected and transported to a central conversion site for anaerobic digestion. The technology to economically produce and convert biomass to methane on a large scale is not currently available, therefore it is necessary to continue conducting research to develop the capability to produce gas in large enough quantities to have an impact on the national energy supply.

To assess the available technology and to identify information gaps, we have developed a systems model, BIOMET *BIO*mass to *ME*Thane (models documented in unpublished reports prepared by J. W. Mishoe *et al.*), that can simulate and report the economic performance of various production and conversion systems and the sensitivity of several factors important to the process. BIOMET provides a framework to design new research programs and to assess the impact of new data as it is developed through research on the impact of commercial scale gas production.

The type of system conceptualized consists of a centralized conversion facility with biomass produced on a regional basis and transported to the facility after harvest. The major components of this system are biomass production and conversion. Associated with each component are the variable costs that account for the inputs and sustenance of operations, and the fixed costs that account for the installation and fixed maintenance of the facilities. To study such a system requires that the output of each component be known as a function of the inputs and that the dynamic performance of each component be known as a function of the unit management. Because the performance of each component depends upon the output of other components it is necessary to study the combined system in order to determine the resultant methane output. For example, weather can greatly influence crop growth; therefore, to assure a sufficient supply of biomass during slow crop growth periods, excess biomass must be accumulated during the high crop growth periods.

BIOMET consists of simulation models of the crop production and conversion components as well as models that describe the economics of crop production, harvesting, transportation and conversion. The implementation of BIOMET is an interactive program designed to allow the user to define the overall system and specify how the system is to be managed. From this, BIOMET will simulate the chosen system. The output includes component performance, overall gas yield, and the levelized cost (defined later) of the gas.

METHODOLOGY

A block diagram of the system level model as defined within BIOMET (Fig. 1) depicts the operating sequences in the model. The first operation in using the model is to define the system configuration by selecting a series of options from menus. The options include crop type selection, field size and location, harvester type and size, transportation type and size, conversion reactor type and size, costing parameters, economic parameters (for the levelized cost model), and gas demand. The menu options have default values, therefore the user need only modify those inputs that need to be changed to satisfy the desired configuration options. BIOMET uses the menu supplied information to select the number of trucks and harvesters for estimating a harvest demand schedule necessary to meet the selected gas demand. With this, the model begins the simulation of the configured system. If any selected component becomes limiting, the resulting effects on total system performance are simulated. The costing models accumulate the cost of events that occur during the simulation. For example, if the

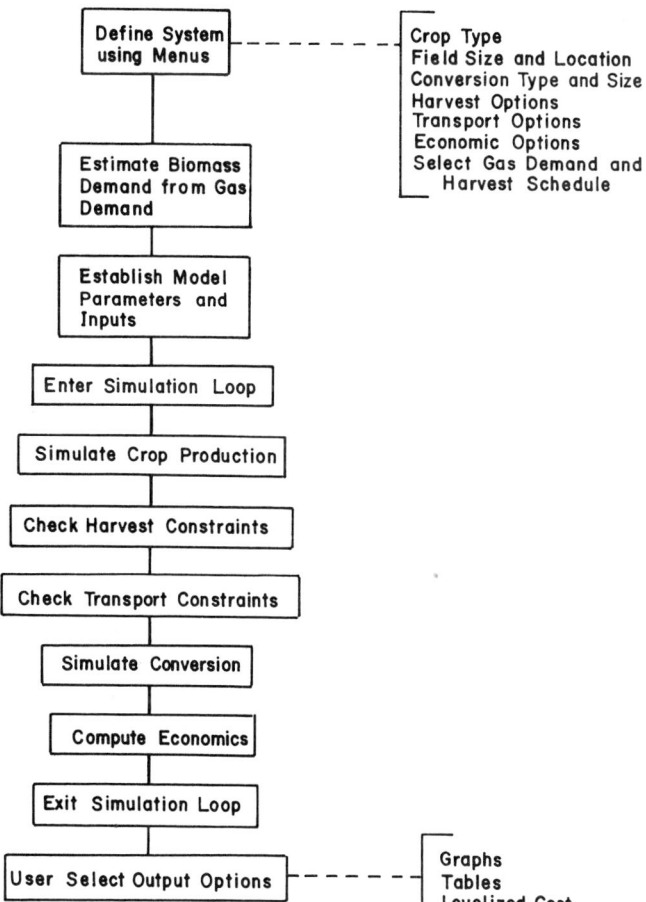

FIG. 1. Block diagram of the steps performed in BIOMET. The options for input are listed to the right.

biomass demanded is not met by the current configuration, only the biomass actually harvested is transported to the conversion site and the transport cost is charged based upon the material actually transported. Generally, because the capital investment is fixed for the system, the unit cost of gas will be higher for the chosen time period because transportation limits the full utilization of the other system components. After the simulation is complete, summary reports list the cost of production and conversion of the biomass to methane. The final unit cost of the methane is presented using levelized cost values over the life of the total system investment. The output options include a graphic presentation of the component response during the simulation. A graphic example is the biomass contained in a given field plotted as a function of time.

BIOMET is currently implemented to meet the needs of a research program which is focused on developing an understanding of how to best grow and convert biomass to methane. Once we adequately understand the system, and incorporate this understanding of the system into the component models, optimal design information will be determined. By adding optimization algorithms to BIOMET the best system parameters can be selected. Studies of the risk factors involved allow determination of reasonable safety factors to assure satisfactory operation even in worst case situations such as extreme weather events.

BIOMET differs from essentially all analysis procedures found in the literature because of the use of process level simulators to represent the performance of selected physical and biological components to supply information for the analysis. This difference is important for several reasons. First, it increases the capability of the analysis procedure to consider factors that cannot be considered otherwise. For example, management of the crop by fertilizer application, harvest scheduling, planting density, etc., are sequential events that define crop growth. However, the timing and sequence of the management inputs greatly influence the response of the crop. The simulation model for the crop integrates the various combinations of these input decisions based upon the current state and sensitivity of the crop. Therefore, by using the simulation model, optimal management strategies are defined from simulation studies and then applied to field conditions. An additional reason for use of the simulation models in the systems analysis is the ability of the model to assess the impact of research information on the performance of the system. The models are structured at the same level of detail as the collected research data. The measured research data are used to compare with the simulation model. This develops confidence that the model does 'mimic' the real system, plus it gives the opportunity for the researcher to determine the response of other variables that were not measured. The sensitivity of the variable can be examined without having to repeat large numbers of experiments in the field. This process of model verification and research is interactive allowing improved experimental design plus model development and validation.

Crop Growth Models

The crop growth modeling uses physiologically based models representing the basic growth processes in response to the crop's environment. The models are designed to respond to solar radiation and temperature to produce biomass. These processes are modified by the model in response to other environmental factors such as plant nutrition and canopy cover. The partitioning of the biomass among various plant parts is described by functions within the model. Component models are included to simulate the cycling of nutrients within the crop environment. For example, the root zone is represented by several zones (Fig. 2), and within each zone a mass balance of nutrients is maintained with crop growth enhanced or stressed, depending upon prevailing conditions.

Water hyacinth (*Eichhornia crassipes* (Mart) Solms) and Napiergrass (*Pennisetum purpureum* Schum.) were the crops used in BIOMET development since they represent promising terrestrial and aquatic feedstock species. The basic processes for both crops are the same; however, the structure and growing conditions differ such that it is necessary to implement the models in a different format.

Water Hyacinth

The basic physiological equation describing water hyacinth growth (Lorber *et al.*, 1984) is given by:

$$dW/dt = (PG - AMRESP) \times CTURN - DETRIT \tag{1}$$

where:

dW/dt = change in biomass dry weight ($g\ m^{-2}$)
PG = gross photosynthate available
AMRESP = maintenance respiration
CTURN = efficiency of conversion (net photosynthate to biomass dry weight)
DETRIT = production of detrital material ($g\ m^{-2}$)

Water hyacinth remains in the vegetative stage during most of its growth cycle; therefore, it is not necessary to define phenological growth stages. To describe photosynthesis the relationship developed between total daily radiation and the potential available photosynthate for crop growth is as follows:

$$\begin{aligned} PGMAX &= 22 \cdot 318 + 0 \cdot 102 \times DLANG &&DLANG > 100 \\ &= 0 \cdot 32 \times DLANG &&DLANG <= 100 \end{aligned} \tag{2}$$

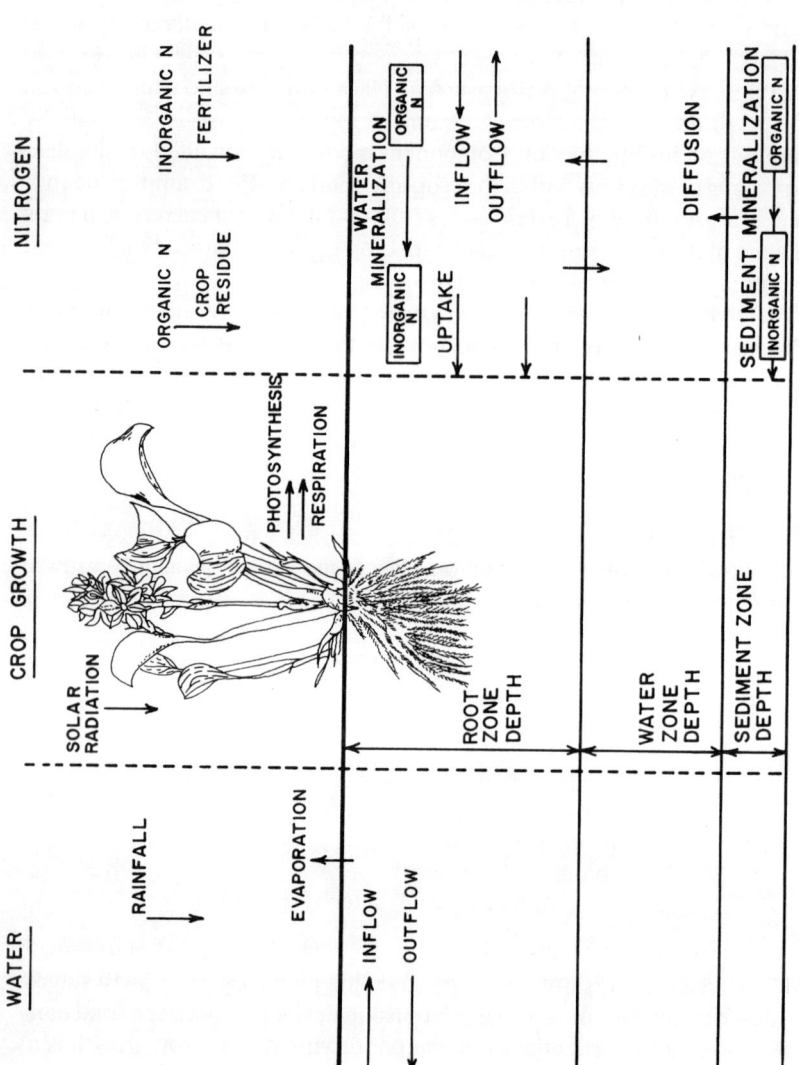

FIG. 2. Example of water hyacinth growth modeling philosophy of BIOMET.

where:

PGMAX = potential available photosynthate
DLANG = daily solar radiation (langleys)

If conditions are optimum for crop growth, the photosynthate available for growth would be equal to the potentially available photosynthate. However, stresses caused by insufficient nutrient or weather conditions will limit the crop's ability to produce and to use photosynthate for growth. Therefore, daily photosynthate available for growth is given by:

$$PG = PGMAX \times f_T \times f_N \times f_P \times f_D \qquad (3)$$

where:

PGMAX = potentially available photosynthate
f_T = temperature limiting function
f_N = nitrogen limiting function
f_P = phosphorus limiting function
f_D = plant density limiting function

The functional relationships used to describe the growth limiting functions are presented in Fig. 3. At 29°C growth was maximal and at temperatures below 13°C growth stopped (Fig. 3A). The effects of freezing temperatures were not considered. The density limiting function, f_D (Fig. 3B) used to describe the effect of canopy density on photosynthesis, showed low densities to possess limited leaf areas so that light was not fully intercepted resulting in reduced photosynthesis.

The functions for nitrogen (N) and phosphorus (P) stress (Figs. 3C and 3D) showed lower plant concentrations of N and P to reduce plant photosynthesis. Plant nutrient uptake is based upon root zone supply and plant demand. In the case of N a luxury uptake is permitted up to 3·5% of plant biomass. Growth reduction occurs when plant N falls below 1·5% of plant biomass.

Maintenance respiration (AMRESP) of eqn (1) is extracted from gross photosynthate (PG) before the remainder (PG − AMRESP) is converted to plant dry matter (PG − AMRESP) × CTURN. Maintenance respiration is estimated as a simple linear function of the crop biomass:

$$AMRESP = CRESP \times BIOM \qquad (4)$$

where:

AMRESP = maintenance respiration (g m^{-2})

CRESP = respiration coefficient
BIOM = biomass dry weight (g m^{-2})

CRESP was calibrated to 0·01, which is similar to the theoretical range of estimates 0·015–0·025 given by Penning de Vries (1975).

Detrital production, DETRIT of eqn (1) occurs when conditions are poor for growth, or when conditions are optimal and death is caused by

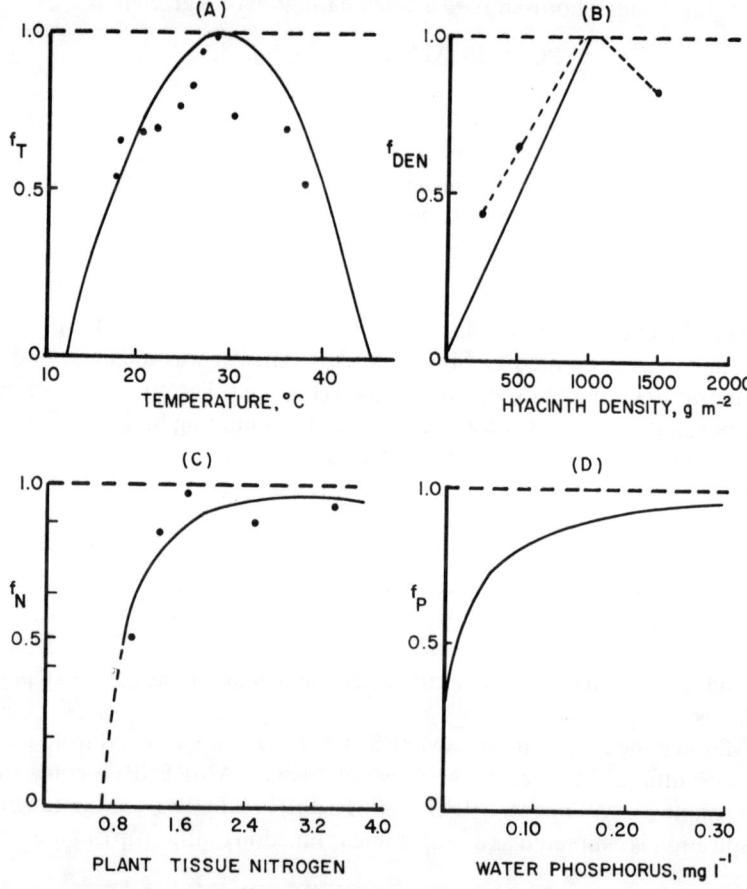

FIG. 3. Limiting functions (Lorber *et al.*, 1984) to potential available photosynthate, Pgmax. (A) air temperature limiting function, (B) density limiting function, (C) plant nitrogen limiting function, (D) plant phosphorus limiting function.

dense, overcrowding of plants. Poor conditions are signaled by gross available photosynthate (PGAVL) being insufficient to meet the respiration demand (AMRESP) and the resulting detrital production is given in eqn (1) as (PG − AMRESP) × CTURN). When conditions are optimal, water hyacinths vegetatively reproduce to cover the water surface and then continue to grow vertically. At high densities, growth continues, but plant material lower in the canopy dies. For the water hyacinth model, detrital production due to high densities occurs above 2400 g m^{-2}.

Water hyacinths will uptake N and P until a maximum plant concentration of each is reached. Plant demand for N and P is based on the attempt to reach this maximum concentration. Plant reservoirs of N and P are maintained throughout the simulation.

The maximum plant concentration for P, CUPP, is considered site specific in the sense that it reflects potential uptake under prevailing conditions. In natural environments, competition from algae and other aquatic plants reduces nutrient uptake by water hyacinths compared to plants grown alone in tanks of nutrient solution. CUPP is set at 0·01 (1·0%) for simulations of tank experiments and 0·005 (0·5%) for simulation of natural environments. The uptake rate for P is described as:

$$PUPT = ACUPP \times (PGAVL \times CTURN) \quad (5)$$

where:

PUPT = daily P uptake of plant (g m^{-2})
PGAVL × CTURN = new plant growth daily (g m^{-2})
ACUPP = CUPP × FP
CUPP = maximum P concentration in the plant
FP = uptake inhibitor for P

Phosphorus is added to the system through inflow water and mineralization. The inflow of P is described as:

$$PINF = GPINF/SAREA \quad (6)$$

where:

PINF = P inflow (g m^{-2})
GPINF = P inflow water (g^{-2})
SAREA = surface area of tank or lake (m^2)

Mineralization of P is described as:

$$PMIN = CMINP \times (2 \times \times ((WTMP - 30.7)/10.0)) \quad (7)$$

where:

PMIN = P mineralized daily
CMINP = P mineralization coefficient
WTMP = water temperature (calculated as average daily air temperature °C)

Uptake rate is equal to PUPT if enough nutrients are available to maintain this uptake demand. When the nutrient level in the water drops, the uptake rate drops accordingly. The water hyacinths can still grow unchecked in nutrient-low waters until the plant concentrations drop below a critical amount. In the model, this level is assumed equal for all growing conditions and environments.

The maximum N concentration in the plant, CNCMAX, is considered constant in all environments in the model. CNCMAX is set to 0·035 (3·5%). Assuming that the biomass present plus the new growth attempt to reach maximum N concentration, the N demand is described as:

$$\text{UPIN} = (\text{PGAVL} \times \text{CTURN}) \times \text{CNCMAX} + \text{BIOM} \times (\text{CNCMAX} - \text{CNCP}) \qquad (8)$$

where:

UPIN = plant demand for N (g m^{-2})
BIOM = biomass dry weight (g m^{-2})
CNCMAX = maximum N concentration in the plant
CNCP = current N concentration in the plant
PGAVL × CTURN = new biomass grown this day (g m^{-2})

If the plant reaches this maximum concentration all excess N uptake is returned to the water. Then the plant will continue to uptake only what is needed to maintain the maximum concentration level.

Plant uptake is limited to the N available in an effective root zone. If insufficient N is available in this zone, plant N concentration drops. The plant experiences nutrient stress if the N concentration falls below a luxury concentration, CNCLUX (currently set at 0·015 (1·5%)). This limitation factor f_N was described earlier.

A mass balance of organic and inorganic (NH_4 and NO_3) N is maintained in the model. The water is divided into three zones, an effective root zone, a water zone and a sediment zone. The inorganic N which the plant uptakes from the root zone may be added to the system in several ways, by entering through the inflow water, by mineralization of organic N, by diffusion from the sediment, and by fertilization. Removal of N by plant uptake is determined by the N demand (eqn 8) and the amount of inorganic N present in the 'effective' root zone. The water hyacinth root length increases or decreases relative to the amount of N available in the water. A large amount of N available causes roots to be very short, and they grow longer when less N is available. The 'effective' root zone is determined by a moving average N concentration in the water and may be deeper than the actual root length. If there is sufficient N available in the effective root zone to meet the plant's requirements, it is taken up by the plant. However, if there is a deficit of N available in the effective root zone, only the N in this root zone is taken. Then, the concentration of inorganic N is recalculated assuming a well-mixed system.

Reddy (1983) grew water hyacinths in 900 liter outdoor tanks in Sanford, FL throughout 1982. The water in the tanks was changed weekly and initial N and P concentrations in the new water were 21 and 3 mg liter^{-1}, respectively. This was found to be more than adequate to supply the plants. Other factors which might have hindered growth such as insect damage were also controlled. These data provided a unique opportunity to calibrate the model of water hyacinth growth under unstressed conditions. The crop was harvested five times during 1982 with measured weekly crop densities. Solar radiation and temperature data were collected at the Sanford site and used in model simulations for generating the observed and calibrated growth curves (Fig. 4).

To measure the response of water hyacinths in 'natural' water Reddy (1983) grew water hyacinths in vexar mesh baskets in a reservoir adjoining Lake Apopka in central Florida near Orlando. Lake Apopka is highly eutrophic with N concentrations approaching 1·0 mg liter^{-1}. The observed and predicted growth curves for this test (Fig. 5) showed model predictions of growth to again match observations well. Also shown (Fig. 5) is the potential growth of water hyacinths during the same period with nutrients non-limiting. Notably, nutrients are still limiting water hyacinth growth in Lake Apopka, despite the eutrophic status of the lake.

Napiergrass
Biomass growth of Napiergrass is based on a carbon balance which includes photosynthesis, respiration, partitioning and senescence. The basic physiological equation describing Napiergrass growth is given by:

$$dW/dt = (PG - R_M) \times E_p - S_L \qquad (9)$$

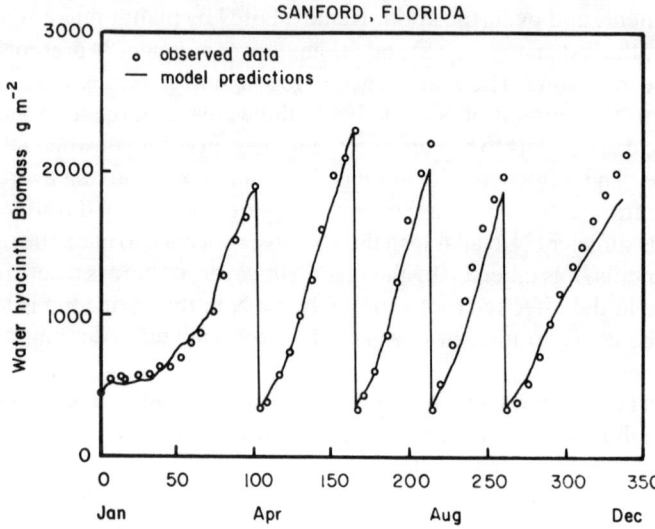

FIG. 4. Validation of water hyacinth growth model using experimental data from Sanford, Florida (data from Reddy, 1983). (Figure from Lorber *et al.*, 1984)

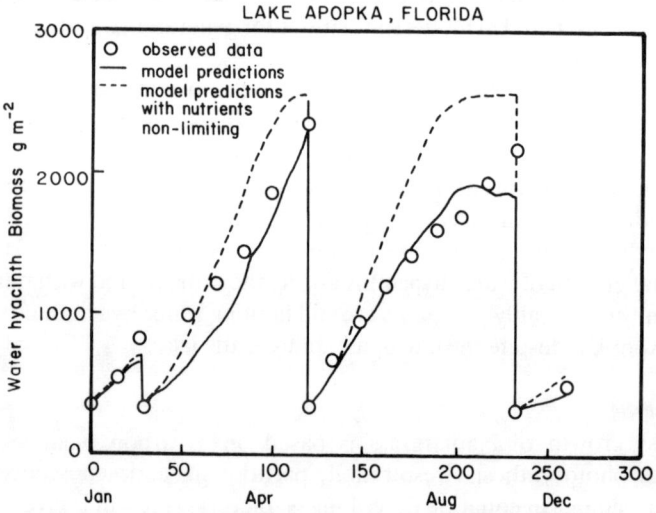

FIG. 5. Validation of water hyacinth growth model using data of water hyacinths growing in a reservoir with water gravity fed from Lake Apopka, Florida (data from Reddy, 1983). (Figure from Lorber *et al.*, 1984)

where:

dW/dt = change in biomass dry weight (g m^{-2})
PG = gross photosynthate available
R_M = maintenance respiration
E_P = efficiency of conversion (net photosynthate to biomass dry weight)
S_L = senescence of leaves (g m^{-2})

Biomass is partitioned into various plant parts: leaves, stem and roots. These partitioning coefficients may vary with phenological stage or water stress. However, it is assumed in BIOMET that Napiergrass remains in the vegetative stage, so limited phenological growth stages are modeled. Governing model equations for new growth into leaves, stems, and roots are:

$$WLDOT = XLF \times PGAVL/ALPGR - SLDOT - XMLDOT \quad (10)$$

$$WSDOT = XSTM \times PGAVL/ALPGR - SSDOT - XMSDOT \quad (11)$$

and

$$WRDOT = XRT \times PGAVL/ALPGR - SRDOT - XMRDOT \quad (12)$$

where:

WLDOT = new leaf growth this day (g m^{-2})
WSDOT = new stem growth this day (g m^{-2})
WRDOT = new root growth this day (g m^{-2})
XLF = leaf partitioning coefficient
XSTM = stem partitioning coefficient
XRT = root partitioning coefficient
PGAVL = daily photosynthate available
ALPGR = conversion efficiency
SLDOT = leaf senescence (g m^{-2})
SSDOT = stem senescence (g m^{-2})
SROOT = root senescence (g m^{-2})
XMLDOT = leaf protein remobilization rate (g m^{-2})
XMSDOT = stem protein remobilization rate (g m^{-2})
XMRDOT = root protein remobilization rate (g m^{-2})

Carbohydrate is supplied from gross photosynthesis (PG) which is based on daily photosynthetically active radiation (PAR), temperature, leaf area index, N content of the leaves, water stress and fertilization. The relationships developed between daily photosynthetically active radiation and the potential available photosynthate for crop growth in the Gainesville, Florida area are:

$$\text{PTSMAX} = \text{PHFAC1} \times \text{PAR}/(\text{PHFAC2} + \text{PAR}) \qquad (13)$$

where:

PTSMAX = potential available photosynthate
PAR = daily photosynthetically active radiation
PHFAC1 = coefficient of equation
PHFAC2 = coefficient of equation

If conditions are optimum for crop growth, the photosynthate available for growth would be equal to the potentially available photosynthate. However, stresses caused by insufficient nutrients or inclement weather will limit the crop's ability to use photosynthate for growth. Therefore, daily photosynthate available for growth is given by:

$$\text{PG} = \text{PTSMAX} \times \text{PGFAC} \times \text{TPHFAC} \\ \times \text{SWFAC} \times \text{AGEFAC} \times \text{SFFAC} \qquad (14)$$

where:

PG = photosynthate available for growth
PTSMAX = potential available photosynthate
PGFAC = leaf area limiting function
TPHFAC = temperature limiting function
SWFAC = soil water limiting function
AGEFAC = N content limiting function
SFFAC = fertilization limiting function

Maintenance respiration is divided into two parts: respiration cost to maintain membranes and ion balances across membranes, and respiration cost due to protein turnover. The equation describing these processes is:

$$R_M = \text{RO} \times \text{TOTWT} + \text{RP} \times \text{PG} \qquad (15)$$

where:

R_M = maintenance respiration
RO = coefficient for membrane maintenance
TOTWT = total plant dry weight
RP = coefficient for maintenance due to protein turnover
PG = photosynthate available for plant growth

The model has been calibrated from data collected in Gainesville, Florida. Additional data are needed to finalize the component N and root growth models.

Conversion

The technology to convert biomass to methane on a commercial scale is poorly understood. This is the area where the most improvement in efficiencies is required to produce gas economically. The approach to modeling the basic conversion processes attempts to describe the best estimate of expected rates for large scale systems. In some cases limited data are available to quantify parameters, therefore it was often necessary to depend upon expert opinion to estimate values. The advantage of using the modeling approach is that it identifies missing information and it provides a method to utilize bench scale data to study commercial scale systems.

A schematic diagram of a proposed commercial scale system (Fig. 6) shows that any number of leachbeds (LB) operated in parallel and sequentially filled with biomass could comprise the system. Leachate from the LB is cycled through any number of parallel packed bed (PB) reactors. The recycle streams pass through the PB and return to the LB. The LBs are operated in batch mode and the PBs are operated continuously.

The LB model equations (unpublished report to the Gas Research Institute by J. Andrews, Rice University, Houston, TX, titled: On the Mathematical Modeling of Biological and Packed Bed Processes for Converting Biomass to Methane—see also Chapter 18) describe the time variability of mass concentrations. The reactant is the biomass expressed in terms of COD equivalents. The biomass consists of two components, the fast biodegradable biomass (BMF) and the slow biodegradable biomass (BMS). Biomass is assumed to be converted first into soluble intermediates (SI) and then into the final product, the fatty acids (FAT) by acid forming bacteria (XA). The main process in the LB is the conversion of biomass to fatty acids by acid forming bacteria. There will be some methane produced by inflowing methane producing bacteria XM.

The governing equations include seven LB state variables, $Y(I)$, which are defined as: slow and fast biodegradable biomass concentration $Y(1)$ and $Y(2)$; soluble intermediate concentration $Y(3)$; fatty acids concentration $Y(4)$; acid forming bacteria concentration $Y(5)$; methane forming bacteria concentration $Y(6)$; and methane production $Y(7)$. The associated daily change of the concentration is characterized by $YP(I)$ and can be

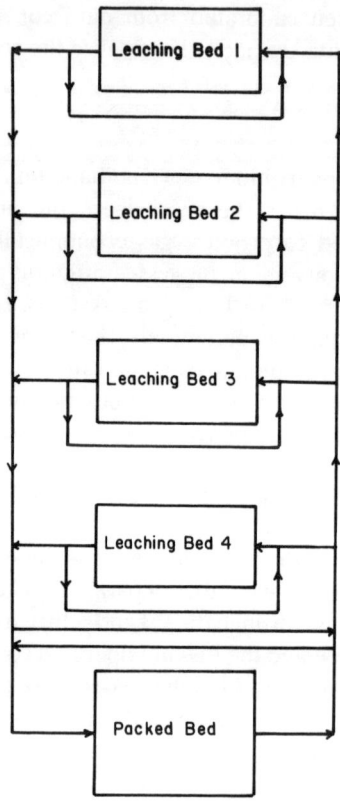

FIG. 6. Conceptional leachbed–packed bed conversion system plan.

computed from the following equations:

$$YP(1) = -K1P \times Y(1) \qquad (16)$$

$$YP(2) = -K1 \times Y(2) \qquad (17)$$

$$YP(3) = RTM1 \times (SIO - Y(3)) + DUM1 - Y(5) \times MUXA/YB \qquad (18)$$

where:

$$MUXA = K2 \times Y(3) \qquad (19)$$

$$DUM1 = (K1 \times Y(2) + K1P \times Y(1))/Y1 \qquad (20)$$

$$YP(4) = RTM1 \times (FATO - Y(4)) + MUXA \times Y(5) \\ \times YA/YB - MUXM \times Y(6)/Y2 \qquad (21)$$

Here MUXM and the retention time (1/RTM1) is given by:

$$\text{MUXM} = \text{MUHAT} \times Y(4)/(KS + Y(4)) \qquad (22)$$

$$\text{RTM1} = \text{FLOW/VULB} \qquad (23)$$

$$YP(5) = \text{RTM1} \times (\text{XAO} - \text{XAOUT}) + Y(5) \times (\text{MUXA} - \text{KD1}) \qquad (24)$$

$$YP(6) = \text{RTM1} \times (\text{XMO} - \text{XMOUT}) + Y(6) \times (\text{MUXM} - \text{KD2}) \qquad (25)$$

The outflowing bacteria concentrations, XAOUT and XMOUT, are assumed to be inversely proportional to the LB retention time and proportional to the corresponding bacteria concentration.

$$\text{XAOUT} = Y(5) \times K6 \times \text{RTM1} \qquad (26)$$

$$\text{XMOUT} = Y(6) \times K6 \times \text{RTM1} \qquad (27)$$

$$YP(7) = \text{VULB} \times YD \times Y(6) \times \text{MUXM}/Y2 \qquad (28)$$

All concentrations are expressed in g CODe liter^{-1}, except for bacteria concentrations which are expressed in g liter^{-1} and the methane production is given in liters at STP.

To account for the effects of temperature, relationships developed by O'Rourke (1968) have been used to modify KS, K2, and MUHAT.

$$KS = KS_{35} \, 10^{0.009\,(35-T)} \qquad (29)$$

$$\text{MUHAT} = \text{MUHAT}_{35} \, 10^{-0.016\,(35-T)} \qquad (30)$$

$$K2 = K2_{35} \, 10^{-0.025\,(35-T)} \qquad (31)$$

where:

T = Temperature of reactor (°C)
$KS_{35}, K2_{35}, \text{MUHAT}_{35}$ = parameter values for $T = 35°C$

The PB system is modeled as two reactors in series, a completely mixed slurry reactor (lower part) and a fixed film reactor (upper part).

The reaction which predominates in the PB is the conversion of fatty acids into methane. Again, the differential equations for the mass balance are expressed in terms of mass concentrations $Y(I)$ and their associated daily time variations $YP(I)$. Since there is no flow of biomass out of the LB the relevant state variables are now acid forming bacteria $Y(1)$, methane producing bacteria $Y(2)$, soluble intermediate concentration $Y(3)$, and fatty acids concentration Y(4). To represent the lower one-third of the PB the following four equations are used:

$$YP(1) = Y(1) \times (MUXA2 - KD1) - (RTI \times XAOUT + RTIW \times Y(1)) \tag{32}$$

$$YP(2) = Y(2) \times (MUXM2 - KD2) - (RTI \times XMOUT + RTIW \times Y(2)) \tag{33}$$

$$YP(3) = Y(1) \times MUXA2/YB + RTI \times (SI1 - Y(3)) \tag{34}$$

$$YP(4) = Y(1) \times MUXA2 \times YA/YB - Y(2) \times MUXM2/Y2 + RTI \times (FAT1 - Y(4)) \tag{35}$$

where MUXA2 and MUXM2 are given by

$$MUXA2 = K2 \times Y(3) \tag{36}$$

$$MUXM2 = MUHAT \times Y(4)/(Y(4) + KS) \tag{37}$$

In the PB the slurry reactor flow (F) is divided into a waste flow (FW) and a fixed film reactor flow ($F - FW$). If we define RFW as the ratio of the waste flow to the recycling stream as:

$$FW = RFW \times F \tag{38}$$

the associated reciprocal retention times are:

$$RTI = F/VB \tag{39}$$

$$RTIW = FW/VB \tag{40}$$

$$RTIB = (F - FW)/VU \tag{41}$$

The upper two-thirds of the PB behaves like a plug-flow reactor. This is approximated by four completely mixed reactors in series with different fatty acids concentrations Y(5) − Y(8). These different concentrations in each subreactor are determined by the following four difference equations.

$$YP(5) = RTIB \times (Y(4) - Y(5)) - K5 \times Y(5) \tag{42}$$

$$YP(6) = RTIB \times (Y(5) - Y(6)) - K5 \times Y(6) \tag{43}$$

$$YP(7) = RTIB \times (Y(6) - Y(7)) - K5 \times Y(7) \tag{44}$$

$$YP(8) = RTIB \times (Y(7) - Y(8)) - K5 \times Y(8) \tag{45}$$

Here it is assumed that the soluble intermediate concentration and the bacteria concentrations are constant through the whole PB. Then the methane production in the PB adds up to:

$$YP(9) = CH42DT + K5 \times YD \times (Y(8) + Y(7) + Y(6) + Y(5)) \times V3 \quad (46)$$

where CH42DT is the methane production in the lower one-third of the PB and is defined as:

$$CH42DT = V2 \times YD \times Y(2) \times MUXM2/Y2 \quad (47)$$

Again the same temperature dependence was assumed for MUXA2, MUXM2, K2 and K5 as for the LB model equations. For the PB the initial values and constants are used assuming that no outflow of bacteria from the PB occurred. Therefore,

$$XAOUT = 0.0 \quad (48)$$

$$XMOUT = 0.0 \quad (49)$$

To demonstrate the performance of the conversion model Figs 7 and 8 present outputs of a single LB and the total methane output from all PBs. The desired output of the LB is the fatty acids. The cyclic nature of the methane output in Fig. 8 results from the configuration and operation of the LB. During the 28 days between refills a sharp increase of fatty acid output occurs followed by a slow decrease (Fig. 7). Because there are four LBs, the frequency of the cycles for the PB follows the total supply of fatty acids.

Because the overall energy balance of the conversion system can affect the cost of methane sufficiently, a model of the temperature control requirements of the reactors was developed. As reported above, the conversion model responds to temperature by increasing methane output at higher temperatures. It was also assumed that methane will be used to supply fuel to heat the reactors. Total heat requirements include (1) replacement of heat loss from the reactor surfaces, (2) replacement of heat loss during recycle of leachate, and (3) the addition of heat to bring the biomass to the operating temperature. Insulation can be added to the reactor wall as required. The selection of a boiler size is based upon worst case heat loads with the assumption the boiler will be 50% efficient. In addition, heat loss from piping and miscellaneous equipment is proportional to the heat loss of the reactor walls.

Results of simulations with our assumptions for a reactor operated at several temperatures (Table 1) indicate the minimum cost of operation

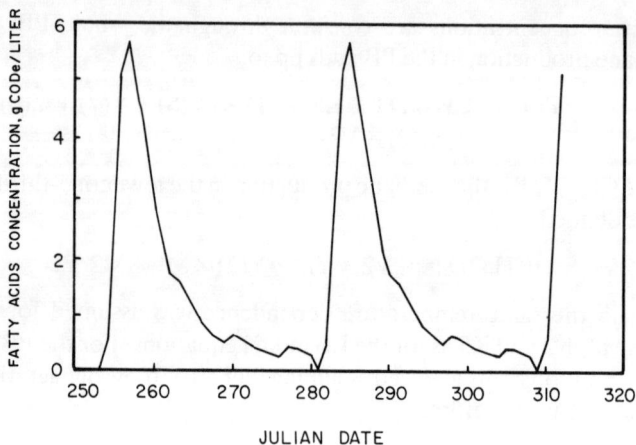

FIG. 7. Simulated fatty acid output from a single leachbed with recycle in a system with four leachbeds and one packed bed.

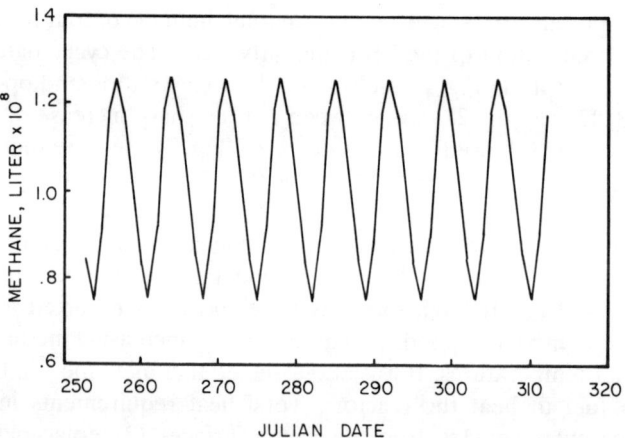

FIG. 8. Simulated output of methane from the leachbed–packed bed system.

occurs when the reactors are not heated. The improvement of reactor stability and minimum risk criteria were not included in the analysis; these factors, however, may be more important than the cost saving gained without heating.

TABLE 1
Simulated Heat Requirements and Cost of Gas at Different Reaction Temperatures Assuming Florida Environmental Conditions

Operation temperature (°C)	Total CH_4 produced (liter)	Gas used for heating (liter)	Total capital cost (levelized) ($ GJ^{-1})
No heating	2·96E+10	0·00	2·17
20	3·00E+10	2·09E+9	2·32
25	3·16E+10	4·01E+9	2·36
30	3·29E+10	6·84E+9	2·49
35	3·41E+10	1·023+10	2·69

Energetics

It is essential to perform energetic analyses of potential biomass/methane systems so that their energy feasibility and positive net energy is assured before commercialization. Energy feasibility may be determined using the measure, Energy Yield Ratio (Fluck and Baird, 1980). Energy Yield Ratio is the quotient of the summation of F, all feedback energy from various sectors of the general economy required to produce methane, and Y, the energy content of the methane produced. Most currently commercially used energy sources exhibit Energy Yield Ratios between 4 and 20, with higher values indicating greater desirability from an energy analysis point of view. Any Energy Yield Ratio greater than unity indicates a positive net energy in that more energy is produced than is consumed by the process. To perform an energy analysis, all direct and indirect energy inputs are identified and quantified, and the summation is compared with the energy produced by the biomass/methane system by forming the Energy Yield Ratio. Direct energy flows include diesel fuel, electric power, etc. Indirect energy flows include the energy required to provide fertilizer, buildings, equipment, labor, etc.

The first generation of energy analyses made thus far within BIOMET are incomplete in that they do not yet include conversion, but only include production, harvesting and transportation. Energy Yield Ratios found thus far range from 4·7 to 9·0 for water hyacinth and from 11·4 to 15·2 for Napiergrass. These values of Energy Yield Ratio are competitive with many alternative energy production systems. However, inclusion of conversion is expected to reduce these values considerably as additional energy inputs are included, especially those for heating reactors. Nevertheless, results so far are encouraging.

Economic Models

The goal of BIOMET is to predict the cost of gas assuming a commercial scale system. To accomplish this, BIOMET includes budget generators and levelized cost economic models that characterize the cost requirements of the proposed conversion and production system. BIOMET will simulate the performance of crop growth, biological conversion, and the heat balance of the reactor, and the remainder of the system is based upon estimated fixed and variable costs.

Total costs of the production of methane from water hyacinth and Napiergrass are simulated in order to:

1. provide an efficient cost accounting method which can be easily assessed and modified; and
2. express system performance with a cost per unit output over the life of the system.

In order to achieve the latter objective, a levelized cost-of-service price (Clark et al., 1982) is used. A constant-dollar, single-valued price per unit gas is calculated which represents the time-varying cost-of-service over the life of the plant. This price, if charged for each unit of gas produced over the life of the plant, would yield the same discounted present value of revenue as would the actual cost-of-service price. The revenue must be sufficient to:

1. amortize debt;
2. cover operating and maintenance costs; and
3. provide a return on both common and preferred equity.

Capital and variable costs of production, harvesting, transportation, and conversion to methane are based on user-defined scenarios.

The generalized equation describing the levelized cost-of-service per unit gas produced is given by:

$$\text{TSYSCL} = (\text{GTPI} + \text{WTPI} + \text{CCT}) \times \text{CCR} + \text{VOMNGL} + \text{VOMWHL} + \text{CVOMCL} \qquad (50)$$

where:

TSYSCL	= total system levelized cost ($ GJ^{-1})
GTPI	= total capital investment for Napiergrass ($)
WTPI	= total capital investment for water hyacinths ($)
CCT	= conversion capital cost ($)
CCR	= capital charge rate (GJ^{-1})

VOMNGL = levelized variable O & M costs for Napiergrass ($GJ^{-1})
VOMWHL = levelized variable O & M costs for water hyacinths ($GJ^{-1})
CVOMCL = levelized variable O & M costs for conversion ($GJ^{-1})

The capital charge rate (CCR) represents the required rate of return on the capital investment. This rate depends on a number of variables including book life, construction period, tax rate, inflation rate, tax life, service factor, design capacity, discount rate and leverage rate. The user can modify any of these values to examine their effect on the levelized cost-of-service price.

Variable costs of production, harvesting, transportation and conversion of Napiergrass and water hyacinths are levelized to account for escalation in the variable O & M costs. Individual cost elements (fuel, land rental, by-product credit, and general O & M) are handled separately. This allows the user to examine the effect of different escalation rates for each major component of O & M costs.

The cost model calculates the initial investment (for biomass feedstock and methane production) prior to simulation based on user-defined scenarios, and then accumulates the variable costs during simulation. The levelized cost model uses these costs to calculate the levelized cost-of-service price. Initial investment and variable costs are calculated separately for Napiergrass production, harvesting and transportation; for water hyacinth production and harvesting; and for conversion of feedstocks into methane.

Water hyacinths costs are based on a winch-boom design fabricated and tested by Bagnall (1986). The user is limited to a choice of four sizes estimated to produce feedstock for 0·1, 0·5, 1·0 and 3·0 × 10^6 GJ conversion plants. Capital investment and variable cost are estimated based on the user-supplied geometry, size, and operating parameters. Napiergrass production, harvesting and transportation costs include both capital investment and variable operating costs. Components of the initial capital investment include: harvester, trucks and initial land preparation. Annual variable cost components include: land rental, crop management (fertilizer, pesticides, etc.), labor, fuel (for trucks and harvesters), and machine operating maintenance.

The economics of the conversion system for Napiergrass is divided into unit processes of biomass storage, LB reactors, leachate storage, PB reactors, gas pressurization, gas clean up, insulation, heat equipment and effluent disposal. To compute the cost of each component, a capital and variable cost curve (Warren *et al.*, 1985) was used that described cost as a function of design capacity and actual usage.

System Simulation

BIOMET was used to simulate the growth of water hyacinths for a period of 70 days exposed to various temperatures (Fig. 9), with 15°C, the lowest temperature simulated, being the point where biomass accumulation essentially stops. Growth rates increased as temperatures increased up to a maximum rate at 30°C. For these growth rates to occur, N in the amounts shown must be available for uptake from the water. If this level cannot be supplied directly from sources such as lake sediment or lake inflow, then N

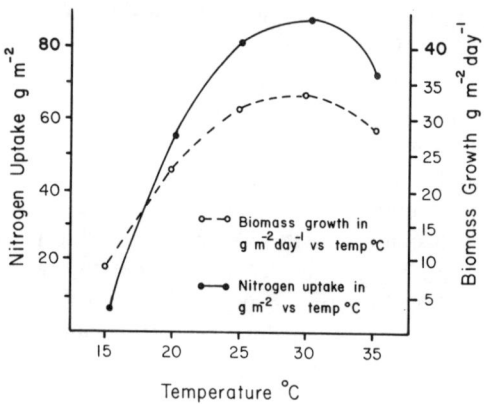

FIG. 9. Simulated response of water hyacinth growth and nitrogen uptake in response to varying environmental temperatures.

as foliar applications may be needed. The decision to add N must be based upon the need to maintain the potential growth rates. However, it may not be economically or environmentally feasible to apply fertilizer to an aquatic system.

To examine the increase in cost due to increased transport distance, BIOMET was used to simulate example cases by varying the distance from the conversion site of a single 3200 ha field. The conversion cost, number of harvesters and production cost were held constant and only the number of trucks and cost of operation were varied. The increased capital cost to add more trucks was a minor increase in cost as compared to the increased variable cost (Table 2). For the shortest distance of 0·1 km the levelized production cost was $2·11 GJ^{-1} using six trucks. Where the field was

TABLE 2
A 3200 ha Napiergrass Farm Various Distances from the Conversion Site

Distance (km)	Number of trucks	Biomass harvested (t ha^{-1})	Distance traveled (10^3 km)	Capital[a] cost ($ GJ^{-1})	Variable[a] cost ($ GJ^{-1})	System[b] cost ($ GJ^{-1})
0·1	6	31·9	60	0·12	1·99	6·50
10	12	31·9	1 212	0·12	2·24	6·75
20	19	31·9	2 376	0·14	2·48	7·01
30	25	31·9	3 540	0·15	2·73	7·26
60	44	31·9	7 031	0·18	3·46	8·02
100	70	31·9	11 686	0·21	4·42	9·02

[a] The capital and variable cost included are for biomass production.
[b] Includes a constant cost of $4·39 GJ^{-1} for conversion.

TABLE 3
Simulations of a 3200 ha Napiergrass System 20 km from the Conversion Site

Number of trucks	Biomass harvested (t ha^{-1})	Distance traveled (10^3 km)	Capital[a] cost (GJ^{-1})	Variable[a] cost (GJ^{-1})	System[b] cost (GJ^{-1})
9	13·5	567	0·12	4·05	10·04
12	20·1	1 102	0·12	3·06	8·23
15	26·3	1 778	0·13	2·60	7·32
19	31·9	2 376	0·14	2·48	7·01
31	31·9	2 376	0·15	2·49	7·03

[a] The capital and variable cost included are for biomass production.
[b] Includes a constant cost of $4·39 GJ^{-1} for conversion.

100 km away, 70 trucks were required and the production cost increased to $4·63 GJ^{-1}, however, of this increase only $0·09 GJ^{-1} was due to increase in capital cost. These results indicate that it is important to design improved transport systems for the longer distances.

In the same example the number of trucks was varied to determine the cost sensitivity of the methane gas. The standard case chosen was the 3200 ha field located 20 km from the conversion site. The simulations performed included examples with the truck numbers from 9 to 31 which resulted in a cost variance from $70·01 GJ^{-1} for the case of 19 trucks and $10·04 GJ^{-1} for the case with 9 trucks (Table 3). The reason for the high

TABLE 4
Simulations with Different Harvest Schedules Varying from High in Summer and Low in Winter to Low in Summer and High in Winter

Summer/winter (%)	Field size (ha)	Number of trucks	Number of harvesters	Capital[a] cost ($ GJ^{-1})	Variable[a] cost ($ GJ^{-1})
75/25	3 250	59	6	0·22	3·15
64/36	3 405	49	5	0·20	3·18
50/50	3 815	49	5	0·20	3·58
36/64	3 985	59	6	0·22	3·68
25/75	4 225	69	7	0·25	3·84

[a] The capital and variable cost included are for biomass production.

cost when the number of trucks was decreased was due to the resultant decrease in harvested biomass because of limited transportation. From a cost viewpoint it is important not to limit available transportation.

The results of simulations for a Napiergrass system with different harvest schedules are presented in Table 4. Biomass productivity is greater in the summer months, therefore, by varying the rate of harvest, there is an effect on total annual yield from a given production area (Table 4). For the simulation, gas and biomass demand were varied from high in the summer and low in the winter to low in the summer and high in the winter. The winter months were defined as October through March. The total farm size was varied to allow the biomass supply to meet the desired gas production schedule. The schedules with 25% of the biomass harvested in the summer required the largest field size and had the greatest cost. The lowest biomass production cost occurred when most of the harvest occurred during the summer months. Considerations such as a variable gas market will also influence the value of gas produced during different months of the season. The large capital investment in the conversion facility encourages constant rates of gas production.

SUMMARY

The development of BIOMET is an on-going activity that has not been completed. Nevertheless, BIOMET is a very useful tool for integrating new research information and determining the impact on the overall performance of the conceptualized systems. Additional work will focus on the improvement of process conversion models. The conversion of the biomass to methane represents about 60–70% of the total cost of the gas. Improving this technology will greatly reduce the cost of methane. Several problems have been identified in the management of the biomass crop. One of these is the unknown effect of multiple harvests. Currently, we are using the crop models to study response to various management options to prevent major yield losses. Plans are to investigate the risk factors associated with major events causing crop loss. As determined in this chapter, to be cost effective the system must not be limited by a lack of biomass. Protection from biomass shortages caused from events such as freezing temperatures must be designed into the system.

Results indicate that methane produced from biomass can be cost competitive with other energy sources. However, the continuation of focused research efforts is required to develop the necessary technologies and the procedures to manage these technologies.

REFERENCES

Bagnall, L. O. (1986). Harvesting systems for aquatic biomass. In: *Biomass Energy Development*, Smith, W. H. (ed.), pp. 259–73. Plenum Press, New York.
Clark, C. E., Francer, R. B. and Requlinski, S. G. (1982). A levelized cost of service price formula, pp. A2–A17. DFI Project 1267. Decision Focus, Incorporated, Palo Alto, California.
Fluck, R. C. and Baird, C. D. (1980). *Agricultural Energetics*. AVI Publishing Company, Incorporated, Westport, Connecticut, 192 pp.
Hayes, T. D., Biljetina, R., Reddy, K. and Chynoweth, D. P. (1986). An integrated wastewater energy production system. *Energy From Biomass and Wastes X*. Washington, D.C. (in press).
Lorber, M. N., Mishoe, J. W. and Reddy, K. R. (1984). Modeling and analysis of waterhyacinth biomass. *Ecological Modeling* 24, 61–7.
Mishoe, J. W., Lorber, M. N., Peart, R. M., Fluck, R. C. and Jones, J. W. (1984). Modeling and analysis of biomass production systems. *Biomass* 6, 119–30.
O'Rourke, J. T. (1968). Kinetics of anaerobic waste treatment at reduced temperatures. Ph.D. dissertation, Stanford University, Stanford, California.
Penning de Vries, F. W. T. (1975). The cost of maintenance processes in plant cells. *Annals of Botany* 39, 77–92.
Reddy, K. R. (1983). Waterhyacinth production systems in nutrient-rich waters. In: *The Methane from Biomass and Waste Program Annual Report for 1982*, Smith, W. H. (ed.), pp. 30–2. Institute of Food and Agricultural Sciences, University of Florida, Gainesville FL.
Warren, C. S., Bruderly, D. E., Angelieri, M., Bilello, L. J., Bucalo, S., Finger, G. W., Hart, R., Hinton, S. W., Newman, J. R. and Vinzant, J. W. (1984). The

Methane from Biomass and Waste Program—Task I: Evaluation of the Lake Apopka Natural Gas District. Contract No. 5080-323-0423(E), Reynolds, Smith and Hills Architects-Engineers-Planners, Incorporated, Task Report (January). Gas Research Institute, Chicago, Illinois, 288 pp.

Warren, C. S., Hinton, S. W. and Knott, S. J. (1985). The Methane from Biomass and Waste Program—Task IV: Equipment Costs Handbook for Biomass and Waste Systems. Contract No. 5080-323-0423(E), Reynolds, Smith and Hills Architects-Engineers-Planners, Incorporated, Vols 1 and 2—Draft Final Report (October). Gas Research Institute, Chicago, Illinois, 212 pp.

4
Conceptual Design of a Commercial Biomass-to-Methane System

T. D. HAYES
Gas Research Institute, Chicago, Illinois, USA

C. S. WARREN and S. W. HINTON
Reynolds, Smith and Hills Architects-Engineers-Planners, Inc., Jacksonville, Florida, USA

INTRODUCTION

The development of a biomass-to-methane (BTM) system requires an answer to the fundamental question of whether the technology could ever be economically competitive, even if reasonable research goals are achieved. In order to answer this question, researchers, modelers and engineers must work together to: (1) create an educated vision of what a biomass-to-methane system might physically and economically look like given today's technology, (2) establish research goals in light of their understanding of scientific principles, and (3) quantify incremental improvements in cost that could be realized if key goals are met.

Our purpose is to present the results of a conceptual design and preliminary cost analysis study conducted in 1982, just one year following the initiation of the GRI/IFAS Methane from Biomass Program. At that time, such a study was needed to answer a number of questions of critical importance to the systems modeling effort and research planning; these included the following:

- At what cost could pipeline quality methane be produced from a BTM system given what might be possible to do with today's technology?
- At what cost could pipeline quality methane be produced if key BTM system performance goals were achieved through research?
- Could BTM systems ever become competitive with other sources of natural gas without requiring major breakthroughs in technology?

In order to address the above issues, the study focused on the conceptual design and cost analysis of BTM systems for a specific site in Florida, the Lake Apopka Natural Gas District (LANGD). The scope of the study included a survey of the availability of land and aquatic areas for the production of high-yielding biomass crops, conceptual layout and design of BTM systems as if they were to be implemented in the LANGD, and a preliminary technical and economical evaluation of each BTM system examined. A variety of sources of information was used in this study. Primary sources included preliminary data from this program on energy crop production and anaerobic digestion, expert opinion from researchers in and external to the program, the literature, site specific data from the Lake Apopka Natural Gas District, and vendors of the major types of equipment utilized in the conceptual design. Where information was needed but not available, assumptions were clearly specified and utilized in the study.

The unique aspect of this study was that design engineers and researchers pooled their expertise to conceptually design and evaluate BTM systems as applied to a specific region of Florida (LANGD) of certain known logistical, political, regulatory, topographical and commercial characteristics. The conceptual nature of the study, however, established the accuracy of the estimated costs at +30% and -20%. The study, therefore, should not be viewed as a reliable guide for detailed design and facility costing.

The analysis summarized here has been used by this program as a rough indicator of the technical and economic status of the integrated BTM concept and the potential for cost improvement through certain research achievements (Warren et al., 1984). Equally important, the study has been used to calibrate computer models (see Mishoe et al., Chapter 3) that simulate the performance and costs of BTM systems. Together with the modeling effort at IFAS, the engineering analysis presented in this chapter has served as an effective tool in quantifying achievements in biomass research in terms of cost improvements and in the planning of effective research directions.

APPROACH

The general steps used to perform this conceptual analysis were the following: (1) select an area of a geographical site of application; (2) using the recommendations from IFAS researchers, identify the most promising energy-crop species for the selected site in terms of potential productivity and convertibility to methane; (3) assess the availability of productive land and aquatic areas to support BTM systems; (4) conceptually design BTM systems based on current and projected research achievements; (5) estimate mass and energy flows through each system; (6) determine size (capacity) and quantity of each of the major pieces of equipment and unit processes; and (7) perform economic and cost sensitivity analyses for each system using conceptual costing techniques (Peters and Timmerhaus, 1980) and financing commonly used by the gas industry (Clark et al., 1982). The generic BTM flow scheme that served as the focus of this study is shown in Fig. 1. Detailed features of this flow scheme were developed for each specific system selected for evaluation.

The area of Florida selected as the site of the study was the Lake Apopka Natural Gas District (LANGD) near Orlando, encompassing West Orange County and South Central Lake County (Fig. 2). This area, identified by the gas company that provides gas to the region, was chosen by virtue of a number of important attributes: (1) the availability of land and water areas suitable for the production of terrestrial and aquatic biomass; (2) an existing pipeline distribution system serving a gas market averaging 3200 GJ day^{-1}; and (3) an interest of LANGD directors in the potential of BTM technology in providing an alternate, stable supply of methane to its customers.

The district contains large tracts of cropland and extensive areas of lakes. The critical question addressed was how much of these areas could be made available for BTM systems given the current commercial use, environmental regulations and zoning restrictions of the locale. Since the district's economic base is principally agricultural, the farming community received considerable attention from this study. Most of the land in the district is in citrus production; however, the Zellwood area on the north side of Lake Apopka is also one of the state's major vegetable-producing areas. Since numerous agricultural processing plants exist in the district, transporting crops to a central processing facility, as would be envisioned for a BTM system, is a well-accepted operation in the farming community. Other factors related to farmer acceptance of energy cropping such as adaptability to new management practices, rate of return, and land application of conversion process effluents, were considered in the study's effort to determine the amount of land that might be dedicated to energy cropping and the economic conditions that would provide sufficient incentives for farmers to participate in energy farming ventures. These conditions were incorporated into the financial assumptions used in energy farming economics.

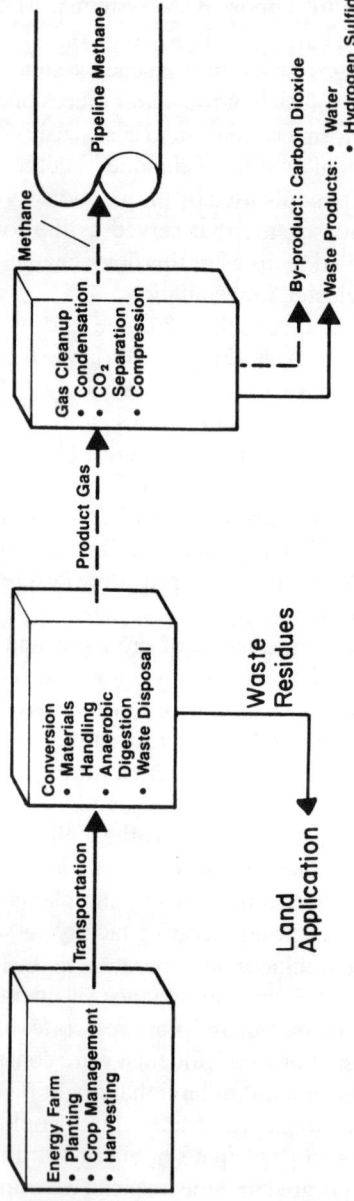

FIG. 1. Elements of a biomass to methane system. (——— Flow of solids and liquids; ——— flow of gases.)

Conceptual Design of a Commercial Biomass-to-Methane System

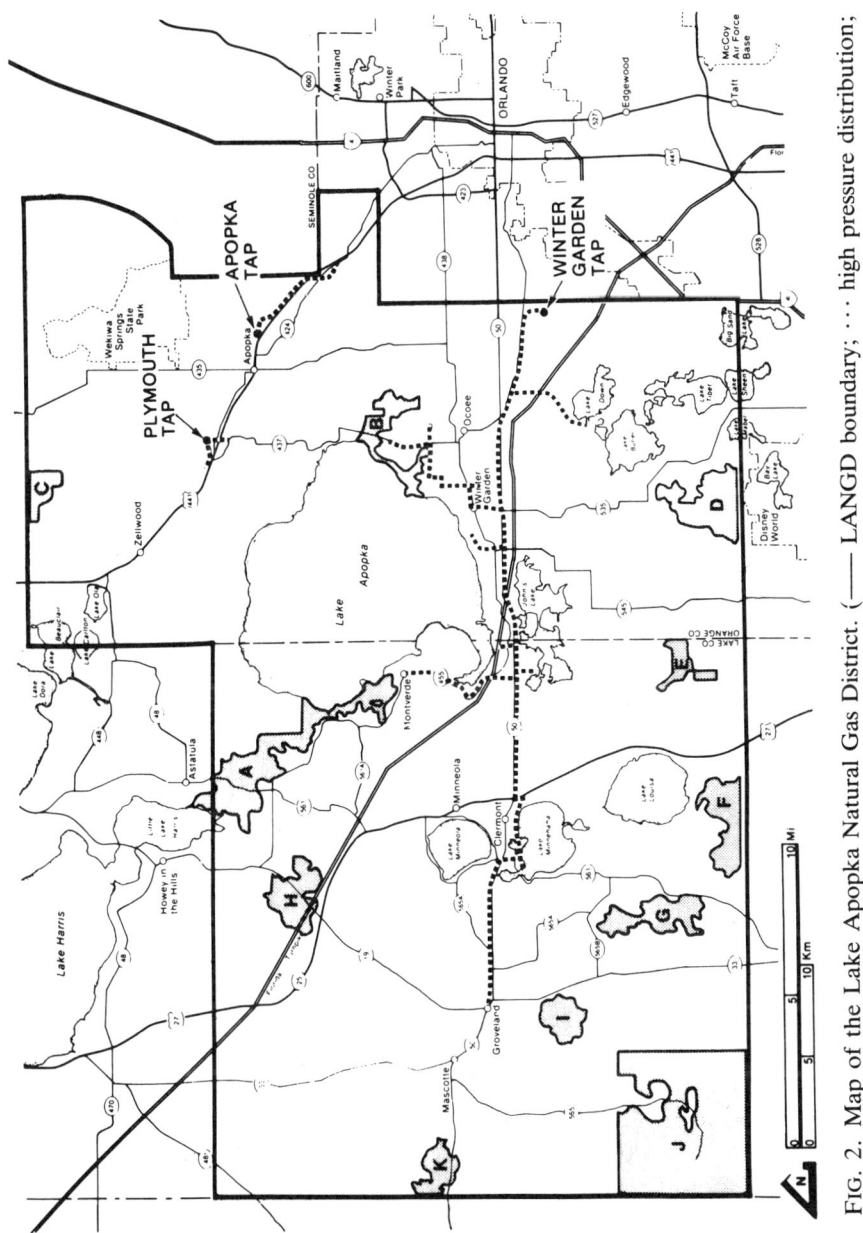

FIG. 2. Map of the Lake Apopka Natural Gas District. (——— LANGD boundary; ⋯ high pressure distribution; areas A–K are potentially favorable biomass production sites. Sources: LANGD, 1982; ESE, 1983.)

Napiergrass and water hyacinth were selected as the biomass energy crops of emphasis for the LANGD on the basis of preliminary production yield data from small plots and bioassay information which indicated favorable ultimate biodegradabilities for these biomass species in anaerobic digestion. At the time of this study, IFAS researchers were achieving nearly 50 dry Mg ha^{-1} yr^{-1} in the region of central Florida in the production of water hyacinth in small lake water test channels and more than 30 dry Mg ha^{-1} yr^{-1} in the production of Napiergrass in small field plots (0·1 ha). Although more research was needed to develop optimum energy crop management methods for these species, water hyacinth and Napiergrass were considered to be among the fastest growing species that could be produced at an acceptable level in the LANGD. These two biomass species together could effectively utilize the terrestrial and aquatic areas potentially available for energy cropping.

Conceptual designs for water hyacinth and Napiergrass systems were then prepared as a collective effort involving IFAS researchers, engineers from Reynolds, Smith and Hills and invited consultants. The Napiergrass and water hyacinth systems were considered under two technology applications scenarios: (1) base case design or the conceptual design of the system based on what might be achieved using today's technology; and (2) advanced case design based on the achievement of key research and development goals related to production and conversion. The scope of the systems costing portion of the study, then, was based on the evaluation of four cases. Assumptions used in the analysis of these four BTM systems are summarized (Table 1) and discussed here.

Napiergrass BTM Systems

Napiergrass BTM systems, base and advanced cases, both assumed that the biomass energy crop would be produced on productive cropland, since preliminary calculations indicated that the achievement of very high biomass yields on good quality farmland would result in significantly lower per-megagram costs of biomass than possible with marginal land.

In the conceptual BTM system designs for Napiergrass, planting cost is based on the establishment of plantlets at a row spacing of 1·0 m on each hectare of energy farm because at the time of the study there were no high-yielding Napiergrass hybrids that were capable of producing seed. This planting cost, however, was distributed over a 7-year cycle due to the perennial nature of Napiergrass production. Fertilizers are applied at rates of 134 kg of nitrogen, 17 kg of phosphorus and 34 kg of potash ha^{-1} yr^{-1} and lime addition at a rate of 740 kg ha^{-1} yr^{-1} is assumed. On the basis of

TABLE 1
Assumptions Used in the Analysis of Baseline and Advanced BTM Systems Sized for a Net Methane Output of 3200 GJ day^{-1}

	Low-solids baseline	Low-solids advanced	High-solids baseline	High-solids advanced
Feedstock production and harvesting				
Feedstock type	Hyacinth/pennywort	Hyacinth/pennywort	Napiergrass	Napiergrass
Annual yield (dry Mg ha^{-1})	50	75	30	45
Area required (ha)	2 520-Lake Apopka	1 780-Lake Apopka	4 210-sites within LANGD	2 150-contiguous site
Harvesting method	Hydrilla harvester	Winch-boom	Silage harvester	Silage harvester
Nutrients added	No	Yes	Yes	Yes
Biomass transportation				
Method	Barges/trucks	Slurry pipeline, 5% solids	Truck/trailer	Truck/trailer
Average distance (km)	7·8	1·7	40·5	4
Conversion system				
Acids stage	Slurry reactor	Slurry reactor	Slurry reactor	Leaching-bed reactor
Biogas stage	Packed-bed reactor	Fixed-film reactor	Packed-bed reactor	Packed-bed reactor
Methane yield (m^3 kg^{-1} VS)	0·44	0·28	0·44	0·38
Feedstock requirements (dry Mg day^{-1})	345	365	345	265
Methane content (%)	60	60	60	85
Lime requirements (Mg day^{-1})	55	0	55	9
Heat input requirements	Yes	No	Yes	No
Biogas production (m^3 day^{-1})	207 000	143 000	207 000	101 000
HRT (days) acid prod./methane prod.[b]	1·5/1·5	3·0/3·0	1·5/1·5	—/1·5
Reactor volumes (m^3)				
Acid phase	11 300	16 000	11 300	9 970
Methane phase	18 200	25 200	18 200	12 600
Gas cleanup and compression				
Cleanup method	Kryosol process	Kryosol process	Kryosol process	Kryosol process
Compression (10^6 Pa)[a]	Two-stage 0·1 → 1·7	Two-stage 0·1 → 1·7	Two-stage 0·1 → 1·7	Two-stage 0·1 → 1·7
Net methane output (GJ^{-1} day^{-1})	3 200	3 200	3 200	3 200

[a] Compression was designed to 1·7 × 10^6 Pa for integration into LANGD's distribution system; compression to 6·2 × 10^6 Pa for transport pipeline integration involves an added stage.
[b] Hydraulic retention time in days of the acid production and methane production stages of the conversion system.

the history of Napiergrass production in south and central Florida as a forage crop under natural climate, no irrigation was used in the conceptual BTM designs. Cultivation is performed using conventional mechanized agricultural equipment and harvesting is accomplished using a two-row harvester driven by a 90–110 kW tractor. This system blows the cut and ground material into a forage wagon which follows the harvester. Full wagons are then taken to a transfer station on the farm where the chopped Napiergrass in the wagons is unloaded by a bottom conveyor into 10 Mg trailers. Since Napiergrass is produced and harvested continuously throughout the year in southern Florida, no storage was employed in the BTM systems, though storage would no doubt be required for cold climate BTM systems where harvesting would be restricted to the summer and fall.

The biomass is then trucked directly from each of the energy farms to the feeding system of a two-phase anaerobic digestion system which biologically converts the biomass to a biogas mixture comprised of methane and carbon dioxide. Moisture and carbon dioxide are removed from biogas using the Kryosol process which performs a cryogenic gas separation based on differential condensation under low temperature and elevated pressure of 1.7×10^6 Pa. In each of the BTM system concepts, the resultant gas stream, containing 97% methane, is inserted into the LANGD natural gas distribution system; excess gas not used by LANGD is compressed to 6.2×10^6 Pa for injection into the Florida Gas Transmission Co. pipeline.

Undigested solids (effluent) from the anaerobic digestion facility are dewatered with a filter press to a 30% total solids cake and returned to the biomass production areas through the use of sound land application procedures; this practice conserves nutrients and soil-building organic matter and may have a positive long-term effect on erosion control and sustained biomass yields. Wastewater generated by the conversion process is disposed of through spray irrigation of land areas near the conversion facility.

Principal differences between the base and advanced Napiergrass BTM systems relate to planting technology and the biomass yields achieved in energy farming and to the design and performance of the conversion systems. The base case production yield was the yield obtained from IFAS experimental Napiergrass plots by 1982 amounting to 30 dry Mg ha^{-1} yr^{-1}. IFAS researchers estimated that yields could be increased by 50% through optimization of management practices and through conventional breeding to develop Napiergrass varieties of improved resistance to drought, pests and disease; hence, a sustainable yield of 45 dry Mg ha^{-1} yr^{-1} was assumed for the advanced Napiergrass energy farm. It was also assumed that costs for the base case planting technique could be reduced by 50% through the development of a new artificial seed planting technique. This concept would involve the economic production of plant embryos in large numbers from individual cells through the use of tissue culture; these plant embryos would be fluidized in a nutrient gel matrix and rapidly planted in the earth using a fluid drilling technique.

Although it was assumed for both cases that Napiergrass would be produced by private farms under contract to the conversion facility, the location of those farms relative to the conversion facility and the resultant transportation distances differ between the base and advanced cases. In the base case, farms were distributed throughout the LANGD and the average distance to the conversion facility is about 40 km. In the advanced case, it assumed that agreements could be established with farms in a contiguous area surrounding the conversion facility and that the average distance to the facility is reduced to 4 km.

Conversion technologies selected for the base and advanced Napiergrass BTM systems were also markedly different. At the time of this study, it was well known that the anaerobic digestion of many lignocellulosic materials resulted in low rates of conversion and low methane yields (Buswell, 1936; McCarty *et al.*, 1979). The complex structure of lignocellulose limits the accessibility of the plant cell wall to microbial attack, while the lignin content in the cell wall acts as a physical barrier to efficient degradation (Klein, 1972). Since the feedstock for the BTM system conversion facility was to be purchased at a significant cost, it was considered important that the anaerobic digestion process for both the base and advanced case be aimed at achieving high methane yields. Because of this requirement, emphasis in the base case design was placed on the pretreatment of the biomass feedstock.

A number of pretreatment methods were considered for the base case; the most noteworthy alternatives are summarized by Levy *et al.* (1986) including size reduction, alkali treatment, delignification by solvents, steam explosion, acid hydrolysis, cellulose solvent extraction, and enzymatic treatment. Caustic treatment of cellulosics is the oldest and best known method of increasing digestibility. This technique, often referred to as the mercerizing process, causes disruption of crystallites and amorphous material, a significant degree of depolymerization, an increase of the internal surface area, and hydration and swelling of the cellulose. By 1980, alkali treatment had been tested extensively to increase the biodegradation of straw (Woodman and Evans, 1947; McFarlane and Pfeffer, 1981; Colleran *et al.*, 1982), municipal refuse (Gossett *et al.*, 1976), and animal manure. Several good reviews of the caustic pretreatment are in the literature (Ashare and Wilson, 1979; Shuler, 1980). In general, caustic pretreatment consists of soaking biomass in sodium hydroxide for 1–24 h; maximum digestibility improvements are usually achieved at dosages of 0·1–0·15 g (2·5–3·8 meq) of sodium hydroxide per gram of cellulosic. Under these conditions, conversion efficiencies are increased to as high as 70–90% for certain types of forage biomass materials (Ashare and Wilson, 1979). Although sodium hydroxide is the most commonly used base for the alkaline pretreatment process, other chemicals, such as lime, can be used to control pH, to disrupt the lignocellulosic complex and improve digestibility.

Anaerobic digestion processes tested in conjunction with pretreatment have mainly been limited to the single stage CSTR and the two-phase digestion process. Two-phase digestion was selected for the base case because of reported kinetic advantages over single stage reactors (Ghosh *et al.*, 1975). In this process, complex biomass organics are converted to volatile acids and other soluble intermediates in the acetogenic phase followed by conversion of the intermediate soluble compounds (volatile acids) to methane and carbon dioxide in the methane formation phase. Solids retention times recommended for a CSTR used to perform the acetogenic first phase conversion of sewage sludge range from 12 to 72 h (Eastman and Ferguson, 1981). Use of an attached film packed bed process as the methane formation phase has allowed overall two-phase retention times to be decreased to as low as 3 days (Norrman and Frostell, 1977).

The conversion concept selected for the base case Napiergrass BTM system combines chemical pretreatment (lime) with the two-phase digestion process. Pretreatment conditions selected for the base case Napiergrass conversion concept include grinding, soaking of biomass feedstock for 24 h in a solution of lime (CaO) at a concentration of 0·2 g (3·6 meq) per dry gram of biomass solids at 35°C, and neutralization to pH 7·5. The neutralized slurry is held in a storage tank for 36 h and the partially solubilized biomass feed stream with a total solids concentration of about 5% is introduced into the acid-phase continuous stirred tank reactor (CSTR) at a 36-h retention time followed by a 36-h retention time packed bed. Retention times for the entire digestion system total 3·0 days and a conversion efficiency of 85% with a methane output of 0·44 m^3 kg^{-1} of organic feed is assumed. These conditions and performance levels selected for base case conversion appear optimistic compared to the solids retention times (> 10 days) and conversion efficiencies associated with conventional high-rate digestion of feedstocks that are not pretreated, yet the short retention times were chosen for the purpose of allowing a good evaluation of lime pretreatment. The raitonale was that the use of the lowest retention times for the reactors would give the technology the benefit of the doubt in terms of performance and would result in the lowest-cost possible for the lime pretreatment concept as a base case process; any reductions below that base case cost through advanced digestion research would then represent an indisputable improvement in technology. As will be seen in the results, however, the capital cost of the conversion facility is low compared to the total BTM system cost; increasing total reactor retention time from 3 to 10 days would only result in a modest additional cost of $0·30 GJ^{-1}, less than 3% of the total gas cost for the base case BTM systems.

The advanced case Napiergrass conversion scheme eliminated pretreatment and utilized a different type of two-phase digestion system consisting of a leaching bed biomass hydrolysis reactor followed by a packaged bed methane digester. Performance results of this process tested by European investigators at the bench scale on straw and grass clippings have been reported by Colleran *et al.* (1982) and the conditions and performance described for the experiments on grass clippings were used to conceptualize the advanced case conversion system for Napiergrass. This two-phase system, consisting of a leaching bed at an 18-day solids retention time (SRT) followed by a packed bed digestion of the leachate at a 3-day SRT achieved an 80% conversion of grass organics to methane at 30°C. This two-phase system theoretically makes it possible to increase the methane content of the gas produced from the methane phase through operation of the reactor system as a pH-swing carbon dioxide absorption/desorption gas separations system (Hayes and Isaacson, 1986); the use of this concept could result in the production of 80–90% methane gas directly from the methane formation reactor allowing substantial savings in gas cleanup costs.

In the advanced Napiergrass conversion concept, chopped biomass with a 35% total solids content is dumped without dilution into large open concrete tanks of 106 m diameter × 9 m high. Sufficient number of leaching beds is provided to allow a 60-day SRT: 5 days to fill the bed, 30 days reaction time, 20 days for draining and dewatering, and 5 days for unloading the effluent solids. Water is sprayed over the material and as the Napiergrass is hydrolyzed by acid-phase bacteria, soluble products (including volatile acids) are removed by the liquid leaching through the bed. The leachate is then transferred to a packed bed digester operated at an SRT of 3 days where solubilized organic matter is converted to an 85% methane gas product at a methane yield of 0·38 m^3 kg^{-1} organic feed. Operating temperatures assumed for this process were in the range of

25–30°C or within the ambient temperatures of Central Florida; no process heating, therefore, was assumed for this digestion scheme. Although this methane yield is lower than assumed for the base case (0·44 m^3 kg^{-1} organic feed), the principal advantages of this system over the base case are noteworthy: (1) insignificant costs for chemicals (lime); (2) methane enrichment in the biogas product reduces gas cleanup costs; and (3) less lime sludge requiring disposal.

Water Hyacinth BTM Systems

Lake Apopka, with its 13 400 ha of aquatic surface area, was central to the conceptual design of base case and advanced water hyacinth BTM systems. In both cases it was assumed that this lake would be used to support an aquatic energy farm consisting of co-cropping water hyacinth with pennywort: water hyacinth predominantly produced from March through November and pennywort from December through February. This is the natural pattern of hyacinth and pennywort production in central Florida when these two biomass species are allowed to be co-produced on the same lake areas. Nutrient balances performed on the lake indicated that sufficient nitrogen and phosphorus was contained in the water and sediments to support more than 30 years of energy farm activity. Since the lake is eutrophic, an added benefit of lake restoration can potentially be realized.

In the energy farm concept for both the base and advanced cases, hyacinth and pennywort, henceforth collectively referred to by the term 'water hyacinth', are grown on one large confined area in Lake Apopka covering thousands of hectares of water surface. Because the biomass species produced are aquatic weeds that are controlled at great expense in Florida, a floating containment barrier would be installed around the biomass production area. Management of the hyacinths mainly consists of pest and disease control and proper harvesting. A harvest cycle of approximately 3 months was chosen with different sections of the farm being harvested each day. Thousands of megagrams of the wet biomass feedstock (95% water, 5% solids) are delivered every day to the shore where they are collected and transported less than 2 km to the conversion facility. The hyacinth solids are then fed directly into a two-phase anaerobic digestion process and converted to biogas which is upgraded to pipeline quality using previously discussed gas processing and compression conditions. Digested solids are dewatered and disposed of via land application; wastewater is eliminated through spray irrigation of nearby fields.

Differences in the base case and advanced case hyacinth BTM systems relate mainly to production yields achieved, mechanized equipment used for hyacinth harvesting, the transportation mode of the hyacinth from the lake to the conversion facility, and the digestion technologies used for hyacinth conversion. The base case hyacinth production yield was the yield achieved in test channels under conditions that simulated lake conditions; by 1982, yields of about 50 dry Mg ha^{-1} yr^{-1} were achieved in these small aquatic growth systems with no fertilizer addition. For the advanced case, IFAS researchers estimated that yields could be increased by 50% through optimization of crop management. Such crop management techniques include a better definition of the harvesting schedule needed to maintain densities optimum for crop growth, improved pest and disease control and foliar applications of small amounts of nutrients and plant growth hormone (e.g. gibberellic acid).

Harvesting in the base case design was performed with commercially available harvesters and barge equipment originally designed and used for hydrilla weed control. Though this equipment was incorporated into the base case design concept, the system can be labor intensive and slow because of the low biomass capacities of the barges and the limited barge velocities used for moving the biomass from the lake to the shore. Barges are transported to shore and unloaded by a bottom conveyor into trucks for transport to the conversion facility.

Harvesting for the advanced case system was simplified through the use of a large floating boom collector which maneuvers over the lake controlled by four shore-based traction winches (Fig. 3); harvesting could then be performed by one person operating the controls from a 10-m high control tower. The floating plants would be forced to shore and lifted to a shore-based conveyor by a floating flail harvester. The biomass would be shredded and pumped through a slurry pipeline to the conversion facility.

The base case water hyacinth conversion concept is the same base case Napiergrass digestion scheme employing lime pretreatment except that no dilution of the feedstock was needed to adjust the feed concentration to 5%. Because both hyacinth and Napiergrass are largely comprised of lignocellulose and hemicellulose, the reactor conditions and the assumed conversion performances for the two base cases are identical; the methane yield for base case hyacinth conversion is assumed to be about 0·44 m^3 kg^{-1} organic feed.

The advanced case hyacinth digestion system also utilizes a two-phase system consisting of an acetogenic CSTR followed by a packed bed but differs from the base case in that no chemical pretreatment is used, no heat

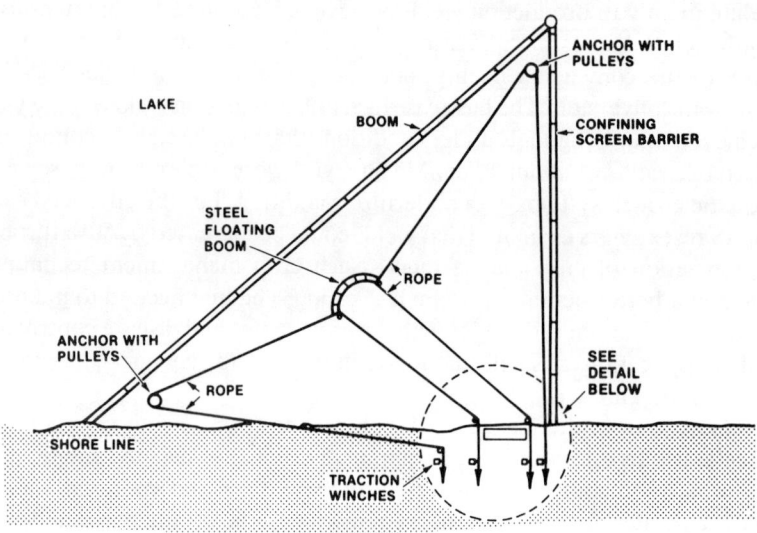
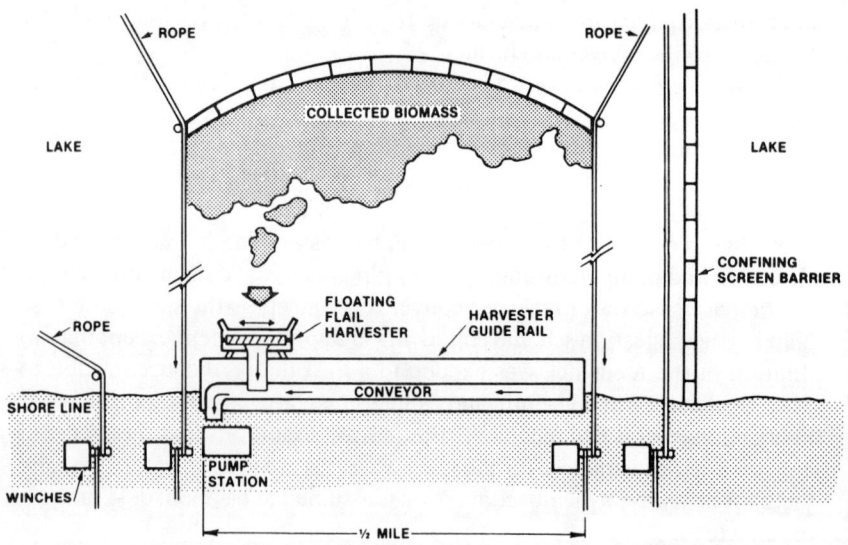

FIG. 3. Active boom-winch harvester concept. (Source: RSH, 1983.)

inputs are assumed so the reactor temperature ranges from 25 to 30°C, and only intermittent mixing is applied to the acetogenic reactor. These improvements are made possible through recycle of the acetogenic reactor effluent solids (separated from the liquid fraction by centrifugation) back to the acetogenic reactor to increase solids residence time and through advances in kinetic control of the process stemming from research on reactor microbial ecology and enhancement of biological hydrolysis. This two-phase system operates at a 4-day SRT for the acid phase and a 3-day SRT for the methane phase, achieving a methane yield of $0.28\,\text{m}^3\,\text{kg}^{-1}$ organic feed. This methane yield is lower than what is assumed for the base case employing caustic pretreatment but higher than methane yields of 0.15–$0.20\,\text{m}^3$ reported in the literature for hyacinth digestion with a CSTR (Ghosh et al., 1975; Vaidyanathan et al., 1985). Again, the assumptions made in the base and advanced conversion cases are well suited to an evaluation of the cost effectiveness of pretreatment applied to energy crop digestion.

Analysis Methods

The conceptual design and costing of the BTM systems were first performed at a net methane output equal to the average daily natural gas demand of the LANGD or $3200\,\text{GJ}\,\text{day}^{-1}$. In order to estimate the amount of biomass required to achieve this output for each BTM system, mass and energy flow diagrams were constructed. Simplified mass flow diagrams for the base case and advanced BTM systems (Fig. 4) together with key design parameters were used to estimate the sizing and number of unit processes comprising the subsystems (production, transportation, conversion and gas cleanup) of each BTM system.

The project results and mass flow analyses were then translated into system costs; for each BTM system these costs were represented by 54 capital and O&M cost categories. The average daily amounts of biomass were used to compute the areas of energy farm required; area requirement and the biomass output requirement defined energy farm capital and operating costs. Calculation of transportation costs required a knowledge of the location of the biomass production areas relative to the conversion facility. Potential Napiergrass production areas identified in this study are on the map of the LANGD (Fig. 2). In evaluating Napiergrass systems, the amounts of biomass required were compared to the amounts and locations of Napiergrass that could be grown in the LANGD. Potential sources were assumed and the conversion facility was located in a location that would minimize Napiergrass transportation costs. In the analysis of hyacinth BTM systems, the conversion facility was always located in close proximity

Base Case
Water Hyacinth or Napiergrass

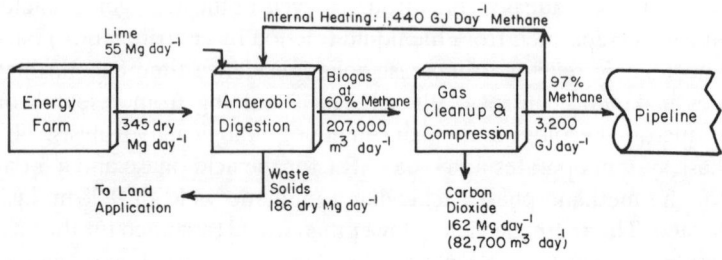

Advanced Case
Napiergrass

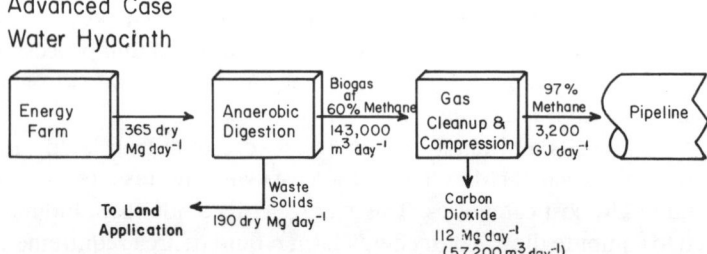

Advanced Case
Water Hyacinth

FIG. 4. Flow diagrams for Napiergrass and water hyacinth BTM systems sized for a net methane output of 3200 GJ day^{-1}.

to the lake, usually less than 2 km. From this information, transportation equipment capital and operating costs were calculated. Capital and operating costs for the conversion facility (including anaerobic digestion and gas cleanup) were computed largely on the basis of mass and energy flows and process design criteria. Cost curves were constructed for each unit process and used to relate process throughput capacity requirements

TABLE 2
Basic Cost Assumptions

Item	Value
Lime	$78 Mg^{-1}
Electricity	$0·04 kWh^{-1}
Oil (No. 1)	$0·28 liter^{-1}
Personnel (includes overheads @ 50%)	$14 h^{-1}
Insurance	0·7% of capital
Maintenance materials	
Conversion equipment	2% of capital
Mobile farm equipment	8% of capital
Fertilizer	
Elemental N	$564 Mg^{-1}
Elemental P	$458 Mg^{-1}
Elemental K	$329 Mg^{-1}
Dolomite	$30 Mg^{-1}
Cultivation (Napiergrass systems)	$33.30 ha^{-1}
Pest control (Napiergrass systems)	$46.20 ha^{-1}
Land purchase cost	
Farm	$2 500 ha^{-1}
Conversion facility	$16 200 ha^{-1}

to capital and annual operating and maintenance (O&M) expenditures. Conceptual costing procedures used for the evaluation of the conversion facilities included obtaining the capital and operating costs of major pieces of equipment based on vendor information, calculating approximate costs of supporting structures and ancillary equipment from vendor cost manuals (Richardson, 1981) and using factors to estimate various indirect costs such as contingency (20%), engineering (6%), contractor overhead and profit (25%) and spare parts (1%) (Peters and Timmerhaus, 1980). Capital and O&M costs (in 1982 dollars) were first calculated for BTM systems sized for a net methane output of 3200 GJ day^{-1} (matched to the LANGD gas demand). Costs were also calculated for plant sizes of 3, 0·5 and 0·1 times the LANGD size to determine the effect of scale on BTM system economics.

Annualized, constant dollar (1982) costs, referred to in this analysis as 'levelized costs', were computed using a financial model commonly employed by the gas industry for economic feasibility evaluations and using financial assumptions based on municipal (public) and private ownership (Clark et al., 1982). Costs were based on municipal ownership of the conversion facility. Basic cost assumptions are presented in Table 2 and

TABLE 3
Basic Assumptions used for the Levelized Cost Calculations

Variable	Investor-owned[a]	Publicly owned[a]
Allowed AFUDC rate	0·108	0·092
Book life	30 years[b]	30 years[b]
Construction period	2 years	2 years
Current-dollar discount rate	0·113	0·113
Constant-dollar discount rate	0·05	0·05
Current-dollar construction cost escalation rate	0·06	0·06
Current-dollar recurring cost escalation rate	0·06	0·06
Current-dollar fuel escalation rate	0·08	0·08
Fraction financed by debt	0·65	0·65
Fraction financed by common equity	0·30	—
Fraction financed by preferred equity	0·05	—
Fraction financed by nonborrowed funds	—	0·35
Inflation rate	0·06	0·06
Effective after-tax current-dollar cost of capital	0·078	0·092
Current-dollar return to debt	0·092	0·0812
Current-dollar return to common equity	0·145	—
Current-dollar return to preferred equity	0·092	—
Current-dollar return to nonborrowed funds	—	0·113
Combined State and federal tax rate	0·50	0
Tax life	15 years	—
Working capital fraction	0·125	0·125
Investment tax credit	0·100	0·0
Base year for cost estimation	1981	1981
Initial year for plant operation	2000	2000

[a] Financed by debt plus equity or nonborrowed funds.
[b] 10 years-mobile farm equipment.

financial assumptions are summarized in Table 3. The BTM system costs arising from these assumptions were compared to the GRI projections of gas price trends (GRI) as a measure of cost competitiveness; these projections indicated that gas from BTM systems coming into the market beyond the year 2000 could be cost competitive with other sources of supplemental

TABLE 4
Comparison of Grass BTM System Performance Goals with the Current Accomplishments of Biomass Research

Performance parameter	Projection of this Study (goals)	Research status
Biomass yield (dry Mg ha^{-1} yr^{-1})[a]	50	61
Anaerobic digestion		
Methane yield (m^3 kg^{-1} versus added)[b]	0.38	0.34
Organic loading rate (kg m^{-3} −day)[b]	9.6	4.83
Methane content in biogas (%)[c]	85	80

[a] Experimental Napiergrass production plots of 0.1 ha size.
[b] Steady state data from a 4 m^3 vertical flow reactor fed with an energy grass (623 × Rio Sorghum) at a solids concentration of 7.6%. This reactor is located at the Walt Disney World Resort Complex near Orlando, FL.
[c] Data from a two-phase digester operated on grass clippings (Colleran et al., 1982).

gas if the plant gate gas price was in the range of $5–$6 GJ^{-1}. The results to follow revealed the potential of each BTM system in meeting this economic criterion.

It must be reiterated that the assumptions used to arrive at the advanced case costs were not predicated on breakthroughs in research; the assumptions regarding the performance of the BTM system were considered achievable with existing knowledge and many of these assumptions were adopted as goals for the IFAS program. Since 1982, good progress has been made in meeting these goals as shown in Table 4. Opportunities for further reductions in cost through long-term fundamental research are discussed in Chapter 27.

RESULTS

Information from the survey on LANGD land use indicated that only a small fraction (less than 10%) of the total potential biomass production areas would be required to support the natural gas demand of the LANGD, 3200 GJ day^{-1} or 1.2 × 10^6 GJ yr^{-1} (Fig. 5). This suggests that large scale energy farming could be conducted in this region with minimal disruption to existing land use patterns, agriculture and the community. The study also found that if 80% of the total potential biomass production

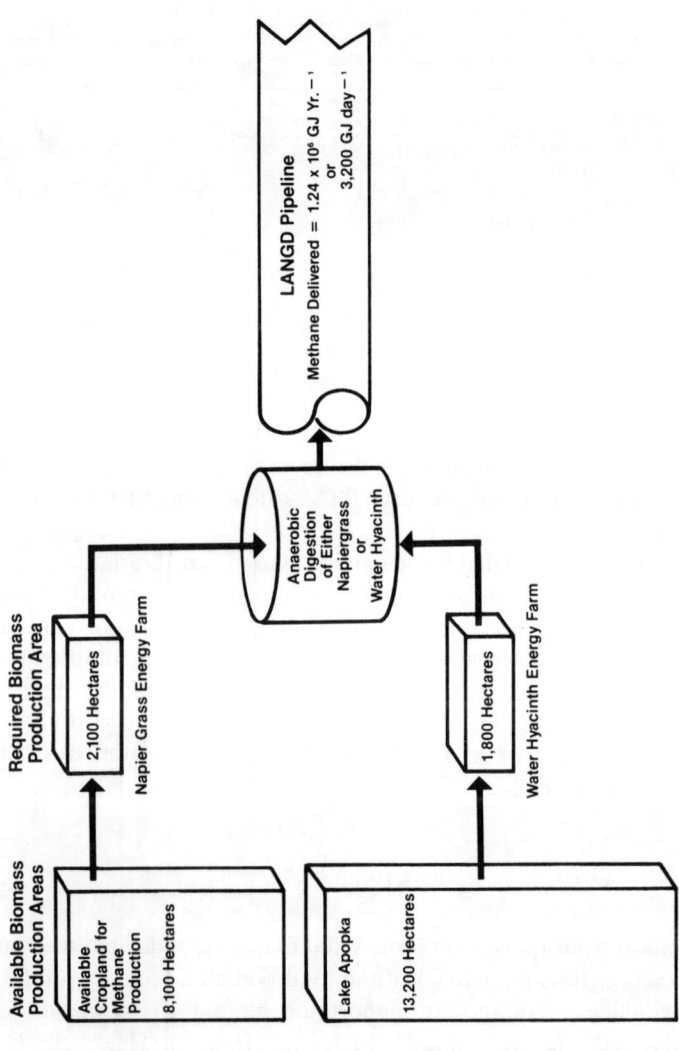

FIG. 5. Regional biomass availability to satisfy the Lake Apopka Natural Gas District methane demand.

areas were utilized for energy farming, nearly 14×10^6 GJ yr^{-1} of methane could be produced. This level of energy farming was, however, considered to be unlikely; the largest output considered in this analysis for a BTM system was $3 \cdot 6 \times 10^6$ GJ yr^{-1}.

BTM Systems Economics

Capital and operating cost estimates (in 1982 dollars) for the four BTM systems sized to deliver 3200 GJ day^{-1} of methane for the pipeline (Table 5) show greatest capital cost in every BTM system to be associated with the anaerobic digestion facility. The levelized costs (inflation corrected amortized costs) of all of the BTM systems show (Table 6) that the levelized gas costs from the base cases for Napiergrass and water hyacinth both exceed $10 GJ^{-1} and are clearly not competitive in price. Under advanced case assumptions, however, the prices of gas from Napiergrass and water hyacinth BTM systems were cost competitive at $5.35 and $6.05 GJ^{-1}, respectively; because of these results, most of the assumptions used for the advanced case BTM systems were adopted as research goals for the program. To date, many of these goals have already been achieved.

Base case and advanced case component costs (levelized costs, $ GJ^{-1}) were compared (Fig. 6). A number of observations in the costs can be made that are common to all of the BTM systems evaluated by this study. First, the highest capital cost is incurred by the conversion component and the lowest capital cost is allocated to transportation. The costs also showed that the highest O&M costs were generally observed in the conversion and production components; in nearly every case, with the exception of the advanced water hyacinth case, the sum of the conversion and production O&M exceeded over half of the total cost of the BTM system. The high use of labor and chemicals (lime pretreatment of biomass) in the base cases and the continued dependence on manual labor for the handling and movement of biomass in the advanced cases were the main reasons why O&M contributions were highest in the conversion and production components.

A comparison of production cost information between the base and advanced cases reveals that a substantial reduction in cost did occur but the reasons for this reduction differ between Napiergrass and water hyacinth. In Napiergrass production, increasing yield from 30 to 45 Mg ha^{-1} yr^{-1} decreased the land area for production and crop management and harvesting time. Simplification of the planting technique reduced manpower requirements to establish the Napiergrass plots. These advances were the major factors in decreasing Napiergrass production costs. On the other hand, since no cost was assumed for the rental of lake area used for

TABLE 5

Capital and Annual Operating Costs for the Major Elements of the BTM Systems Sized to Deliver 3200 GJ day^{-1} of Pipeline Quality Methane

Cost element	Costs (millions of 1982 dollars)	
	Capital	Annual O&M[a]
Napiergrass		
Base case		
Production and harvesting	2·5	2·99
Transportation	0·8	1·18
Conversion	31·8	2·52
Gas cleanup and compression	8·3	0·82
Total	43·4	7·51
Advanced case		
Production and harvesting	1·4	1·60
Transportation	0·2	0·24
Conversion	21·5	1·63
Gas cleanup and compression	5·0	0·48
Total	28·1	3·95
Water hyacinth		
Base case		
Production and harvesting	13·7	2·14
Transportation	4·9	2·38
Conversion	32·5	2·59
Gas cleanup and compression	8·3	0·82
Total	59·4	7·93
Advanced case		
Production and harvesting	12·1	0·66
Transportation	1·4	0·16
Conversion	15·9	1·26
Gas cleanup and compression	6·4	0·66
Total	35·8	2·74

[a] Include fuel and land rental (Napiergrass only).

hyacinth production (environmental benefits were assumed to pay for the use of the lake), increased hyacinth yield had an effect only on the size and cost of the hyacinth containment structure. It was the use of the boom winch harvesting concept that was chiefly responsible for the 60% reduction in the hyacinth production cost. Development of this hyacinth

TABLE 6

Summary of Levelized Costs of BTM Systems Sized to Deliver 3200 GJ day^{-1} of Pipeline Quality Methane

Cost element	Costs in 1982 dollars GJ^{-1}		
	Capital	O&M	Total
Napiergrass			
Base case			
Production and harvesting	0·40	2·90	3·30
Transportation	0·15	1·15	1·30
Conversion	2·00	2·20	4·20
Gas cleanup and compression	0·50	0·90	1·40
Sum	3·05	7·15	10·20
Advanced case			
Production and harvesting	0·20	1·55	1·75
Transportation	0·05	0·25	0·30
Conversion	1·60	1·60	3·20
Gas cleanup and compression	0·30	0·50	0·80
Sum	2·15	3·90	6·05
Water hyacinth			
Base case			
Production and harvesting	1·80	2·10	3·90
Transportation	0·60	2·30	2·90
Conversion	2·00	2·30	4·30
Gas cleanup and compression	0·50	0·90	1·40
Sum	4·90	7·60	12·50
Advanced case			
Production and harvesting	0·85	0·70	1·55
Transportation	0·10	0·20	0·30
Conversion	1·10	1·30	2·40
Gas cleanup and compression	0·40	0·70	1·10
Sum	2·45	2·90	5·35

harvesting concept then became one of the objectives of the program (see Chapter 8).

Likewise, a comparison of transportation costs between the base and advanced cases shows that a large reduction in cost did occur, but again the reasons for this reduction for Napiergrass are different from those for

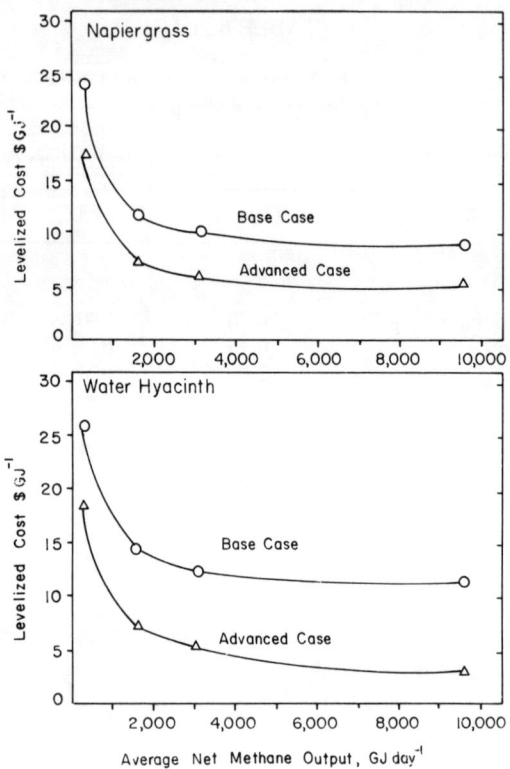

FIG. 6. Comparison of base case and advanced case component costs for Napiergrass and water hyacinth—BTM systems sized to deliver 3200 GJ day^{-1} of pipeline quality gas.

water hyacinth. In the case of Napiergrass, the cost reduction was due to the effect of decreasing the average trucking distance from 40 km to 4 km by changing from dispersed farms located all over the LANGD to establishing biomass delivery contracts only with nearby farms comprising a nearly continuous biomass production sector surrounding the BTM conversion facility. Hyacinth transportation cost reductions, however, were achieved by shifting from trucking in the base case to the advanced case concept of making a slurry of the water hyacinth feedstock at the lake shore and pumping the slurry to the conversion facility. These costs indicated that for transportation distances of less than 2 km, a slurry pipeline can be cost effective in the movement of aquatic biomass of a high water content

Conceptual Design of a Commercial Biomass-to-Methane System 73

from the energy farm to the conversion facility. Resolving technical uncertainties associated with this method became one of the experimental tasks of the IFAS program; these uncertainties included slurry rheology, frictional losses in laminar fluid flow through pipes, and solids separation behavior of hyacinth mixtures (see Chapter 8).

Conversion costs for the base and advanced BTM systems for both Napiergrass and water hyacinth indicated the feasibility of a tradeoff between the added expenditure of chemicals for pretreatment versus the potential costs benefits gained in reducing reactor volumes (i.e. increasing conversion rate) and in the achievement of higher methane yields. A comparison of the levelized conversion costs between the base and advanced cases for both feedstocks indicated that even under optimistic assumptions regarding the effectiveness of lime pretreatment, anaerobic digestion without pretreatment can be substantially lower in cost (25–45%) provided research goals for advanced conversion can be achieved. These results pointed the research program away from experimentation with chemical pretreatment and towards biotechnological and computer control strategies for the enhancement of the digestion process performance.

The economic analysis also showed that if high methane content gas (80–85% methane) could be produced directly from the digester, a significant reduction of nearly 50% could be achieved in the costs of gas cleanup to pipeline quality ($>97\%$ methane). This observation can be made in comparing the upgrading cost for the 65% methane gas produced by the base case Napiergrass system with the cost of upgrading the 85% methane produced in the advanced Napiergrass BTM system. These potential gas cleanup savings underscored the importance of pursuing the research on two-phase digestion systems that could be manipulated in operation to achieve improved separations of gases and a high energy gas produced directly from the digestion process.

Not obvious from the cost estimates are the effects that the performance of one BTM system component can have on another. For example, biomass yield not only affects the cost of production, but also determines the degree of regional dispersion of the biomass feedstock which, in turn, affects the costs of transportation. Methane yield affects not only the cost of conversion, but also makes a large impact on the amount of biomass required for a desired net methane output, transportation costs, production area required and the amount of effluent waste solids generated by the conversion facility requiring disposal. Process heating and heat recovery affects energy generation efficiency of the conversion facility which, in turn, affects the amount of biomass required, etc. Quantifying these sensitivities is beyond our scope and purpose, but the conceptual costs presented here were used to calibrate BTM systems models which were used to perform tradeoffs and sensitivity assessments. These models are discussed in Chapter 3.

Effect of Scale

As with many crop production and processing operations, the size of the BTM system has a pronounced effect on gas production economics. Levelized gas costs calculated for each of the four BTM systems at sizes of 3·0, 1·0, 0·5 and 0·1 times the methane demand of the LANGD are presented in Fig. 7. As shown in this figure, increasing the scale of operation from 320 to 9600 GJ day^{-1} reduced the levelized cost of gas production substantially for all cases. Base case costs decreased from over $24 GJ^{-1} to the $9–12 GJ^{-1} range and advanced case costs decreased from over

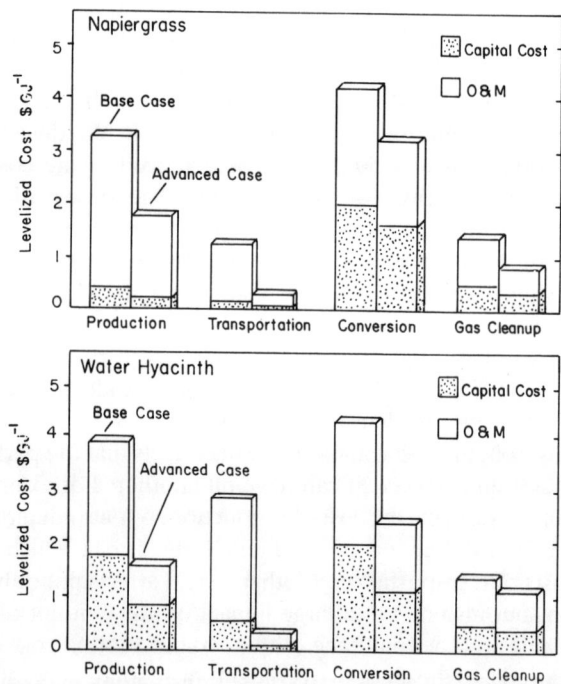

FIG. 7. Effect of scale on the levelized cost of methane production from base case and advanced Napiergrass and water hyacinth BTM systems.

$17 GJ^{-1} to the $5–6 GJ^{-1} range as net methane output increased from 320 to 9600 GJ day^{-1}. These plots seem to indicate that an economy of scale for all systems is reached around the size of the LANGD gas demand; the costs of the 3200 GJ day^{-1} BTM systems were all within 15% of the lowest systems costs represented in this analysis by the BTM scale of 9600 GJ day^{-1}. It would be expected, however, that at some extremely large size beyond the highest capacity shown here, costs would ultimately rise with increasing scale due to the burdensome transportation requirements of very large systems.

SUMMARY

This analysis, conducted early in the program, indicated that if key research goals could be achieved, advanced biomass to methane (BTM) systems could be designed to achieve the large scale production (> 3200 GJ day^{-1}) of pipeline quality gas at competitive prices in the range of $5–6 GJ^{-1}. In the judgement of the researchers the technical achievements needed to reach this cost range would not require fundamental breakthroughs in technology. As seen in this analysis, BTM systems designed on the basis of conventional agriculture and anaerobic digestion concepts are clearly not competitive with anticipated prices of supplemental gas. Research, however, plays a critical role in bringing about the improvements necessary to drive down biomasss derived gas costs to the competitive range.

For the specific site of the Lake Apopka Natural Gas District, many favorable factors point to the potential practical and economic feasibility of supplying nearly all of the region's gas demand from biomass resources upon achievement of key research goals that would make it possible to implement advanced BTM technology. It is also important that an economy of scale for both Napiergrass and hyacinth BTM systems is reached at around the magnitude of net gas output required by LANGD (3200 GJ day^{-1}).

Since its completion, this study has been used extensively in the development of BTM systems analysis models capable of performing more detailed engineering tradeoffs and sensitivity analyses. Information from these models has been invaluable in identifying research that will make a substantial contribution toward the goal of developing a cost effective biomass to methane technology.

REFERENCES

Ashare, E. and Wilson, E. H. (1979). Analysis of digester design concepts. Department of Energy Report No. C00-2991-42, pp. 81–88. Available from the National Technical Information Service, Springfield, California.

Buswell, A. M. and Hatfield, W. D. (1936). Anaerobic fermentation. Bulletin No. 32, Division of the State Water Survey, Urbana, Illinois.

Clark, C. E., Fancher, R. B. and Regulinski, S. G. (1982). A levelized cost of service price formula. Decision Focus, Incorporated, Palo Alto, California.

Colleran, E., Barry, M. and Wilkie, A. (1982). The application of the anaerobic filter design to biogas production from solid and liquid agricultural wastes. *Symposium Proceedings of Energy from Biomass and Waste VI*, p. 443–481. Institute of Gas Technology, Chicago, Illinois.

Eastman, J. A. and Ferguson, J. F. (1981). Solubilization of particulate organic carbon during the acid phase of anaerobic digestion. *Journal of the Water Pollution Control Federation* **53**, 352.

Ghosh, S. and Klass, D. L. (1981). Methane production by anaerobic digestion of water hyacinth. In: *Fuels from Biomass and Wastes*, pp. 129–149. Ann Arbor Science Publishers Incorporated, Ann Arbor, Michigan.

Ghosh, S., Conrad, J. R. and Klass, D. L. (1975). Anaerobic acidogenesis of wastewater sludge. *Journal of the Water Pollution Control Federation* **47**, 30.

Gossett, J. M., Healy, J. B. Jr., Stuckey, D. C., Young, L. Y. and McCarty, P. L. (1976). Heat treatment of refuse for increasing anaerobic biodegradability. Report No. NSF/RANN/SE/AER-74-17940-A01/PR/75/4, Department of Civil Engineering, Stanford University, Stanford, California.

Hayes, T. D. and Isaacson, H. R. (1986). Advanced concepts for methane enrichment in anaerobic digestion. *Proceedings of the 21st Intersociety Energy Conversion Engineering Conference*, pp. 205–212. American Chemical Society, Washington, D.C.

Klein, J. A. (1972). Anaerobic digestion of solid wastes. *Compost Science* **13**(1), 6.

Levy, J., Tong, X., Lang, M., Bedard J., and McCarty, P. L. (1986). Thermochemical pretreatment of lignocellulosic biomass for increasing anaerobic biodegradability to methane. Solar Energy Research Institute Report No. SERI/STR-231-2780. Available from the National Technical Information Service, Springfield, Virginia.

McCarty, P. L., Young, L. Y., Owens, W. F., Stuckey, D. C. and Colberg, P. J. (1979). Heat treatment of biomass for increasing biodegradability. *Proceedings of the Third Annual Biomass Energy Systems Conference*. Solar Energy Research Institute Report No. SERI/TD-33-285, Gidden, Colorado.

McFarlane, P. N. and Pfeffer, J. T. (1981). Biological conversion of biomass to methane. Solar Energy Research Institute Report No. SERI/TR-98357-1. Available from the National Technical Information Service, Springfield, Virginia.

Norrman, J. and Frostell, B. (1977). Two-stage anaerobic treatment. *Proceedings of the 39th Purdue Industrial Waste Conference*, Vol. 32, p. 387. Ann Arbor Science Publishers, Incorporated, Ann Arbor, Michigan.

Pavlostathis, S. G. and Gossett, J. M. (1985). Alkaline treatment of wheat straw for increasing anaerobic biodegradability. *Biotechnology and Bioengineering* **37**, 334.

Peters, M. S. and Timmerhaus, K. D. (1980). *Plant Design and Economics for Chemical Engineers*, pp. 147–224. McGraw-Hill, New York.

Richardson (1981). Process plant estimating standards. The rich rapid construction cost estimating system. Richardson Engineering Services, Incorporated, San Marcos, California.

Shuler, M. (1980). *Utilization and Recycle of Agricultural Wastes and Residues*, pp. 45–46. CRC Press, Boca Raton, Florida.

Vaidyanathan, S., Karadou, K. M., Shroff, K. C. and Mahajan, S. P. (1985). Biogas production in batch and semicontinuous digesters using water hyacinth. *Biotechnology and Bioengineering* **27**, 905.

Warren, C. S., Bruderly, D. E., Angelieri, M., Bilello, L. J., Bucalo, S., Finger, E. W., Har, R., Hinton, S. W., Newman, J. R. and Vinzant, J. W. (1984). The methane from biomass and waste program: evaluation of the Lake Apopka Natural Gas District. Final Report No. GRI84/0015.1. Available from Gas Research Institute, Chicago, Illinois.

Woodman, H. E. and Evans, R. E. (1947). The nutritive value of fodder cellulose from wheat straw. *Journal of Agricultural Science* **37**, 202.

5
Energy Crop Development

W. H. SMITH
Center for Biomass Energy Systems, University of Florida—IFAS, Gainesville, Florida, USA

and

J. R. FRANK
Gas Research Institute, Chicago, Illinois, USA

INTRODUCTION

Screening, Evaluation and Improvement

Domestic crops now grown by farmers superbly meet the requirements for food, feed, fiber. Plant quality, culture methods, and materials handling, processing, preservation and distribution have been so well developed that crops are easily overproduced, despite the fact that the amount of land in crops has remained nearly constant in this century. These crops were developed for environments without input constraints (fertilizer, pest protection, irrigation) to meet consumer needs such as taste, morphological quality, and nutrition. In general, for energetic, economic, and sometimes plant chemistry reasons, the crops used for food, feed and fiber are not suitable as energy crops. Therefore, biomass crops and the procedures to manage them must be developed for their energy values just as the others were designed for food, feed and fiber values.

Growing crops deliberately for their energy commodity value is a new concept; although the mechanics of crop selection and improvement are well established. The process involves the selection of promising species among the various plant resource groups (grasses and herbaceous, woody, root and tuberous, freshwater and marine aquatics, halophytes, and hydrocarbon producers) for their potential as abundant producers of biomass with modest production inputs. We identified ten characteristics that differentiate potential biomass crops from those developed for food, feed and fiber (Smith, 1983). These mainly relate to crops that produce high yields of usable biomass without large cultural inputs in an array of soils and environments. Potentially useful selections were field tested in a variety of environments reflecting Florida's diversity—near tropical to temperate climatic conditions, wet to droughty soils, aquatic to terrestrial environments—to identify adaptable, highly productive selections with desirable genetic variability and tolerances to potential stresses. About 350 species/cultivars/varieties in over 100 genera were screened through this process and the most promising ones selected for further evaluation (Smith and Frank, 1984). In addition, each species was assayed for its *in vitro* fermentability and plant chemical properties as they relate to plant part, age, morphology and environmental conditions in which the crop was grown. These data are being related to methanogenic characteristics of the biomass.

Once high-yielding species showing promise as energy crops were identified, area availability including regional/national adaptability was ascertained, and cropping opportunities and agronomic characteristics were determined. The latter objective included determination of optimum plant density, planting dates, harvest period and frequency and cultural inputs to optimize energy yield per unit area, time, and invested energy. All species were not investigated with equal intensity. For example, the most promising terrestrial (*Pennisetum purpureum* (L.) Lam., Napiergrass) and freshwater aquatic (*Eichhornia crassipes* (Mart) Solms, water hyacinth) species were rigorously researched for the purpose of evolving crop growth models compatible with an overall systems model (biomass to methane—BIOMET, see Chapter 3) to determine the production parameters to which methane gas cost was most sensitive. An initial application (Frank *et al.*, 1984) of these growth models revealed the sensitivity of methane gas cost on the yield per unit area for a crop (Fig. 1). This analysis using realistic assumptions showed the importance of targeting crops or cropping sequences that could result in annual harvests exceeding 25 Mg ha^{-1} or about twice the yield of conventional food crops. The intent is to arrive at procedures to calibrate the models of BIOMET for any crop when the minimal array of parameters has been described. The production and conversion modeling process in the integrated analysis system (including the progress toward generic crop growth models) is discussed in Chapter 3).

Many of the highly promising energy crop species do not reproduce sexually. Costly vegetative propagation has traditionally been the method available for increasing the size of such crops. To address these problems, tissue culture propagation has been utilized with two model species types—Napiergrass and related grasses, and sweet potato (*Ipomoae batatas* (L.) Lam.) and related tuberous crops. Species among both types

Energy Crop Development

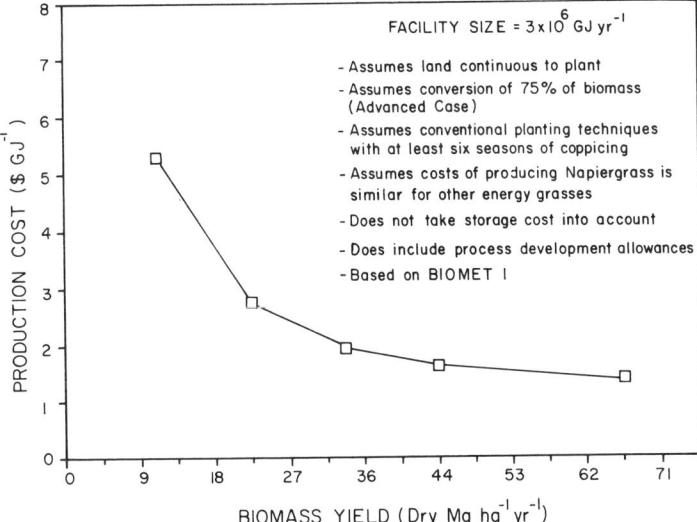

FIG. 1. The effects of biomass yield on production costs of energy grasses.

proved amenable to tissue culturing. Tissue culturing provides the opportunities to obtain sufficient amounts of material for field tests in months rather than years; to develop techniques of protoplast isolation and subsequent somatic fusion and genetic transformation; and to explore mechanical planting of somatic embryos either in mass-produced containers or suspended in fluidized gels used as a carrier for the 'artificial seeds'. In addition, genetic variability among promising species is being exploited by conventional selection, breeding techniques and by advanced biotechnological techniques.

The sequence of chapters that follows describes the results from the application of this screening, evaluation and improvement process for development of biomass feedstocks for methane generation. In summary, several promising species have been selected and the technology for increased growth and genetic improvement has been advanced.

REFERENCES

Frank, J. R., Hayes, T. and Smith, P. H. (1984). The production of methane from biomass in the US: Economic, tradeoffs, and prospects. In: *Energy from Biomass*, Palz, W., Coombs, J. and Hall, D. O. (eds.), pp. 484–8. Elsevier Applied Science, London.

Smith, W. H. (1983). Energy from biomass: A new commodity. In: *Agriculture in the 21st Century*, Rosenblum, J. W. (ed.), pp. 61–9. John Wiley, New York.

Smith, W. H. and Frank, J. R. (1984). Comparative biomass yields of energy crops. In: *Energy from Biomass*, Palz, W., Coombs, J. and Hall, D. O. (eds.), pp. 323–9. Elsevier Applied Science, London.

6
Model Crop Systems: Sorghum, Napiergrass

G. M. PRINE
Agronomy Department, University of Florida—IFAS, Gainesville, Florida, USA

L. S. DUNAVIN, B. J. BRECKE
Agricultural Research and Education Center, University of Florida—IFAS, Jay, Florida, USA

R. L. STANLEY
North Florida Research and Education Center, University of Florida—IFAS, Quincy, Florida, USA

P. MISLEVY, R. S. KALMBACHER
Agricultural Research and Education Center, University of Florida—IFAS, Ona, Florida, USA

and

D. R. HENSEL
Agricultural Research and Education Center, University of Florida—IFAS, Hastings, Florida, USA

INTRODUCTION

The tall grasses, such as corn (*Zea mays* L.), sorghum (*Sorghum bicolor* (L.) Moench.), Napiergrass (*Pennisetum purpureum* Schum.), and sugarcane (*Saccharum* spp.), are among the most efficient plants in converting solar energy to biomass. All have the efficient C_4 metabolic system that returns 4–5 units of energy for each unit used (Bassham, 1983) and grow especially well at the high growing season temperatures in temperate and subtropical climates. The perennials grow year round under tropical climates. Disadvantages of these plants are slow growth at cooler temperatures, actual top growth death of perennial plants by freezing temperatures, or death of the entire plant where soil temperatures go below freezing. Annual plants (e.g. sorghum) complete their life cycle before winter, therefore they can be grown further north than the perennials. Under long warm growing conditions, as in Florida, these tall-growing grasses and members of the *Saccharum* and *Erianthus* are quite superior to most other land plants in producing annual yields of biomass (Prine and Mislevy, 1983; Smith and Frank, 1985; Chapter 3 in this volume).

Sorghum and Napiergrass were selected as superior crops on which to concentrate energy research efforts and to develop crop production models (see Chapter 3). Sorghum is more widely adapted than corn to stressful soil conditions, such as low fertility, drought or temporary flooding, and can produce multiple crops by ratoon growth without re-seeding. Napiergrass, a perennial, develops faster and produces more biomass than sugarcane in subtropical climates with a 7·5–9 month growing season. In tropical climates, where both crops can grow for 12 months, sugarcane biomass yields greatly exceed Napiergrass (Alexander, 1985). However, Napiergrass grows better in less fertile and droughty soils than sugarcane. Development of energy sugarcane accessions, such as Louisiana L79-1002, may challenge Napiergrass as a high biomass producer in the subtropics (Giamalva *et al.*, 1984) after an establishment year. Research is already under way in Florida to study this possibility.

This chapter covers sorghum developmental research on Agricultural Research Centers near Jay, Quincy, Hastings and Ona, Florida, and Napiergrass grown near Jay, Quincy, Gainesville and Ona, Florida. Research was initiated in 1981 and continued through 1985. Other research to develop tall grasses as biomass feedstocks for methane is included in Chapters 9, 10 and 16.

SORGHUM

Sorghum is a tropical grass with abundant genetic variation within the species. Some sorghum types are tall with large stems, and other types are shorter with smaller stems and relatively narrow leaves. Agronomic cultivars have been developed fitting various characteristics to the desired use, such as grazing, silage, grain and syrup production. The high yield potential, particularly of the taller cultivars, modest nitrogen requirements and relatively low water requirements make sorghum a prime candidate for biomass exploitation. Many of the present cultivars with high biomass potential are either tall with medium to low grain production or are relatively tall with a high grain production. In both cases, the potential for lodging is usually high. This problem can result in harvesting difficulty, damage to lodged material, and possible loss of or reduced ratoon harvests. In Florida, sorghum is seeded annually, with the production of 2–3 harvests per year; however, the high-yielding sweet sorghums may produce only one good crop when harvested in the late dough stage.

Cultivar Yields

Various cultivar trials have been conducted during the five-year period (1981–85) at Jay, Quincy, Ona, Gainesville and Hastings, Florida. These trials have included entries of sweet or syrup-type, forage or silage-type, grain or combine-type sorghums, the grazing-type sorghum × sudangrass hybrids, and various crosses between types. Cultivar trials included various commercial cultivars and experimental single-cross hybrids from the Sorghum Breeding Program at Texas A&M University. Trials with these latter entries have been conducted at Jay, Quincy and Hastings since 1983. Biomass yields from leading entries among the Texas A&M crosses were higher at Quincy and Jay than at Hastings. At Jay in 1983, TX623 × Pickett-3 sorghum had a two-harvest yield of 35·6 Mg ha^{-1} dry biomass. In 1984, Atlas × Rio yielded 28·9 Mg ha^{-1}, and in 1985, TX623 × Mn 1500 yielded 29·4 Mg ha^{-1}. Single-harvest yields at Quincy were 20·2 Mg ha^{-1} green biomass for TX623 × Pickett-3 in 1983, 13·8 Mg ha^{-1} dry biomass for TX623 × 82CS796 in 1984, and 10·3 Mg ha^{-1} dry biomass for Atlas × Mn 1500 in 1985. At Quincy, Rio had the highest two-year average biomass yield. Biomass yields for the hybrids can be compared to yields presented in Table 1 for the leading commercial sorghum cultivars at AREC, Jay, from 1981 through 1985. Five of the highest yielders were sweet sorghums, seven were forage sorghums, and two were experimental single crosses. Data reported by Stanley and Dunavin (1986) give a three-year comparison of two forage sorghums and two sweet sorghums. The sweet types showed a general yield advantage; however, in 2 of the 3 years, the ratoon harvests from the forage types were generally superior.

The grazing-type sorghum × sudangrass hybrids were tested in multiple-harvest trials. The McCurdy cultivar, Sweet M, produced the highest biomass yield obtained during the five-year period in 1982, yielding 21·6 Mg ha^{-1} dry biomass. The Northrup-King cultivar, Sordan 79, harvested four times in 1984, yielded 14·3 Mg ha^{-1} biomass and was placed in a two-harvest trial in 1985 where it yielded 17·4 Mg ha^{-1}, as compared to a four-harvest yield in 1985 of 7·0 Mg ha^{-1} of dry biomass. Since these trials

TABLE 1
Production of Highest Yielding Sorghum Entries at AREC, Jay, 1981–85, in a System of Two Harvests per Year

Sorghum cultivar	Year	Dry biomass ($Mg\ ha^{-1}$)
Wray sweet sorghum	1982	45·2
911 forage sorghum	1981	44·5
NK367 forage sorghum	1981	41·0
FS451 forage sorghum	1981	40·2
MN1500 sweet sorghum	1982	38·6
Theis sweet sorghum	1981	36·8
FS25 forage sorghum	1981	36·5
Red Top Kandy Xtra forage sorghum	1984	35·8
TX623 × Pickett-3	1983	35·6
M81E sweet sorghum	1982	35·6
FS25E forage sorghum	1984	34·3
Cow Vittles forage sorghum	1983	29·9
Rio sweet sorghum	1984	29·9
TX623 × MN1500	1985	29·4

were separate, they are not directly comparable; however, the implication is that the biomass yields from grazing types can be increased considerably if they are cut less frequently like the other sorghums. The multiple-harvest use might be better suited when a continual supply of biomass is necessary.

Fertilization

Several fertility experiments were conducted at various Research Centers during the five-year period. Data for a trial conducted at Ona from 1982 to 1984, on a Smyrna fine sand (sandy siliceous hyperthermic typic Haplaquod), showed that 'normal' fertilization (224–49–186 kg ha^{-1} N–P–K initially and 168–25–93 kg ha^{-1} N–P–K to regrowth) produced significantly higher yields of biomass than did 'half normal' fertilization (Table 2). Double fertilizer treatments increased average yields by less than one-third. There was no difference in the response to fertilizer between forage and sweet types of sorghum.

A second experiment was conducted at AREC, Jay, from 1982 through 1985, utilizing the sweet sorghum M81E. In this trial, conducted on a Tifton sandy loam (Plinthic Paleudult) soil, the initial fertilization each year was 45–59–112 kg ha^{-1} (N–P–K). Two rates of side-dressed nitrogen (78 and 156 kg ha^{-1} N) were main plots, two rates of potassium

TABLE 2
Biomass Yield from Sweet Sorghum (M81E) and Forage Sorghum (Pioneer® 911) at Two Fertility Levels, AREC, Ona, 1982–84

Fertilizer level (N–P–K)	Yield of dry biomass (Mg ha^{-1})	
	Sweet sorghum[a]	Forage sorghum
Normal 224–49–186 at planting and 168–25–93 kg ha^{-1} for regrowth	25·8a	26·2a
½ normal	22·2b	19·5b

[a] Means within a column followed by the same letter are not significantly different at the 0·05 level of probability.

TABLE 3
Biomass Yield of M81E Sweet Sorghum as Affected by Two Levels Each of Nitrogen, Potassium and Seed, AREC, Jay, 1982–85

Treatment	Dry biomass yield (Mg ha^{-1})
Fertilizer level (kg ha^{-1})	
224 side-dressed ammonium nitrate	22·5
448 side-dressed ammonium nitrate	24·1
112 K in initial fertilization	23·2
224 K (initial + side-dress)	23·4
Seeding rate (kg ha^{-1})	
3·4 seed	22·4
11·2 seed	24·3

Seeding rate yield difference was significant ($P = 1·92\%$). No other interaction was significant at $P = 5\%$ level of probability

(112 kg ha^{-1} K in initial fertilization and initial + 112 kg ha^{-1} side-dressed) were subplots and two rates of seeding (3·4 and 11·2 kg ha^{-1}) were subplots. No significant interaction between experimental variables and neither N nor K significantly affected yield (Table 3). However, the higher seeding rate did increase biomass yield by 7·8% over the low seeding rate.

An experiment was conducted at Quincy in 1983 on a Leon fine sand (Aeric Haplaquod) soil to study five levels (0, 50, 100, 200, 400 kg ha^{-1}) of nitrogen (N), three levels (0, 44, 88 kg ha^{-1}) of phosphorus (P), and four levels (0, 83, 166, 332 kg ha^{-1}) of potassium (K). Although the highest

TABLE 4
Biomass Yields of Sorghum (Green Weight) in Fertility Trial at AREC, Quincy in 1983

Nutrient level (kg ha^{-1})			Biomass yield (Mg ha^{-1})		
N	P	K	Total (GWT)	Stalk (GWT)	Grain (DWT)
0	88	332	22·2	18·1	0·74
50	88	332	49·4	42·2	1·18
100	88	332	48·2	42·5	2·24
200	88	332	63·6	54·9	2·83
400	88	332	67·7	61·2	2·96
400	0	332	63·9	55·0	3·02
400	44	332	69·6	62·7	2·86
400	88	0	65·1	58·7	2·74
400	88	83	65·7	57·7	2·79
400	88	166	60·1	54·1	2·99

yield of total green biomass (69·6 Mg ha^{-1}) was obtained at 400–44–332 kg ha^{-1} (N–P–K) (Table 4), little yield improvement occurred at N rates greater than 200 kg ha^{-1}. Phosphorus and K soil reserves appeared adequate since little response to these fertilizers was recorded.

Quality of Biomass

Bioassay results of biomass convertibility to methane and fermentability have been reported by Shiralipour and Smith (1984), Bjorndal and Moore (1986) and in other chapters in this volume. These data indicate some variability between the sorghum types as to methane potential, with some indication of higher potential from sweet sorghum. There are also indications that variations in fermentability and methanogenesis are occurring among plant parts. Grain and leaves are more fermentable than stalks and roots and young tissue is more convertible than old tissue. The grain content (Table 5) varied among several sorghum cultivars and hybrids. Also, plant age and plant structure can be manipulated should these prove to be important in conversion.

Crop Management

Sorghum responds to varied management practices such as seeding rates and fertility. At the Ona AREC, sorghums were found to support high numbers of stubby root, lesion, and stunt nematodes; however, the nematicide, Nemacur 15G, applied at 6·7 kg ha^{-1} on the soil surface at the

TABLE 5
Average Grain Content of Several Sorghum Types and Crosses, AREC, Jay, 1982–84

	Grain content (%)		
Sorghum cultivar	First harvest	Ratoon harvest	Season total
M81E sweet sorghum	19	12	17
MN1500 sweet sorghum	9	3	7
Titan R forage sorghum	43	28	34
Red Top Kandy Xtra forage sorghum	38	14	28
TX623 × Pickett-3	20	17	22
Atlas × TX430	30	28	33
TX623 × Rio	34	38	35
Rio sweet sorghum	17	21	18

end of March each year failed to significantly increase sorghum yields or quality, although slightly higher yields were obtained where the nematicide was applied.

Seeding rates can also affect biomass yields. Sweet sorghums utilized for syrup generally are seeded at $3.4\,\text{kg ha}^{-1}$, while a seeding rate of $11.2\,\text{kg ha}^{-1}$ is recommended for forage sorghum. In a comparison of ten sorghums at two seeding rates at AREC, Jay in 1983 (Stanley and Dunavin, 1986), higher yields of biomass were consistently obtained at the $11.2\,\text{kg ha}^{-1}$ seeding rate. Yields at the $3.4\,\text{kg ha}^{-1}$ seeding rate were 41–89% of those at the higher rate.

Experiments at AREC, Jay, in 1983, examined the effect of row spacing and seeding rate on biomass production of sorghum (Stanley and Dunavin, 1986). The sorghum hybrid TX623 × Rio was the test sorghum, and row widths of 0.46 and 0.91 m were used with seeding rates of 11.2 and $3.4\,\text{kg ha}^{-1}$. Only seeding rate had a significant effect on yield, with about 63% as much dry biomass produced at the lower seeding rate.

Three legumes (Dixie crimson clover, Yuchi arrowleaf clover, and Cahaba White vetch) were grown at Jay from 1982 to 1985 during the cool season, followed by M81E sweet sorghum grown during the warm season. The objective was to determine sorghum biomass yield as affected by harvested legumes followed by no-till planted sorghum compared with legumes plowed under and followed by sorghum planted on a conventional seedbed. Sorghum produced about 93% as much dry biomass when the legumes were harvested as when the legumes were plowed under. The relationship of higher sorghum yields where legumes were incorporated held true for each of the three legumes with highest sorghum yields following the crimson clover. Side-dressing the sorghum with ammonium nitrate at 224 kg ha^{-1} of N produced higher yields than the rate of 112 kg ha^{-1} of N.

Several plant-growth regulators were tested on sweet sorghum at AREC, Jay over a four-year period (Table 6). Results indicate none of the

TABLE 6
Effect of Selected Plant Growth Regulators on Juice Content, Brix (% w/w of Sugar in Juice), and Dry Biomass Yield of Sweet Sorghums—4-year Average

Plant growth regulator	Rate (kg ha^{-1})	Juice (%)	Brix (%)	Dry biomass (Mg ha^{-1})
Glyphosate	0·1	26	13	19
Glyphosine	3·0	26	14	21
Ethephon	0·3	27	13	21
Mefluidide	0·2	28	13	22
Control	—	26	13	20

TABLE 7
Effect of Selected Herbicides on Weed Control and Sweet Sorghum Dry Biomass Yield—3-year Average

Herbicide	Annual grass control (%)	Annual broadleaf control (%)	Dry biomass yield (Mg ha^{-1})
Metolachlor + atrazine	83	67	20·7
Alachlor + atrazine	100	84	23·6
Pendimethalin + atrazine	100	76	20·5
Terbutryn	85	60	15·7
Propazine	40	78	17·7
Atrazine + oil	95	97	22·4
Metolachlor + propazine	100	75	19·7
Control	0	0	14·6

plant growth regulators evaluated caused dramatic changes in the parameters measured. The mefluidide treatment, however, appeared to increase the juice content and dry-biomass yield without affecting the brix content.

Herbicide treatments except propazine alone, provided excellent control of annual grass (Table 7). Alachlor + atrazine and atrazine + oil provided superior control of broadleaf weeds compared to the other herbicide treatments and resulted in the highest dry-biomass yields.

Sorghums were tested at AREC, Hastings to make use of residual fertilizer following potato crops. In 1983, average dry biomass yields were 1·8, 3·0, 2·8, 2·6 and 2·5 Mg ha^{-1} for 10 sorghum entries cut at 60, 74, 88, 102 and 116 days, respectively, following planting. Dry-biomass yields in 1984 were 5·6, 10·1 and 6·2 Mg ha^{-1} from 10 entries cut 61, 90 and 121 days, respectively, after planting.

NAPIERGRASS

Napiergrass, or elephantgrass, is a long-lived, tall, perennial warm-season grass that matures earlier than sugarcane. In subtropical climates where the warm growing season is shorter than 9 months, it may produce higher biomass than sugarcane. Alexander (1985), in Puerto Rico, found Napiergrass outyielded energy sugarcane when cut at 6-month intervals, but sugarcane was quite superior in yield to Napiergrass when cut at 12-month intervals. Under Florida conditions where the frost-free season is 7·5–9 months long, Napiergrass often makes higher biomass yields than sugarcane, winter-kills less often, and ratoons for more years. Most Napiergrass accessions in the first observation plots, established in the fall of 1981 at Gainesville, are still in good condition. A Napiergrass (N-51) has grown as an escape on the Georgia Coastal Plain Station at Tifton for over 35 years.

The effective life of Napiergrass plantings under high production and annual harvesting in Florida has not been determined yet, but 6–10 years seems realistic, except where some abnormal condition such as flooding occurs. Frequent harvesting makes Napiergrass more susceptible to winter kill (Calhoun and Prine, 1985) and shortens the expected stand life. Early plantings of Napiergrass at Gainesville and Ona showed this tall grass to be higher-yielding than most other herbaceous plants. Perennial grasses such as Napiergrass or sugarcane appear to be superior to annual tall grasses such as sorghum, because annual planting is avoided. Adequate stand establishment of many of the seeded annuals can be difficult to obtain if planted year after year on the same soil. It may be necessary to rotate Napiergrass also, but only after 6 or more years have passed. A shortcoming of Napiergrass and sugarcane compared to annually-seeded grasses is their need for vegetative propagation. Large quantities of stems are required so that only 10–15 ha can be planted from a hectare of planting material. This problem may be solved through the use of biotechnology techniques (see Chapter 9).

Genotype Selection

Once Napiergrass was identified as a potential biomass crop, many sources of germplasm were located in subtropical US. No new germplasm was

TABLE 8
Average Biomass Yields of Napiergrass Accessions at Three Locations during the Period of 1983–85

Napiergrass accession	Dry biomass yield ($Mg\ ha^{-1}$)			
	Average annual yield			Three-location average
	Jay	Quincy	Gainesville	
PI 300086	40·5	31·2	35·8	35·8
N 51	45·0	31·0	31·4	35·8
N 43	35·5	38·9	30·4	34·9
Merkeron	44·7	40·7	30·3	38·6
PP 13	33·7	33·0	28·8	31·8
N 13	31·6	44·8	32·7	36·3
PP 19	23·3	33·5	29·0	28·6
N 40 (PI 388894)	—	26·7	30·8	—
PP 3	—	37·5	27·9	—
N 109	—	36·9	28·1	—
PP 9	39·8	—	26·5	—
B 16	31·3	—	—	—
N 19	22·1	—	—	—
N 22	36·4	—	—	—
Years studied (N rate in year kg ha^{-1})[a]	1983 (24) 1984 (168) 1985 (168)	1983 (75) 1984 (168) —[b]	1983 (200) 1984 (168) 1985 (168)	

[a] Nitrogen applied in 4–1–2 ratio of N–P_2O_5–K_2O in spring.
[b] Stands were thinned so severely by January 1985 freeze that yields were not taken in 1985.

imported because of concern for introducing some unknown pest. Small collections of germplasm were found at several IFAS research stations in Florida, the USDA Plant Materials Center at Brooksville, Florida, and the Georgia Coastal Plain Station at Tifton, Georgia. We also collected wild ecotypes that had escaped from Florida plantings made in the 1930s and 1940s, particularly around Lake Okeechobee. In each case we selected only the highest-yielding accessions; so, the forty Napiergrass accessions in our first observation nursery were all better biomass producers. From this nursery we selected Napiergrass accessions for trials at Ona, Gainesville, Quincy and Jay, locations exhibiting large variations in climate and soils.

Napiergrass accession PI 300086 was selected as the control in accession trials and for many early experiments, because of increased availability of planting material at the USDA SCS Plant Materials Center at Brooksville, FL, where it had performed well. Several high-yielding genotypes performed well at north Florida locations of Jay, Quincy and Gainesville (Table 8). PI 300086 had been one of the highest-yielding accessions each season. This accession is the latest-maturing and most resistant to lodging, but it was greatly thinned by the exceptionally severe freezes of December, 1983 and January, 1985 in north Florida locations (Table 9). Thus, PI 300086 cannot be the principal Napiergrass genotype for energy use in north Florida, though it can be used in warmer peninsular Florida. Merkeron, N-51, N-43 and N-13 accessions yielded well in north Florida and showed cold tolerance and the potential to grow as perennials throughout much of the southeast. Tests are now in place to determine the northern limit beyond which Napiergrass would need to be grown as an annual.

Clipping Frequency and Cold Tolerance

In 1983 and 1984, PI 300086 Napiergrass was harvested at 6, 8, 12 and 24 week harvest intervals (Calhoun and Prine, 1985) on a Sparr sandy soil near Gainesville. Shoot yields for both seasons increased as the length of harvest interval increased (Fig. 1). The biomass yield for the year when cut at 6 week intervals was only half of the 40 Mg ha^{-1} harvested at the single 24 week cutting.

Reduced yield at shorter harvest intervals was also found by other researchers (Capiel and Ashcroft, 1972; Mwakka, 1972; Omaliko and Obioha, 1981). Because the highest biomass yields are from single-season harvests, biomass storage will be necessary to assure a year-round feedstock. Where higher quality biomass is needed, harvests of Napiergrass at shorter intervals will be necessary. Shiralipour and Smith (1984) found

TABLE 9
Lodging and Winter Survival Values for Napiergrass at Different Locations and Seasons

Napiergrass accession	Lodging rating Jay[a]	Lodging rating Gainesville[b]	Number per plot Plants Quincy	Number per plot Tillers Jay	Quincy Live plants	Quincy Live tillers	Gainesville Stand survival[c] %	Gainesville Tiller vigor[d]
PI 300086	0	0·25	7	6	0·5	0·6	45	7·8
N 51	2	5·0	20	145	2·5	12·7	75	8·8
N 43	1	4·5	24	158	4·3	27·5	88	9·1
Merkeron	3	5·0	27	125	4·8	35·5	90	9·5
PP 13	0	2·3	16	56	3·1	20·4	83	7·5
N 13	2	2·5	25	112	4·2	34·5	78	9·3
PP 19	3	3·5	22	89	4·9	41·2	98	9·3
N 40	—	4·0	12	—	1·4	5·4	75	9·3
PP 3	—	0·75	22	—	3·3	10·8	93	9·8
N 109	1	1·0	16	—	1·8	5·2	85	8·3
PP 9	2	3·8	—	81	—	—	93	8·8
B 16	0	—	—	4	—	—	—	—
N 19	1	—	—	91	—	—	—	—
N 22	0	—	—	48	—	—	—	—
Quincy Common	—	—	25	—	4·6	16·0	—	—
Date of rating	11/13/85	11/23/83	4/26/84	3/23/84	4/30/85	4/30/85	3/10/85	3/10/85

[a] Lodging rating at Jay: 0 = None to 3 = severe.
[b] Lodging rating at Gainesville 0 = None to 10 = 100% of plants lodged.
[c] New shoots compared to old stems from last season.
[d] Tiller vigor = 0—dead to 10 = longest and most vigorous tillers.

younger Napiergrass plants to produce more methane per kg than older plants. Frequent harvests bring on additional problems, such as winter kill and need for additional fertilization. Calhoun and Prine (1985) found severe winter kill of PI 300086 Napiergrass following the harvest season at the 6, 8 and 12 week harvest intervals in both winters. Napiergrass shoots are killed by freezing temperatures, so the plants must regenerate from

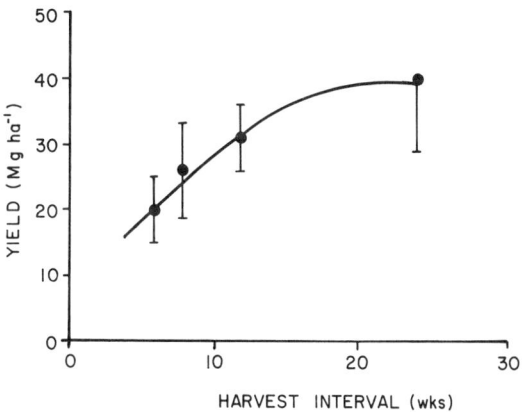

FIG. 1. Regression of combined 1983 and 1984 dry biomass yield (Y) of PI 300086 Napiergrass on harvest interval (X) at Gainesville, FL: $Y = 4·3 + 3·15X - 0·07X^2$, $R^2 = 0·51$. Error bars represent 95% confidence interval.

below-gound stems, rhizomes and roots. Cutting frequency affects the food reserves in these underground plant parts and the ability of the plants to survive frosts and below-freezing temperatures. Cold and warm periods are intermittent during winters in the subtropics; therefore, regrowth often starts between cold periods. Periodic low temperatures kill young shoot growth generated from food reserves before sufficient growth has occurred for the production/accumulation and return of food reserves to root storage. After each freeze-kill of regrowth, the food reserves are progressively lower and the plant is weaker. Some plants can get too weak to recover in spring. In the winters of 1983–84 and 1984–85, the soil in north Florida sometimes froze several inches deep, and the underground plant parts were killed by being frozen directly. In a normal winter, most winter kill of Napiergrass plants apparently occurs after a gradual loss of food reserves.

Growth Analyses

Calhoun (1985) performed a growth analysis of PI 300086 Napiergrass grown near Gainesville. The Napiergrass was harvested at 4 week intervals over the two growing seasons of 1983 and 1984; a regression equation determined relating average shoot dry biomass accumulations over the two seasons to time in days (Fig. 2) beginning after growth on March 15, though some shoots had emerged earlier both seasons. The Napiergrass achieved an average maximum dry biomass of 41·5 Mg ha^{-1} (41·5 g m^{-2}) after 240 days of growth. Growth rate during the linear growth phase (44–215 days) was 22·7 g m^{-2} day^{-1}. The pattern of dry biomass accumulation of plant components by the Napiergrass for both the 1983 and 1984 seasons was for leaf biomass to become maximum in mid-summer while stem biomass continued to increase until fall (Fig. 3). Maximum leaf area index (LAI) was observed in early August and reached 12·4 in 1983 and 7·2 in 1984. Leaf number per tiller (a measure of developmental rate) was highly correlated with growing degree days $R^2 = 0.94$ and equal to 0·12 leaves day^{-1}. Data from these growth analyses were used to calibrate the Napiergrass portion of the BIOMET models (see Chapter 3).

Fertilization

Fertilization trials have been conducted mainly on Napiergrass with one single harvest per season. Nitrogen needs have usually been adequate at rates up to 224 kg ha^{-1} when applied with adequate P and K and other elements (Tables 10–12). The fertilizer requirement will go up if the crop is harvested more than once a season, and possibly as stands grow older and fewer nutrients are available from the soil. The application of phosphate (P_2O_5) and potash (K_2O) in a 4–1–2 ratio of N–P_2O–K_2O appears adequate. However, better evaluation of P and K needs should be determined in future research. No minor-element deficiency symptoms were noted, but these may become important on some Florida soils, particularly after a number of years of biomass removal. The excellent biomass yields of Napiergrass with no fertilizer at Ona and Gainesville (Tables 10 and 12), and with modest nitrogen rates at Jay (Table 11), gives an indication of the ability of Napiergrass to remove nutrients to support impressive growth from relatively infertile soils.

Propagation

Napiergrass is normally propagated vegetatively. In the spring, old plants consisting of many tillers can be broken into plants of one to several tillers. This destroys the original plant and can transmit soil-borne diseases,

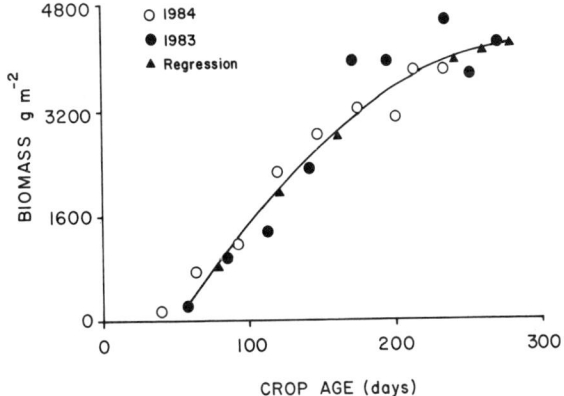

FIG. 2. Shoot dry biomass accumulation of PI 300086 Napiergrass at Gainesville, FL, in 1983 and 1984, where 0 days = March 15. Regression equation: $W = -1692 + 39 \cdot 05t - 0644t^2$; $R^2 = 83$. (○ 1984; ● 1983; ▲ regression.)

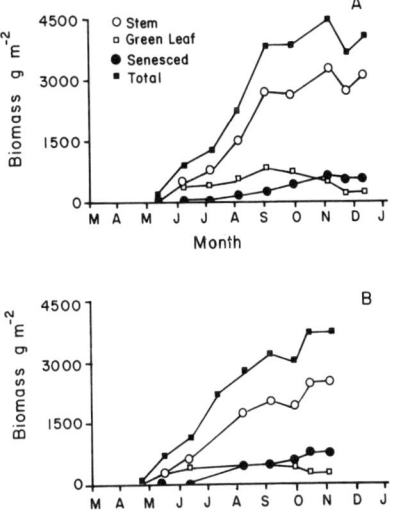

FIG. 3. Dry biomass accumulation of plant components by PI 300086 Napiergrass at Gainesville, FL, in 1983 (A) and 1984 (B). (○ Stem; □ green leaf; ● senesced; ■ total.)

TABLE 10
Effect of Fertilizer Levels and Applications on Dry Biomass Yields of PI 300086
Napiergrass Harvested once Annually, ARC, Ona, 1984

Fertilizer $(N-P_2O_5-K_2O)$ $(kg\ ha^{-1})$	Application[a] 1	2	Dry biomass yield ($Mg\ ha^{-1}$)		
			1984	1985	Two-year average
No fertilizer	—	—	24·9	30·1	27·5
a. 168–16–62	X	—	55·6	60·2	57·9
b. 168–22–83	X	—	45·2	47·1	46·2
c. 168–44–166	X	—	40·3	41·6	41·0
d. 168–44–166	—	X	50·4	52·0	51·2
Average medium N			47·9	50·2	49·1
e. 336–22–83	X	—	60·5	50·3	55·4
f. 336–44–166	X	—	54·2	72·0	63·1
g. 336–44–166	—	X	43·0	43·7	43·4
Average high N			52·6	55·3	54·0
Average for 1 application			51·3	54·2	52·8
Average for 2 application			46·8	47·9	48·4
Average for low PK			53·7	52·5	53·1
Average for high PK			47·0	52·3	49·7
Average for all fertilizers			50·0	52·4	51·2

[a] When two applications were made, the total was divided into two equal treatments applied on April 20, 1984 (first application) and June 21, 1984 (second application).

insects and nematodes. The most common method is to plant stem pieces of various lengths. In Florida, it is usually late July before stems are tall and mature enough to make good seed pieces. Woodard et al. (1985) studied planting methods and dates for Napiergrass during the year at two locations (Table 13). Satisfactory summer plantings were made from late July to early September. The success of plantings in November and early December depends upon climatic conditions following planting. For example, an early freeze can destroy all planting material. The most troublesome problem is warm periods when new shoots grow, followed by a killing freeze. Napiergrass planted in the summer makes considerable growth during the planting season and is ready for full production the next season (Fig. 3). The bottom nodes on the plant stem of Napiergrass make

TABLE 11
Biomass Yield of Three Napiergrass Accessions at Three Fertility Rates, AREC, Jay, 1984 and 1985

Accession	Dry biomass yield for rates averaged over two years (Mg ha^{-1})			Dry biomass yield averaged over fertility rates (Mg ha^{-1})		
	Low[a] rate	Medium[a] rate	High[a] rate	1984	1985	Two-yr avg
N-43	23·7	21·6	22·2	19·9	25·1	22·5
N-51	19·6	24·3	21·0	19·5	25·6	22·6
Merkeron	25·6	25·3	23·6	19·9	29·3	24·8
Average	23·0	23·7	22·3	19·8	26·8	23·3

[a] Fertility rates included 56, 112 and 224 kg N ha^{-1} in a 4–1–2 ratio of N–P$_2$O$_5$–K$_2$O.

TABLE 12
Dry Biomass Yields of PI 300086 Napiergrass under Different Fertilization Levels on Sparr Fine Sandy Soil at the Dairy Research Unit near Gainesville over Four Growing Seasons

Nitrogen rate (4–1–2 ratio N–P$_2$O$_5$–K$_2$O) (kg ha^{-1})	Dry biomass yields (Mg ha^{-1})				Four-year average
	1982	1983	1984	1985	
Single application in spring					
0	31·8	21·3	31·4	27·7	28·1
56	34·8	33·4	39·3	32·8	35·1
112	40·9	38·5	39·9	34·6	38·5
224	49·0	46·1	55·9	43·3	48·6
448	42·1	36·7	57·2	48·5	46·1
Split application (½ in spring + ½ in mid-season)					
224	41·6	42·1	38·4	33·7	39·0
448	56·8	37·2	50·5	39·6	46·0
672	48·5	40·5	47·9	44·5	45·4

the best planting material, but all hard stem portions can be used. Three and four node seed pieces are normally planted horizontally in soil at about 5 cm deep in summer, and slightly deeper in winter. Drought occurring before new germinating seed pieces have developed roots will kill the new

TABLE 13
Emergence and Winter Survival of PI 300086 Napiergrass as Affected by Planting Date at the Beef Research Unit and Lake Panasoffkee

Planting number	Planting date in 1982	Surviving hills (hills plot^{-1})[a]			
		BRU Count made		Lake Panasoffkee Count made	
		1/15/83	7/12/83	1/15/83	7/12/83
1	4 July	7·2b[b]	7·2cd	7·5c	7·5c
2	24 July	9·2a	9·2ab	9·2ab	9·2ab
3	14 Aug.	9·2a	9·0ab	9·7a	9·7a
4	4 Sept.	9·0a	8·5abc	10·0a	10·0a
5	25 Sept.	9·5a	7·0cd[d]	10·0a	9·5ab
6	16 Oct.	9·7a	4·2e[d]	10·0a	9·5ab
7	6 Nov.	9·5a	5·5de[c]	10·0a	10·0a
8	27 Nov.	8·7a	9·7a	9·5a	10·0a
9	18 Dec.	2·7c	8·0bc	7·7bc	8·5bc
10	8 Jan. (1983)	—	9·7a	—	10·0a

[a] Each plot contained 10 four-node stem cuttings which were placed horizontally in a furrow and covered 10–13 cm deep.

[b] Means followed by the same letter within counting date are not significantly different at the 5% level according to the Duncan's Multiple Range Test (DMRT).

[c,d] Hill counts were significantly lower on 7/12/83 than on 1/15/83 within planting dates at the 5%[c] and 1%[d] level, respectively.

plants. Weak new plants that develop during warm periods in winter are sometimes killed by freezes. Mechanization in harvesting planting material and actual planting will be needed for planting large areas of Napiergrass. It is possible that machinery used in harvesting and planting sugarcane seed pieces will adapt to Napiergrass.

Harvesting and Storage

Napiergrass harvested for energy use will probably only be harvested once each year. This means that the biomass will need to be stored until needed. Where the biomass can be stored as silage without affecting its use appears promising. Preliminary experiments with making silage from PI 300086 Napiergrass indicate that mature plants chopped directly from the field ensiled with no difficulties. Napiergrass is also being cut, solar dried and roll baled at the Ona Agricultural Research and Education Center. The feasibility of both techniques needs further research since both dry and ensiled biomass should have specific benefits.

SUMMARY

Sorghum is a dependable annual C_4 grass crop that is adapted to many soils and can be grown as an energy crop over much of the continental USA. The biomass yields of highest yielding sorghums (ca 40 Mg ha^{-1} yr^{-1}) often approach those of the perennial tall grasses. Typical annual yields in the region are usually 20–30 Mg ha^{-1}. In Florida and warmer southeastern USA, sorghum biomass yields are below those of perennial tall-growing grass species such as Napiergrass, sugarcane and others which often produce 40 60 Mg ha^{-1} yr^{-1}. Sorghum matures much earlier than the tall perennial grasses so it can furnish biomass earlier. Also, multiple harvests and early fall harvesting does not affect survival as it does with Napiergrass and sugarcane.

In warm southeastern USA, Napiergrass and its hybrids (*Pennisetum* spp.) comprise energy crops that will grow throughout the 7·5–9 months. The *Pennisetums* may be challenged by the more cold hardy of the energy canes such as L79-1002. Where the warm growing season is longer than 9 months, the energy cane and sugarcane will probably yield more biomass than the *Pennisetums*. Future research on tall perennial grasses should involve selecting the most winter hardy genotypes available. Breeding and biotechnology of the *Pennisetums* should develop plants having not only higher biomass yields with low input, but higher quality of the biomass for methane production, better winter survival, and propagation from seeds or tissue-culture derived plantlets. Continued research will determine water and nutrient needs and better methods of planting and harvesting the tall perennial grasses.

Storage of the tall-grass biomass for year-round use should receive priority in future research. Success of the tall-grasses biomass program hinges on whether we can store the biomass produced cheaply with an acceptable loss of quality and quantity of the harvested product. Hay and silage experiments already in progress should answer these questions.

REFERENCES

Alexander, A. G. (1985). *The Energy Cane Alternative*. Elsevier Science Publishers, Amsterdam, Netherlands, 507 pp.

Bassham, J. A. (1983). Biomass production potential. *Proceedings of Soil and Crop Science Society of Florida* **42**, 2–8.

Bjorndal, K. A. and Moore, J. E. (1986). Prediction of fermentability of biomass feedstocks from chemical characteristics. In: *Biomass Energy Development*, Smith, W. H. (ed.), pp. 447–54. Plenum Press, New York.

Calhoun, D. S. (1985). Elephantgrass performance in a warm temperature environment. M.S. thesis, University of Florida, 78 pp.

Calhoun, D. S. and Prine, G. M. (1985). Response of elephantgrass to harvest interval and method of fertilization in colder subtropics. *Proceedings of Soil and Crop Science Society of Florida* **44**, 111–15.

Capiel, M. and Ashcroft, G. L. (1972). Effect of irrigation, harvest interval and nitrogen on the yield and nutrient composition of Napiergrass (*Pennisetum purpureum*). *Agronomy Journal* **64**, 396–8.

Giamalva, M. J., Clark, S. J. and Stein, J. M. (1984). Sugarcane hybrids as biomass. *Biomass* **6**, 61–8.

Mwakka, E. (1972). Effect of cutting frequency on productivity of napier and Guatemala grasses in western Kenya. *East Africa Agricultural and Forestry Journal* **37**, 206–11.

Omaliko, C. P. and Obioha, F. C. (1981). Yield and quality of herbage harvested under various rainy-season and dry-season managements. *Agronomy Journal* **73**, 1081–3.

Prine G. and Mislevy, P. (1983). Grass and herbaceous plants for biomass. *Proceedings of Soil and Crop Science Society of Florida*, **42**, 8–12.

Shiralipour, A. and Smith, P. H. (1984). Conversion of biomass into methane gas. *Biomass* **6**, 85–92.

Smith, W. H. and Frank, J. R. (1985). Comparative biomass yields of energy crops. In: *Energy from Biomass*, Palz, W., Coombs, J. and Hall, D. O. (eds.), pp. 323–29. Elsevier Applied Science, London.

Stanley, R. L., Jr. and Dunavin, L. S. (1986). Potential sorghum biomass production in North Florida. *Biomass Energy Development*, Smith, W. H. (ed.), pp. 217–26. Plenum Press, New York.

Texas Agricultural Experiment Station. (1985). GRI Annual Report, April 1984 to March 1985. Sorghums for methane production. Gas Research Institute, Chicago, Illinois.

Woodard, K. R. (1985). Techniques in the establishment of elephantgrass (*Pennisetum purpureum* Schum). M.S. thesis, University of Florida, 90 pp.

Woodard, K. R., Prine, G. M. and Ocumpaugh, W. R. (1985). Techniques in the establishment of elephantgrass (*Pennisetum purpureum* Schum). *Proceedings of Soil and Crop Science Society of Florida* **44**, 216–21.

7

Water Hyacinth (*Eichhornia crassipes* (Mart) Solms) Biomass Cropping Systems: I. Production

K. R. REDDY
Central Florida Research and Education Center, University of Florida —IFAS, Sanford, Florida, USA

INTRODUCTION

Water hyacinth (*Eichhornia crassipes* (Mart) Solms), a floating aquatic macrophyte, is one of the most prominent aquatic plants found throughout tropical and subtropical areas. Under natural conditions, water hyacinths are widely distributed in eutrophic lakes, reservoirs and streams. Nutrients discharged from urban, industrial and agricultural developments often hasten the onset of eutrophic conditions. Large acreages of aquatic plants in natural water bodies suggest that these plants can be grown on a large scale, if economical harvesting methods and utilization opportunities can be developed.

Rapid growth of water hyacinth in nutrient-rich waters makes this plant an ideal candidate for use in treating polluted waters while producing abundant biomass and improving water quality (Boyd, 1969; Yount and Crossman, 1970; Stowell *et al.*, 1981; Reddy *et al.*, 1983; Reddy and Sutton, 1984; Reddy and DeBusk, 1985a). The resulting biomass can be used for production of gaseous fuels (Wolverton and McDonald, 1981; Shiralipour and Smith, 1984), feed (Bagnall *et al.*, 1974), fiber and other products (Nolan and Kirmse, 1974).

This report summarizes experiments conducted to develop a model aquatic plant system using water hyacinth as a possible feedstock for producing methane from biomass. Research was conducted: (1) at the Central Florida Research and Education Center in Sanford, Florida; (2) at eutrophic Lake Apopka, near Zellwood, Florida in experimental channels; and (3) at the Walt Disney World wastewater treatment facility in

*Florida Agricultural Experiment Station Journal Series No. 7644.

experimental channels. Other related physiological research was conducted at the University of Florida's main campus in Gainesville. Detailed experimental procedures and most results have been published previously elsewhere, as cited.

SELECTION

Freshwater aquatic macrophytes useful in biomass production can be categorized into three groups: floating, submersed, and emergent (Reddy et al., 1983). The economic success of biomass production depends foremost on high year-round growth rates and nutrient assimilation by the plants. Other important plant characteristics include: (1) possession of photosynthetic surfaces that efficiently utilize solar energy; (2) tolerance to a wide variety of wastewaters; (3) requirement of few additional energy inputs; (4) resistance to pests and diseases; (5) tolerance to environmental stresses such as freezing temperatures and wind; and (6) adaptability to management. These criteria have been used in screening aquatic plants for biomass production.

Selected aquatic plants were cultured for a period of 1 year in outdoor microcosm ponds (concrete tanks with 1000 liters volume) under nutrient

TABLE 1
Comparative Biomass Yields of Selected Aquatic Plants Cultured under Nutrient Nonlimiting Conditions

Aquatic plant	Cumulative annual yield (Mg (dry wt) ha^{-1} yr^{-1})	Highest yield during active growing season [a] (Mg (dry wt) ha^{-1} yr^{-1})
Water hyacinth	105	235
Water lettuce	72	146
Pennywort	43	87
Salvinia	40	67
Azolla	11	22
Duckweek	18	45
Giant duckweed	11	22
Elodea	13	47
Cattails (shoots)	45	—
Bulrush (shoots)	34	—

[a] Annual biomass yield calculated from highest periodic yield during the year.

nonlimiting conditions (Reddy and DeBusk, 1984, 1985a,b) to determine relative growth rates and nutrient-removal potential. Water hyacinth outperformed all other plants evaluated (Table 1) both in terms of biomass yield and nutrient removal (Reddy and DeBusk, 1985a), although winter performance was poor compared to pennywort. Two species of pennywort, i.e. *Hydrocotyle umbellata* L. and *Hydrocotyle ranunculoides* L., were identified as more cold-tolerant than any other plant evaluated. Biomass yields of emergent plants such as *Typha* spp., *Scripus* spp. and *Phragmites* spp. were lower than for water hyacinth, water lettuce and pennywort, but were higher than for other aquatic plants tested. Thus, water hyacinth is a highly promising aquatic plant for biomass production systems in tropical and subtropical eutrophic freshwaters. It was selected for further evaluation in order to develop a database on the biological, chemical and physical factors regulating its growth and nutrient-removal potential.

MORPHOLOGY OF WATER HYACINTHS

Water hyacinth belongs to the family Pontederiaceae and has a unique morphology (Center and Spencer, 1981; Gopal and Sharma, 1981). The plant is free-floating, consisting of a rhizomatous stem, a rosette of leaves, fibrous roots and stolons (Fig. 1). The rhizome (stem) consists of an axis with several short internodes. These internodes develop offshoots and bear both leaves and roots. The elongated internodes develop into stolons, and at the end of each stolon a new plant is produced.

The morphology of water hyacinth plants can be altered by cultural and management practices, nutrient supply, temperature and solar radiation. In natural systems, water hyacinths with varying leaf size, root and shoot lengths, and rhizome size can occur. For example, water hyacinth plants obtained from 30 different Florida locations showed a wide range of physical differences (Oki *et al.*, 1985a). When these plants were cultured under identical conditions with nutrients nonlimiting, physical differences among the various plant types were significantly reduced. Original plant vigor still existed after 6 months of culture. Plants initially larger in size produced larger ramets, compared to initially smaller plants which produced smaller ramets. Genetically, it appears that there is only one species of water hyacinth in Florida, since isozyme patterns of water hyacinth populations do not differ (Wain and Martin, 1980; Oki *et al.*, 1985a).

Although water hyacinth reproduces vegetatively, the plants are also

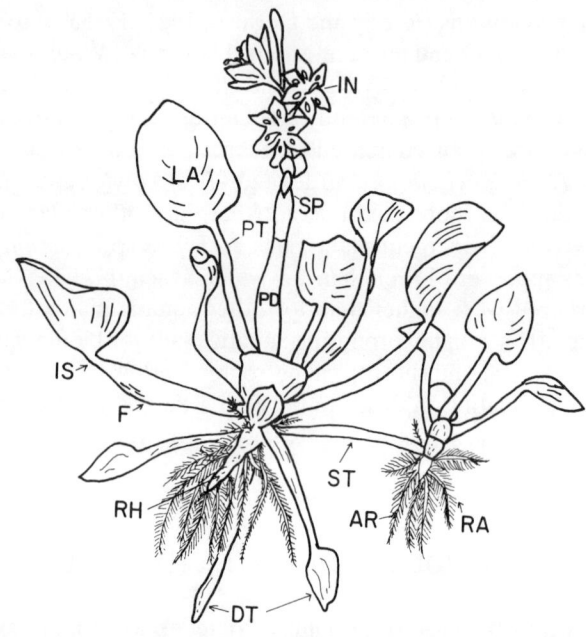

FIG. 1. A generalized diagram of a water hyacinth plant. The major morphological structures are (AR) adventitious roots; (RA) root hairs; (RH) rhizome; (ST) stolon; (DT) detritus tissue attached to the plant; (F) float; (IS) leaf isthmus; (PT) leaf petiole; (PD) peduncle; (SP) spathe; (LA) leaf lamina; (IN) inflorescence.

capable of sexual reproduction (Barrett, 1980). Seeds lie dormant at the sediment–water interface, then germinate when favorable conditions exist. The seeds have been reported to remain viable for several years (Matthews, 1967).

NUTRIENT REQUIREMENTS

Carbon

The upper limit to potential productivity of a plant is set by its ability to photosynthetically incorporate CO_2 into organic compounds. Water hyacinth is commonly considered to possess a C_3 carbon pathway. The plant does not possess Krantz's anatomy but, like many other aquatic plants, it shows unusual anatomical adaptations such as extensive lacunal air-space systems (Sculthorpe, 1967; McNaughton and Fullem, 1970;

Patterson and Duke, 1979). Despite its C_3 pathway, this plant has shown relatively high net photosynthetic rates, including values up to 40 mg CO_2 $dm^{-2} h^{-1}$ (Patterson and Duke, 1979). In conformity with high photosynthetic rates, dry matter yields of up to 64 g m^{-2} day^{-1} (234 Mg ha^{-1} yr^{-1}) have been reported (Reddy and DeBusk, 1984). Water hyacinth growth rates have been found to increase to 0·040 specific growth rate day^{-1} (2·5 times), when ambient CO_2 content of the surrounding air was increased ten-fold (to 3300 ppm) during photosynthetic periods. Spencer and Bowes (1986) reported even greater increases in carbon fixation in CO_2 enriched atmospheres.

Nitrogen

Critical levels of N, uptake and translocation of N, effect of N on plant morphology, and effect of varying sources of N were investigated in microcosm experimental tanks containing nutrient-enriched medium (Reddy and Tucker, 1983; Reddy and DeBusk, 1984). The resultant growth rate (Figs. 2(A,B)) was directly proportional to the N concentration of the culture medium, up to 10 mg N $liter^{-1}$ and an N loading rate of 416 mg N m^{-2} day^{-1}. Thus, N can be limiting when water hyacinths are cultured in waters containing less than 10 mg N $liter^{-1}$. Others have also shown that an increased rate of N supply significantly increases growth rates of water hyacinth (Ower et al., 1981; Sato and Kondo, 1981). Plant shoot/root ratios inversely vary with N concentration of the culture medium (Fig. 2(A)). Plant tissue N content also increases with increased levels of N in the culture medium, with a maximum tissue N level of 40 mg N g^{-1} of plant tissue being observed at an N concentration of 50 mg N $liter^{-1}$. Similarly, growth rates increase until plant tissue N content reaches 16 mg N g^{-1}; further increases in tissue N content do not increase plant growth rates significantly (Fig. 2(B)). Critical plant tissue N is 7 mg N g^{-1} (Fig. 2(B)). Below this value, net increase in growth eventually approaches zero.

Water hyacinth growth and productivity are also influenced by the type of N present in the water (Tucker, 1981; Reddy and Tucker, 1983). Net productivity is highest in water containing equal amounts of NH_4^+ and NO_3^-, and decreases progressively for waters containing NH_4^+, NO_3^- and urea, respectively. The shoot/root ratio was in the range of 4·2–6·5 for treatments with $NH_4^+ + NO_3^-$, NH_4^+ or NO_3^-, and of 3·7 for the treatment involving urea. Contribution of root biomass to total biomass yield is inversely related to the level of plant-available N in the culture medium (Reddy and Tucker, 1983). Ash content of the plant tissue is lower for plants cultured in

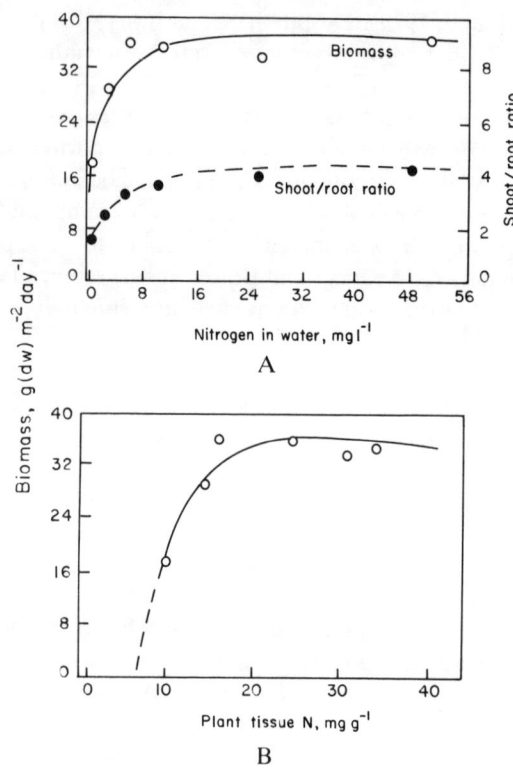

FIG. 2. (A) Effect of nitrogen concentration of the culture medium on biomass yield of water hyacinth; (B) relationship between plant tissue N and biomass yield of water hyacinths.

solutions containing NH_4^+ than plants grown in NO_3^- solution. Shoots generally contain less mineral matter than roots (Reddy and Tucker, 1983). Others (Shiralipour et al., 1981) have shown that N supplied through foliar application of urea produces more biomass yield of water hyacinth than does N supplied in the water as $(NH_4)_2CO_3$ or KNO_3. Although biomass yields in our experiments were higher for the treatment receiving both NH_4^+ and NO_3^-, plants assimilated NH_4^+ first, followed by NO_3^-. This was later confirmed by Oki et al. (1985b) and Reddy et al. (1986a), using ^{15}N to determine the preferential uptake of NH_4^+ and NO_3^- by water hyacinth and egeria (Egeria densa), respectively.

Plant morphology can be altered considerably by N levels. Plants cultured under nutrient nonlimiting conditions had high shoot/root ratios

TABLE 2

(a) Morphological Characteristics of Water Hyacinth with Varying Shoot/Root Weight Ratios

Shoot/root weight ratio	Leaf	Petiole	Rhizome	Roots	Root volume (cm^3)	Length Root (cm)	Length Shoot (cm)	Shoot/root length ratio	Leaf	Petiole	Rhizome	Roots
	(% of the total biomass yield)								(% dry matter)			
<2	17·4	38·9	8·5	35·2	464	67·4	48·8	0·72	14·60	6·41	5·81	3·72
2–4	25·7	49·7	4·8	19·8	304	43·6	52·4	1·20	13·36	5·73	5·60	3·60
>4	34·6	47·9	5·2	12·3	204	30·2	51·5	2·55	11·34	5·06	5·44	3·77

(b) Translocation of ^{15}N in Plant Parts of Water Hyacinth with Varying Shoot/Root Weight Ratios

Shoot/root ratio	Leaf	Petiole	Rhizome	Roots	Leaf	Petiole	Rhizome	Roots	Total	Leaf	Petiole	Rhizome	Roots
	(% N)				(^{15}N uptake mg/container)					(% of total ^{15}N uptake)			
<2	2·91	1·51	1·40	1·10	60·4	100·0	18·2	27·7	206·3	29·3	48·5	8·8	13·4
2–4	3·46	1·68	1·32	1·32	87·7	120·3	6·8	22·2	237·0	37·0	50·8	2·9	9·4
>4	4·49	2·59	3·93	1·98	78·9	89·1	10·4	22·3	200·7	39·3	44·4	5·2	11·1

with up to 85% of their yield as leaves and petioles. In contrast, plants cultured in nutrient-limiting conditions had low shoot/root ratios with leaves and petioles contributing only about 53% of the total yield (Table 2a). Percent dry matter in plant tissue was in the order: leaf > petiole > rhizome > roots. Up to 85% of the added ^{15}N was translocated into photosynthetic plant parts (leaves and petioles) for plants with high shoot/root ratios, while only 73% of the ^{15}N was translocated into photosynthetic tissues for plants with low shoot/root ratios (Table 2b). Rapid translocation of N was probably due to the high photosynthetic rates of plants with high shoot/root ratios.

Phosphorus

Phosphorus (P) is the major plant nutrient which next most frequently limits the growth of water hyacinth. Phosphorus limitation of plant growth not only depends on the P concentration of the culture medium, but also on the N/P ratio of the water. Water hyacinths were cultured in water containing varying levels of P in microcosm experimental tanks (1000 liters) placed outdoors to determine P requirements (Figs. 3A, B). Responses to P additions were observed up to a concentration of $1 \cdot 06$ mg P liter^{-1} ($80 \cdot 2$ mg P m^{-2} day^{-1}); further increases in P did not increase growth rate. Extrapolation of results from Fig. 3A suggests that biomass yields of water hyacinth approach zero when P concentration of the culture medium is about $0 \cdot 02$ mg liter^{-1}. This is in agreement with the results of others (Haller et al., 1970 and Knipling et al., 1970).

Maximum biomass yield was observed when plant tissue P was about 4 mg P g^{-1}, while yields approached zero when P content of the tissue was 1 mg P g^{-1} (Fig. 3B). Increasing the P concentration of the culture medium increased both N and P contents of the plant tissue.

Potassium

Potassium (K) seldom limits the growth of water hyacinth although plants collected from various locations in Florida contained an average of $38 \cdot 1$ mg K g^{-1} of tissue, indicating high requirements for K. The growth rates of water hyacinth were maximum at 12 mg K liter^{-1}; a further increase in K levels did not improve biomass yields significantly (Fig. 4A). In eutrophic lake water K levels were found to be in the range of 10–25 mg liter^{-1}, while in sewage effluents K levels were found to be less than 10 mg liter^{-1}. Though potassium does not likely limit water hyacinth growth in lake waters, deficiency symptoms have been observed during the summer months when plants were cultured in sewage effluent at the Walt

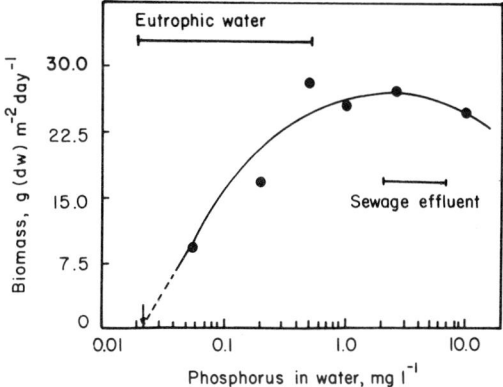

FIG. 3. (A) Effect of phosphorus concentration of the culture medium on biomass yield of water hyacinth.

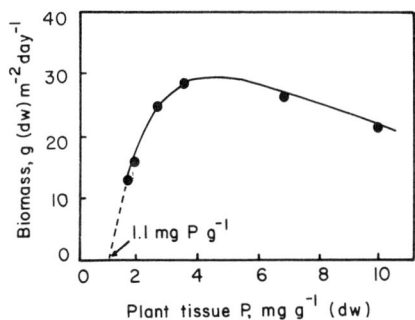

FIG. 3. (B) Relationship between plant tissue P and biomass yield of water hyacinth.

Disney World site. In microcosm tanks, water hyacinth yields increased until plant tissue levels reached to 48 mg K g^{-1}; further increase in tissue K did not improve yields (Fig. 4B). Plant tissue K was found to increase with increased levels of K in the culture medium.

Sodium

Biomass yields of water hyacinth decreased significantly with increased additions of Na to the medium. Plants did not survive at a sustained Na concentration >1500 mg liter^{-1}. Haller et al. (1974) concluded that water

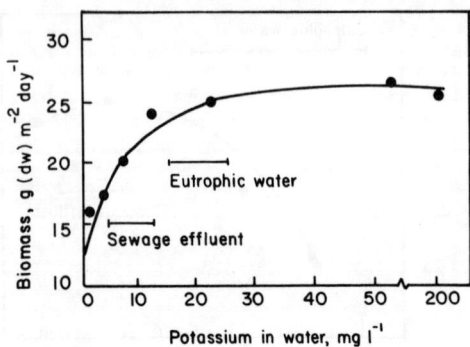

FIG. 4. (A) Effect of potassium concentration of the culture medium on biomass yield of water hyacinth.

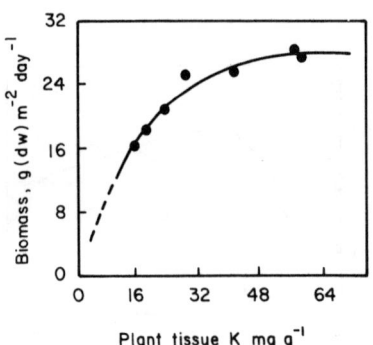

FIG. 4. (B) Relationship between plant tissue K and biomass yield of water hyacinth.

hyacinth will not live in waters with a sustained salt concentration of 2500 mg liter^{-1} (equivalent to about 1000 mg Na liter^{-1}). A similar trend was also observed between conductivity of the water and biomass yields. Plants did not survive in water with a substantial electrical conductivity >8000 μmhos cm^{-1}, indicating that direct use of waters with high ionic strength (such as digester effluents) could be toxic to hyacinth.

Calcium and Magnesium

Most of Florida's water bodies and domestic wastewaters contain adequate concentrations of Ca and Mg to support maximum growth of water hyacinth, so deficiencies are rare. Tissue concentrations of water hyacinth plants collected at various locations in Florida averaged 16·6 mg Ca g^{-1} and 5·6 mg Mg g^{-1}, respectively (Penfound and Earle, 1948).

Trace Elements

The critical levels of trace elements needed to obtain maximum growth of water hyacinth are still unknown. Water hyacinths cultured in NO_3^--rich waters exhibited chlorosis, even though N was present in adequate levels. Upon addition of chelated iron (Fe), however, the chlorosis symptoms disappeared (Reddy, 1983). The Fe requirements of water hyacinth were investigated in 60 liter tubs containing varying levels of Fe-EDTA; one set of tubs contained NO_3^- as the N source, while a second set contained NH_4^+ as the N source. Higher levels of Fe were needed when the culture medium contained NO_3^- as compared to NH_4^+ (Fig. 5). The critical level for Fe

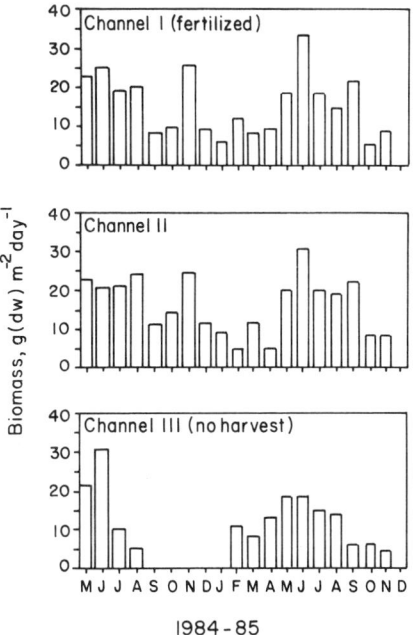

FIG. 5. Seasonal changes in biomass production of water hyacinth cultured in Lake Apopka water. Channel I, plants fertilized 3 times during the summer at a rate of 10 kg N and 3 kg P ha^{-1} per application, applied as foliar spray. Channels II and III were not fertilized. Channels I and II were maintained at a water hyacinth density of 15–20 kg (fresh wt) m^{-2}, while channel III was operated with no harvest.

appears to be around 1 mg Fe liter^{-1} when NH_4^+ is present as the N source; a much higher level of Fe (perhaps 5 mg liter^{-1}) is required when NO_3^- is present instead. Nitrogen was added at a rate of 20 mg liter^{-1} and replenished once every 7 days. These results confirm the Fe requirement during enzymatic reduction of NO_3^- in the plant. Iron can play a significant role when water hyacinths are cultured in wastewaters rich in NO_3^- and low in Fe. This situation is common in secondary sewage effluents, when Fe deficiency significantly reduces plant growth. The critical role of this and other trace elements in water hyacinth growth and management needs further research.

Growth Regulators

The effectiveness of gibberellic acid (GA_3), a plant growth regulator, in improving water hyacinth biomass yields was evaluated in controlled experiments (Oki and Reddy, 1985). The critical concentration of GA_3 was in the range of 0·03–0·05 mg liter^{-1} (Table 3a). Application of GA_3 also increased shoot length and altered other morphological characteristics such as petiole length and leaf area. The length to width ratio of newly produced petioles increased markedly with the increase in GA_3 concentration of the culture medium. However, maximum leaf area was recorded at a GA_3 concentration of 0·3 mg liter^{-1}. The highest concentration of GA_3 (0·5 mg liter^{-1}) induced profuse flowering and vegetative propagation, which normally are not observed in natural populations. Fertilizer and GA_3 applied as foliar spray on water hyacinth plants cultured in eutrophic Lake Apopka water increased biomass yields by about 36% compared to plants receiving no GA_3 spray (Table 3b). The role of growth regulators in improving biomass yields and nutrient utilization by aquatic plants is still not well known.

ENVIRONMENTAL FACTORS

Water hyacinth adapts to light exposures. Under artificial light conditions of 90, 320 and 750 $\mu E\ m^{-2}\ s^{-1}$, Patterson and Duke (1979) observed photosynthetic rates of 14, 27 and 29 mg $CO_2\ dm^{-2}\ h^{-1}$, respectively. However, at full sunlight (2000 $\mu E\ m^{-2}\ s^{-1}$) rates were much higher (34 mg $CO_2\ dm^{-2}\ h^{-1}$) and at the highest intensity biomass yield of water hyacinth was maximum. Under central Florida conditions, maximum solar radiation (6500 kcal m^{-2} day^{-1}) occurred during May, when maximum growth rates as high as 64 g (dry wt) m^{-2} day^{-1} have been recorded resulting in a solar energy conversion efficiency of 4·3%.

TABLE 3

(a) Effect of Gibberellic Acid (GA_3) on Biomass Yield and Morphological Characteristics of Water Hyacinth (Oki and Reddy, 1985). Values in Parentheses Represent Percent Increase in Yield over Control

GA_3 (mg liter^{-1})	Height (cm)	Leaf area (cm^2)	Number of new rosettes per tub	Number of inflorescences per tub	Net productivity (g (dw) m^{-2} day^{-1})
0·00	31·2	93·1	50·0	1·0	12·6 (0)
0·01	35·3	106·9	50·0	8·3	18·1 (44)
0·03	37·4	116·7	54·0	8·3	19·1 (51)
0·05	40·9	114·5	55·3	15·7	19·4 (54)
0·10	36·3	97·6	58·7	17·8	15·5 (23)
0·50	38·2	65·0	77·3	22·7	12·4 (−2)

(b) Effect of Foliar Application of Gibberellic Acid and Fertilizer on Morphology of Water Hyacinth Grown in Lake Apopka Water. Values are Means of Three Replications at the End of the Experiment. Values in Parentheses Represent Percent Increase over Control. Study Period 7/13/84–8/10/84

Foliar spray treatment	Height (cm)	Length to width ratios Petiole	Length to width ratios Lamina	Leaf area (cm^2)	Net productivity (g (dw) m^{-2} day^{-1})
Control	26·3	19·1	0·9	83·9	34·1 (0)^2
Fertilizer[a]	40·4	22·1	0·9	135·9	39·9 (17)
GA_3 0·05 ppm	53·9	62·0	1·3	112·0	35·2 (3)
Fertilizer + GA_3 0·03 ppm	61·8	57·6	1·1	148·5	46·3 (36)
Fertilizer + GA_3 0·05 ppm	61·9	62·7	1·2	148·6	45·7 (34)

[a] Application rates: N = 3947 mg m^{-2}, P = 460 mg m^{-2}, K = 1184 mg m^{-2}.

In dense stands, water hyacinth leaves grow upright rather than horizontally over the water surface. This architecture provides greater leaf surface area for photon capture and increased photosynthetic efficiency, suggesting that high growth density maximizes solar energy interception. However, the poor relationship ($R^2 = 0.48$) between solar radiation and growth rates suggests that interacting factors regulate growth. Center and Spencer (1981) reported a strong relationship between leaf area index of water hyacinths and solar radiation, but no data relating growth rate to solar radiation.

Temperature

The relationship between water hyacinth growth and mean daily temperature was significant ($R^2 = 0.54$), though with considerable scatter of data points. The optimal temperature for maximum growth of water hyacinth ranged from 27 to 30°C, with maximum temperature tolerance of up to 40°C. At constant temperatures but under low light conditions (Knipling et al., 1970), and at optimal temperature under light-saturated conditions, significant interactions occurred between light, temperature and plant growth.

Water hyacinth plants, sensitive to low temperatures, showed no net increase in growth under field conditions when average daily temperatures dropped below 10°C (Reddy and Bagnall, 1981). Temperatures sustained below 0°C can be lethal to the plants. We exposed water hyacinth plants to freezing temperatures (0°C) for 1, 2, 3, 4 and 6 nights (each night for a period of 12 h) along with daytime temperatures of 10°C. All parts of the plants were completely killed when exposed to freezing temperatures for 6 consecutive nights. In the remaining treatments, leaves and petioles were damaged but recovered and resumed normal growth within a few days. Water hyacinth plants cultured in outdoor tanks were exposed to a freeze in January 1982, for three consecutive nights (-1.7, -3.3 and -2.8°C) (Reddy, 1984). Shoots and petioles of the plants were killed, but the plants resumed normal growth in about 2 weeks.

Low temperatures can limit the growth of water hyacinth biomass for energy in other areas with cooler climates. Our research has shown that more cold-tolerant plants such as pennywort, when grown with water hyacinth, can help resolve the problem of poor biomass supply during the winter. Future research is needed to determine whether genetic and nutritional factors can be used to induce greater cold tolerance into water hyacinth plants.

SOURCES OF CULTURE MEDIUM

Nutrient-enriched effluents for water hyacinth growth include eutrophic natural waters, municipal sewage effluents, agricultural drainage waters and effluents from industrial operations. In effluent streams, water hyacinth can be grown in channels or reservoirs, with facility size depending on type of effluent, residence time and degree of nutrient removal desired while sustaining hyacinth growth.

Lake Water

Many lakes have become enriched from human activity. In Florida, Lake Apopka (12 500 ha surface area), a highly eutrophic lake sustained by drainage from adjacent farming operations on organic soils and municipal and industrial discharges, was investigated. Plant-available N and P were low in the eutrophic lake water compared to other types of enriched waters (Table 4). Inorganic N and P discharged into Lake Apopka are contained in algal biomass, thus decreasing their availability to higher plants. Once adequate water hyacinth populations are established in portions of the lake, algae should be shaded out and any inorganic forms of N and P discharged to the lake should be readily available to water hyacinth. Other nutrients such as K, Ca and Mg appear to be nonlimiting. Micronutrients can potentially limit the growth of water hyacinth but, at present, critical levels of micronutrients needed for maximum growth of water hyacinth remain unknown.

Agricultural Drainage Water

Drainage from 7500 ha of farmed organic soils adjacent to Lake Apopka can contain significant concentrations of plant-available N and P (Reddy *et al.*, 1982). A similar situation exists in south Florida, where about 1 million

TABLE 4
Chemical Composition (mg liter^{-1}) of Culture Media Used to Culture Water Hyacinth (Reddy *et al.*, 1982, 1985; Reddy, 1984)

Parameter	Agricultural drainage water	Eutrophic Lake Apopka water	Primary sewage effluent	Secondary sewage effluent	Nutrient medium	Methane digester effluent[a]
BOD$_5$	15·0	5·0	246·0	9·0	—	>2500
TKN	4·5	4·6	29·2	3·9	—	326
Ammonium N	0·7	0·1	23·0	1·8	10·5	289
Nitrate N	1·1	0·1	0·0	1·5	10·5	—
Total P	0·7	0·3	6·6	0·7	—	12
SRP	0·6	<0·0	5·3	0·4	3·1	—
Potassium	19·0	15·0	8·4	9·4	24·0	123
Calcium	57·0	55·0	26·4	27·3	20·0	23
Magnesium	26·0	24·0	5·6	5·6	4·8	14
Iron	<0·1	<0·1	0·3	0·3	0·6	—
Manganese	<0·1	<0·1	0·04	0·03	0·3	—

[a] Effluent obtained from an anaerobic digester receiving water hyacinth as a feedstock.

ha of organic soils are located adjacent to Lake Okeechobee (600 ha in surface area). The lake receives drainage waters through back-pumping. A potentially large acreage of drainage canals and retention ponds used to either transport or hold drainage water is also available for water hyacinth biomass production, while concurrently improving drainage water quality (Reddy and Bagnall, 1981; Reddy et al., 1982).

Municipal Sewage Effluents
Another source of nutrients to culture water hyacinth (Wooten and Dodd, 1976; Wolverton and McDonald, 1979; Reddy et al., 1985) is provided by municipalities in the sun-belt states, which are seeking alternative, biological wastewater-treatment systems in order to reduce costs. Both primary and secondary sewage effluents (Table 4) contain an adequate supply of plant-available N and P. Primary sewage effluent contains predominantly NH_4^+, while secondary sewage effluent usually has NO_3^- as its predominant form of inorganic N. In secondary sewage effluent, high NO_3^- levels associated with high pH can also induce Fe deficiency, resulting in chlorosis. Limiting nutrients can be added either to the effluent or directly sprayed on the leaves. Primary sewage effluent has high concentrations of Biological Oxygen Demand (BOD), which can create oxygen stress on plants at high loading rates, but can be avoided by aeration at the inflow of the channel.

Other Effluents
Feedlot runoff, swine lagoon effluent, and methane digester effluent usually contain both high BOD and high salt concentration, which may be lethal to water hyacinth plants if grown directly in the effluent. Dilution to reduce salt and BOD levels is often necessary before water hyacinth can be grown. Limited information is available on the feasibility of growing water hyacinth in undiluted effluents.

Results presented by several researchers (Wolverton and McDonald, 1975a,b; Tatsuyama et al., 1979; Cooley et al., 1979; Muramoto and Oki, 1983) indicate that water hyacinth can even be grown on wastewaters high in heavy metals and toxic organics (Wolverton and McKnown, 1976).

Although water hyacinth can be grown in a wide variety of nutrient-enriched wastewaters, many such waters may not contain nutrients in optimal ratios to achieve maximum plant growth. Future research should determine optimal nutrient ratios as well as limiting nutrient concentrations in wastewaters, and should develop methods to apply corrective treatments without deteriorating water quality.

GROWTH RATES AND BIOMASS YIELDS

Summaries follow of our field investigations to determine seasonal variations in dry biomass yields of water hyacinth cultured in: (1) nutrient medium, (2) agricultural drainage water, (3) eutrophic lake water, (4) primary and secondary sewage effluents, and (5) methane digester effluents.

Nutrient Medium

Water hyacinths were cultured in a nutrient medium to determine potential biomass yields in microcosm tanks placed outdoors under nutrient nonlimiting conditions (Reddy, 1984; Reddy and DeBusk, 1984). Growth rates were maximum during the months of May and June, when solar radiation was highest and humidity was less than 70%. Considering the seasonal variations in solar radiation, average annual solar energy conversion efficiency was calculated at 3·12% (Reddy, 1984). Average annual biomass yield was found to be 106 Mg ha^{-1} yr^{-1}, with about 50% of the total yield obtained during May–August, 30% during September–December, and 20% during January–April. During January–February, only 3% of the total yield was obtained.

Agricultural Drainage Water

Water hyacinth was also cultured in retention reservoirs (1240 m^2) to utilize nutrients from agricultural drainage water. Detailed experimental descriptions and results were reported in a series of papers (Reddy and Bagnall, 1981; Reddy et al., 1982; Reddy, 1984). Water hyacinths grown for 3 years in these waters (1979–81) showed an average biomass yield of 50–61 Mg ha^{-1} yr^{-1}. Maximum growth rates of up to 46 g m^{-2} day^{-1} occurred during the summer and the average solar energy conversion efficiency was 1·8% for these water hyacinths.

Eutrophic Lake Water

For a period of 4 years (1982–85), water hyacinth was cultured in eutrophic Lake Apopka water to determine the feasibility of improving lake water nutrient levels in order to grow water hyacinth (Reddy, 1984; Reddy et al., 1984; DeBusk et al., 1985). During 1982 and 1983, earthen reservoirs (1240 m^2, depth 1 m) were gravity fed with Lake Apopka water at a residence time of 1·5 days. During 1984 and 1985, four channels (6 m wide, 66 m long and 0·6 m deep) were constructed with concrete-block side walls and calcareous clay bottoms. These channels were gravity-fed with lake water to obtain a residence time of 1·5 days. Three channels were stocked with plants, while one channel was operated without plants. Channel I was operated at constant plant density (750–1000 g m^{-2}) and received foliar fertilization (June, August and September, each time at a rate of 10 kg N and 3 kg P ha^{-1}) during the summer season; channel II was also operated with constant plant density but with no fertilization; channel III contained no plants; and channel IV was operated with variable plant density and no harvest. Water chemistry was monitored twice a week, at both the inflow and outflow of each channel. Growth rates and plant chemistry were monitored every two weeks.

During 1982 and 1983, average dry biomass yield was 50 and 40 Mg ha^{-1} yr^{-1}, respectively, while no special management was practiced. Hyacinth was harvested only when plant density reached maximum with no additional growth being recorded. Seasonal distribution of yield followed solar radiation (Reddy, 1984). Annual dry biomass yields during 1984–85 were in the range of 55–65 Mg ha^{-1} yr^{-1} (Fig. 5). Maximum growth rates (38 g m^{-2} day^{-1}) were observed during the month of May, when incident solar radiation was maximum. Although water hyacinth plants responded to foliar application of N and P during the active growing season, increase in biomass yield was not sustained for long periods. Increase in biomass yield due to foliar application during June was about 30%. During the month of January 1985, all photosynthetic tissue was killed but the plants recovered from living roots and rhizomes and resumed normal growth within 4–6 weeks.

Sewage Effluent

In a prototype water hyacinth-based wastewater-treatment system currently in operation at Walt Disney World Resort Complex near Orlando, Florida (Reddy, 1984; Reddy et al., 1985), dry biomass yield of water hyacinth cultured in primary sewage effluent averaged 52 Mg ha^{-1} yr^{-1} during 1981–84. These biomass yields may not represent the true yield potential of hyacinth cultured in sewage effluent, since these systems were not optimized with respect to biomass yields. Experimental channels were operated during 1983–84, at varying residence times (3, 6, 12 and 24 days) using primary sewage effluent. Biomass yields were not affected by residence time of the primary sewage effluent, although yields recorded during this period were lower (36–45 Mg ha^{-1} yr^{-1}) than had been obtained in previous years. An unusually hard freeze during the winter, high canopy temperatures and high humidity during July–September, and insect damage were all contributing factors to the low biomass yields.

Water hyacinth biomass annual yields and seasonal distribution in secondary sewage effluent were the same as those reported for hyacinth cultured in primary sewage effluent (Reddy et al., 1985) and in other types of waters.

Methane Digester Effluents

Conversion of biomass in anaerobic digesters results in a waste product which must be disposed in an environmentally safe manner. Digester effluent can be potentially treated, and its nutrients recycled, by growing water hyacinth and feeding the resultant biomass back to the digester. Only limited information is available on the performance of water hyacinth in digester effluents. We have conducted two greenhouse studies to determine the growth of water hyacinth cultured in diluted and undiluted digester effluent obtained from anaerobic digesters fed with water hyacinth feedstock (Reddy and Tucker, 1983; Moorhead, 1986). Growth rates of water hyacinth were in the range of 19–34 g m^{-2} day^{-1}, rates comparable to those obtained in sewage effluent (Fig. 6). Ammonia concentration of the digester effluent significantly affected the growth of water hyacinth (Fig. 6). Plant growth was not affected adversely when NH_4^+ concentration of the effluent was less than 104 mg N liter^{-1}, and the plants did not survive at a NH_4^+ concentration >200 mg N liter^{-1}.

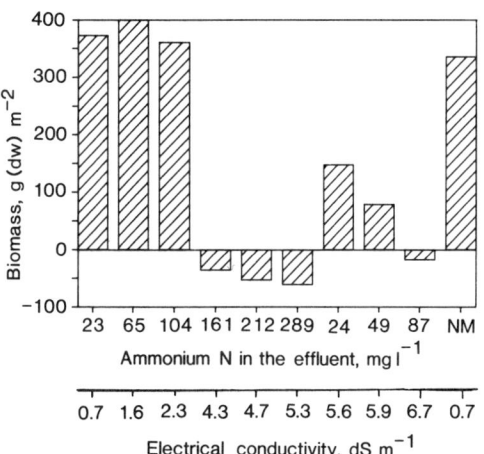

FIG. 6. Effect of ammonium N concentration on water hyacinths cultured in methane digester effluent (NM = nutrient medium).

NUTRIENT SUPPLY

Water hyacinth cultured in either natural systems such as eutrophic lakes and streams, or in artificial systems such as retention ponds or channels, derive nutrients from a number of sources. These include nutrients in the water itself, nutrient release from underlying sediments or detritus plant tissue trapped in water hyacinth mats or deposited on the bottom, and dinitrogen fixation by epiphytic algae present on the water hyacinth plants. Limiting nutrients may also be supplied through foliar sprays.

Foliar Fertilization of Nutrients

Under certain conditions, it may be necessary to apply limiting nutrients in order to increase biomass production and enhance the uptake of problem nutrients. For example, our studies have shown that both N and P can limit water hyacinth growth in eutrophic lake waters. Plants responded to foliar fertilization of N and P during the summer growing season, with a significant increase in biomass yields (DeBusk et al., 1985). Strong evidence emerged that P was a limiting nutrient in Lake Apopka water, since the application of N and P to water hyacinth increased net biomass yields by 51%, compared with only a 6% increase resulting from application of N alone. Use of ^{15}N-enriched urea in the foliar spray revealed that only 62–74% of the foliar-applied N was assimilated into plant tissue. When micronutrients were applied through foliar fertilization of water hyacinth cultured in primary sewage effluent, biomass yields increased significantly.

Sediments

In natural systems, sediments can play a significant role in supplying nutrients to water hyacinth, but in artificial systems this source of nutrients may not be significant. Since Lake Apopka was used as a model system for growing hyacinths, we (Reddy et al., 1986b) determined the nutrient-release potential of Lake Apopka sediments for water hyacinth to be 42–124 mg N m^{-2} day^{-1}. Fundamental processes involved in nutrient release need to be evaluated in order to quantify this source of nutrients under sustained production of water hyacinths in lakes.

Detritus

During water hyacinth biomass production, detritus plant tissue can be generated in the photic zone through the aging of plant parts, overcrowding or freeze damage and eventually deposited at the sediment–water interface. Some of this detritus can also be trapped in the root zone.

In a managed system, detritus comprised about 10–12% of the total biomass yield (DeBusk et al., 1983; Moorhead, 1986). Upon decomposition, this detritus releases nutrients utilizable by the live plants. Water hyacinths growing on eutrophic Lake Apopka water (Reddy et al., 1986c) contained 65% of the plant biomass as live tissue, 25% as dead plant tissue attached to the plant, and about 10% as detritus deposited on the sediment surface. Frequent harvesting of plants for biomass utilization will significantly reduce the production of detritus. The decomposition rate of detritus is faster in the root zone ($k = 0.011$–$0.014\,\text{day}^{-1}$) than at the sediment–water interface ($k = 0.006\,\text{day}^{-1}$). About 21–45% of the detritus N was released when decomposition occurred in the root zone, while only 11–17% was released when decomposition occurred at the sediment–water interface (Reddy et al., 1986c). Decomposition of detritus was independent of detritus content over the range 10–28 mg N g^{-1} of tissue. Decomposition rates were also higher when detritus P content was >2.2 mg P g^{-1}. The significance of detritus production and decomposition, and the basic pathways involved in decomposition and nutrient release of detritus to support water hyacinth are topics for future research.

Nitrogen Fixation

Although water hyacinth plants do not themselves have the capability of fixing atmospheric N_2, several N_2-fixing algae and bacteria are capable of growing on various plant parts. A few studies (Purchase, 1977; Nayak et al., 1979) have indicated that N_2-fixing bacteria such as *Azotobacter* sp. and *Azospirillum* sp. are capable of growing on water hyacinth plants. Acetylene reduction activity of water hyacinth plants cultured in Lake Apopka water has been measured under laboratory conditions. The N_2-fixing potential of epiphytic algae present on various plant parts of water hyacinth was 1.7 mg N kg^{-1} day^{-1}. No identification was made of the type of algae present on these water hyacinth plants. Currently, studies are underway to quantify the significance of N_2 fixation in water hyacinth cultures.

MANAGEMENT PRACTICES AND SYSTEMS

Sustained biomass production throughout the year is controlled largely by the cultural methods used for management of the system. Plant density, harvesting frequency, and monoculture versus polyculture have each been investigated.

Plant Density and Harvesting Frequency

A hypothetical growth curve (Fig. 7) as a function of time reveals three phases of biomass accumulation if surface area available remains fixed during the growing period. Phase I of the growth curve represents an initial lag followed by exponential growth. During this phase plant density is usually low, and net biomass yield per unit area is also low, but specific growth rates (percent increase in biomass per day in relation to initial density) are high. Phase II represents a linear growth phase, where biomass yields are highest and plant densities are in an intermediate range. Phase III represents a nonlinear growth phase once more, where biomass yields are low and plant densities are highest compared to the other two phases. In the linear phase of the growth curve, plant densities are optimum for achieving maximum biomass yields. Growth curves of water hyacinth and optimal plant densities were also found to be influenced by seasons (DeBusk et al., 1981; Reddy and DeBusk, 1984). During cooler months, optimal plant density was in the range of 500–800 g m^{-2} while, during warmer months, optimal density was in the range 500–2000 g m^{-2}.

A much higher optimal density range was observed for hyacinth grown under nutrient nonlimiting conditions, though a narrow range was observed for plants grown in a nutrient-limited system (Lake Apopka water). Under both conditions, the growth rate of hyacinth was maximum in the plant density range of 500–1000 g m^{-2}. Depending on biomass

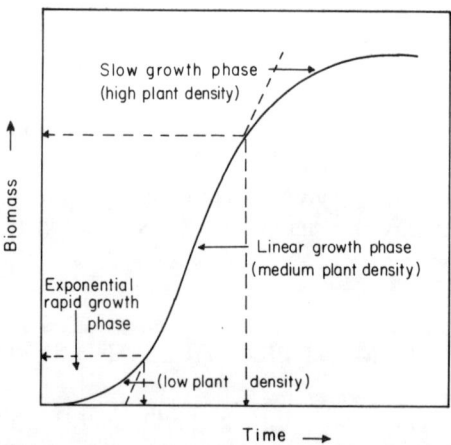

FIG. 7. Hypothetical growth curve of water hyacinths showing typical changes in plant density during growing season.

requirements, plants normally can be allowed to grow until they reach the higher end of the linear phase, and then harvested back to the lower end of the linear phase.

Biomass yield of water hyacinth was 19 g m^{-2} day^{-1} in a system where plants were not harvested, compared to 24 g m^{-2} day^{-1} when harvested at maximum plant densities with no net growth, and 32 g m^{-2} day^{-1} when harvested once every 2 weeks in order to maintain optimal densities.

Monoculture versus Polyculture Systems

Water hyacinth production is significantly affected by cold temperatures. Temperatures at or below the freezing point for several hours can result in the damage of photosynthetic plant tissue. To sustain biomass yields, cold-tolerant aquatic plants may be grown along with water hyacinth. Earlier, performance of two cold-tolerant pennywort varieties were compared in monoculture. Later, in outdoor microcosms, we evaluated the performance of water hyacinth and pennywort for a period of 1 year under nutrient nonlimiting conditions. Water hyacinth growth (Fig. 8) after the

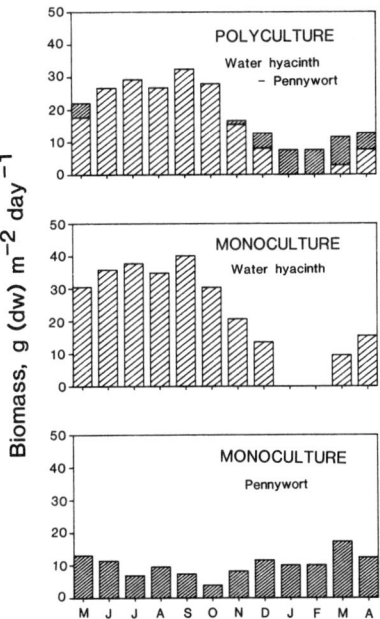

FIG. 8. Growth rates of water hyacinths and pennywort cultured in monoculture and polyculture systems.

January 1982 freeze, and yields were near zero both during January and February. Pennywort growth was not affected (growth rate of 8 g (dry wt) m^{-2} day^{-1}). During the winter months, pennywort occupied the major part of each tank by pushing water hyacinth plants to one side; during warmer months this situation was reversed. Although pennywort appears promising for the winter months, limited information is available on its growth characteristics and nutrient requirements in order to optimize growth. Pennywort plants are intertwined and are hard to separate from one another, so harvesting methods developed for water hyacinth may not be suitable for the harvesting of pennywort. The ratio of water hyacinth/pennywort standing crop needs to be monitored in order to optimize year-round biomass production. Polyculture systems may play a role in water hyacinth-based sewage treatment where year-round plant growth and water treatment are essential.

Water Hyacinth Production in Lakes and Streams

Management strategies are needed for a continuous supply of biomass if water hyacinths are to be grown in lakes as a feedstock for methane production. Biomass supply per unit area is variable because water hyacinth growth is dependent on solar radiation and temperature. The critical months with very little or no growth are December, January and February. During this period either the methane digester should be supplied with other feedstock such as pennywort; or the digester design should be based on average annual biomass yield, and follow a special harvesting strategy involving management of the system at varying plant densities (Reddy, 1984). Harvesting constant amounts of biomass every day will automatically result in varying plant densities since growth rates are different each month. For example, if the biomass harvest is lower than the biomass yield in a particular month, then there will be a net accumulation of biomass in the system, thus resulting in a higher plant density. This situation usually occurs during an active growing season. On the other hand, if biomass harvested is greater than biomass yield during the month, then there will be a net decrease in standing crop, thus resulting in lower plant density. This strategy stores surplus biomass during active growing months, and harvests *in situ* stored biomass during slow-growing months in order to meet the biomass demand.

Water Hyacinth Production in Sewage Treatment Systems

The practical application of water hyacinth production in sewage effluent depends on the degree of water treatment which hyacinths can provide to the effluent. Water hyacinth biomass production in sewage effluent offers a limited source of feedstock for methane digesters and can be potentially used as well with other sources of wastes available from small communities. Our studies have shown (Reddy et al., 1985) that water hyacinths can be used successfully to treat primary sewage effluent in order to meet secondary treatment standards, and to treat secondarily treated effluent in order to meet tertiary treatment standards. Application of plant growth regulators increases biomass yields significantly, while concurrently improving water quality. Increasing biomass yields not only enhances the improvement of water quality, but also reduces the area needed for treatment systems.

Although a number of studies (Cornwell et al., 1977; McDonald and Wolverton, 1980; Reddy and Sutton, 1984; Reddy et al., 1985) have been conducted on the use of water hyacinths in sewage treatment, specific information on design criteria is still lacking. Few data are available on the effects of water depth, loading rates, harvesting frequency and special optimization techniques such as application of limiting nutrients, and aeration, to reduce BOD stress on plants and to improve water quality. These data are needed in order to develop economically feasible, optimized biomass production systems.

BIOCHEMICAL AND PHYSICO-CHEMICAL TRANSFORMATIONS IN A WATER HYACINTH PRODUCTION SYSTEM

The presence of water hyacinths decreases the oxygen content of the water (Rai and Munshi, 1979; Reddy, 1981), thus altering the metabolic activity of microorganisms present in the root zone. A dense cover of water hyacinths increases the dissolved CO_2 level of the water. The water column under water hyacinth mats, can be divided into three zones (Fig. 9A) based on O_2 content. Zone I is the root zone (rhizosphere) area where water hyacinths transport oxygen through the leaves and petioles into the roots; subsequently, the oxygen not used during respiration is released into the adjacent environment. This process has been demonstrated for other aquatic plants by Armstrong (1964). No data are available on the potential of water hyacinth for transporting oxygen from shoots to the root zone. Oxygen in the root zone can play a significant role in regulating several carbon and nitrogen transformations. For example, the pumping of oxygen by water hyacinth can significantly decrease the BOD of primary sewage

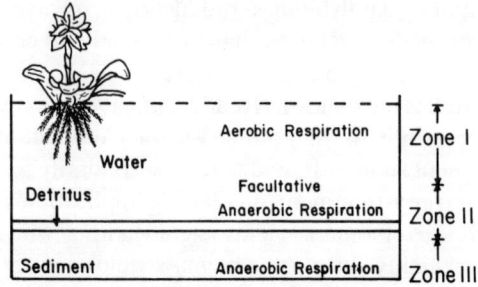

FIG. 9. (A) Schematic presentation of a water hyacinth system showing the zones with different microbial respiratory activities.

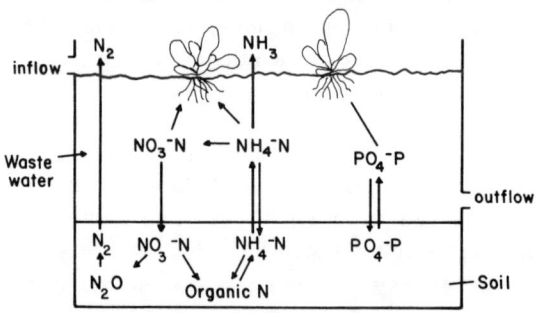

FIG. 9. (B) Schematic presentation of the nitrogen and phosphorus transformations functioning in water hyacinth systems.

effluent (Reddy et al., 1985). Oxygen concentration in the water column decreases with depth, approaching near zero in zone II. In this zone, microbial metabolism switches from aerobic to facultative anaerobic respiration, where NO_3^- is used as an electron acceptor during respiration. Nitrate entering this zone is reduced to N_2 and probably lost subsequently. In zone III, anaerobic microbial respiration probably dominates the system, where sulfates are reduced by bacterial respiration. A detailed review of these processes for aquatic systems is presented by Fenchel and Jorgensen (1977).

A number of N transformations are known to be functioning in an aquatic system (Fig. 9B), but only limited information is available on the significance of these processes with respect to the fate of N in water hyacinth systems (Reddy, 1983). In a lake system, N transformations in the underlying sediment can play a significant role in supplying N to the plant. Nitrogen in water hyacinth systems is present in organic and inorganic

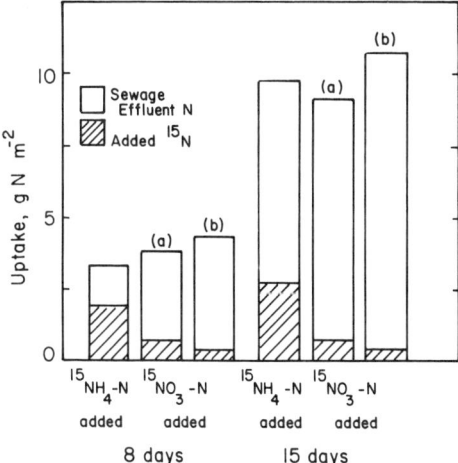

FIG. 10. Plant uptake of labeled NO_3 and NH_4 from sewage effluent, indicating the significance of nitrogen transformations.

forms. In a eutrophic lake such as Lake Apopka, most of the N is tied up in organic form, and this form of N can be converted into inorganic forms and then utilized by plants. Unfortunately this process is poorly understood.

Denitrification (conversion of NO_3^- to gaseous end products such as N_2O and N_2) can play a major role when water hyacinths are used for polishing secondarily treated wastewaters. In a water hyacinth system containing secondarily treated sewage effluent, about 80% of the added $^{15}NO_3$ was lost through denitrification (Fig. 10), with only 18% being recovered in plant tissues. Denitrification can potentially occur in the underlying sediment, in the water column if devoid of oxygen, and in anoxic sites of the root zone. For example, detritus trapped in the root zone provides an ideal environment for denitrification. Future research should be directed toward quantitatively determining biochemical and physico-chemical processes and the role of these processes in improving water quality during plant biomass production.

QUALITY OF PLANT BIOMASS

Nutrient composition of the culture medium can significantly influence the quality of the biomass produced by affecting allocation of carbon among plant parts. Plants cultured in nutrient-poor waters have elongated root systems and shoot/root ratios are usually in the range of 1–2, while plants cultured in nutrient-rich waters tend to have shallow roots with a shoot/root ratio in the range of 4–6. An increased proportion of root biomass lowers overall quality of the biomass since roots usually decompose slowly and produce less methane (Shiralipour and Smith, 1984). Nitrogen concentrations of the culture medium and plant density appear to be inversely related to root length.

Chemical Composition

The quality of plant biomass is also affected by nutrient assimilation and by the resulting chemical composition of water hyacinth tissue. Water hyacinth tissue N and P composition was monitored throughout the year to determine seasonal variability for plants cultured in eutrophic Lake Apopka water and in sewage effluents. Nitrogen and P content of the tissue was higher during winter than summer months (Table 5). Slightly lower concentrations during the summer months were due to dilution as a result of more biomass being produced per unit area. Nitrogen concentration of the plant tissue was in the range of 14–22 mg N g^{-1}, while P content was in

TABLE 5
Nitrogen and Phosphorus Concentrations of Water Hyacinth Plants Grown in Various Culture Mediums

Culture medium	Plant tissue (mg g^{-1})	
	Nitrogen	Phosphorus
Agricultural drainage water	20·5 ± 1·8	3·5 ± 0·9
Eutrophic Lake Apopka water		
a. Managed/fertilized	16·3 ± 2·0	1·8 ± 0·6
b. Managed/no fertilization	16·8 ± 2·7	1·8 ± 0·7
c. Unmanaged/no harvest	16·4 ± 2·8	1·7 ± 0·4
Primary sewage effluent		
a. 3 days retention time	37·2 ± 3·6	12·9 ± 2·9
b. 6 days retention time	36·1 ± 3·6	12·9 ± 2·9
c. 12 days retention time	33·3 ± 3·2	12·4 ± 2·9
d. 24 days retention time	33·5 ± 3·4	12·3 ± 2·7
Secondary sewage effluent	20·5 ± 3·8	7·9 ± 2·2
Digester effluent	22·4 ± 2·1	9·8 ± 1·5
Nutrient medium	39·5 ± 2·1	9·5 ± 1·5

Fertilizer applied as foliar spray three times during the summer months, at a rate of 10 kg N and 3 kg P ha^{-1}.

the range of 1–3 mg P g^{-1}. Foliar fertilization with N and P three times a year did not increase plant tissue N and P levels significantly. Nitrogen and P contents of plants cultured in primary sewage effluent at Walt Disney World have ranged from 33 to 37 mg N g^{-1}, while P concentrations have ranged from 12 to 13 mg P g^{-1}. Plants cultured in secondary sewage effluent had N and P contents of 20 and 8 mg g^{-1}, respectively.

High N and P ratios for plants cultured in Lake Apopka water suggest that P is limiting the plant growth. Nitrogen to P ratios of the lake water were higher (10–25) than the ratios observed for plant tissue (6–16), indicating that some P was probably released from the underlying sediments or from the decomposition of detritus. A much narrower ratio of N to P was observed for plants cultured in primary and secondary sewage effluent, indicating that both N and P were not limiting the growth of hyacinth. Water hyacinth cultured under nutrient nonlimiting conditions has an N/P ratio of 3–5 (Sato and Kondo, 1981; Reddy and Tucker, 1983).

Plant Uptake

Nutrient removal through plant uptake is directly related to the growth rates of plants (Rogers and Davis, 1972; Boyd, 1976). Nitrogen and P removal rates by water hyacinths cultured in Lake Apopka water were 0·9–6·5 kg N ha^{-1} day^{-1} and 0·06–0·74 kg P ha^{-1} day^{-1}. Removal rates were lowest during winter and highest during summer months. Annual average removal rates were approximately 850 kg N ha^{-1} yr^{-1} and 80 kg P ha^{-1} yr^{-1}. Higher removal rates were observed for plants cultured in primary and secondary sewage effluent, with annual removal rates of 1726 and 1193 kg N ha^{-1} yr^{-1}, while P-removal rates were 387 and 205 kg P ha^{-1} yr^{-1}, respectively (Reddy et al., 1985). Water hyacinth grown in agricultural drainage water removed 730 kg N ha^{-1} yr^{-1} and 159 kg P ha^{-1} yr^{-1} (Reddy et al., 1982). A wide range of N and P uptake rates was reported from studies conducted at other locations (Boyd, 1976; Reddy and Sutton, 1984). Most of the reported values were based on short-term measurements and were then subsequently extrapolated to an annual basis. Thus, the values reported represent potential removal rates rather than actual removal rates for a given set of climatic conditions.

WATER QUALITY IMPROVEMENT

Use of water hyacinth for biomass energy farms in freshwater lakes and streams depends on the plant's capacity to utilize nutrients in order to concurrently promote growth and improve water quality. Two case studies involving the use of water hyacinth for the production of biomass while improving water quality of eutrophic lake water and sewage effluents will be discussed below.

Lake Water Treatment Using Water Hyacinths

For a period of 18 months, experimental channels were used to determine the effectiveness of water hyacinth for improving Lake Apopka water quality (discussed in earlier sections). Several water quality parameters were monitored both at the inflow (Lake Apopka water) and the outflow (water leaving hyacinth channel) sites; only changes in N, P and chlorophyll-a concentrations, key indicators of the impact of water hyacinth on lake water quality, will be presented here.

Total N concentration of the lake water varied considerably, with values of 2·5–8·1 mg N liter^{-1}. It occurred primarily as organic, rather than readily available forms (Table 6), with an average value of 4·35 mg N liter^{-1}. During a 36-h residence time in a channel containing water hyacinth, total N concentrations decreased by about 50%. Foliar

TABLE 6
Chemical Composition of the Influent (Lake Apopka Water) and Effluent Leaving Experimental Channels Containing Water Hyacinth Plants. Hydraulic Retention Period = 36 h. Experimental period, June 1984–December 1985

System	TKN	TP	Chlorophyll-a
	(mg liter^{-1})		(mg m^{-3})
Influent (Lake Apopka water)	4·35	0·27	55·1
	±1·04	±0·06	±10·8
Effluent			
1. Channel with water hyacinth (foliar fertilized and frequent harvesting)	2·79	0·14	15·4
	±0·69	±0·05	±7·9
2. Channel with water hyacinth (frequent harvesting)	2·68	0·15	11·2
	±0·68	±0·05	±5·2
3. Channel with water hyacinth (no harvesting)	3·00	0·16	10·9
	±0·88	±0·06	±5·9
4. Channel with no plants	4·06	0·23	43·0
	±0·87	±0·08	±20·8

application of urea during the months of May, July and September did not alter resultant quality of the water. Removal rates of N were significantly lower for the channel which contained no plants, or for the channel where water hyacinths were not harvested for a period of 1 year. Maximum P concentration (0·85 mg P liter^{-1}) of the lake water was in August, while the minimum value (0·2 mg P liter^{-1}) was observed in October. This was similar to the N pattern (Table 6). Average total P concentration was 0·27 mg P liter^{-1}.

Chlorophyll-a concentration of the lake water also showed trends similar to those for total N and P (Table 6), with an average value of 55 mg m^{-3}. Water hyacinths were effective in decreasing chlorophyll-a concentrations of the lake water.

The beneficial effects of water hyacinths in improving water quality are clear. Nitrogen removal rates, calculated from the reduction in mass loading, indicate that about 50–60% of the incoming N was removed, while up to 80% of the incoming P was being removed. Significant removal of both N and P was due to the settling of particulate matter, especially algal cells of the lake water. This caused a sediment build-up of as much as 30 cm over a period of 1 year. This type of situation will not arise when plants are cultured directly in the lake, however, because water hyacinth plants shade the algae. In the experimental channels, lake water rich in algal cells was allowed to flow through each channel at a relatively rapid rate (36 h residence time), resulting in a rapid build-up of sediment. Frequent harvesting appears beneficial in the improving of water quality. Although the results presented are favorable with respect to improving water quality, long-term effectiveness of the system is not known.

Sewage Effluent Treatment

Evaluations of the efficiency of water hyacinth biomass production with respect to use of nutrients from primary and secondary sewage effluents revealed that water hyacinth plants were also effective in BOD removal, with about 80% reduction in 3 days (Table 7a). Nitrogen and P removal were both very poor, however water hyacinths were effective in reducing N and P concentrations of secondary sewage effluent (Table 7b). Nitrate removal was better than 90%, while NH$_4$ removal was about 80% and TKN was reduced by about 50%. Both biotic and abiotic processes likely accounted for removal of BOD, N and P from the secondary effluent. Effective removal of BOD from primary sewage effluent by water hyacinth is due to: (1) filtration by the roots, (2) sedimentation of solids, and (3) aerobic oxidation of carbon in the root zone. Detailed discussions of

TABLE 7

(a) Chemical Composition (mg liter^{-1}) of the Influent and Effluent of a Water Hyacinth Channel Receiving Primary Sewage Effluent (Values Presented are Averages for the Period August 1983–October 1984)

		Effluent			
		Residence time (days)			
Parameter	Influent	3	6	9	24
Total solids	348·6	259·9	256·1	250·1	261·3
	±31·5	±28·6	±23·4	±28·4	±54·9
BOD$_5$	199·6	44·1	34·2	34·1	28·4
	±29·9	±17·4	±23·1	±12·6	±11·0
TKN	29·2	25·6	23·9	19·2	17·8
	±3·7	±3·8	±4·3	±4·1	±4·2
NH$_4^+$–N	23·0	21·9	21·1	16·7	15·4
	±3·7	±4·3	±3·8	±3·3	±5·9
TP	6·6	5·6	5·7	5·0	5·2
	±0·6	±0·9	±0·9	±0·8	±0·9

(b) Chemical Composition (mg liter^{-1}) of the Influent and Effluent of a Water Hyacinth Channel Receiving Secondary Effluent with a 3-day Residence Period (Values Presented are Averages for the Period August 1983–October 1984)

Parameter	Influent	Effluent
Total solids	239·9 ± 22·1	217·7 ± 15·4
TKN	3·9 ± 1·8	1·7 ± 2·1
NH$_4$–N	1·8 ± 1·4	0·3 ± 0·4
NO$_3$–N	1·5 ± 0·5	0·1 ± 0·04
TP	0·7 ± 0·2	0·3 ± 0·2
Ortho-P	0·41 ± 0·2	0·2 ± 0·2

nutrient-removal processes and of the effectiveness of water hyacinth-based sewage treatment are found elsewhere (Wooten and Dodd, 1976; DeBusk et al., 1983; Reddy et al., 1985).

SUMMARY AND CONCLUSIONS

The results of a series of experiments conducted on a model aquatic biomass crop, water hyacinth, are summarized. Initial screening of aquatic plants as biomass feedstocks for methane production revealed water hyacinth to be the most productive aquatic macrophyte, with maximum potential yields of 234 Mg ha^{-1} yr^{-1} under nonlimiting conditions. Natural populations of water hyacinth showed considerable morphological variability but, genetically, it appears that there is only one species at present in Florida.

High photosynthetic rates increased water hyacinth growth rates by 2·5 times when the CO_2 content of the air was enriched. Maximum biomass yields were observed when the culture medium contained 20 mg N liter^{-1}, 1·1 mg P liter^{-1}, and 12 mg K liter^{-1}. Further increase in the levels of these nutrients did not increase yields but did improve the plant composition. Iron was found to be the critical micronutrient controlling biomass yields when water hyacinth was cultured in NO_3^--rich wastewaters. Application of growth regulators (GA_3) increased biomass yields of water hyacinth 30–50%.

Growth rates in relation to environmental variables such as temperature and solar radiation revealed that freezing temperatures up to 6 nights successively during the winter months caused the death of photosynthetic tissue. However, the plants recovered after 2–4 weeks and resumed normal growth. Polycultures of water hyacinth and the cold-tolerant pennywort sustained levels of year-round biomass production at higher levels than did monocultures.

Water hyacinth can be grown successfully in a variety of nutrient-enriched waters. Maximum growth rates of up to 64 g m^{-2} day^{-1} were observed in May during periods of highest solar radiation incidence when plants were cultured under nutrient nonlimiting conditions; average biomass yield of this system was 106 Mg ha^{-1} yr^{-1}. Biomass yields of 50–65 Mg ha^{-1} yr^{-1} were observed when plants were cultured either in amended or unaltered agricultural drainage water or eutrophic lake water. Similar yields were also obtained in primary or secondary sewage effluent.

Water hyacinth grown for biomass improved agricultural drainage water, sewage effluent, or eutrophic lake water. In eutrophic Lake Apopka water, N and P loading were decreased by 50 and 70% using water hyacinth. The BOD of primary sewage effluent decreased by 90% using water hyacinth plants, while P and NO_3^- concentrations of secondary sewage effluent were decreased by 50 and 90%, respectively.

Although the results presented in this chapter provide a comprehensive database on morphological and physiological characteristics, nutrient requirements, cultural techniques and performance under field conditions, a number of data gaps exist which may limit the application of hyacinth biomass systems either for lakes, streams or sewage effluents with concurrent water treatment.

Fundamental understanding remains necessary on: understanding and regulation of the fundamental processes of photosynthesis; mineral nutrition; nutrient cycling within-plant and between plant/water/sediment; stress tolerances to cold, toxins and pests; growth regulation; and genetic variations. Such understanding is essential if we are to reliably design and operate such systems.

Field data are also needed on performance of water hyacinth when cultured directly in lakes, including resultant impact on water quality, fish and other nontarget biota. Optimization tecniques for a lake–hyacinth system are likely to differ from those for a water hyacinth-based sewage treatment system.

Use of water hyacinth biomass farms to provide sewage treatment is gaining the acceptance of regulatory agencies. Deliberate growth of hyacinth biomass in eutrophic lakes for either establishing biomass energy farms or for restoration of lakes is also attracting increased attention. Needed is a strong database and demonstration projects to convince public and regulatory agencies to use this system. The potential to generate a desirable energy form such as methane, is also improving the acceptability of such systems.

ACKNOWLEDGEMENTS

The technical assistance of W. F. DeBusk, J. C. Tucker, F. H. Hueston, R. Fowler, M. Fisher, and K. K. Moorhead, who participated in various phases of the investigations, is gratefully acknowledged. The author also acknowledges the excellent contributions of Dr Yoko Oki, a visiting scientist from Japan. Critical review of the manuscript by Dr B. L. McNeal is also gratefully acknowledged. Thanks goes as well to Miss Brenda Clutter for her patience and diligence in the preparation of this manuscript.

REFERENCES

Armstrong, W. (1964). Oxygen diffusion from the roots of some British bog plants. *Nature, Loud.*, **204**, 801–2.

Bagnall, L. O., Baldwin, J. A. and Hentges, J. F. (1974). Processing and storage of water hyacinth silage. *Hyacinth Control Journal* **12**, 73–9.

Barrett, S. C. H. (1980). Sexual reproduction in *Eichhornia crassipes*. *Journal of Applied Ecology* **17**, 101–12.
Boyd, C. E. (1969). Vascular aquatic plants for mineral nutrient removal from polluted waters. *Economic Botany* **23**, 95–103.
Boyd, C. E. (1976). Accumulation of dry matter, nitrogen and phosphorus by cultivated water hyacinths. *Economic Botany* **30**, 51–6.
Center, T. D. and Spencer, N. R. (1981). The phenology and growth of water hyacinths (*Eichhornia crassipes* [Mart] Solms) in a eutrophic north-central Florida lake. *Aquatic Botany* **10**, 1–32.
Cooley, T. N., Martin, D. F., Durden, W. C., Jr and Perkins, B. D. (1979). A preliminary study of metal distribution in three water hyacinth biotypes. *Water Research* **13**, 343–8.
Cornwell, D. A., Zoltek, J. Jr, Patrinely, C. D., des Furman, T. and Kim, J. I. (1977). Nutrient removal by water hyacinth. *Journal of Water Pollution Control Federation* **49**, 57–65.
DeBusk, T. A., Ryther, J. H., Hanisak, M. D. and Williams, L. D. (1981). Effects of seasonality and plant density on the productivity of some freshwater macrophytes. *Aquatic Botany* **10**, 133–42.
DeBusk, T. A., Williams, L. D. and Ryther, J. H. (1983). Removal of nitrogen and phosphorus from wastewater in a waterhyacinth-based treatment system. *Journal of Environmental Quality* **12**, 257–62.
DeBusk, W. F., Reddy, K. R. and Tucker, J. C. (1985). Management strategies for water hyacinth production in a nutrient-limited system. In: *Biomass Energy Development*, Smith, W. H. (ed.), pp. 275–86. Plenum Press, New York.
Fenchel, T. M. and Jorgensen, B. B. (1977). Detritus food chains of aquatic ecosystems. The role of bacteria. *Advances in Microbial Ecology* **1**, 1–58.
Gopal, B. and Sharma, K. P. (1981) *Water Hyacinth*, Hindasia Publishers, New Delhi, India, 128 pp.
Haller, W. T., Knipling, E. B. and West, S. H. (1970). Phosphorus absorption by and distribution in water hyacinths. *Proceedings of Soil Crop Science Society of Florida* **30**, 64–8.
Haller, W. T., Sutton, D. L. and Barlow, W. C. (1974). Effects of salinity on growth of several aquatic macrophytes. *Ecology* **55**, 891–5.
Knipling, E. B., West, S. H. and Haller, W. T. (1970). Growth characteristics, yield potential and nutrient content of water hyacinths. *Proceedings of Soil Crop Science Society of Florida* **30**, 51–63.
Matthews, L. J. (1967). Seedling establishment of water hyacinth (*Eichhornia crassipes*). *PANS* **13**, 7–8.
McDonald, R. C. and Wolverton, B. C. (1980). Comparative study of wastewater lagoon with and without waterhyacinth. *Economic Botany* **34**, 101–10.
McNaughton, S. J. and Fullem, L. W. (1970). Photosynthesis and photorespiration in *Typha latifolia*. *Plant Physiology* **45**, 703–7.
Moorhead, K. K. (1986). Nitrogen cycling in an integrated 'Biomass for Energy' system. Ph.D. dissertation, University of Florida, Gainesville, Florida.
Muramoto, S. and Oki, Y. (1983). Removal of some heavy metals from polluted water by water hyacinth (*Eichhornia crassipes*). *Bulletin of Environmental Contamination and Toxicology* **30**, 170–7.

Nayak, D. N., Swain, A. and Rao, V. R. (1979). Nitrogen fixing *Azospirillum lipoferum* from common weeds associated with rice and aquatic ecosystems. *Current Science* **48**, 866–7.

Nolan, W. J. and Kirmse, D. W. (1974). The paper making properties of water hyacinth. *Hyacinth Control Journal* **12**, 90–7.

Ogwada, R. A. (1983). Growth, nutrient uptake and nutrient regeneration by selected aquatic macrophytes. M. S. thesis, University of Florida, Gainesville, Florida.

Oki, Y. and Reddy, K. R. (1985). Response of water hyacinth to exogenously supplied gibberellic acid (GA_3). Paper presented at 25th annual meetings of the Aquatic Plant Management Society, July 21–24, 1985. Vancouver, British Columbia, Canada.

Oki, Y., Reddy, K. R. and Wain, R. P. (1985a). Morphological variations and isoenzyme patterns of water hyacinths in Florida. *Weed Research (Japan)* **30**, 191–2.

Oki, Y., Nakagawa, K. and Reddy, K. R. (1985b). Uptake and translocation of ^{15}N in water hyacinth. *Proceedings of the 10th Asian-Pacific Weed Science Society Conference*, pp. 317–24.

Ower, J., Cresswell, C. F. and Bate, G. C. (1981). The effects of varying culture nitrogen and phosphorus levels on nutrient uptake and storage by the water hyacinth, *Eichhornia crassipes* (Mart) Solms. *Hydrobiologia* **85**, 17–22.

Patterson, D. T. and Duke, S. O. (1979). Effect of growth irradiance on the maximum photosynthetic capacity of water hyacinth (*Eichhornia crassipes*). *Plant and Cell Physiology* **20**, 177–84.

Penfound, W. T. and Earle, T. T. (1948). The biology of the water hyacinth. *Ecological Monographs* **18**, 447–72.

Purchase, B. S. (1977). Nitrogen fixation associated with *Eichhornia crassipes*. *Plant Soil* **46**, 283–6.

Rai, D. N. and Munshi, J. D. (1979). The influence of thick floating vegetation (water hyacinth: *Eichhornia crassipes*) on the physico-chemical environment of a fresh water wetland. *Hydrobiologia* **62**, 65–9.

Reddy, K. R. (1981). Diel variations in physico-chemical parameters of water in selected aquatic systems. *Hydrobiologia* **85**, 201–7.

Reddy, K. R. (1983). Fate of nitrogen and phosphorus in a wastewater retention reservoir containing aquatic macrophytes. *Journal of Environmental Quality* **12**, 137–41.

Reddy, K. R. (1984). Water hyacinth biomass production in Florida. *Biomass* **6**, 167–81.

Reddy, K. R. and Bagnall, L. O. (1981). Biomass production of aquatic plants used in agricultural drainage water treatment. In: *1981 International Gas Research Conference Proceedings*, pp. 376–90. Government Institute Inc., Rockville, Maryland.

Reddy, K. R. and DeBusk, W. F. (1984). Growth characteristics of aquatic macrophytes cultured in nutrient-enriched water. I. Water hyacinth, water lettuce, and pennywort. *Economic Botany* **38**, 229–39.

Reddy, K. R. and DeBusk, W. F. (1985a). Nutrient removal potential of selected aquatic macrophytes. *Journal of Environmental Quality* **14**, 459–62.

Reddy, K. R. and DeBusk, W. F. (1985b). Growth characteristics of aquatic macrophytes cultured in nutrient-enriched water. II. Azolla, duckweed and salvinia. *Economic Botany* **39**, 200–8.

Reddy, K. R. and Sutton, D. L. (1984). Water hyacinths for water quality improvement and biomass production. *Journal of Environmental Quality* **13**, 1–8.

Reddy, K. R. and Tucker, J. C. (1983). Growth and nutrient uptake of water hyacinth. I. Effect of nitrogen source. *Economic Botany* **37**, 236–46.

Reddy, K. R., Campbell, K. L., Graetz, D. A. and Portier, K. M. (1982). Use of biological filters for agricultural drainage water treatment. *Journal of Environmental Quality* **11**, 591–5.

Reddy, K. R., Sutton, D. L. and Bowes, G. E. (1983). Biomass production of freshwater aquatic plants in Florida. *Proceedings of Soil and Crop Science Society of Florida* **42**, 28–40.

Reddy, K. R., DeBusk, W. F. and Bagnall, L. O. (1984). Water hyacinth biomass production in eutrophic lake water. In: *1984 International Gas Research Conference Proceedings*, pp. 630–44. Government Institute, Inc., Rockville, Maryland.

Reddy, K. R., Hueston, F. M. and McKim, T. (1985). Biomass production and nutrient removal potential of water hyacinth cultured in sewage effluent. *Journal of Solar Energy* **107**, 128–35.

Reddy, K. R., Tucker, J. C. and DeBusk, W. F. (1986a). The role of submerged macrophytes in water quality improvement (submitted for publication).

Reddy, K. R., Jessup, R. E. and Rao, P. S. C. (1986b). Nitrogen dynamics in a eutrophic lake sediment (submitted for publication).

Reddy, K. R., DeBusk, W. F. and Fowler, R. (1986c). Detritus production and decomposition of water hyacinths in eutrophic lake water (submitted for publication).

Rogers, H. H. and Davis, D. E. (1972). Nutrient removal by waterhyacinths. *Weed Science* **20**, 423–8.

Sato, H. and Kondo, T. (1981). Biomass production of water hyacinth and its ability to remove inorganic minerals from water. I. Effect of the concentration of culture solution on the rates of plant growth and nutrient uptake. *Japanese Journal of Ecology* **31**, 257–67.

Sculthorpe, C. D. (1967). *The Biology of Aquatic Vascular Plants*, Edward Arnold, London, 610 pp.

Shiralipour, A. and Smith, P. H. (1984). Conversion of biomass into methane gas. *Biomass* **6**, 85–94.

Shiralipour, A., Garrard, L. A. and Haller, W. T. (1981). Nitrogen source, biomass production, and phosphorus uptake in water hyacinth. *Aquatic Plant Management* **19**, 40–3.

Spencer, W. and Bowes, G. (1986). CO_2 enrichment increases biomass production of water hyacinth. Paper presented at the Conference on Research and Applications of Aquatic Plants for Water Treatment and Resource Recovery, July 20–24, 1986. Orlando, Florida.

Stowell, R., Ludwig, R., Colt, J. and Tchobanoglous, G. (1981). Concepts in aquatic treatment system design. *Journal of Environmental Engineering* **107**, 919–40.

Tatsuyama, K., Egawa, H., Yamamoto, H. and Nakamura, M. (1979). Sorption of heavy metals by the water hyacinth from the metal solution. III. Some experimental conditions influencing sorption. *Weed Research (Japan)* **24**, 260–3.

Tucker, C. S. (1981). The effect of ionic form and level of nitrogen on the growth and composition of *Eichhornia crassipes* (Mart) Solms. *Hydrobiologia* **83**, 517–22.

Wain, R. P. and Martin, D. F. (1980). A genetic comparison of three forms of water hyacinth. *Journal of Environmental Science and Health* **15**, 625–33.

Wolverton, B. C. and McDonald, R. C. (1975a). Water hyacinths and alligator weeds for removal of lead and mercury from polluted waters. NASA Technical Memo Number TM-X-72723, National Space Technology Laboratories. Bay St. Louis, Mississippi.

Wolverton, B. C. and McDonald R. C. (1975b). Water hyacinths and alligator weeds for removal of silver cobalt and strontium from polluted waters. NASA Technical Memo Number TM-X-72727, National Space Technology Laboratories. Bay St. Louis, Mississippi.

Wolverton, B. C. and McDonald, R. C. (1979). Water hyacinth (*Eichhornia crassipes*) productivity and harvesting studies. *Economic Botany* **33**, 1–10.

Wolverton, B. C. and McDonald, R. C. (1981). Energy from vascular plant wastewater treatment systems. *Economic Botany* **35**, 224–32.

Wolverton, B. C. and McKnown, M. M. (1976). Water hyacinths for removal of phenols from polluted waters. *Aquatic Botany* **2**, 191–201.

Wooten, J. W. and Dodd, J. D. (1976). Growth of water hyacinths in treated sewage effluent. *Economic Botany* **30**, 29–37.

Yount, J. L. and Crossman, R. (1970). Eutrophication control by plant harvesting. *Journal of Water Pollution Control Federation* **42**, 173–83.

8
Water Hyacinth (*Eichhornia crassipes* (Mart) Solms) Biomass Cropping Systems: II. Harvesting and Handling

L. O. BAGNALL

Agricultural Engineering Department, University of Florida—IFAS, Gainesville, Florida, USA

INTRODUCTION

Water hyacinths, although productive and chemically and biologically attractive as a digester feedstock, are difficult to harvest and transport because of their bulk and the environment in which they grow. Aquatic biomass harvesting and transporting systems have been devised, but their labor requirements are high and productivity low and they do not appear to be economically feasible for commercial production of aquatic biomass (Bagnall, 1986).

In this program, design criteria and components to harvest and handle water hyacinth biomass were researched for developing systems with higher productivity and lower cost, labor, and energy requirements. Included were determination of mechanical, hydrodynamic and aerodynamic properties of plant mats, interactions of machinery with plants, and performance of processing and handling equipment.

IN-SITU PROPERTIES OF WATER HYACINTHS

To optimize the design of systems to harvest free-floating water hyacinths, the interactions of the individual and massed plants with restraints, barriers, machinery, wind and current must be predictable. The literature about the properties of water hyacinth in any context is sparse (Bagnall, 1974, 1982; Cifuentes and Bagnall, 1976). In-situ behavior of water hyacinth plants and mats of plants is important to the design of any mechanical water hyacinth management system. Some managers have an intuitive grasp of this behavior, but if systems are to be designed to be operated by unskilled labor or robots, the behavior must be quantified.

Physical-Mechanical Properties

Mats of water hyacinth may be loaded in lateral compression, tension and shear. When loaded, they may remain intact, fragment, rotate about vertical and horizontal axes, and/or slide over or under adjacent mats. All of these behaviors affect the capacity and efficiency of the systems with which they are harvested. Systems that require that the plants be supported by their own buoyancy while being transported in the water are especially sensitive to these behaviors.

Compression properties of water hyacinth mats were determined using two frame sizes. Mats of water hyacinth grown on wastewater were compressed in a 90×90 cm frame (Sivakumaran and Bagnall, 1984). The forces used were much higher than are likely to be encountered in a harvesting system. Force, displacement and time were measured by a digital data acquisition system. Mats of water hyacinth at a wastewater site and at a lake water site were compressed in a 3×3 m frame with forces which should be within the range encountered in harvesting systems. The larger frame was used for scaling and to introduce rotating and over-riding behaviors which could not be produced by the small frame. Instrumentation of the large frame was similar to that of the small frame.

The data on compression characteristics of the mats (Fig. 1) were regressed to a form similar to that found in the bulk compression tests. In tests with high-density samples in the large frame, rotation about a horizontal axis and subduction were observed.

On the basis of these results, it appears to be feasible to design a compacting rake which should be sufficiently more efficient than current types to justify the additional complexity.

Aerodynamic Properties

Water hyacinth plants and mats move about freely with the wind and harvesting system designs have been proposed to use the wind as part of a gathering mechanism. The mats are also compacted, to some unknown extent, by the wind. Forces induced by the wind may be a significant factor in rafting plants to removal sites.

Relatively undisturbed water hyacinth mats, 1.2×2.4 m, were lifted from a wastewater pond and transferred to a short, variable-speed wind tunnel (Bagnall, 1984). The mat was contained in a board frame with a

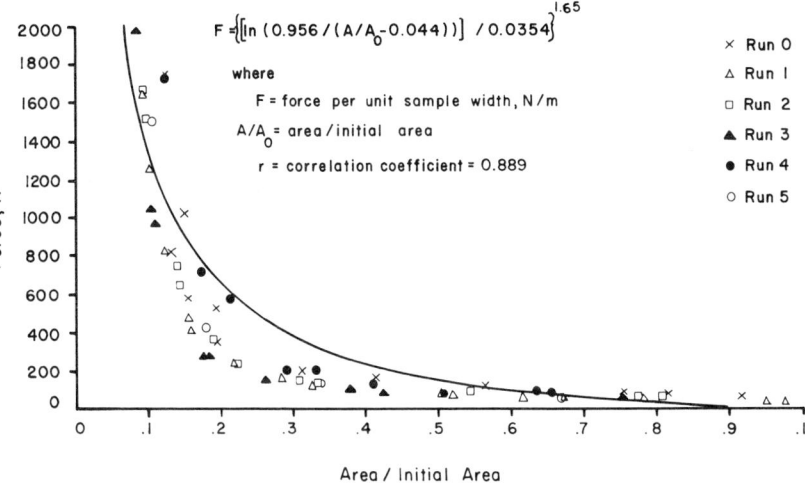

FIG. 1. In-situ compression properties of 1 m² water hyacinth mats.

minimal projection above the water and restrained by a load cell. Wind speed was varied and measured while the force was measured by the load cell. Sample size was reduced sequentially to observe the effect of size so that a model could be constructed.

The aerodynamic tests shown in Fig. 2 revealed that drag forces are small, relative to mechanical and hydrodynamic drag forces, but under sustained wind, the plants can move a considerable distance and compact appreciably. Aerodynamic drag should not be a major force affecting in-water transportation.

Hydrodynamic Properties

Hydrodynamic drag forces, coupled with application of mechanical force and constraint, cause water hyacinth mats to rotate, disintegrate and over-ride. It is the major force acting to retard movement of plants across the water and, hence, the greatest energy sink in the harvesting system.

Mats of water hyacinth were prepared as for the aerodynamic drag tests (Bagnall, 1984). They were placed in an inner frame suspended from an outer, floating frame by strain-gaged beams. The frame was towed through the water at various speeds by a winch and the force and position monitored continuously by a digital data acquisition system.

The results of the hydrodynamic tests are shown in Fig. 3 and the coefficients, a and b, for the regression in the form $F = aV^b$ in Table 1.

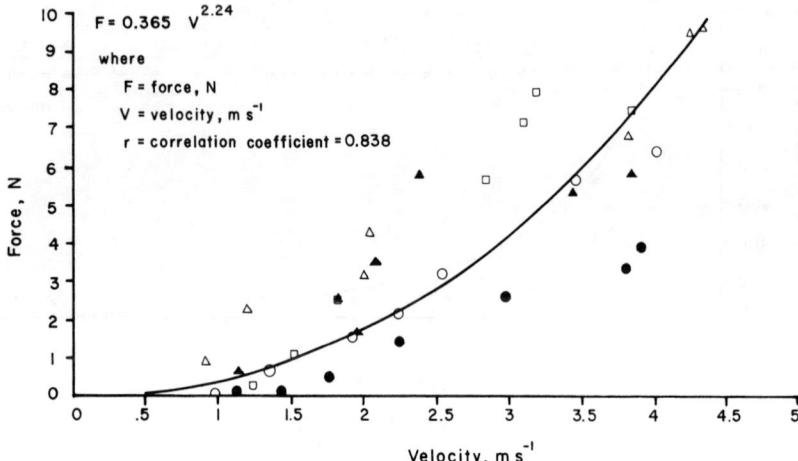

FIG. 2. Aerodynamic drag of water hyacinth mats.

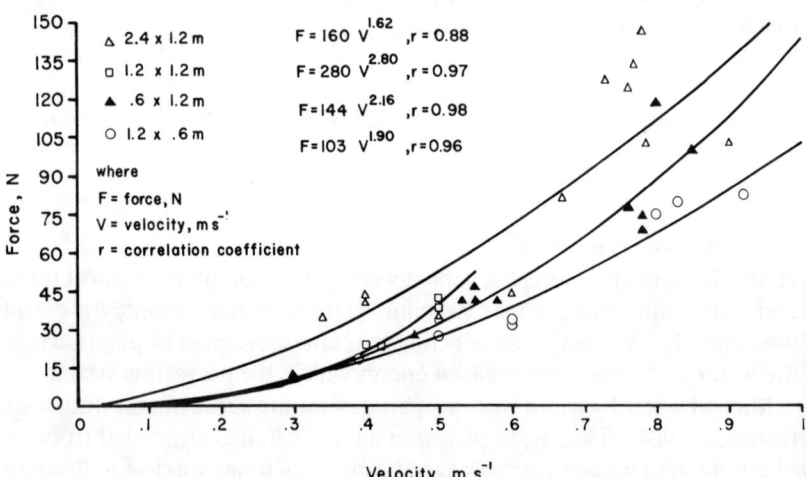

FIG. 3. Hydrodynamic drag of water hyacinth mats.

TABLE 1
Hydrodynamic Drag Coefficients ($F = aV^b$)

L	W	a	b	r
2·4	1·2	160	1·62	0·88
1·2	1·2	280	2·80	0·97
0·6	1·2	144	2·16	0·98
1·2	0·6	103	1·90	0·96

Drag force increased approximately quadratically with speed. Drag force appeared to increase with sample size and root length. In some of the higher-speed tests with the larger samples, the mat rolled about a lateral horizontal axis.

HARVESTING SYSTEMS

Water hyacinth harvesting consists of gathering, severing, moving overwater, elevating, chopping and overland transportation, not necessarily in that order. Numerous devices, sequences and combinations may be appropriate. Most water hyacinth harvesting systems are immobile; the plants are transported over-water before they are elevated. Most submersed plant harvesting systems are mobile; the plants are elevated before being transported over-water. Water hyacinth systems are not usually mobile because the volume of plant material to be moved is overwhelming. However, if the volume can be reduced and the over-water transportation made more efficient, mobile systems may be developed. A hybrid system may combine the best features of the mobile and immobile systems: move the plants in the water to a confined pool, then elevate them with a short-range mobile chopper and continuous transportation system (Bagnall, 1986).

Zellwood Harvesting System

An immobile harvesting system, based on the concept proposed in the report by Warren et al. (1984), was designed, built and installed in a 0·4 ha pond R1 at Zellwood. The system consisted of a rake or gathering boom, a takeout conveyor, a chopper and an elevator, arranged as shown in Fig. 4.

The plants were gathered by a 6 m wide rake of light steel and PVC construction. Two rakes were designed. The first (Fig. 5), was a straight

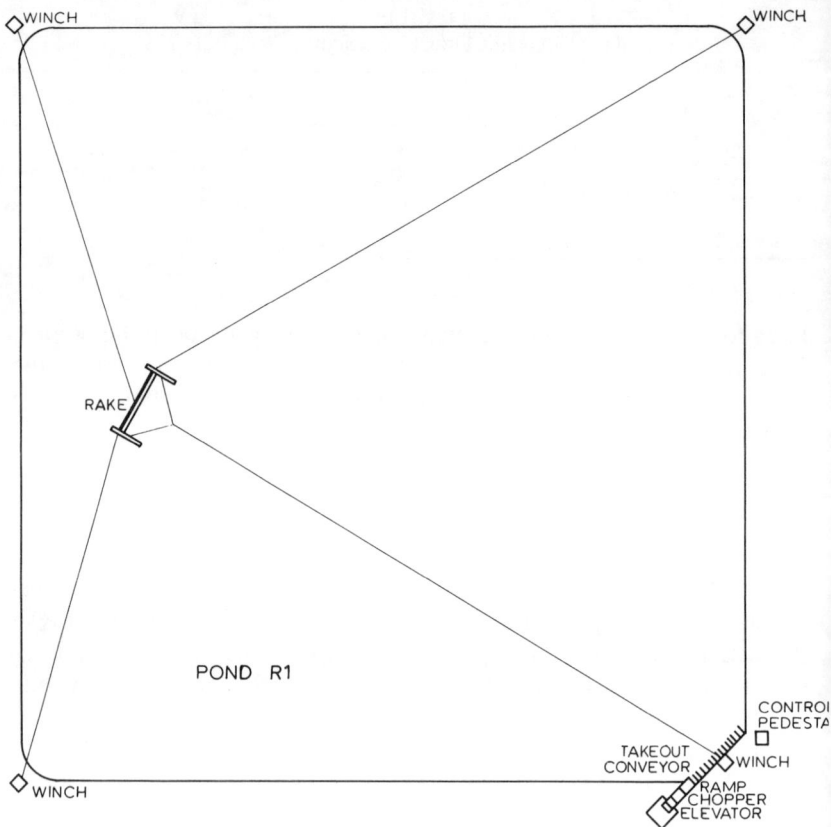

FIG. 4. Zellwood R1 immobile harvesting system.

rigid rake which tilted up to clear the water hyacinth on the rearward stroke. The second rake design, more readily expandable to large spans, folded laterally to present a narrow section on the rearward stroke. The rake was maneuvered at 0.5 m s^{-1} through light lines by four fractional-kilowatt electric winches at the corners of the pond. The winches were controlled by switches on the control pedestal near the takeout.

The takeout flights stripped a 60 cm stream of water hyacinth plants from the face of the gathered mat and fed it onto an elevating flat-wire-belt conveyor which fed the chopper. The takeout conveyor (Fig. 6) was 6 m wide and of a unique, cantilevered-flight design. The flights were tined and

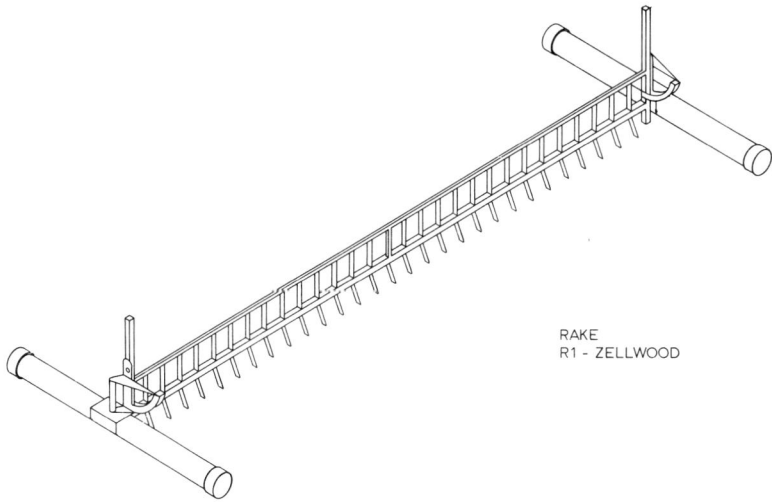

FIG. 5. Rake used in Zellwood R1 immobile harvesting system.

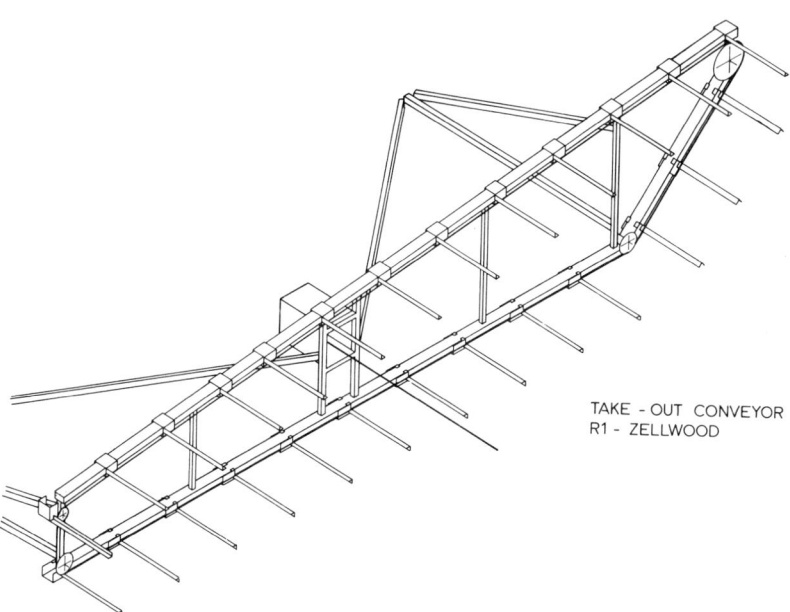

FIG. 6. Takeout conveyor used in Zellwood R1 immobile harvesting system.

were spaced at approximately twice their length. The flights were driven at approximately 0.5 m s^{-1} by a hydraulic motor, the valve for which was at the control pedestal.

The harvested plants were chopped by a 60 cm wide, 40 cm diameter flail chopper operating at 1800 rpm which was mounted on the frame of the flat-wire-belt conveyor. The chopped plants were elevated to storage or transportation by a chain-and-flight farm elevator.

The design capacity of the Zellwood system, based on harvesting the projected weekly production in 2 h, was 20 Mg h^{-1}. Observed sustained capacity never exceeded 2 Mg h^{-1}. Fuel consumption by the mobile power system was 1.5 liters h^{-1}, but power actually delivered to the harvesting system components was much lower. Capacity was limited by excessive spillage from the rake, poor transfer from the rake to the takeout conveyor, excessive spillage from the takeout conveyor and major stoppages at the transition from the takeout to the live ramp.

The cable-operated rake was energy-efficient and could be maneuvered to almost any point in the pond, but the lines had to be constantly monitored to prevent tangling with plants, machinery and each other. The single cable-powered rake left substantial gaps in the supply to the takeout which restricted system capacity.

As the rake was drawn in, small and large mats broke off and eddied out of the path of the rake. Usually what arrived at the takeout was the area between the rake frame and a line connecting the points of the pontoons. Reducing the speed of the rake would reduce spillage, but would probably reduce capacity. The straight rake discharged fairly cleanly but contained a small mat. The laterally folding rake held a larger mat, but the mat tended to stay in the rake when it arrived at the takeout. Rake cycle times were approximately as predicted, but rake capacity was below prediction because there was no surcharge.

Many of the water hyacinths captured by the takeout flights rotated out of the path of the flights. Many of the plants that remained in the path of the flights rolled under the flights, especially if the roots were long, the tops short and the water level low. During some tests, there were only 25 cm of water below the takeout and the roots were often longer than 50 cm. At the transition, where the takeout and ramp lifted the plants from the water, the plants, especially if they had rolled into a large ball, would jam between the takeout and the ramp. The flight bars should be at the surface, with the tines extending below. The ramp should be submerged far enough that the horizontally moving plants engage the sloped surface above the idler radius. Clearance between the takeout and the ramp should be adjustable, preferably automatically, so that traction is adequate to move the plants but spacing is flexible enough to accommodate large, entangled masses.

The flail chopper worked well. When overloaded, the flails fold back and the rotor occasionally jammed, allowing poorly chopped or unchopped plants to pass. The likely solutions are to use heavier flails, higher speed and more power.

Combine Harvesting System

The mobile combined harvester–chopper (Fig. 7) was self-contained and could move to the water hyacinths, gather, elevate, chop and transport them. Fixed site harvesting was possible if the plants were moved to it. A 90 cm swath was gathered into a 30 cm wide cylinder/shearbar chopper. The chopped plants were elevated to subsequent processing or transportation.

Plants were gathered by a combination lateral opposed screw feeder and reel. The screws forced a 90 cm wide swath into the path of the 30 cm wide reel, which crowded the plants rearward, then upward into the chopper feeding rolls. When the plants were bunched under the gatherer, it rode

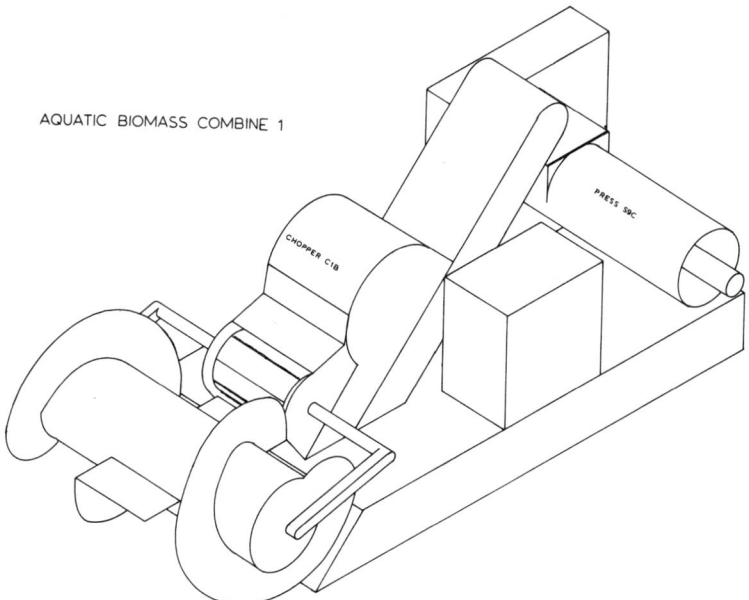

FIG. 7. Water hyacinth combine 1.

over them. Nominal peripheral speed of the flight tips was 0.5 m s^{-1}. The front of the barge was sloped to ease the inward and upward passage of the plants to the chopper.

The gathered plants were elevated by the rear sweep of the gatherer over the front surface of the barge and into the cleated chopper feed rolls. The feed rolls completed the elevation to the shearbar.

The plants were finely chopped by a 30 cm wide, 30 cm diameter cylinder/shearbar chopper operating at 600 rpm. The chopped product was discharged downward into an elevating conveyor which dropped it on the barge deck for storage or transportation.

Capacity of the combine was below expectations, due to poor and irregular feeding by the gathering-elevating system. Performance of the gatherer was adversely affected by the increased width of the barge behind it, requiring that some of the water hyacinth be fed in directly from the end. The plants bunched behind the gatherer and were elevated to the chopper in large masses which the chopper feed rolls could not handle. Some of the plants then carried over the top, where they jammed on the support frame cross brace, bending the gatherer reel flights, and were thrown out in front of the gatherer.

To be effective in dense water hyacinths, the chopper should be at least 60 cm wide and the convergence of the gatherer should not be much greater than 2:1. The feeding reel gatherer was effective when allowed to float upward to climb over large loads and driven slowly enough to obtain penetration and traction. With all aquatic plants, the elevating angle should not exceed 30°. The cylinder/shearbar chopper, in the commercially available sizes, is generally inadequate for high loading rates of the randomly oriented water hyacinths. Unless a fine, uniform chop is needed. some other chopping device should be used. If a cylinder/shearbar chopper is to be used, the throat should be at least 60 cm wide and 30 cm high and the feed rolls should be extremely aggressive.

Handling System

The function of the handling system was to prepare the harvested product for transportation and to transport it from the harvesting site to the digester. The system may be as simple as transporting intact plants by truck or as complicated as liquefication of the plants with subsequent pipeline transportation.

Volume and Particle Size Reduction

Intact water hyacinths have a typical bulk specific volume of $12 \text{ m}^3 \text{ Mg}^{-1}$ (Bagnall, 1982). Reduction of this volume is an obvious way to enhance the efficiency of any type of handling system. Specific volumes approaching 1 m^3 Mg^{-1} can be achieved by compaction or particle size reduction. Particle size reduction has the further advantage of improving the fluidity and of preparing a product more suitable for microbial digestion.

Three cylinder-shearbar choppers and a flail chopper were designed specifically for use on water hyacinth. They had lighter cutting elements than typical forage harvesters because of the relative lack of strength of the plant material. Two of the cylinder/shearbar choppers had 30 cm diameter by 30 cm wide cylinders and the third had a 40 mm diameter by 60 mm wide cylinder. The 30 cm choppers had cleated and smooth feed rolls to control flow to the cylinder and the 60 cm chopper had upper and lower feed aprons. All operated with cylinder speeds of about 600 rpm and discharged the chopped product downward. The flail chopper rotor was 40 cm in diameter by 60 cm wide and was driven at 1800 rpm.

The small choppers chopped 1·9 Mg h^{-1} and required an energy input of 0·87 J g^{-1}. The large chopper chopped 5·0 Mg h^{-1}. The product from the small choppers was fine and uniform and was 3·8 times as dense as the intact water hyacinth feedstock. The flail chopped up to 15 Mg h^{-1} and required an energy input of 0·4 J g^{-1}. Specific energy requirement increased as speed increased, as mat thickness increased and as throughput decreased. As speed increased, the product became finer, more uniform and denser.

An alternative method of particle size reduction, producing smaller particle sizes, is extrusion by forcing the material through constricted passages with high pressure gradients and shear stresses. Structural disintegration in extrusion extends down to the cellular level which, in addition to producing a fine, fluid product, exposes and releases internal cell constituents for access by digesting microorganisms. Screw extruders are commonly used in the food, feed, plastics and chemical industries to grind, blend, cook and shape moist or plastic materials.

The extruder/presses used on water hyacinth were simpler than most of those used in industry. They consist of modified heavy conveying screws in pipe casings with restraining teeth or bars projecting into the path of the extruded material. The extruder propels the product from the inlet to the outlet, shearing and compressing it as it goes. Extruder/presses from 100 to 300 mm in diameter and pitch with length approximately proportional to diameter were used. Speeds ranged from 30 to 90 rpm. Feeding throats were over two pitches in length. Two or three restrictors, in addition to the discharge restrictor, were used in all of the extruder/presses. Most of the extrusion reduction was done in screw presses designed to remove over 50% of the water from water hyacinth (Bagnall, 1980). In order to do this, large segments of the casing wall were removed and the casing was lined with perforated steel so that the removed water could drain. One screw extruder/feeder was designed to extrude and pump the product without separation.

The 150 mm screw press extruded 0.6 Mg h^{-1}. The 150 mm screw feeder extruded 0.9 Mg h^{-1} without supplementary restriction and 0.3 Mg h^{-1} with maximum supplementary restriction. The 230 mm screw press extruded 1.6 Mg h^{-1} and had a specific energy requirement of 7 J g^{-1}. The products from the screw presses were a fibrous press cake and a juice or press liquor. The products were mixed to produce a pumpable product. The juice, freed in the screw feeder, was not allowed to separate and was re-absorbed at the discharge. The only screw feeder product that was pumpable without supplementary water was that produced with the maximum restriction. Re-pressing the product in the same or smaller device further reduced the particle size and produced a readily pumpable product, but at any energy input of approximately 10–12 J g^{-1}.

Transportation

Harvested water hyacinth can be transported from the harvester to the digester in batches or continuously. Traditionally, batch transportation by truck has been used when water hyacinth has been transported long distances. Trucks allow flexibility in the location of both the harvester and the utilization site, and can be used elsewhere when the harvesting system is not in operation. In a commercial biomass/biogas operation, this flexibility is of little value. The volume that can be carried in a single truckload is small relative to the daily requirement of a large biogas plant.

An alternative to batch transportation is continuous transportation of fluidized product. One method of fluidization is to add enough water to suspend the coarse plant fragments and pump with high-volume pumps, then re-cycle the suspending water. Another method is to free some of the water in the plants, reduce the particle size, and suspend the fine particles in the freed water. The advantages of the latter method are that there is no return water line, the diluting water doesn't have to be managed, and the lines, pumps, motors and energy requirement can be much smaller.

The best type of pump to use for this application is some variation on the screw pump, capable of handling fluids with high solids concentrations and delivering them at high pressure. Progressing cavity pumps, consisting of a tubular metal impeller in an elastomeric liner, are widely used in waste management and readily available in a wide range of configurations.

Two 100 mm progressing cavity pumps, one by Roper and the other by Allweiler, were used to evaluate requirements for pumping extruded water hyacinth. The Roper pump was driven by a variable speed electric drive and was connected to a pipeline. The Allweiler pump was driven at 425 rpm by an electric motor and was used to pump into a continuously expanding reactor.

Performance of the pumps depended greatly on the quality of the feedstock. Soft, wet (<5% dry matter), finely divided (double-extruded) water hyacinth could be pumped without supplementary water at rates in excess of 10 000 kg h^{-1} whereas fibrous, dry (>5% dry matter), coarsely divided (single-extruded) water hyacinth required over 100% supplementary water and could not be pumped at rates greater than 1000 kg h^{-1}. A major problem with the coarsely divided product was that plant rhizomes, which had passed through the extruder nearly intact, would lodge in the inlet throat of the pump. Increasing pump speed increased throughput, not only by simply increasing running displacement, but by increasing the suction head on the intake.

The 100 mm pump should be the smallest size considered for pumping extruded water hyacinth. An agitated hopper should be used to enhance feeding and permit pumping a less highly prepared product. Larger pumps will have the advantage, in addition to larger capacity, of being less sensitive to particle size. Single-stage pumps should be used, with transfer stations as appropriate, rather than multi-stage, high-pressure pumps to reduce the probability of separation, plugging and bursting of pipes.

SUMMARY

Mechanical, aerodynamic and hydrodynamic properties of water hyacinth relevant to the design of harvesting systems were determined and nonlinear regressions relating force, compaction, and velocity were found. Two water hyacinth harvesting systems, one immobile and the other mobile, were designed, built and tested. Each system has unique advantages and components could be used interchangeably between types of systems. Factors affecting pipeline transportation of water hyacinth, including size reduction and pumping, were studied. Capacity and efficiency of pump increased and supplementary water requirement decreased as size reduction energy increased.

REFERENCES

Bagnall, L. O. (1974). Mechanical properties of mature water hyacinth stems. Southern Association of Agricultural Scientists/Southeast region ASAE.
Bagnall, L. O. (1980). Intermediate-technology screw presses for dewatering aquatic plants. ASAE paper 80-5044. American Society of Agricultural Engineers.
Bagnall, L. O. (1981). Extrusion pulping aquatic plants for protein production. ASAE paper 81-1526. American Society of Agricultural Engineers.
Bagnall, L. O. (1982). Bulk mechanical properties of waterhyacinth. *Aquatic Plant Management* **20**, 49–53.
Bagnall, L. O. (1984). Hydrodynamic characteristics of water hyacinth plants. ASAE paper 84-5030. American Society of Agricultural Engineers.
Bagnall, L. O. (1984). Energy requirements and capacities for chopping aquatic plants. Aquatic Plant Management Society.
Bagnall, L. O. (1986). Harvesting systems for aquatic biomass. In: *Biomass Energy Development*, Smith, W. H. (ed.), pp. 259–73. Plenum Press, New York.
Bagnall, L. O. and Petrell, R. J. (1985). Mechanical, hydrodynamic and aerodynamic properties of water hyacinth mats. Aquatic Plant Management Society.
Cifuentes, J. and Bagnall, L. O. (1976). Pressing characteristics of waterhyacinth. *Aquatic Plant Management* **14**, 71–5.
Sivakumaran, K. and Bagnall, L. O. (1984). In-situ mechanical properties of water hyacinth. ASAE Paper 84-5029. American Society of Agricultural Engineers.
Stewart, J. S. III. (1972). Energy and flow requirements for chopping water hyacinths. M.E. thesis, University of Florida, Gainesville, Fl.
Warren, C. S., Bruderly, D. E., Angelieri, M., Bilello, L. J., Bucalo, S., Finger, G. W., Hart, R., Hinton, S. W., Newman, J. R. and Vinzant, J. W. (1984). Evaluation of biomass to methane systems in the Lake Apopka natural gas district. Reynolds, Smith and Hills (GRI84/0015·1).

9
Tissue Culture of Gramineous Biomass Species

K. RAJASEKARAN,[a] I. K. VASIL[a] and S. C. SCHANK[b]
[a] *Botany Department,* [b] *Agronomy Department, University of Florida—IFAS, Gainesville, Florida, USA*

INTRODUCTION

Several members of the grass family (Gramineae) are efficient biomass producers. More prominent among these are sugarcane (*Saccharum* spp.), Napiergrass (*Pennisetum purpureum* Schum.) and Napiergrass hybrids, sorghum (*Sorghum bicolor* (L.) Moench.) and bahiagrass (*Paspalum notatum* Flugge). Most are grown for fodder, grain or other food/feed values. Some selections among the grasses are also showing promise as biomass energy crops (Burton, 1986). Genetic improvement of grass species by classical breeding is time consuming, laborious and is often restricted by factors such as sterility, incompatibility and poor seed set. Plant tissue culture techniques allow rapid multiplication of an elite variety or a novel hybrid and offer the potential of crop improvement by using various methods. These include procedures for embryo rescue, homozygous diploid plants obtained from microspore-derived haploids, selection of useful mutants and variants, somatic hybridization and genetic transformation (Vasil, 1984). These advanced techniques are being exploited to provide opportunities to rapidly improve and propagate biomass selections.

Since 1980 much progress has been made in tissue culture of grass species which were previously considered recalcitrant. Reproducible plant regeneration systems have been established for most of the important species and it has been shown that somatic embryogenesis, rather than shoot morphogenesis, is the predominant pathway of regeneration in grass species (Vasil, 1983, 1985). We provide below a summary of work from our laboratory on pearl millet (*Pennisetum americanum* (L.) K. Schum.), Napiergrass (elephantgrass; *Pennisetum purpureum* Schum.) and

Napiergrass hybrids (*P. americanum* × *P. purpureum*), which have the efficient C4 photosynthetic pathway and produce high amounts of biomass (Smith and Frank, 1985).

PLANT REGENERATION FROM CALLUS CULTURES

Embryogenic callus cultures have been established with several grass species including genotypes of *P. americanum*, *P. purpureum* and *P. americanum* × *P. purpureum* hybrids (Table 1) now under improvement for biomass energy. Basal portions of the innermost 3–4 young leaves, or segments of immature inflorescences, or immature embryos, are used to initiate cultures on Murashige and Skoog's (1962) medium supplemented with growth regulators and coconut milk (Vasil and Vasil, 1984a). Two types of callus are produced within 2–3 weeks of culture initiation. Embryogenic callus is white to pale yellow in color, compact and nodular in appearance, and is comprised of small, densely cytoplasmic cells with prominent nuclei and several starch grains. On the other hand, non-embryogenic callus is friable and contains loose cells which are often elongated and highly vacuolated. Embryogenic callus arises only from specific tissues in the explants (Vasil, 1985). Continued proliferation of embryogenic callus can be achieved by subculturing every two weeks. Nonembryogenic callus does not respond well to subcultures and must be discarded. Somatic embryos develop rapidly from embryogenic callus cultures when the concentration of 2,4-dichlorophenoxyacetic acid (2,4-D) is lowered (0·5 mg liter^{-1}). Subsequent germination of somatic embryos into green plantlets can be accomplished by transfer to basal nutrient medium lacking growth regulators and artificially illuminated growth chambers. All plantlets are able to survive transfer to soil without any need for special acclimatization procedures.

The efficiency of embryogenic callus formation depends upon several factors. The developmental stage of the explants is considered one of the most important factors. Establishment of embryogenic callus cultures has been possible, so far, from only young, immature plant parts such as immature inflorescences or embryos and young leaves. The well-developed, mature embryos, inflorescences or leaves rarely form callus which often is nonmorphogenic or produces only roots. A clear developmental gradient exists in the explants with regard to embryogenic capacity. The genotype of the donor plant also has been reported to influence the capacity for regeneration in many grass species (e.g. Hodges *et al.*, 1986). This is in contrast to the reports on somatic embryogenesis in several genotypes, both inbreds and hybrids, of 13 different species, where no strong influence of genotype was observed (Vasil, 1985). Even in those instances where genotypic differences exist, it should be possible to overcome these by selection of appropriate media and suitable developmental stages of the explant.

Exogenous growth regulators are added to the culture medium to obtain the formation of embryogenic callus. In tissue cultures of most grass species 2,4-D is the only major growth regulator needed to elicit embryogenic response. However, hormonal control of somatic embryogenesis will be better understood by analysis of endogenous levels of growth regulators present in the explants and callus cultures. The basal region of young leaves of *Pennisetum purpureum*, which is developmentally the youngest part and is highly embryogenic when compared to the more mature middle and distal parts of the leaves, has been shown to contain higher levels of both indoleacetic acid (IAA) and abscisic acid (ABA) (Rajasekaran *et al.*, 1987). Explants from mature leaves, which do not form callus, contain relatively low levels of endogenous IAA and ABA. Of the cytokinins analyzed only 2-isopentenyladenine (2iP) is present in leaves and appears not to be correlated with embryogenic competence. In comparison to nonembryogenic callus, the embryogenic callus consistently contains higher levels of endogenous IAA and ABA. Non-embryogenic callus contains higher levels of cytokinins (zeatinriboside, dihydrozeatinriboside, 2iP and isopentenyladenosine) than embryogenic callus. Hormonal differences between the two different callus types are maintained during regular maintenance for over a year by subculture at 3–4 week intervals. Callus cultures maintained without subculture lose their embryogenic competence after about 4 weeks and contain only insignificant levels of IAA and ABA (Rajasekaran *et al.*, 1987).

Optimization of plant regeneration from long term embryogenic callus cultures of Napiergrass can be achieved by subculturing at a biweekly interval (Table 2). Infrequent subculturing (4 weeks or longer) and increased culture age greatly increase the formation of soft and necrotic nonembryogenic callus and severely reduce plant regeneration (Chandler and Vasil, 1984a; Chandler *et al.*, 1984). Albino plantlets often are produced in long-term cultures and the ratio of albino to green plants increases with increasing culture age. Nevertheless, it is possible to

TABLE 1
Tissue Culture of *Pennisetum* spp.

Pennisetum spp.	Variety/ genotype	Parentage/source	Source of explant	References
P. americanum (pearl millet)	Gahi 3 (cms)	Tifton, GA	Immature inflorescence	Vasil and Vasil, 1981a Botti and Vasil, 1984, Swedlund and Vasil, 1985 Vasil and Vasil, 1981a, 1982b
			Immature embryo Shoot primordia from mature embryos Cell suspension Protoplast	Botti and Vasil, 1983 Vasil and Vasil 1981b, 1982a Vasil and Vasil, 1980
	—	23A × J993	Mesocotyl[a]	Rangan, 1976
	IP 3128 IP 3676 IP 4122 IP 5313	ICRISAT, India	Mesocotyl[a]	Subbarao and Nitzsche, 1984
	Tift 28 Tift 23DB Tift 18DB Tift 23B Gahi HMP-559 7203 Senegal Bulk	Tifton, GA	Mesocotyl[a]	Subbarao and Nitzsche, 1984
	WC-C75		Seedling roots/ Immature embryos	Nabors *et al.*, 1983

Species	Genotype	Source	Explant	Reference
P. purpureum (Napiergrass or elephantgrass)	PP1, 2, 12, 13, 15	Selections from Merkeron Menlo		Haydu and Vasil, 1981
	PP3, 4	Selections from Cameron		
	PP17, 18, 19, 20, 21	Selections from Cachoeiro do Itapemirim, Brazil		
	PP5	Selection from TA25		
	PP8	Selection from Merkeron		
	PP9	Selection from PI 201209		
	PP16	Selection from TA143		
	PP14, 22	Unknown origin		
	?		Young leaves	Haydu and Vasil, 1981
	?		Anthers	Haydu and Vasil, 1981
			Young leaves	Chandler and Vasil, 1984a, Chandler et al., 1984
	PP12, 13		Immature inflorescence	Wang and Vasil, 1982
	PP6, 9, 13, 15		Immature inflorescence	Chandler et al., 1984
	N75		Young leaves	
	PI 300086		Immature inflorescence	Chandler et al., 1984
			Young leaves	Rajasekaran and Vasil (unpubl.)
	PP12, 13		Cell suspension	
			Protoplast	Vasil et al., 1983
	?		Immature inflorescence[a]	Bajaj and Dhanju, 1981
P. americanum × P. purpureum (hybrid Napiergrass)	N5	Tift 239DA cms × Merkeron	Immature inflorescence	
			Immature embryo	Vasil and Vasil, 1981a
	Sel. 3, 4, 5	23 DA × N75	Young leaves	
			Immature inflorescence	Chandler et al., 1984

P. purpureum genotypes PP1 to 22 and hybrid Napiergrass selection Nos. 3, 4 and 5 are from a collection maintained by Dr S. C. Schank, University of Florida.
[a] Regeneration is by shoot morphogenesis; all others by somatic embryogenesis.

TABLE 2
Effect of Subculture Interval on Proportion of Embryogenic Callus Pieces which Continue to Produce Embryogenic Callus in Napiergrass. Mean of at Least Five Replicates (after Chandler and Vasil, 1984a)

Subculture interval (weeks)	Proportion of subcultured embryogenic callus pieces continuing to produce embryogenic callus (%)
2·4	40
4·1	21
6·0	5

TABLE 3
Rate of Regeneration *in vitro* of Green Plants of Napiergrass from a Single Leaf Explant (Albinos Excluded). An Average of 50 mg Embryogenic Callus was Initially Produced and it was Multiplied by Subculture at 2 Week Intervals (after Chandler *et al.*, 1984)

Weeks after initiation of culture	Number of green plants which could be regenerated
8	93
24	12 300
26	24 600

regenerate approximately 25 000 green plants from a single leaf or inflorescence explant in seven months (Table 3; Chandler and Vasil, 1984a; Chandler *et al.*, 1984).

REGENERATION OF PLANTS FROM CULTURED PROTOPLASTS

Leaf mesophyll cells have been used as an ideal source for culture of protoplasts in several dicotyledonous species. Similar attempts with mesophyll protoplasts of grasses have not been successful (Vasil, 1983, 1984). Embryogenic cell suspension cultures have been established in several grass species (Vasil and Vasil, 1984b), including pearl millet and Napiergrass (Table 1). The suspension cultures contain up to 2×10^7 cells ml^{-1} and are fast growing with a mean cell doubling time of 27–30 h (Karlsson and Vasil, 1986). Protoplasts isolated from embryogenic cell suspension cultures of *Pennisetum americanum*, *P. purpureum*, and *Panicum maximum* (Vasil and Vasil, 1980; Lu *et al.*, 1981; Vasil *et al.*, 1983) have been cultured successfully to produce somatic embryos and plants. Unfortunately, in no instance could the protoplast-derived plants be grown to maturity in soil. Therefore, the recent report of regeneration of green plants from protoplast cultures of sugarcane (Srinivasan and Vasil, 1986) is encouraging.

Somatic Hybridization

The availability of totipotent protoplasts is an important prerequisite for somatic hybridization, which is useful for transfer of nuclear and/or cytoplasmic traits between sexually incompatible, distantly related species. Intergeneric somatic hybridization has been demonstrated recently in the Gramineae by using protpolasts isolated from embryogenic cell suspension cultures. Somatic hybrid cell lines of *Panicum maximum* + *Pennisetum americanum* (Ozias-Akins *et al.*, 1986) and *Pennisetum americanum* + *Saccharum officinarum* (Tabaeizadeh *et al.*, 1986) have been obtained following protoplast fusion. The fusion products have been shown to be hybrids by examination of the electrophoretic banding pattern of isozymes and by restriction analysis of total DNA. These findings, along with recent reports on transformation of protoplasts of *Zea mays* (Fromm *et al.*, 1986), *Triticum monococcum* (Lorz *et al.*, 1985) and *Lolium multiflorum* (Potrykus *et al.*, 1985), indicate that genetic engineering of grass species for incorporation of desirable traits is feasible.

SELECTION FOR SALT TOLERANCE *IN VITRO*

One of the potential applications of tissue culture techniques in plant improvement is the production of salt-tolerant plants through selection of salt-tolerant cell lines and subsequent regeneration of plants from such lines (e.g. Nabors *et al.*, 1980). Several embryogenic cell lines of *Pennisetum americanum*, capable of growth in liquid suspension cultures containing up to 1·16% NaCl, were isolated (Rangan and Vasil, 1983). These suspensions were grown for more than a year in the presence of salt and were found to retain their salt-tolerance after several subcultures in the absence of salt. Somatic embryogenesis was inhibited at higher concentrations of NaCl and consequently, no plants were regenerated. Similar attempts with leaf-derived embryogenic callus of *P. purpureum*

TABLE 4
Growth of Salt-tolerant (Selected at 1·25% NaCl) and Non-tolerant Callus Lines of Napiergrass in Salt Medium (after Chandler and Vasil, 1984b)

Concentration of NaCl in medium (%)	Growth (% FW of control)	
	Nontolerant	Tolerant
0	100	100
0·25	80	259
0·50	61	231
0·75	61	191
1·00	37	91
1·25	35	123
1·50	35	77
2·00	29	41

resulted in the identification and selection of tolerant callus capable of growth at 1·25–2·0% NaCl (Chandler and Vasil, 1984b). Growth of tolerant callus was better than nontolerant callus in media containing different levels of NaCl (Table 4). After 30–40 weeks under selection, the tolerant callus became necrotic, but healthy callus was recovered after transfer to a salt-free medium. Unfortunately, plants regenerated from such callus were sensitive to salt irrigation. It is likely that repeated subculture in NaCl medium induced physiological adaptation of the cell or callus lines to salt and therefore no stable mutants were selected.

UNIFORMITY OF EMBRYOGENIC CALLUS CULTURES AND REGENERATED PLANTS

Tissue cultures of grass species, like those of other plants, are prone to chromosomal, genetic and molecular changes (Bayliss, 1980; Orton, 1984). These changes occur during callus formation or during the cell proliferation stage (Larkin and Scowcroft, 1981). Barbier and Dulieu (1980) have demonstrated in tobacco that mosaicism in the original explants also may be responsible, at least in part, for the occurrence of variability in callus cultures and regenerants. Swedlund and Vasil (1985) reported the presence of polyploid and aneuploid cells in the inflorescence explants of *P. americanum*. Embryogenic callus obtained from these inflorescence explants contained predominantly diploid cells after 1 month and 6 months in culture (92% and 76%, respectively). According to these authors there was a slight drift toward aneuploidy and polyploidy in long-term callus cultures. However, among approximately 100 regenerated plants, nearly all were diploid (99%) indicating that there is a strong selection in favor of plant regeneration from cytogenetically normal cells. No gross morphological changes were observed in any of the chromosomes of the regenerated plants. Similar results have been reported previously with plants regenerated from embryogenic cultures of *Panicum maximum* (Hanna *et al.*, 1984). Recent cytological analysis of nearly 600 regenerated plants of a triploid, sterile *Pennisetum* hybrid (*P. americanum* cv. 'Tift DA23' × *P. purpureum* 'N75'; selection No. 3) indicated stability of triploid status (3 × = 21) in all the regenerated plants except two which were hexaploid (6 × = 42) (Rajasekaran *et al.*, 1986).

Three- to four-year-old embryogenic cell suspensions of *Panicum maximum* and *Pennisetum purpureum* were analyzed for nuclear DNA content by cytophotometry and flow cytometry and were found to contain more than 75% diploid cells (Karlsson and Vasil, 1986). It has been suggested that plants regenerated from embryogenic callus cultures are cytologically normal and stable because embryogenic cells appear to have selective advantage during morphogenesis (Swedlund and Vasil, 1985).

We have compared the morphological features of several hundred plants regenerated from tissue culture (TC) of *Pennisetum* hybrid selection No. 3 to an equal number of vegetatively propagated controls (V) for two years. Care was taken to keep the age and shoot growth of the TC and V plants uniform prior to field planting. Two replicated field trials were conducted with 224 plants each of TC and V plants in the first field (Field 1) and 300 each of TC and V plants in the second field (Field 2). Vegetatively propagated plants did not show any morphological variability. On the other hand, there were 23 phenotypic variants in the TC-regenerant population of 524 plants (4% variation), most of them being dwarf with narrow and erect leaves. Morphological analyses indicated similarities in characters among the phenotypic variants indicating that they may have arisen from only a few variant cell lines. Statistical analysis (Duncan's Multiple Range Test at $P = 0.05$) showed that the 23 variants could be grouped into less than nine morphological groups (Rajasekaran *et al.*, 1986). This analysis confirmed that the variation among plants regenerated from embryogenic callus cultures was minimal (Hanna *et al.*, 1984; Swedlund and Vasil, 1985) and also indicated that the total number of variant plants observed among the TC population does not always correspond to the number of independent mutational events which may have occurred in the callus cultures. Results from this study indicated that plant regeneration from embryogenic callus cultures can provide a useful alternative for the rapid, reliable propagation of hybrid Napiergrass which is conventionally multiplied by cuttings.

BIOMASS YIELD OF TISSUE CULTURE-DERIVED PLANTS

Biomass yield was recorded from both tissue culture-derived and vegetatively propagated plants of the Napiergrass hybrid selection No. 3. Tissue culture-derived plants yielded significantly higher biomass than vegetatively propagated plants at first harvest approximately 80–100 days after planting or regrowth (Tables 5 and 7). The difference in biomass yield between plants of tissue culture origin and vegetatively propagated controls was not significant at the second harvest. Nevertheless, the annual biomass yield was higher from tissue culture-derived plants (Table 5) and this may be due to the earlier establishment and the formation of significantly more tillers than the vegetatively propagated plants (Tables 6 and 7). Method of propagation, either tissue culture or vegetative, did not affect the percent dry matter content (Tables 6 and 7). It is likely that the increased tiller production observed in tissue culture-derived plants is a result of culture conditions such as changes in nutritional or endogenous hormonal status or the presence of well established vascular contact between shoot and primary root system prior to transfer to soil, as compared to adventitious roots in the vegetatively propagated plants (Rajasekaran *et al.*, 1986).

SUMMARY

Regeneration of plants through somatic embryogenesis is a common phenomenon in tissue cultures of grass species. Embryogenic callus cultures are useful for rapid clonal multiplication because of the relative cytological stability and uniformity of regenerated plants. Embryogenic cell suspension cultures are, by far, the only source of regenerable protoplasts. The success in plant regeneration from tissue cultures of *Pennisetum* and other important grass species, and the related development of techniques for genetic transformation, provide a potentially useful opportunity for the genetic improvement of these grass species as biomass energy crops.

TABLE 5
Biomass Yield of Hybrid Napiergrass Selection No. 3 (Field 1). Transplanted 13 June 1984 into 8 Field Replications, each Containing 28 Plants

Method of propagation	Biomass yield (kg ha^{-1})					
	1984			1985		
	Harvest 1 (80)^a	Harvest 2 (75)	Total (155)	Harvest 1 (108)	Harvest 2 (102)	Total (210)
Tissue culture	5 426a^b	4 808a	10 234a	8 207a	8 859a	17 066a
Vegetative	3 288b	4 506a	7 794b	5 813b	8 669a	14 482a

^{a} Days to harvest from planting/regrowth are given in parentheses.
^{b} Column means followed by the same letter are not significantly different at $P = 0.05$ according to Duncan's Multiple Range Test.

TABLE 6
Tiller Number and Percent Dry Matter Content of Hybrid Napiergrass Selection No. 3 (Field 1)

Method of propagation	Mean number of tillers		% Dry matter	
	1984	1985	1984	1985
Tissue culture	53a^a	42^a	17·0a	21·6a
Vegetative	32b	32b	18·2b	21·7a

^{a} Column means followed by the same letter are not significantly different at $P = 0.05$ according to Duncan's Multiple Range Test.

TABLE 7
Biomass Yield of Hybrid Napiergrass Selection No. 3 (Field 2). Transplanted 14 July 1984 into 8 Field Replications, each Containing 25 Plants. Due to Late Planting, only one Harvest was Made on 13 Nov 1984, 120 days after Planting. All the Plants in this Field were Killed in 1984 due to Freezing Temperatures

Method of propagation matter	Biomass yield (kg ha^{-1})	Number of tillers produced	% Dry matter
Tissue culture	7 616a^a	43a	20·8a
Vegetative	4 106b	20b	20·1a

^{a} Column means followed by the same letter are not significantly different at $P = 0.05$ according to Duncan's Multiple Range Test.

ACKNOWLEDGEMENTS

Parts of the research work reported in this chapter were supported by the Monsanto Company, St. Louis, Missouri.

REFERENCES

Bajaj, Y. P. S. and Dhanju, B. S. (1981). Regeneration of plants from callus cultures of Napiergrass (*Pennisetum purpureum*). *Plant Science Letters* **20**, 343–5.
Barbier, M. and Dulieu, H. L. (1980). Effets genetiques observes sur des plantes de Tabac regenerees a partir de cotyledons par culture *in vitro*. *Annals de Amelioration des Plantes* **30**, 321–44.
Bayliss, M. W. (1980). Chromosomal variation in plant tissues in culture. In: *Perspectives in Plant Cell and Tissue Culture*, Vasil, I. K. (ed.). *International Review of Cytology Supplement* **11A**, 113–44.
Botti, C. and Vasil, I. K. (1983). Plant regeneration by somatic embryogenesis from parts of cultured mature embryos of *Pennisetum americanum* (L.) K. Schum. *Zeitschrift fur Pflanzenphysiologie* **111**, 319–25.
Botti, C. and Vasil, I. K. (1984). Ontogeny of somatic embryos of *Pennisetum americanum*. II. In cultured immature inflorescences. *Canadian Journal of Botany* **62**, 1629–35.
Burton, G. W. (1986). Biomass production from herbaceous plants. In: *Biomass Energy Development*, Smith, W. H. (ed.), pp. 163–71. Plenum Press, New York.
Chandler, S. F. and Vasil, I. K. (1984a). Optimization of plant regeneration from long term embryogenic callus cultures of *Pennisetum purpureum* Schum. (Napiergrass). *Journal of Plant Physiology* **117**, 147–56.
Chandler, S. F. and Vasil, I. K. (1984b). Selection and characterization of NaCl tolerant callus from embryogenic cultures of *Pennisetum purpureum* Schum. (Napiergrass). *Plant Science Letters* **37**, 157–64.
Chandler, S. F., Rajasekaran, K. and Vasil, I. K. (1984). Large scale propagation of Napiergrass and giant Napiergrass by tissue culture. In: *Proceedings of the 1984 International Gas Research Conference*, pp. 561–6. Government Institute, Inc., Rockville, Maryland.
Fromm, M. E., Taylor, L. P. and Walbot, V. (1986). Stable transformation of maize after gene transfer by electroporation. *Nature, Loud.* **319**, 791–3.
Hanna, W. W., Lu, C. and Vasil, I. K. (1984). Uniformity of plants regenerated from somatic embryos of *Panicum maximum* Jacq. (Guineagrass). *Theoretical and Applied Genetics* **67**, 155–9.
Haydu, Z. and Vasil, I. K. (1981). Somatic embryogenesis and plant regeneration from leaf tissues and anthers of *Pennisetum purpureum*. *Theoretical and Applied Genetics* **59**, 269–73.
Hodges, T. K., Kamo, K. K., Imbrie, C. W. and Becwar, M. R. (1986). Genotype specificity of somatic embryogenesis and regeneration in maize. *Bio/Technology* **4**, 219–23.

Karlsson, S. and Vasil, I. K. (1986). Growth, cytology and flow cytometry of embryogenic cell suspension cultures of *Panicum maximum* Jacq. and *Pennisetum purpureum* Schum. *Journal of Plant Physiology* **123**, 211–27.

Larkin, P. J. and Scowcroft, W. R. (1981). Somaclonal variation—a novel source of variability from cell cultures for plant improvement. *Theoretical and Applied Genetics* **60**, 197–214.

Lorz, H., Baker, B. and Schell, J. (1985). Gene transfer to cereal cells mediated by protoplast transformation. *Molecular and General Genetics* **199**, 178–82.

Lu, C. Y., Vasil, V. and Vasil, I. K. (1981). Isolation and culture of protoplasts of *Panicum maximum* Jacq. (Guineagrass): Somatic embryogenesis and plantlet formation. *Zeitschrift fur Pflanzenphysiologie* **104**, 311–18.

Murashige, T. and Skoog, F. (1962). A revised medium for rapid growth and bioassays with tobacco tissue cultures. *Physiologia Plantarum* **15**, 473–97.

Nabors, M. W., Gibbs, S. E., Bernstein, C. S. and Meis, M. E. (1980). NaCl-tolerant tobacco plants from cultured cells. *Zeitschrift fur Pflanzenphsyiologie* **97**, 13–17.

Nabors, M. W., Heyser, J. W., Dykes, T. A. and DeMott, K. J. (1983). Long duration, high-frequency plant regeneration from cereal tissue cultures. *Planta* **157**, 385–91.

Orton, T. J. (1984). Somaclonal variation: Theoretical and practical considerations. In: *Genetic Manipulation in Plant Improvement—16th Stadler Genetics Symposium*, Gustafson, J. P. (ed.), pp. 427–68. Plenum Press, New York.

Ozias-Akins, P., Ferl, R. J. and Vasil, I. K. (1986). Somatic hybridization of *Pennisetum americanum* and *Panicum maximum* (Gramineae). *Molecular and General Genetics* **203**, 365–70.

Potrykus, I., Saul, M., Petruska, J., Paszkowski, J. and Shillito, R. D. (1985). Direct gene transfer into protoplast of a graminaceous monocot. *Molecular and General Genetics* **199**, 183–8.

Rajasekaran, K., Schank, S. C. and Vasil, I. K. (1986). Characterization of biomass production, cytology and phenotypes of plants regenerated from embryogenic callus cultures of a *Pennisetum americanum* × *P. purpureum* (hybrid triploid Napiergrass). *Theoretical and Applied Genetics* **73**, 4–10.

Rajasekaran, K., Hein, M. B., Davis, G. C., Carnes, M. G. and Vasil, I. K. (1987). Endogenous growth regulators in leaves and tissue cultures of *Pennisetum purpureum* Schum. *Journal of Plant Physiology* **130**, 13–25.

Rangan, T. S. (1976). Growth and plantlet regeneration in tissue cultures of some Indian millets: *Paspalum scrobiculatum* L., *Eleusine coracana* Gaertn., and *Pennisetum typhoideum* Pers. *Zeitschrift fur Pflanzenphysiologie* **78**, 208–16.

Rangan, T. S. and Vasil, I. K. (1983). Sodium chloride tolerant embryogenic cell lines of *Pennisetum americanum* (L.) K. Schum. *Annals of Botany* **52**, 59–64.

Smith, W. H. and Frank, J. R. (1985). Comparative biomass yields of energy crops. In: *Energy from Biomass*, Palz, W., Coombs, J. and Hall, D. O. (eds.), pp. 323–9. Elsevier Applied Science, New York.

Srinivasan, C. and Vasil, I. K. (1986). Plant regeneration from protoplasts of sugarcane (*Saccharum officinarum* L.). *Plant Physiology* **126**, 41–8.

Subbarao, M. V. and Nitzsche, W. (1984). Genotypic differences in callus growth and organogenesis of eight pearl millet lines. *Euphytica* **33**, 923–8.

Swedlund, B. and Vasil, I. K. (1985). Cytogenetic characterization of embryogenic callus and regenerated plants of *Pennisetum americanum* (L.) K. Schum. *Theoretical and Applied Genetics* **69**, 575–81.

Tabaeizadeh, Z., Ferl, R. J. and Vasil, I. K. (1986). Somatic hybridization of sugarcane (*Saccharum officinarum* L.) and pearl millet (*Pennisetum americanum* (L.) K. Schum.). *Proceedings of National Academy of Science USA* **83**, 5616–19.

Vasil, I. K. (1985). Somatic embryogenesis and its consequences in the gramineae. In: *Genetic Engineering in Eukaryotes*, Lurquin, P. and A. Kleinhofs (eds.), pp. 233–52. Plenum Publishing Corporation, New York.

Vasil, I. K. (ed.) (1984). *Cell Culture and Somatic Cell Genetics of Plants, Vol. 1, Laboratory Procedures and their Applications*, Academic Press, Orlando.

Vasil, I. K. (1985). Somatic embryogenesis and its consequences in the gramineae. In: *Tissue Culture in Forestry and Agriculture*, Henke, R. R., Hughes, K. W., Constantin, M. P. and Hollaender, A. (eds.), pp. 31–47. Plenum Publishing Corporation, New York.

Vasil, V. and Vasil, I. K. (1980). Isolation and culture of cereal protoplasts. II. Embryogenesis and plantlet formation from protoplasts of *Pennisetum americanum*. *Theoretical and Applied Genetics* **56**, 97–9.

Vasil, V. and Vasil, I. K. (1981a). Somatic embryogenesis and plant regeneration from tissue cultures of *Pennisetum americanum* and *P. americanum* × *P. purpureum* hybrid. *American Journal of Botany* **68**, 864–72.

Vasil, V. and Vasil, I. K. (1981b). Somatic embryogenesis and plant regeneration from suspension cultures of pearl millet (*Pennisetum americanum*). *Annals of Botany* **47**, 669–78.

Vasil, V. and Vasil, I. K. (1982a). Characterization of an embryogenic cell suspension culture derived from inflorescences of *Pennisetum americanum* (pearl millet; Gramineae). *American Journal of Botany* **69**, 1441–9.

Vasil, V. and Vasil, I. K. (1982b). The ontogeny of somatic embryos of *Pennisetum americanum* (L.) K. Schum.: in cultured immature embryos. *Botany Gazette* **143**, 454–65.

Vasil, V. and Vasil, I. K. (1984a). Induction and maintenance of embryogenic callus cultures of Gramineae. In: *Cell Culture and Somatic Cell Genetics of Plants, Vol. 1*, Vasil, I. K. (ed.), pp. 36–42. Academic Press, Orlando.

Vasil, V. and Vasil, I. K. (1984b). Isolation and maintenance of embryogenic cell suspension cultures of gramineae. In: *Cell Culture and Somatic Cell Genetics of Plants, Vol. 1*, Vasil, I. K. (ed.), pp. 152–8. Academic Press, Orlando.

Vasil, V., Wang, D. and Vasil, I. K. (1983). Plant regeneration from protoplasts of *Pennisetum purpureum* Schum. (Napiergrass). *Zeitschrift fur Pflanzenphysiologie* **111**, 233–9.

Wang, D. Y. and Vasil, I. K. (1982). Somatic embryogenesis and plant regeneration from inflorescence segments of *Pennisetum purpureum* Schum. (Napier- or elephantgrass). *Plant Science Letters* **25**, 147–54.

10
Breeding Grasses for Improved Biomass Properties for Energy

S. C. SCHANK

Agronomy Department, University of Florida—IFAS, Gainesville, Florida, USA

and

L. S. DUNAVIN

Agricultural Research and Education Center, University of Florida—IFAS, Jay, Florida, USA

INTRODUCTION

Agnes Chase (1948), renowned agrostologist, wrote in the preface of *Grass, the Yearbook of Agriculture*:

'Of all plants the grasses are the most important to man. All our breadstuffs—corn, wheat, oats, rye, barley and rice and sugarcane are grasses. Bamboos are grasses, and so are the Kentucky bluegrass and creeping bent of our lawns, the timothy and redtop of our meadows. Grasses have been so successful in the struggle for existence that they have a wider range than any other plant family; they occupy all parts of the earth and exceed any other in the number of individuals.'

The grass family contains about 620 genera and 10 000 species. Their uses have been categorized by Lawrence (1951) as follows: food for humans, shelter, feed for domestic animals, industrial uses, turf, ornamentals, soil conservation and dune stabilization, and biomass for energy production.

The grasses have greater variation than other families of plants; thus, they would be the most malleable in the hands of a plant breeder. Important properties desired in a biomass plant must be carefully evaluated. Characteristics, such as high dry matter yield, regeneration properties, plant morphology, seasonal growth patterns, tissue chemical composition, stress tolerance, lodging resistance and the ability to associate with nitrogen-fixing bacteria and/or legumes are traits that can be manipulated genetically. Desirable traits of biomass crops and how they differ from other agricultural crops were outlined by Smith (1983). Superior strains of grasses for specialized uses have been the goal of plant breeders for a long time. Hanson and Carnahan (1956) have appraised problems confronting the grass breeders and have summarized techniques and methods available for making improvements. Sound objectives were emphasized since they determine the ultimate success of the breeding program. In this program, large quantities of convertible biomass at a low cost is the overall objective.

Comparative biomass yields of energy crops were reviewed by Smith and Frank (1985) and it was concluded that among grasses some of the most promising genera were *Pennisetum* (Napiergrass), *Saccharum* (sugarcane) and *Sorghum* (sorghum). These grasses possess the more efficient C-4 photosynthetic pathway that results in 4–5 units of energy for each unit used in production. In Florida, biomass yields of the first two grasses have approached 50–70 Mg ha^{-1}, and yields of 20–45 Mg ha^{-1} were recorded for sorghum. This chapter reviews selection and breeding work done on these genera of grasses since this program was initiated 5 years ago to develop biomass feedstocks for methane generation.

PENNISETUM SELECTION AND BREEDING

Napiergrass is generally propagated vegetatively since the plants are either sterile or the seeds are very small and lack the vigor for successful establishment in field plantings. Pearl millet, on the other hand, has large seed. The two species hybridize readily and the F_1 progeny have perenniality, vigor, high quality, high yield, and great immunity to diseases such as rust. Using pearl millet as the female parent has several advantages as outlined by Jauhar (1981). Despite the suggestion twenty years ago (Powell and Burton, 1966) that the production of hybrid seed could be accomplished in a frost-free environment, the commercialization of the hybrid between *Pennisetum americanum* (L.) K. Schum. and *Pennisetum purpureum* Schum. has not been achieved. Excellent summary papers describe the hybridization process in *Pennisetum* and the subsequent genetic variation (Muldoon and Pearson, 1979; Jauhar, 1981; Dujardin and Hanna, 1984).

The primary objective of this breeding research on *Pennisetum* is to evaluate new interspecific hybrids of *P. purpureum* × *P. americanum* for biomass production and to determine the potential of genetic improvement of these Napiergrass (elephantgrass) hybrids. This work builds on some twelve years of experience with several *Pennisetum* introductions and chromosomal evaluations of eight hybrids and chromosomal counts on three Napiergrass varieties (Schank *et al.*, 1975). Pearl millet (*P. americanum*), an annual forage grass that has been used in the southern US for decades, hybridizes readily with Napiergrass. The F_1 hybrids of this cross are triploids and therefore sterile. However, with clonal selection, many F_1 genotypes with potentially desirable properties have been identified and evaluated.

In 1982, dwarf Napiergrass (N75) was used as the male parent and dwarf pearl millet was used as the female parent in a series of hybridizations. The purpose of these experiments was to improve the already highly successful N75. From over 2000 hybrid genotypes grown in 1982, selection was made for rapid regrowth, high forage quality and persistence. The frequency of dwarf offspring was extremely low from this dwarf × dwarf parentage. Six of the dwarf hybrids, however, were selected and were planted in a replicated test in 1983 at a nearby University of Florida site for comparison with four other crosses. We identified a genotype from this trial, calling it 'Selection No. 3', which had certain distinct superior characteristics such as leafy upright growth, high leaf to stem ratio and high *in vitro* organic matter digestibility. It survived the winters of 1983, 1984 and 1985 very well, and is a triploid perennial plant which has 'semi-dwarf' genes but is sexually sterile. Selection No. 3 has been propagated vegetatively with planting material provided to many collaborating producers from Latin America in 1985.

Interspecific hybrids between a dwarf pearl millet, 23DA (female parent) and dwarf Napiergrass N75 (male parent) were included in 1983 studies. Over 1500 seedlings from that cross were evaluated and selections of the most vigorous dwarf types were chosen for further testing. In a genetic study from some of the seedlings above, the dwarf × dwarf cross produced 670 tall plants compared to 54 dwarf plants (close to a 15:1 ratio).

Hybridization further expanded the genetic base in 1983 because 19 different crosses were used to look for genetic differences in vigor and tolerance to waterlogging. This is considered an important trait because of the many wet soil sites in Florida and the southeastern lower coastal plain. A large amount of variation in siblings from the same two parents was noted. Some of the plants survived four cycles of intense waterlogging in 1983. Further testing of these plants is underway in order to form a basis for increased use of Napiergrass hybrids on wet or waterlogged sites.

In 1984, the hybridization scheme included making crosses of pearl millet to various male Napiergrasses. Selection PI 300086, which Prine (1985) identified from among several varieties as having a high biomass yield potential average of 48·3 Mg ha^{-1}, in three years of testing, was used in many of these crosses. A total of 119 crosses were made in the glasshouse in 1984. Seed was germinated in February and March and approximately 12 000 seedlings were transplanted to the field in late April. Some hybrids showed promising genetic potential with dry biomass yields ranging from 19·5 Mg ha^{-1} (Selection No. 95) downward to 4·1 Mg ha^{-1} (line N75) (Schank, 1986). The hybrids were harvested in July, so the short-term harvest yields should not be confused with full-season yields. Two selections yielded over 19 Mg ha^{-1} dry biomass yield; 19 selections yielded 15–19 Mg ha^{-1}; 45 selections yielded 10–15 Mg ha^{-1}; and 12 yielded 4·1–10 Mg ha^{-1}.

Many *Pennisetum* selections and/or crosses are sterile; thus, vegetative propagation and rapid multiplication of desired genotypes is important to a biomass energy crop program. The problems associated with vegetative propagation can be circumvented by using tissue culture techniques. Rajasekaran *et al.* (1986) compared the biomass production, cytology and phenotypes in plants regenerated from vegetative cuttings with plants from embryogenic callus cultures of *P. americanum* × *P. purpureum* Napiergrass hybrids (see Chapter 9). Vegetatively propagated plants did not show any morphological variability. Among the tissue culture plants there were 23 phenotypic variants in the regenerated population of 524, most of them being more dwarf and later in flowering than the other regenerants. Two of the variants were hexaploid, which are now being tested for seed production, vigor and other agronomic traits.

The ability of different crosses of Napiergrass to differentially withstand cold winters is an important selection criterion. A rating scale of 0 (no survival) to 7 (70% survival) was used to evaluate various crosses exposed to a winter freeze. Despite a loss of 62% of the plants, the survivors (Table 1) show that sufficient genetic variation is present in these interspecific hybrids to greatly aid the selection of cold tolerant biomass types.

The cold tolerance of N75 is equal to that of Merkeron, one of the most cold-tolerant Napiergrass cultivars (Hanna, 1986), which has survived for nine winters at Tifton, GA, where air temperatures fell to −22°C (−8°F) in January, 1985. Both Merkeron and N75 have been used as the male parent in pearl millet × Napiergrass hybridizations.

Initially, emphasis was placed on having vigorous dwarf types for forage, but with the need for higher biomass production many tall hybrids have now been included in tests. The chemical make-up of plant parts of tall

TABLE 1
Winter Hardiness of *Pennisetum* Hybrids—1984–85. Rating Scale: 0 = no Survival, 1 = 10% Survival, 5 = 50% Survival, 7 = 70% Survival. Pearl Millet (Female Parent) is Listed First

	Winter hardiness rating							
Hybridization	0	1	2	3	4	5	6	7
23A × 17	19	19	1					
23A × 300 086	62	15	5	1				1
Gahi3 × 300 086	117	35	8	1				
Gahi3 × 14	62	3	1					
Gahi3 × 10	41	60	16	1	4	2	2	
Gahi3 × 2	34	40	4	5	2			
23A × 10	44	4						
Totals	379	176	35	8	6	2	2	1
Percentages	62%	29%	5%	1·3%	1%	0·03%	0·03%	0·02%

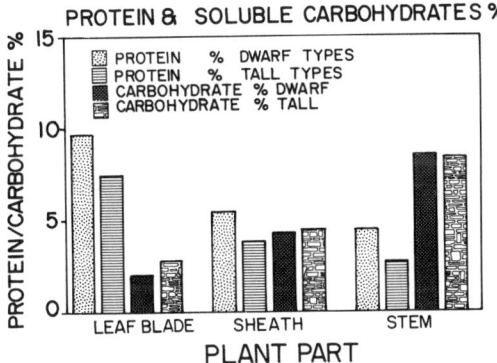

FIG. 1. A comparison between dwarf and tall Napiergrass hybrids in protein content and in nonstructural (soluble) carbohydrates. Plant tissue was separated into the various components, leaf blades, sheaths and stems prior to the analysis. Dwarf plants were selections 3 and 4, and tall plants were selections 10-2 and 10-6, grown at Hague, FL., 1983.

versus dwarf types (Fig. 1) shows contrasts in protein and soluble carbohydrate contents of leaves, sheaths and stems of these differing genotypes. Clearly, the dwarf plants were significantly higher in protein content than the tall genotypes and the leaves were highest in protein content (9·5%) compared to the stems (2·6%). The nonstructural carbohydrate content

TABLE 2
Characterization of Dwarf versus Tall Napiergrass Genotypes—Percent of Plant in the Different Plant Part Components

Genotype	Leaves	Sheath	Rind	Pith	Infl.	Plant total dry wt (g)
N-75 (dwarf)	54·4A[a]	14·3A	25·4D	5·8B	0·0B	79·5
Sel. 3 (dwarf)	37·2B	6·7B	41·0C	10·0B	5·0A	150·6
Gahi3 + 300 086	23·8C	1·8C	61·4B	13·1A	0·0B	690·5
Gahi3 × 16	23·2C	1·6C	60·4B	12·1A	2·7AB	273·8
23A × 16	18·8C	2·8C	61·2B	12·9A	4·3A	250·9
Gahi3 × 10	16·3CD	1·6C	63·9AB	16·7A	1·7B	728·2
23A × 300 086	16·1CD	1·3C	69·2AB	12·2A	1·2B	821·8
Gahi3 × 14	7·5D	0·9C	75·8A	14·7A	0·5B	631·4

[a] Means followed by different letters are significantly different at the 5% level according to Duncan's Multiple Range Test.

TABLE 3
Chemical Components of *Pennisetum* Hybrids

Plant part	Characteristics of plant parts			
	IVOMD %	ADF %	Lignin %	NDFA %
Stem int. (pith)	69·5A[a]	29·4D	4·5C	50·9C
Leaves	56·0B	38·4C	5·0C	72·8B
Sheaths	44·9C	49·0B	8·4B	78·7A
Inflorescences	44·5C	38·9C	7·6B	72·9B
Stem ext. (rind)	28·0D	62·0A	15·2A	83·2A

[a] Means followed by different letters are significantly different at the 5% level according to Duncan's Multiple Range Test.

(soluble) of the corresponding tissues ranged from 2·0% in the leaves to 8·3% in the stems but varied little in similar parts from tall and dwarf selections. The greatly reduced yields by dwarfs may limit their usefulness.

Morphological characterization of the tall versus dwarf plants (Table 2) shows dwarf plants to have less stem material, higher leaf content, and consequently a much higher protein content. The stem component was carefully separated into the rind and pith, with clear evidence that in the tall genotypes, the rind portion can make up 60–75% of the plant weight.

The rind is highly lignified (Table 3). This adversely impacts fermentability and may affect the evolution of methane gas (see Chapter 20). Continuing analyses are evaluating the trade-off of biomass yield, allocation of biomass among plant parts, plant chemical composition, plant age, harvesting, handling and storage, and other factors for providing plant breeders and crop producers guidelines for obtaining the best Napiergrass feedstock for methane production (see Chapter 3).

SORGHUM SELECTION AND BREEDING

Sorghum (*Sorghum bicolor* (L.) Moench), cultivated on 6 million ha in the US, ranks fourth among cereal grains in the world in total production. It is grown in areas of the world where the climate is unsuitable for corn or rice. Because of its C-4 photosynthetic pathway, sorghum has a high productivity in most environments including dry or drought-affected areas. Chynoweth (1983) has identified some sorghum lines as high level producers of methane by using laboratory bioassays.

Variation among species within the genus is great, and like corn, plant breeders have been increasing both grain and stover production. Miller and Monk (1984) have been able to use the variability in the genus to tailor genotypes to fit various agronomic situations. Plant breeders have shortened internodes so that mechanical harvesting is possible. They also have increased stalk juiciness and sugar content for syrup production and increased leafiness and tillering for grazing use of sorghum. Work has recently been aimed toward production of large quantities of biomass for energy purposes with sorghum. Cultivars, which have been developed with high biomass potential, are usually either very tall or they are relatively tall with a high production of grain. In both cases, the potential for lodging is usually large.

At Florida, we have capitalized on the plant breeding efforts of Miller and Monk (1984) and have evaluated many of the newer cultivars and lines as they have been released. Sorghum cultivar trials were conducted at Jay, Hastings and Quincy, Florida. The highest average total yield of dry sorghum biomass produced at Jay in 1984 was by the silage sorghum cultivar Red Top Kandy Xtra which yielded 35·8 Mg ha^{-1} in two harvests (Table 4). Other top yielders were FS25E, MN1500, and M81E. Ten entries from Texas were planted at Jay, Quincy and Hastings, Florida. At Hastings, the entries were harvested at 60, 90 and 121 days after planting following a spring potato crop (see Chapter 6). When harvested 90 days

TABLE 4
Biomass Yield of Sorghum (AREC, Jay, 1984)

Cultivar	Dry biomass (Mg ha^{-1})		
	First harvest	Ratoon	Total
Red Top Kandy Xtra	19·90	15·86	35·76
FS25E	21·86	12·40	34·26
MN1500	24·82	7·04	31·86
M81E	21·12	8·98	30·10
Rio	18·48	11·41	29·89
Atlas × Rio	15·98	12·95	28·93
TX 623 × Atlas	16·25	12·47	28·72
TX 623 × Rio	16·16	12·30	28·46
Gator Chop	15·70	11·93	27·63
TX 623 × Pickett-3	13·25	14·37	27·62
TX 623 × 82cs 796	15·81	11·76	27·57
F80	17·33	10·12	27·45
Titan R	16·12	11·18	27·30
Atlas × TX 432	14·25	12·13	26·38
F70S	13·92	11·73	25·65
Cow Vittles	14·71	10·74	25·45
83S	14·16	10·84	25·00
Atlas × TX 430	14·06	10·57	24·63
235F	13·32	10·77	24·09
Atlas	16·21	7·59	23·80
TX 623 × TX 432	13·97	5·99	19·96
LSD (t ·05)	3·51	3·37	5·81

Planted: April 6, 1984. Harvested: as each cultivar reached late dough: July 13–Aug. 21 and Oct. 10–Nov. 14. Soil: Red Bay sandy loam (Rhodic Paleudult). Fertilizer: 560 kg ha^{-1} 8–24–24 preplant; 224 kg ha^{-1} ammonium nitrate on May 14 and following the first harvest.

after planting, the entry Atlas × Rio (from Texas) and Beef Builder forage sorghum produced similar yields (6 Mg ha^{-1}). After 120 days, Beef Builder had the highest yield (14·9 Mg ha^{-1}). Data on grain biomass from the Quincy and Jay locations are compared in Table 5. Some of the cultivars performed differently at the two locations. For example, Atlas × Rio had the highest grain yield at Jay (7·4 Mg ha^{-1}), while TX 623 × Rio had the highest yield (3·2 Mg ha^{-1}) at the Quincy location (Table 5).

Production data in Florida (Stanley and Dunavin, 1986) indicate that the sorghums are well adapted to the north Florida area. Total biomass production of the forage (or silage) types was as high as 35·8 Mg ha^{-1} after two

TABLE 5
Grain Biomass at Jay and Quincy, Florida of Sorghum Cultivars from Texas A&M, 1984

Cultivar	AREC, Jay, FL. first harvest (Mg ha^{-1})	NFREC, Quincy, FL. single harvest (Mg ha^{-1})
Atlas × Rio	7·35	2·44
Atlas × 623 × TX 432	5·87	2·47
Atlas × TX 432	5·70	2·45
TX 623 × Atlas	5·36	1·21
TX 623 × Rio	5·33	3·18
Atlas	5·19	1·36
Atlas × TX 430	5·20	2·33
TX 623 × 82cs796	4·58	2·58
TX 623 × Pickett-3	3·58	2·43
Rio	2·96	0·96
Average	5·11	2·14

Considerable bird damage occurred on some cultivars.

harvests with grain content of 33%. When forage and sweet types were compared in the same experiment over a 3-year period, the sweet types produced more than the forage types. The recent entries from the Texas A&M research program have been evaluated for 2 years in north Florida, and when harvested twice during a season highest average yield was 31·6 Mg ha^{-1}. When compared to the older sweet types, none of the Texas entries have produced as much total biomass in Florida as the old syrup types. These studies show how genetic selection has been used to identify the best sorghum lines for future testing in this project.

Texas data also generally indicate that sweet sorghums not only yield higher fresh and dry biomass, they also have higher nonstructural carbohydrates (Brix %) and are taller. They are later in maturity and lower in percent dry weight. The principal advantages of grain sorghums include high percent dry weight and early maturity. However, they lack the high biomass production and nonstructural carbohydrates, generally present in sweet sorghum.

The hybrids of grain and sweet sorghums (high energy sorghums) attempt to capitalize on some of the advantages of both types. Biomass production, nonstructural carbohydrates and height are increased as compared to grain types while maturity is delayed only slightly. The high energy sorghums have a faster rate of production than either grain or sweet types of sorghum (Miller and Monk, 1984).

SUGARCANE SELECTION AND BREEDING

Sugarcane (*Saccharum officinarum* L.) is an effective collector of solar energy (Alexander, 1985), storing the energy in the structure of new tissues (fiber) preferentially to accumulating it as soluble sugars. Both fiber and fermentable sugars are produced in abundance. Sugarcane grown as a renewable energy crop requires changes in agronomic practices to maximize yield. This involves improved land preparation; selection of high-biomass varieties; increased fertilization, especially nitrogen; adequate irrigation; and inclusion of cane tops and leaf trash in the tonnage calculations. Samuels (1984) states that when the best agronomic practices are used, yields of up to 280 Mg ha^{-1} of green biomass (112 Mg ha^{-1} dry matter) have been achieved in Puerto Rico. In most sugarcane growing countries of the Caribbean, the average cane yield is presently about 75 Mg ha^{-1} green weight (18 Mg ha^{-1} dry weight) when calculated for a 12-month plant crop (Samuels, 1986).

The problems and potentials of intergeneric hybridization in *Saccharum* breeding work has been summarized by Grassl (1962) and Alexander (1985). The following grass genera, under suitable conditions, may be used as parents in the hybridization of sugarcane: *Erianthus*, *Eccoilopus*, *Eriochrysis*, *Imperata*, *Spodiopogon*, *Miscanthus*, *Narenga*, *Sclerostachya* and *Sorghum*. All 9 genera have been successfully hybridized with *Saccharum*. *Bambusa* (bamboo) and *Zea* (corn) also cross with some difficulty. More recently, Ozias-Akins et al. (1985) have used protoplast fusion to hybridize *Saccharum* and *Pennisetum*.

Each year, a large number of individual seedlings from crosses made at the US Sugarcane Field Station, Canal Point, Florida, are grown as single stools at the US Sugarcane Field Station, Houma, Louisiana, as reported by Hebert and Henderson (1959). They state that the literature contains little information on studies related to the performance of agronomic and chemical characters in sugarcane, especially in regard to their relative performance in single stools and clones. There has been an extensive breeding program in progress in Louisiana since 1919 and many new sugarcane varieties have resulted from this work (Hebert and Henderson, 1959). Tai and Miller (1986) have very recently reported their evaluation of crosses of commercial lines of sugarcane with F1 *Saccharum spontaneum*

Breeding Grasses for Improved Biomass Properties for Energy 179

lines. They evaluated 50 clones of sugarcane (complex hybrids of *Saccharum* spp.) for cold tolerance and found that the estimates of narrow sense heritability of cold tolerance in sugarcane ranged from 12 to 26 in a combined analysis. Although the heritability is low, some genotypes were superior to others in cold tolerance according to the rating on percent green tissue of total leaf area (Tai and Miller, 1986).

Sugarcane stocks used in the GRI/IFAS biomass project originated from the USDA sugarcane station. Mislevy *et al.* (see Chapter 16) report dry matter yields of 15 energy canes and one *Erianthus* (formerly named *Ripidium*). The genetic variation found in these lines in Brix % (range 6–24%), yield, physiology, lodging, response to N fertilization, and other agronomic traits indicates very large genetic differences between lines. Since selection is such a powerful genetic improvement tool, it is believed that not only Brix, but many other *Saccharum* characteristics may be modified through conventional plant breeding methods. Continued cooperation in the evaluation of sugarcanes is expected with the USDA Sugarcane Field Stations. Screening studies conducted in Florida indicate that certain sugarcane (*Saccharum* and *Erianthus*) species may be equal or possibly more productive in biomass yield than Napiergrass. In addition, percentage sugar solids or Brix was 12·0 for promising sugarcane entries, approximately double that found in Napiergrass. Disease or insect problems did not appear to be a problem with the energy cane selections.

Alexander (1985) states:

'. . . it is impossible to overstate the significance of sugarcane's ability to hybridize with its tropical grass peers, both closely and distantly related. This facility has already figured in its natural evolution and in the improvement of select clones for the production of sugar. However, it has far greater significance still for future cane industries requiring multiple products from the same plant.'

He adds that much more can be done to perfect production technologies, increase yields, and lower costs through breeding of superior new progeny. The available germplasm is vast, and is practically untouched. Alexander (1985) concludes:

'. . . the unused intergeneric cross capability is probably the best possible tool for creating the desired new hybrids.'

Chu (1982) estimated that ultimate yields in the order of 157–168 Mg ha^{-1} yr^{-1} for energy canes are possible with 'third generation' hybrids bred and selected specifically for their biomass attributes. However, more work remains in progeny development, cultural management, and harvest technology before such yields can become common in commercial biomass energy operations.

SUMMARY

Of all the genotypes of tropical or subtropical grass species evaluated, the highest biomass yields (20–50 Mg ha^{-1}) have been reported for *Saccharum* (energy cane), *Pennisetum* (Napiergrass), and *Sorghum* (forage sorghums) selections. Additional breeding and selection work is necessary with all three genera to obtain the genotypes that have traits desirable as feedstocks for methane. Sufficient genetic variation is present in each of the three genera to continue to make progress through conventional plant breeding methodology. Interspecific hybridization techniques will allow further genetic variation to be expressed and exploited in the biomass programs. Breeding goals will focus on selections having high biomass yields comprised of chemical constituents convertible to methane that can be grown in energy efficient cropping systems.

REFERENCES

Alexander, A. G. (1985). *The Energy Cane Alternative*. Sugar Series 6, Elsevier Applied Science, New York, 509 pp.

Chase, A. (1948). The meek that inherit the earth. In: *Grass, the Yearbook of Agriculture*, pp. 8–15. US Government Printing Office, Washington, D.C.

Chu, T. L. (1982). Development of second- and third-generation energy cane varieties. Symposium Proceedings: Fuels and Feedstocks from Tropical Biomass. Caribe Hilton Hotel, San Juan, Puerto Rico.

Chynoweth, D. P. (1983). Gasification of land-based biomass. Final Report No. GRI-82/0058. Gas Research Institute, Chicago, Illinois.

Dujardin, M. and Hanna, W. W. (1984). Microsporogenesis, reproductive behavior, and fertility in five *Pennisetum* species. *Theoretical and Applied Genetics* **67**, 197–201.

Grassl, C. O. (1962). Problems and potentialities of intergeneric hybridization in a sugar cane breeding programme. *Proceedings of Congress ISSCT, Mauritius* **11**, 447–56.

Hanna, W. W. (1986). Notice of release of dwarf Tift N75 Napiergrass germplasm USDA, ARS and Georgia Agric. Exp. Stn Memo.

Hanson, A. A. and Carnahan, H. L. (1956). Breeding perennial forage grasses, Technical Bulletin 1145. US Government Printing Office, Washington, D.C., 116 pp.

Hebert, L. P. and Henderson, M. T. (1959). Breeding behavior of certain agronomic characters in progenies of sugarcane crosses at the United States Sugarcane Field Station, Houma, Louisiana, Technical Bulletin 1194. US Government Printing Office, Washington, D.C., 54 pp.

Jauhar, P. P. (1981). Cytogenetics of pearl millet. *Advances in Agronomy* **34**, 407–79.

Lawrence, G. H. M. (1951). Gramineae (grass family). In: *Taxonomy of Vascular Plants*, pp. 387–91. The Macmillan Company, New York.

Miller, F. R. and Monk, R. L. (1984). Breeding and development. In: *Sorghums for Methane Production*, Hiler, E. (ed.). Annual Report, Gas Research Institute, Chicago, Illinois.

Muldoon, D. K. and Pearson, C. J. (1979). The hybrid between *Pennisetum americanum* and *Pennisetum purpureum*. *Herbage Abstracts* **49**, 189–99.

Ozias-Akins, P., Tabaeizadeh, Z. and Vasil, I. K. (1985). Somatic hybridization in the Gramineae. Abstract in the *First International Congress Plant Molecular Biology*, Vol. 1, p. 44. University of Georgia, Athens, Georgia.

Powell, J. B. and Burton, G. W. (1966). A suggested commercial method of producing an interspecific hybrid forage in *Pennisetum*. *Crop Science* **6**, 378–9.

Prine, G. M. (1985). Leucaena and elephantgrass as energy crops in colder subtropics. *5th Annual Solar and Biomass Workshop*, pp. 149–52. Atlanta, Georgia.

Rajasekaran, K., Schank, S. C. and Vasil, I. K. (1986). Characterization of biomass production and phenotypes of plants regenerated from embryogenic callus cultures of *Pennisetum americanum* × *P. purpureum* (hybrid Napiergrass). *Theoretical and Applied Genetics* (in press).

Samuels, G. (1984). Potential production of energy cane for fuel in the Caribbean. *Energy Progress* **4**, 249–51.

Samuels, G. (1986). Growing sugarcane as a renewable energy crop. *Proceedings of Soil Crop Science Society of Florida* **45**, 103–5.

Schank, S. C. (1986). Production of warm season grasses for biomass. In: *Energy from Biomass and Waste Symposium*. Institute of Gas Technology, Chicago, Illinois (in press).

Schank, S. C., Mendes de Souza, R. and Miranda, R. M. (1975). Chromosomal hybrids between *Pennisetum americanum* and *P. purpureum*. *Abstracts Southern Agricultural Workers Conference*, New Orleans, Vol. 2, p. 4.

Smith, W. H. (1983). Energy from biomass: a new commodity. In: *Agriculture in the Twenty-First Century*, Rosenblum, J. W. (ed.), pp. 67–8. John Wiley, New York.

Smith, W. H. and Frank, J. R. (1985). Comparative biomass yields of energy crops. In: *Energy from Biomass, 3rd European Community Conference*, Palz, W., Coombs, J. and Hall, D. O. (eds.), pp. 323–9. Elsevier Applied Science, New York.

Stanley, R. L., Jr and Dunavin, L. S. (1986). Potential sorghum biomass production in north Florida. In: *Biomass Energy Development*, Smith, W. H. (ed.), pp. 217–26. Plenum Press, New York.

Tai, P. Y. P. and Miller, J. D. (1986). Genotype × environment interaction for cold tolerance in sugarcane. *Proceedings of Congress ISSCT, Indonesia* (in press).

Vasil, I. K. (1985). Somatic embryogenesis and its consequences in the gramineae. In: *Tissue Culture in Forestry and Agriculture*, Henki, R. R., Hughes, K. W., Constantin, M. P. and Hollaender, A. (eds.), pp. 31–47. Plenum Press. New York.

Vasil, V. and Vasil, I. K. (1981). Somatic embryogenesis and plant regeneration from tissue cultures of *Pennisetum americanum* and *P. purpureum* hybrid. *American Journal of Botany* **68**, 864–72.

11

Development of Artificial Seeds of Sweet Potato for Clonal Propagation through Somatic Embryogenesis*

D. J. CANTLIFFE, J. R. LIU and J. R. SCHULTHEIS

Vegetable Crops Department, University of Florida—IFAS, Gainesville, Florida, USA

INTRODUCTION

Propagation of most domestic crop species is via seed. Several species including: *Solanum tuberosum* L. (potato), *Saccharum* spp. (hybrid sugarcane), *Manihot esculenta* Crantz (cassava), *Pennisetum purpureum* Schum. (Napiergrass), *Ipomoea batatas* (L.) Lam. (sweet potato) are asexually propagated and among those species possessing potential as biomass crops because of their rapid quantitative production of convertible carbohydrates.

Asexual methods of propagation may be necessary in order to insure trueness-to-type for a particular species. Seed to seed propagation can be costly and time consuming when hybrid cultivars are involved. Problems with current methods of asexual propagation include (1) length of time required to get plant material which will survive stressful conditions, (2) production of disease-free stock, (3) spacial requirements for mass production of plant materials, (4) high labor demands to produce and place plants in the field, and (5) high costs of maintaining vegetative stock materials.

The ability to economically mass produce disease free plant materials by vegetative processes and then to place those materials directly in the field has been limited to only a few species. The number of intermediate transfer-planting steps required to insure propagule survival once the

*Florida Agricultural Experiment Station Journal Series No. 7579.

plantlet is removed from tissue culture has had limited success with many species. Direct mass transfer of tissue culture derived plantlets to the field to date has been impossible. True clones by tissue culturing could greatly increase the number of plants produced from stock plant materials and reduce the production time if the clones could be successfully transferred to the field with a minimum of handling procedures.

At present, successful direct field transfer is limited by, among other problems, development of large numbers of singulated somatic embryos of uniform size which develop rapidly upon sowing. Induction of embryogenesis *in vitro* is dependent on the media used (Tazawa and Reinert, 1969); and the source of tissue and stage of morphological development are also important factors to be considered (Vasil, I. University of Florida, Personal Communication).

The use of embryogenesis as a method of clonal regeneration of plants for commercial production of many biomass crops offers the potential for mass field production with reduced labor costs. Somatic embryos have the potential to be produced by tissue culture, removed and planted directly in the field in a fashion similar to sexually produced embryos. This latter part of the production scheme might be accomplished by the use of such cultural practices as fluid drilling (Bryan *et al.*, 1978; Gray, 1981), or encapsulation (Redenbaugh *et al.*, 1984; Kitto and Janick, 1985a,b).

The use of encapsulation has been limited by problems of embryo survival after desiccation. Fluid drilling would allow the newly formed embryos to be either planted 'as is' without desiccation or germinated, then sown in a fluid gel matrix to which additives such as fungicides and growth regulators may be incorporated (Ohep, 1981). The addition of such compounds around the embryo would protect it during early seedling growth. Possible problems with hormonal regulation of embryo growth could be overcome in the gel solution, should such problems arise, by quantitative additions of growth regulators to the gel. Should storage material be lacking in the cultured embryo, nutrients could also be added to the gel. A new concept now being explored in this program is the addition of mycorrhizae inocula for exploiting the benefits of such symbiotic relations on plantlet survivability.

The primary purpose of the research summarized in this chapter was to produce somatic embryos (synthetic seeds) of the biomass species sweet potato, *I. batatas*, via tissue culture techniques, to investigate direct planting in the field using fluid (gel) sowing techniques. Success with this species in reducing production costs could serve as a model for propagating other promising asexual biomass species.

INDUCTION OF SOMATIC EMBRYOS

The first true shoot apical meristem culture of angiosperms was achieved by Smith and Murashige (1970) who regenerated whole plants from excised apical meristem domes without leaf primordia. Their method was widely employed for production of pathogen-free stocks in many species (Murashige, 1974). *In vitro* somatic embryogenesis in sweet potato has been reported from the anther (Tsai and Tseng, 1979), shoot-tip (Jarret *et al.*, 1984; Liu and Cantliffe, 1984a), leaf, stem and root explant (Liu and Cantliffe, 1984c).

We have demonstrated that apical meristem culture can be used for high frequency embryogenic callus formation in sweet potato. Basically, our results showed that two weeks after culture over 90% of surviving meristem dome explants growing on medium containing $0.5-3.0$ mg liter^{-1} 2,4-dichlorophenoxyacetic acid (2, 4-D) enlarged 2- to 3-fold in diameter (Fig. 1A). After 3–4 weeks the peripheral region of the meristem dome produced smooth-surfaced tissue (Figs. 1B, C). After 4–5 weeks the surfaces were nodulated with 2–15 embryoids at early embryoid developmental stages (Fig. 1D). The nodulated region showed typical characteristics of embryogenic callus in sweet potato (Liu and Cantliffe, 1984b)— pale to bright yellow and compact. After 5–8 weeks numerous embryos arose from the embryogenic callus (Figs. 1E, F), whereas cell layers subjacent to some meristem domes produced a substantial amount of white to pale brown nonembryogenic friable callus. The most distal portion of the meristem dome was also included in the formation of embryogenic callus but always after the peripheral region produced embryogenic callus with embryos (Fig. 1E). After 8 weeks of culture on medium containing $0.5-3.0$ mg liter^{-1} 2,4-D the formation of embryogenic callus was observed from over 90% of the meristem dome explants (Fig. 2). The meristem dome explants cultured on basal medium either produced nonembryogenic friable callus or failed to survive culture. When cultured on medium containing 4.0 mg liter^{-1} 2,4-D, most of them did not continue to grow.

Incorporation of 5% coconut water to medium containing $0-4.0$ mg liter^{-1} 2,4-D promoted the induction of nonembryogenic friable callus from the basal portion of the meristem dome explants. After 2–3 weeks of culture the meristem dome explants were embedded by the nonembryogenic callus.

Somatic embryos have been shown to be initiated in two ways, from single cell origins and from a mass of cells termed a pro-embryonal complex (Pence *et al.*, 1980). In the tissue culture of sweet potato we have

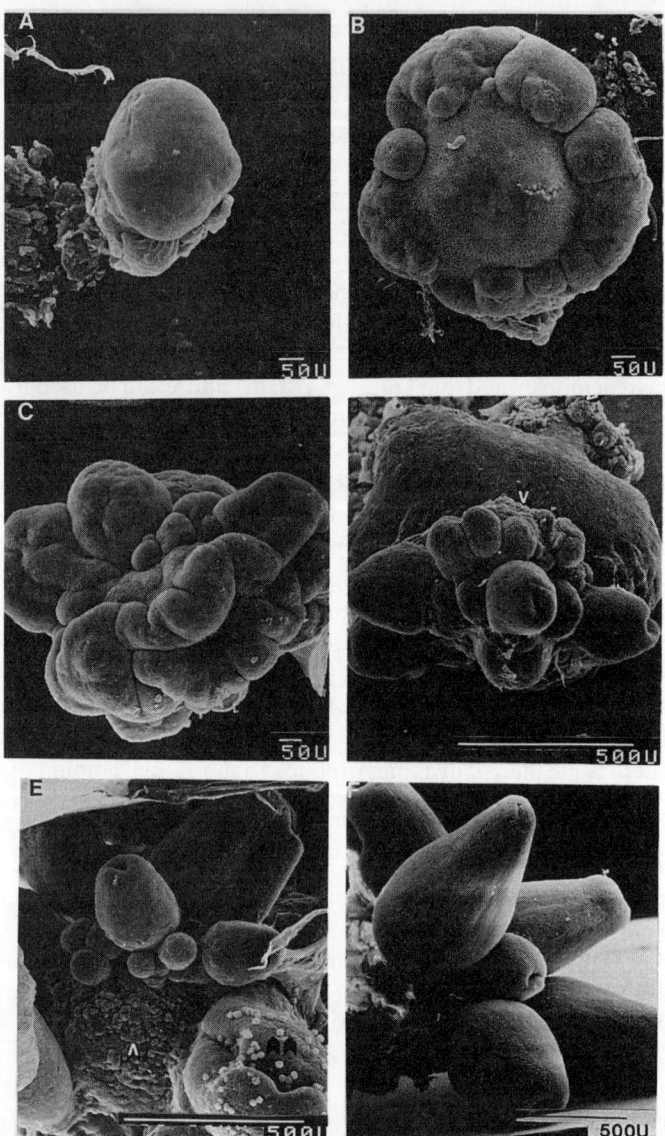

FIG. 1. Scanning electron micrograph of somatic embryogenesis from shoot apical meristem dome explants. (A) Two-week old cultured meristem dome. (B) Four-week old cultured meristem dome showing that the peripheral region produced smooth-surfaced tissue. (C) Same as 'B' but cultured on media containing a higher concentration of 2,4-D ($1 \cdot 0$ mg liter^{-1}). Note the smooth-surfaced tissue proliferated more actively than 'B'. (D) Proliferating immature embryos from embryogenic callus (arrows). (E) Eight-week old cultured meristem dome. Note the distal region (arrow) producing embryogenic callus and the trichomes (arrows). (F) Group of maturing embryos.

Artificial Seeds of Sweet Potato for Clonal Propagation

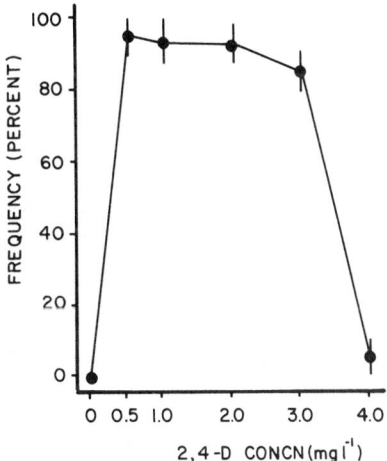

FIG. 2. The frequency of embryogenic callus formation from apical meristem explants cultured on media containing different concentrations of 2,4-D. Vertical bars indicate the standard deviation.

achieved somatic embryogenesis from embryogenic callus generated from root, leaf, petiole and shoot-tip explants, with the most efficient explant coming from the apical dome of the shoot-tip.

Callus, which is morphologically organized and compacted with a yellow pigmentation, has the best embryogenic response to our cultural practices. Somatic embryogenesis occurs when the meristematic condition dedifferentiates to the totipotent state. In other words, when the cell or cellular mass becomes competent and self-sufficient.

In the apical dome of flowering plants there is a very small group of cells located subterminally which are perpetually meristematic. These cells are totipotent. These cells may give the apical dome explant the efficiency observed in producing embryonic tissue. Our on-going research suggests that somatic embryos arise from the apical dome with intervening callus production (Cantliffe and Liu, unpublished). Research addressing single-cell origin of asexual embryos is also in progress.

A brief histological investigation of shoot apical meristem culture showed the stages of callus development. Histologically, a two-layered tunica encloses the corpus (Fig. 3A). The meristem (apical) dome was aseptically excised and plated onto solidified agar media (Fig. 3B). After 2 days in culture (Fig. 3C) the central portion of the explant had become meristematic indicated by the dark histological stain. After 4 days in

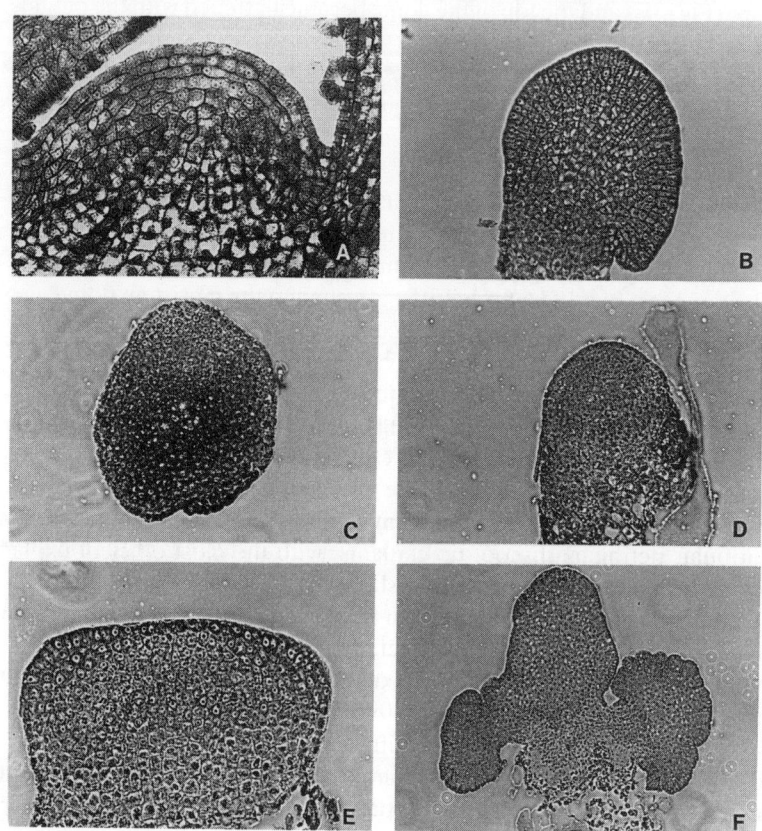

FIG. 3. Histology of somatic embryogenesis in shoot apical meristem culture of 'White Star' sweet potato. (A) Meristem dome before excised (original magnification 400 ×). (B) Meristem dome explant after excision (original magnification 215 ×). (C) After two-day culture (original magnification 232 ×). (D) Four-day old cultured meristem dome (original magnification 175 ×). (E) Eight day divisions in the peripheral region (original magnification 210 ×). (F) 32-day old cultured meristem dome showing numerous embryos arising from embryogenic callus (original magnification 93 ×).

culture (Fig. 3D) relative disorganization in the explant occurred although the epidermis continued as a uniform cover.

The peripheral area of the explant became buttressed by 8 days in culture (Fig. 3E) and the top of the dome became flattened with the bulging of lateral tissue. Multinucleate cells in this region indicated proliferation in a ring-like fashion around the central zone.

By 4 weeks the callus had various outgrowths (Fig. 3F). The radial files on the right resemble embryogenic callus production (Springer et al., 1979), possibly indicating polarity with respect to concentration, physiological or diffusion gradients. The structure on the left of Fig. 3F may be an embryonic mass organizing a single somatic embryo or the proembryonal foundation for future multiple embryo production.

The youngest pair of leaf primordia attached to the meristem dome also produced embryogenic callus. The second and third pair of leaf primordia produced nonembryogenic friable callus which usually covered the meristem dome region prior to any embryogenic callus formation. In some cases the domes of axillary meristem adjacent to the leaf primordia produced embryogenic callus as well. The frequency of embryogenic callus formation declined significantly as the numbers of pairs of leaf primordia attached to the apical meristem explant increased (Fig. 4).

Globular to heart-shaped embryos were transferred onto medium containing 0·03 mg liter^{-1} abscisic acid (ABA) in the dark. They matured into embryos with sporadic rooting (Fig. 5A). On basal medium the maturation of embryos was delayed and they had poorly developed cotyledons. Cotyledonary embryos developed into plants on medium containing 0·02 mg liter^{-1} ABA and 0·02 mg liter^{-1} N(6)-(2-isopentenyl)-adenine (2iP) under light (Fig. 5B).

The apical meristem dome plus the youngest pair of leaf primordia from the cultivar 'GaTG3' were cultured in the same manner. About 80% of the explants also produced embryogenic callus on medium containing 1·0 mg liter^{-1} 2,4-D. Since 'GaTG3' had a smaller meristem dome than 'White Star', excising the meristem dome alone was not possible.

The technique of shoot apical meristem culture in this work may be tedious and time-consuming when compared to leaf, shoot-tip, stem and root explant culture methods previously described for embryogenic callus induction (Liu and Cantliffe, 1983). However, benefits of this method include a dramatic increase in the frequency of embryogenic callus formation. Botti and Vasil (1983) induced embryogenic callus from the shoot apical meristem dome and the two youngest leaf primordia excised from mature embryos or from 4–5 day old seedlings of *Pennisetum purpureum*.

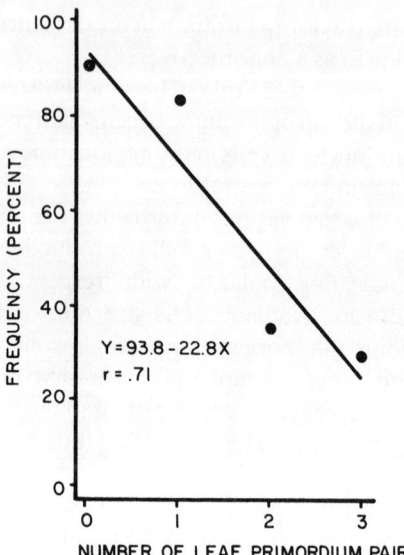

FIG. 4. The frequency of embryogenic callus formation from shoot apical meristems and shoot-tips as related to explant dimension. The slope shows a linear regression between the estimated frequency (Y) and the given number of leaf primordium pair (X). The correlation coefficient (r) is significant at the 1% level.

They realized that the complete removal of mesocotyl tissue from the base of the shoot apex at the time of culture was critical for the formation of embryogenic callus, and for inhibiting the development of nonembryogenic friable callus.

It is believed that the shoot apical meristem preserves the intact genetic composition through mitosis better than any other part of a plant. Some variability pre-exists in the explant such as root and petiole cuttings of *Pelargonium* spp. used to culture *in vitro* (Skirvin and Janick, 1976). Therefore, we may eliminate some unnecessary genetic variation by employing the shoot apical meristem dome as an explant.

Furthermore, the embryogenic compact callus appears to have more karyotypic stability than the nonembryogenic friable callus. Leaf tissue of maize (*Zea mays* L.) produces a higher frequency of tetraploid and octaploid cells in the friable callus than in the compact callus (Freeling *et al.*, 1976). The loss of 'S'-type cytoplasmic male sterility in maize was uniquely associated with the friable callus, whereas the compact callus retained male sterility (Pring *et al.*, 1977). A high uniformity

FIG. 5. (A) Germination of embryos. (B) Plants regenerated via somatic embryogenesis.

in chromosomal normality, pollen stainability and morphological characteristics was observed among plants regenerated from the compact embryogenic callus of Guinea grass (*Panicum maximum* Jacq.) (Hanna *et al.*, 1984). In this and our previous work (Liu and Cantliffe, 1984c) no morphological abnormality was detected from the mature plants regenerated from the compact embryogenic callus. Unfortunately, however, the genomic complexity of sweet potato ($2n = 6\times = 90$) did not allow us a precise cytological study on regenerated plants.

PLANTLET HANDLING

The pathway of plantlet propagation through somatic embryogenesis has been demonstrated for sweet potato (Fig. 5B). Yields of somatic embryos must be increased at all stages (shoot-tip establishment in culture, embryogenic callus production, embryo production, embryo germination, transfer to gel, and seeding). For this a culture medium specific to sweet potato may need to be further defined.

Synchronizaion of embryo development remains a major task. Singulation is associated with this step. The future experimental approach is to obtain, manage, and maintain suspension cultures of single cells to small cell clusters (up to 10 cells) that are embryogenic. Synchronization of development or developmental arrest at desired stages can then be studied, using different incubation conditions and media factors. The successful management of cell suspensions requires the study of cell and cell cluster population growth. At present suspensions of small cell clusters are obtained using a system of sieves and by grinding large clusters.

Gel seeding somatic embryos is a promising direct field sowing technique. Agricultural gels such as Liqua-gel. Natrosol, Viterra 2, and American Colloid (AC) were tested for their phytotoxicity or effect on plant growth (plant length, leaf and root initiation, etc.). The gels were supplemented with inorganic salts, thiamine, HCl, myo-inositol, and 2% sucrose. Viterra and AC gels were toxic to the embryos, over 80% died after 6 days (Table 1). Embryos grew rapidly and normally in Natrosol gel. Sucrose at 1·6% concentration resulted in the highest percentage of normal plantlet formation. Natrosol has promise for use as a gel carrier for fluid drilling of somatic embryos.

Certain growth hormones helped improve normal root and shoot

TABLE 1
Effect of Gel Carrier on Somatic Embryo Survival 6 Days after Placing in Gel

Gel type	% Response		
	Green	Necrosis	Dead
Natrosol	100	0	0
Liqua-gel	40	37	23
Viterra	0	13	87
American Colloid	0	3	97

development of torpedo stage somatic embryos. Plantlet formation was enhanced when NAA was added at concentrations of 0·1–1 μM. When the cytokinin 6-benzyl-aminopurine (BAP) was added at concentrations over 2 μM predominantly shoots developed. Additions of these hormones directly to the gel showed promise to enhance early plantlet development in the fluid drilling system. Cooperative research is in progress with scientists working with N-fixers and mycorrhizae in order to incorporate the potential to reduce fertilizer and other inputs for production of biomass systems produced via somatic embryos.

SUMMARY

Many potentially important species for biomass production cannot be adequately propagated from seed. Attempts to mass produce plants via tissue culture have been limited by a number of intermediate transfer-planting steps to insure plantlet survival once the plantlet is removed from culture. By using embryogenesis, embryos produced in culture might be planted directly in the field by using practices such as fluid drilling.

We have been successful in tissue culturing somatic embryos of sweet potato for the first time. Technology for proliferation of such embryos is at hand. Shoot apical meristem dome explants of sweet potato (cv. White Star) produced embryogenic callus on agar nutrient medium containing 0·5–3·0 mg liter^{-1} 2,4-D. The frequency of embryogenic callus formation was over 90%. The frequency declined sharply as the number of leaf primordium pairs attached to the meristem dome explant increased. The induced callus gave rise to numerous embryoids. When transferred onto hormone free medium, the embryoids readily developed into a torpedo shape before germination. The plantlets were transplanted to soil both in the greenhouse and field, where normal plants developed. The need for a gel which would support normal embryo growth and development was found in Natrosol 250 HHR gel. Further work on synchronization of embryo development will lead to further experimentation with the fluid drilling sowing system.

The perfection of and utilization of embryogenesis and fluid drilling could reduce the cost of producing biomass crops and therefore reduce the cost of the methane produced. Preliminary estimates suggest the fluid drilling of tissue culture produced propagules could involve one-tenth of the cost of conventional planting procedures. As methodologies are developed for direct transfer of sweet potato embryos, transfer systems for other important biomass species can likewise be perfected.

ACKNOWLEDGEMENTS

We thank Ms Jeanne Fischer and Ms Caprice Simonds for assistance with scannning electron micrographs and light microscopic examinations.

REFERENCES

Botti, C. and Vasil, I. K. (1983). Plant regeneration by somatic embryogenesis from parts of cultured mature embryos of *Pennisetum purpureum* Schum. *Zeitschrift fur Pflanzenphysiologie* **111**, 319–25.
Bryan, H. H., Stall, W. M., Gray, D. and Richmond, N. S. (1978). Fluid drilling of pregerminated seeds for vegetable gardening. *Proceedings of the Florida State Horticultural Society* **91**, 88–90.
Freeling, M., Woodman, J. C. and Cheng, D. S. K. (1976). Developmental potentials of maize tissue cultures. *Maydica* **21**, 97–112.
Gray, D. (1981). Fluid drilling of vegetable seeds. *Horticultural Reviews* **1**, 1–27.
Hanna, W. W., Lu, C. and Vasil, I. K. (1984). Uniformity of plants regenerated from somatic embryos of *Panicum maximum* Jacq. (Guinea grass). *Theoretical and Applied Genetics* **67**, 155–9.
Jarret, R. L., Salazar, S. and Fernandez, Z. R. (1984). Somatic embryogenesis in sweet potato. *HortScience* **19**, 397–8.
Kitto, S. L. and Janick, J. (1985a). Production of synthetic seeds by encapsulating asexual embryos of carrot. *Journal of the American Society for Horticultural Science* **110**, 277–82.
Kitto, S. L. and Janick, J. (1985b). Hardening treatments increase survival of synthetically-coated asexual embryos of carrot. *Journal of the American Society for Horticultural Science* **110**, 283–6.
Liu, J. R. and Cantliffe, D. J. (1983). Somatic embryogenesis and plant regeneration in tissue cultures of sweet potato (*Ipomoea batatas* Poir.). *HortScience* **18**, 618.
Liu, J. R. and Cantliffe, D. J. (1984a). Improved efficiency of somatic embryogenesis and plant regeneration in tissue cultures of sweet potato (*Ipomoea batatas* Poir.). *HortScience* **19**, 589.
Liu, J. R. and Cantliffe, D. J. (1984b). Tissue culture propagation development and its application to energy crops. *1984 International Gas Research Conference*, pp. 622–9.
Liu, J. R. and Cantliffe, D. J. (1984c). Somatic embryogenesis and regeneration in tissue cultures of sweet potato (*Ipomoea batatas* Poir.). *Plant Cell Reports* **3**, 112–15.
Murashige, T. (1974). Plant cell and organ culture methods in the establishment of pathogen-free stock. A. W. Pimock Lectures No. 2. Cornell University. Ithaca, New York.
Ohep, J. C. (1981). Incorporation of gel additives in the fluid drilling of pregerminated tomato seeds. M.S. thesis, University of Florida, Gainesville, Florida.

Pence, V. C., Hasegawa, P. M. and Janick, J. (1980). Initiation and development of asexual embryos of *Theobroma cacao* L. *in vitro*. *Zeitschrift fur Pflanzenphysiologie* **98**, 1–14.

Pring, D. R., Levings, C. S. III, Hu, W. W. L. and Timothy, D. H. (1977). Unique DNA associated with mitochondria in the 'S'-type cytoplasm of male-sterile maize. *Proceedings of the National Academy of Sciences* **74**, 2904–8.

Redenbaugh, K., Nichol, J., Kossler, M. E. and Paasch, B. (1984). Encapsulation of somatic embryos for artificial seed production. *In Vitro* **20**, 256–7.

Skirvin, R. M. and Janick, J. (1976). Tissue cultures induced variation in scented *Pelargonium* spp. *Journal of the American Society for Horticultural Science* **101**, 281–90.

Smith, R. H. and Murashige, T. (1970). In vitro development of the isolated shoot apical meristem of angiosperms. *American Journal of Botany* **57**, 562–8.

Springer, W. D., Green, C. E. and Kohn, K. A. (1979). A histological examination of tissue culture initiation from immature embryos of maize. *Protoplasma* **101**, 269–81.

Tazawa, M. and Reinert, J. (1969). Extracellular and intracellular chemical environments in relation to embryogenesis *in vitro*. *Protoplasma* **68**, 157–73.

Tsai, H. S. and Tseng, M. T. (1979). Embryoid formation and plantlet regeneration from anther callus of sweet potato. *Botanical Bulletin of Academic Sinica* **20**, 117–22.

12

Nitrogen Fixation in *Sargassum* Biomass Production Systems

K. T. SHANMUGAM,[a] H. SPILLER[a] and E. J. PHLIPS[b]

[a] *Microbiology and Cell Science Department*, [b] *Fisheries and Aquaculture Department, University of Florida—IFAS, Gainesville, Florida, USA*

INTRODUCTION

The development of biomass energy technologies is based on economic production of large amounts of convertible biomass. Needed are solar energy, water and inorganic nutrients. Nutrients for high productivity are normally sustained with artificial fertilizer additions, a substantial production cost. Estimates from the BIOMET model for Napiergrass indicate that 30% or more of the yearly operating budget for this biomass crop may go to fertilization. More than half the cost of fertilizer is for the reduced nitrogen compounds. Not all agricultural crops involve such high subsidies of nitrogen fertilizer. For example, legumes can be cultivated with minimal or no addition of nitrogen because of their symbiotic association with bacteria which fix atmospheric nitrogen as ammonia. The process of nitrogen fixation by bacteria is an important element in the nitrogen cycle of many ecosystems, including the open ocean (Delwiche, 1981). Estimates of the total input of nitrogen into the world's oceans by nitrogen fixation amount to about 15×10^{12} g N yr^{-1} or about 50% of the total annual input of combined nitrogen (Capone and Carpenter, 1982; Delwiche, 1981). There are several studies on the extent and nature of nitrogen fixation in marine communities (Capone and Carpenter, 1982; Capone and Taylor, 1977; Carpenter, 1972; Carpenter and Cox, 1974; Dugdale *et al.*, 1964; Hanson, 1977; Ryther and Dunstan, 1971; Stewart, 1965).

Nitrogen fixation in marine communities does not guarantee success in providing fertilizer for the biomass crops. In terrestrial agriculture, research efforts have largely focused on the symbiotic rhizobia, or introduction of free-living heterotrophic bacteria, like *Azospirillium* sp. (Kapulnik *et al.*, 1981; Smith *et al.*, 1984) and free-living cyanobacteria

(Rogers and Reynaud, 1982) into soil. Cyanobacteria may be superior to heterotrophic bacteria as an inoculum for nitrogen enrichment because these organisms use sunlight and water as sources of energy and reductant, respectively, and carbon dioxide as the carbon source for growth. Nitrogen fixation by heterotrophs is often limited by the availability of organic carbon compounds.

A number of potential marine biomass crops, including *Sargassum*, are known to have nitrogen-fixing bacteria as epiphytes (Dugdale et al., 1964; Carpenter, 1972; Carpenter and Cox, 1974; Dawes, 1974; Hanson, 1977). Our interest in *Sargassum* as a potential marine biomass crop relates to its successful adaptation to life in the most nutrient-depleted environment of the central north Atlantic Ocean (Dugdale et al., 1964) where the pelagic *Sargassum* is found in abundance. These properties stimulated our studies on the ecology of this process; *in situ*, distribution and physiological characteristics of nitrogen-fixing cyanobacteria associated with *Sargassum* sp. Nitrogen-fixing cyanobacteria have been reported to exist in *Sargassum* communities, but these organisms were not purified to homogeneity for further physiological studies. Although the plant community contains a large number of nitrogen-fixing epiphytes, it is not known whether the fixed nitrogen is directly made available to the plant or the plant benefits from the enhanced nitrogen fertility of the microenvironment. It is essential to determine the physiological properties of the epiphytes to evaluate these characteristics, since all known free-living organisms that fix nitrogen assimilate it for their own growth and do not excrete it into the environment.

In order to evaluate the potential of *Sargassum* as a biomass crop for methane production, we have surveyed the marine coastal environment off the State of Florida for epiphytic nitrogen-fixation capacity associated with *Sargassum*. In addition, the type and distribution of associated cyanobacteria and physiological properties of selected cyanobacteria were also studied. Results of these experiments are presented below. A longer-term goal is to determine if these organisms can replace part or all of the nitrogen needed for biomass crop production.

NITROGEN FIXATION IN *SARGASSUM* COMMUNITIES

Quantity of Nitrogen Fixed

We utilized the acetylene reduction technique to investigate nitrogen fixation in *Sargassum* communities (Phlips et al., 1986), the activity in both the pelagic species *Sargassum fluitans* Borgesen and *S. natans* J. Meyer, and the estuarine, benthic species *S. filipendula* C. Agardh. Nitrogen fixation would represent a significant percentage of the nitrogen required for primary production in the oceanic *Sargassum* community, but a much smaller percentage in the estuarine communities, where other sources of N take precedence (i.e. land and river runoff). Evidence from previous investigations of nitrogen fixation associated with oceanic *Sargassum* indicated the presence of significant activity (Carpenter, 1972; Hanson, 1977). However, the average activities reported were substantially different (i.e. $0 \cdot 02 – 1 \cdot 08 \, \mu g \, N \, m^{-2} \, h^{-1}$, Carpenter (1972) and $4 \cdot 5 \, \mu g \, N \, m^{-2} \, h^{-1}$, Hanson (1977) and seasonal data were not available.

Our two-year study (1984–5) of nitrogen-fixation activity revealed significantly different patterns of activity in oceanic and estuarine *Sargassum* communities. Oceanic communities exhibited strong but variable activity throughout the test period (Fig. 1). There was no apparent seasonal pattern to the variability, and the activity never dropped below a moderate level ($0 \cdot 15 – 0 \cdot 20 \, \mu mol$ ethylene produced $\times$ g *Sargassum* dry wt^{-1} h^{-1}). In contrast, nitrogen fixation in the estuarine community exhibited strong seasonality (Fig. 1). Nitrogenase activity was high in the late spring to early fall, but low or absent in the winter months.

There is substantial nitrogen fixation in both oceanic and estuarine *Sargassum* communities according to our data. The question is, 'How significant is this activity to the N budget of the community?' In the summer of 1984, a series of experiments were performed (Phlips *et al.*, 1986) to compare the rates of acetylene reduction and photosynthesis (measured as oxygen evolution) in order to estimate the potential impact of nitrogen fixation. A C:N ratio of 20 was used to facilitate the comparison (Hanson, 1977). Three stations were tested in June and July, (1) oceanic *Sargassum* from off St. Augustine, Florida, (2) oceanic *Sargassum* from off Miami, Florida, and (3) estuarine *Sargassum* from the mouth of the Homosassa River, Florida. Nitrogen-fixation activity (calculated from acetylene reduction, assuming a 3:1 ratio of acetylene reduction:nitrogen fixation) represented 104, 30 and 5% of the nitrogen required for the rate of photosynthesis observed at the three sites, respectively, a significant nitrogen fixation in the pelagic *Sargassum* communities.

The latter results suggest that nitrogen fixation in pelagic *Sargassum* biomass farms may be able to replace a significant percentage of the need for nitrogen fertilizer. However, if the productivity of *Sargassum* is increased, by phosphate addition, for example, will the rate of nitrogen fixation keep pace with the increased demand for nitrogen? This is a likely

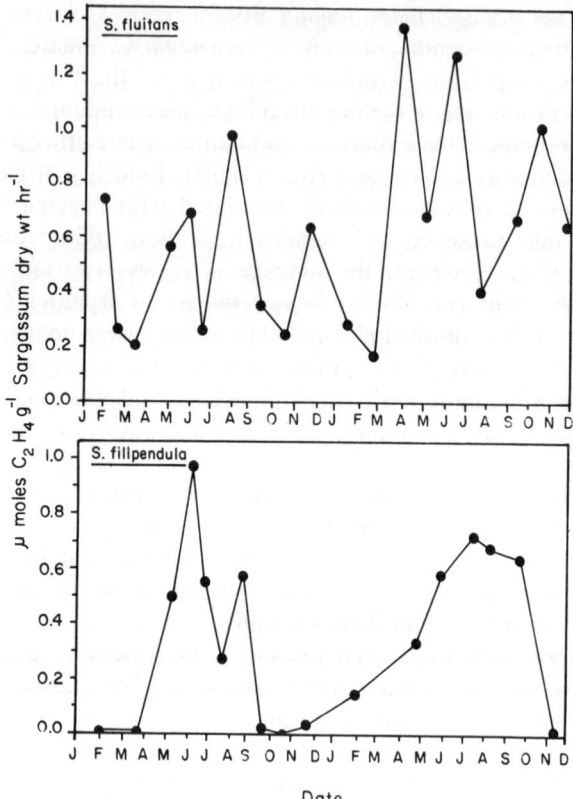

FIG. 1. Nitrogen fixation activity (represented as the rate of acetylene reduction) in oceanic (top) and estuarine *Sargassum* communities (bottom).

possibility since high productivity is desirable, if not essential, to the success of biomass production systems. Therefore, a preliminary investigation was carried out on the effects of phosphate enrichment on nitrogen fixation in both oceanic and estuarine *Sargassum* communities. Incubation of oceanic *Sargassum* in phosphate enriched clear ocean water resulted in an eventual three-fold increase in nitrogen-fixation activity. A similar enrichment of estuarine *Sargassum* samples in coastal water resulted in a less than 20% enhancement. This indicates that there is considerable room for improvement of nitrogen fixation over the levels observed in nature. Thus, *Sargassum* productivity may be linked to a concomitant increase in the amount of nitrogen fixed.

Organisms Responsible for Nitrogen Fixation

The ability to fix nitrogen is restricted to procaryotic organisms. A variety of heterotrophic bacteria, photosynthetic bacteria and cyanobacteria (i.e. blue-green algae) exhibit nitrogen-fixation activity over a range of environmental conditions. Most of these organisms produce significant nitrogenase activity only under anaerobic or microaerobic conditions. A more select group of organisms are able to fix nitrogen under aerobic conditions. These include heterotrophic bacteria like *Azotobacter*, heterocystous cyanobacteria and a few nonheterocystous filamentous and unicellular cyanobacteria. In the generally aerobic *Sargassum* ecosystems, it is likely that most of the nitrogen-fixation activity associated with these communities is attributable to aerobic nitrogen fixers. Furthermore, our experimental results indicate that this nitrogen fixation is strongly light dependent. Rates of nitrogen fixation in the dark are only a small percentage of the rates in the light. This further narrows the organisms responsible for nitrogen fixation to cyanobacteria. Close examination of *Sargassum* samples which exhibit the highest nitrogen-fixation activity revealed visible tufts of cyanobacteria that resemble *Calothrix*. Scanning electron microscopy of *Sargassum* blades shows the nature of these tufts (Fig. 2). In addition, there is a diverse collection of other epiphytic cyanobacteria associated with the surface of *Sargassum*, including *Nostoc*, *Oscillatoria*, members of the poorly defined *Lyngbya*, *Phormidium* and *Plectonema* group and some unicellular species.

For the isolation and purification of the nitrogen-fixing organisms associated with *Sargassum*, material from four major *Sargassum* species was collected and after thorough rinsing with sea water was assayed for nitrogenase activity (Table 1). In *Sargassum* communities, the epiphytic population and nitrogenase activity were highest in older blades as compared to young blades. For the isolation of epiphytic cyanobacteria, blade segments with nitrogenase activity were embedded in sea-water based agar medium containing N_2 as the sole N source and incubated with illumination, at room temperature.

After several weeks, all agar plates were evaluated according to (a) the most dominant cyanobacteria type, (b) the percentage of blades affected by each type, and (c) the frequency of blue-green foci per blade (Table 1). Both benthic *Sargassum* (*S. filipendula* and *S. cymosum*, C. Agardh) and pelagic *Sargassum* (*S. fluitans* and *S. natans*) showed the presence of nitrogen-fixing cyanobacteria and the frequency of leaves affected varied from 75 to 100%. *Dichothrix* occurred on all *Sargassum* species, whereas *Phormidium* (of the LPP-group) (Rippka et al., 1979) and *Oscillatoria*

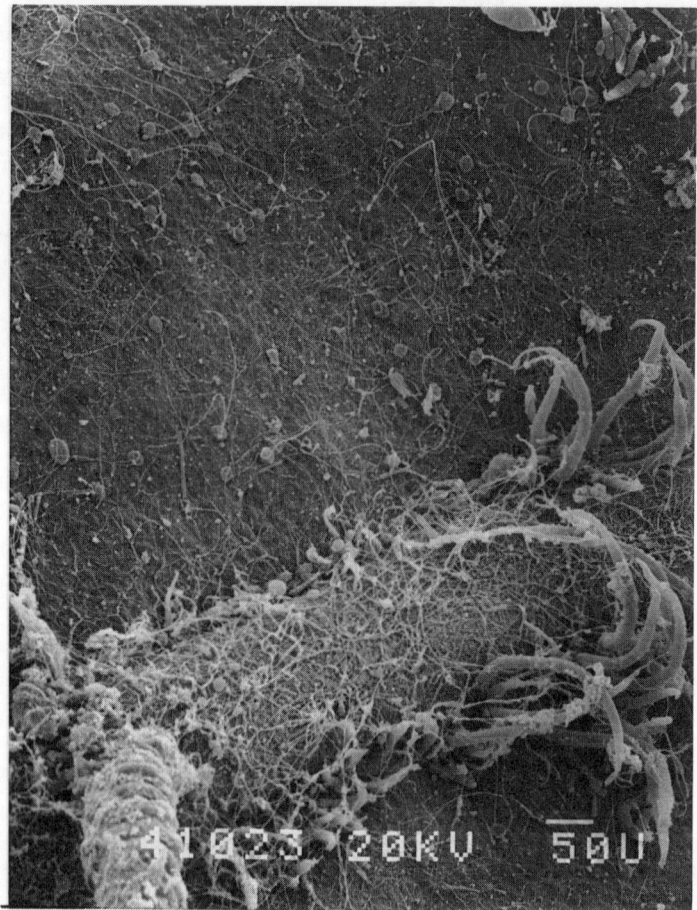

FIG. 2. Scanning electron micrograph of the surface of a *Sargassum* blade, showing upwardly protruding filaments of *Calothrix* sp. (lower right). U indicates μm.

were more dominant on the pelagic species. Unicellular cyanobacteria were a regular occurrence on benthic species, even though they were observed only on 10% of the subcultured blade segments on agar plates. This low value may not represent the real presence in nature, since the unicellular species might have been lost more easily during the frequent washings in the collection process. The frequency of foci per blade, ranging

TABLE 1
Occurrence of Cyanobacteria on *Sargassum* Leaves

Sargassum	No. of cyanobacteria	Most dominant type	Leaves affected (%)	Frequency of foci per leaf
S. filipendula	4	Dichothrix	80	3–15
		unicellular	10	2–15
S. natans	3	Dichothrix	60–100	10–15
		Phormidium		
S. fluitans	3	Branched filamentous		
		possibly Dichothrix	90–100	2–5
		Oscillatoria		
S. cymosum	3	Dichothrix	80	10–15
		unicellular		
		Calothrix		

between 2 and 15, confirms cyanobacteria as a common epiphyte on *Sargassum*.

Isolation of the Cyanobacterial Epiphyte in Pure Culture

Prospective nitrogen-fixing organisms were purified by pour plating as well as by surface plating on media solidified with purified agar. These media contained different carbon compounds as C source with or without inorganic, fixed nitrogen compounds. Cultures were transferred from single colonies, whenever possible, and after serial dilutions of the isolates. Some isolates failed to grow in the absence of inorganic nitrogen compounds or would bleach shortly after reaching the size of a minicolony.

Morphological characterization as well as microscopic studies (Dawes, 1974; Rippka *et al.*, 1979) showed that the cyanobacterial species isolated belonged to these families:

(1) Rivulariaceae, *Dichothrix* Zanardini; *Calothrix fucicola* Agardh from *Sargassum fluitans* and *S. natans*; *Rivularia polyotis* Roth from *S. filipendula* and *S. cymosum*.
(2) Oscillatoriaceae, *Phormidium* Kutzing from *S. cymosum* and several *Oscillatoria* species from *S. filipendula*.
(3) Chroococcaceae, comprising *Synechococcus* Nageli strains isolated from *S. fluitans*, *S. cymosum*, and *S. filipendula*.

We continued our studies with a representative of each group: *Rivularia polyotis*; *Oscillatoria* sp. (a nonheterocystous representative that did not fit the description of *O. erythraea* (Bryceson and Fay, 1981) and a filamentous strain from the LPP-group characterized by clearly defined constrictions between cells of isodiametric shape and a trichome enclosed by a sheath. Two different unicellular cyanobacteria were also purified. One of the two had a spherical shape and originated from *S. fluitans*, and the second which was three times as large and elongated in shape, was obtained from *S. filipendula*. Both strains multiplied by dividing in one plane in successive divisions and lacked a capsule and slime layer when grown in the presence of ammonia. Both strains grew well with bicarbonate as the C source.

Conditions Needed for Expression of Maximum Nitrogenase Activity of the Isolated Organisms

Nitrogenase activity is subject to control mechanisms such as repression by oxygen, the availability of a degradable carbon source and regulation by fixed organic nitrogen compounds. In cyanobacteria, light is a prerequisite for nitrogenase activity to supply ATP produced by cyclic photophosphorylation. The conditions under which nitrogenase activity can be induced vary considerably according to the morphology, light intensity and the availability of a suitable carbon source (Table 2; Rippka et al., 1979). *Oscillatoria* sp. expressed nitrogenase activity only in the presence of fructose, glucose or mannitol, and only when the inhibitor of photosynthetic O_2 evolution, DCMU (3-(3,4-dichlorophenyl)-1,1-dimethyl urea), was also present. In the dark there was no activity. This same observation applied to the representative of the LPP-group, which specifically required fructose and DCMU in the presence of light to support nitrogenase activity (Table 2).

Rivularia sp., found anchored to the cortical layer of the *Sargassum* blade, is a heterocystous and filamentous cyanobacterium that can grow photoautotrophically. Three unicellular cyanobacteria were isolated from the surface of *Sargassum* blades, where they seemed to be loosely adhering to the surface. In these unicellular cyanobacteria, nitrogenase activity was readily detected. In the *Synechococcus* strains tested, both bicarbonate and pyruvate supported nitrogenase activity at high rates. It is possible that the pyruvate effect is mediated through CO_2 released from pyruvate degradation. The specific activity of nitrogenase in pure cultures of different cyanobacteria varied considerably among the strains, the lowest value was found for the filamentous LPP member, the highest in one unicellular strain (Table 2). Considering the generation time of 20 h, under

TABLE 2
Nitrogenase Activity[a] of Cyanobacteria Isolated from *Sargassum*

Organism[b]	C source and condition for maximum activity	Nitrogenase activity
Oscillatoria sp.	Glucose, fructose or mannitol, DCMU, light	7·0
F-22, LPP-group	Fructose plus DCMU, light	2·1
Rivularia sp.	Bicarbonate, light	8·3
Synechococcus sp. 1	Bicarbonate (glucose) light	8·0
Synechococcus sp. 2	Bicarbonate, pyruvate light	17·0

[a] Expressed in μmol ethylene produced h^{-1} mg^{-1} chl with samples that had been fully induced for nitrogenase after previous growth in medium supplied with inorganic nitrogen and CO_2 or bicarbonate as C source.
[b] These are single colony isolates free of contaminating bacteria, as determined by lack of growth in bacterial growth media in the dark.

optimum conditions in the presence of ammonia, for *Synechococcus* sp. strain 1, growth in nitrogen-deficient biotopes would be extremely slow, given the additional requirements for protection against oxygen and the availability of a suitable fixed carbon source for most cyanobacteria attached to *Sargassum*.

Effect of Preformed Organic Compounds on Nitrogenase Activity

Nitrogen fixation requires a supply of reductant and ATP. In cyanobacteria, the metabolic pathways involved in meeting these requirements are carbon dioxide fixation, respiratory breakdown of storage glycogen or the dissimilatory breakdown of fixed carbon compounds supplied externally. Molecular hydrogen also can support nitrogenase activity, in light, at maximum rates. In *Rivularia* sp., where the effect of addition of organic compounds on nitrogen fixation was studied in detail, glucose and pyruvate were found to support nitrogenase activity (Table 3). In these experiments, the nitrogenase activity obtained with H_2 was set at 100% and the effect of various organic compounds on nitrogenase activity in *Rivularia* sp. was compared to this H_2-dependent value. The control sample without any addition still produced half of the maximum activity, indicating that endogenous reserves were mobilized. Bicarbonate addition enhanced nitrogenase activity only slightly over the control value.

TABLE 3
Effect of Carbon Compounds on Nitrogenase Activity in *Rivularia*[a] sp. Isolated from *S. cymosum*

Addition	Nitrogenase activity (% maximum)	
	Light	Light + DCMU
None	50	30
Bicarbonate, 10 mM	62	65
Glucose, 5 mM	72	51
Fructose, 5 mM	45	52
Pyruvate, 12 mM	80	56
Citrate, 10 mM	62	44
Acetate, 10 mM	48	52
Hydrogen, 94%	100	50

[a] Culture grown in limiting-nitrate/CO_2 medium was transferred to fixed-N-free assay medium after exhaustion of nitrate. The activity of the dark controls (minus supplement, pyruvate or bicarbonate addition) was undetectable. Nitrogenase activity of the hydrogen sample was 2·8 μmol h^{-1} mg^{-1} chl.

Pyruvate was a better organic compound, supporting 80% of the maximum activity attainable, followed by glucose with 72%. Citrate also showed a distinct enhancement, to 62%, whereas fructose and acetate addition did not increase nitrogenase activity above the control level. The photosynthesis inhibitor, DCMU, suppressed nitrogenase activity. Since carbon dioxide fixation is inhibited by DCMU, these values might represent the activity supported by endogenous glycogen breakdown plus the dissimilation of carbon externally supplied. The enhancing effect of citrate as well as the decreasing effect of glucose + DCMU addition suggests that hexoses are degraded to the level of CO_2 and then are reassimilated through the Calvin cycle enzymes. For the *Sargassum* community, these findings imply that organic compounds produced from the decay of algal matter can be used to support nitrogen fixation. In this manner, primary production in the nitrogen-deficient marine environment can be enhanced by reutilizing excess carbon released by the *Sargassum* plants (Hanson, 1977).

The Effect of Physical Environment on Nitrogenase Activity and Photosynthetic Oxygen Evolution in *Synechococcus* sp.

The short-term and long-term effects of light on photosynthesis, growth and pigment composition of cyanobacteria are well documented (Bryant, 1981; Van Liere and Walsby, 1982). However, there are only few studies relating nitrogenase activity to light intensity. Since light is a primary component of photosynthesis, the effect of light intensity on nitrogenase activity was determined, using unicellular cyanobacteria because this group of bacteria lacks heterocysts and thus the nitrogenase activity is sensitive to oxygen produced by photosynthesis. Among the unicellular cyanobacteria, only *Gloeocapsa* has extensively been studied with regard to the effect of light on nitrogen fixation (Gallon, 1980). Both nitrogenase and photosynthesis rates (measured as oxygen evolution) were determined at the same light intensities (Table 4). At low light of 3–10 $\mu E\ m^{-1}\ s^{-1}$, where respiration exceeds oxygen evolution, more than 70% of the maximum nitrogenase activity was measured. Light intensity up to 100 $\mu E\ m^{-2}\ s^{-1}$ increased this value to 7·3 units. This light represents only the lower range of photosynthetic oxygen evolution capacity. High light, coupled with high oxygen levels, severely inhibited nitrogenase activity. These data suggest that nitrogen fixation in this unicellular cyanobacteria effectively proceeds only at low light intensity, that is, in shaded areas as on the surface of submerged *Sargassum* blades or on other parts of the plant.

Another unicellular cyanobacterium was found to produce a highly viscous polymer and the nature of this polymer is currently under investigation (Phlips and Hansen, 1987). This strain, designated strain 0011, exhibited high rates of aerobic nitrogen fixation (Table 5) and cell yield (normally in excess of 500 mg liter^{-1}). From the standpoint of environmental physiology, this strain is highly euryhaline (salt tolerant). It is capable of sustained growth, nitrogen fixation and polymer production

TABLE 4
Effect of Light Intensity on Oxygen Evolution and Nitrogenase Activity of a Marine Unicellular *Synechococcus* sp. Strain 1

Light intensity ($\mu E\ m^{-2}\ s^{-1}$)	Net oxygen evolution[a]	Nitrogenase activity[a]
3	−15	0·8
10	−12	3·5
30	0	6·3
100	+18	7·3
200	+50	6·7
450	+96	5·8
1 500	+130	3·0

[a] Expressed as net oxygen balance, with (+) values denoting O_2 evolution and (−) denoting oxygen consumption; net oxygen evolution and nitrogenase activity are expressed as μmol product h^{-1} mg^{-1} chl.

TABLE 5
Salinity Tolerance of Growth, Carbohydrate Production and N_2 Fixation by Unicellular Cyanobacteria (Strain 0011)

Salinity	Cell yield	30-day carbohydrate production ($\mu g\ ml^{-1}$)	Nitrogenase activity[a] Light aerobic	Light anaerobic + DCMU
2·5	0·5	2		
10	2·8	155	70	145
15	3·1	134		
20	4·3	179		
25	4·2	168	160	380
30	3·8	92		
35	3·6	119	65	145
40	3·6	117		
45	3·7	96		
50	3·7	96		
55	3·7	100	32	100
60	3·3	96		

[a] Nitrogenase activity is expressed as μmol ethylene produced $h^{-1}\ mg^{-1}$ chl. Cell yield was determined after 12 days of growth and presented in millions ml^{-1}.

over a wide range of salt concentrations. These characteristics make this strain an excellent candidate for nutrient enrichment of biomass production systems.

SUMMARY

Research results with *Sargassum* support the idea that nitrogen fixation could be used to significantly reduce or replace the need for nitrogen fertilizer in biomass production, at least in certain communities. This capacity has far-reaching implications for the use of cyanobacteria in the enrichment of other biomass production systems. As a step towards this goal, a wide variety of nitrogen-fixing cyanobacteria have been isolated from the *Sargassum* community and have been subjected to tests of their environmental, physiological and biochemical characteristics. However, the isolated organisms failed to excrete any of the ammonia produced by nitrogenase into the environment. Thus, the *Sargassum* community benefits from nitrogen fixation only indirectly through subsequent mineralization of cyanobacteria.

REFERENCES

Bryant, D. A. (1981). The photoregulated expression of multiple phycocyanin species. A general mechanism for the control of phycocyanin synthesis in chromatically adapting cyanobacteria. *European Journal of Biochemistry* **119**, 425–9.

Bryceson, I. and Fay, P. (1981). Nitrogen fixation in *Oscillatoria (Trichodesmium) erythrea* in relation to bundle formation and trichome differentiation. *Marine Biology* **61**, 159–66.

Capone, D. G. and Carpenter, E. J. (1982). Nitrogen fixation in the marine environment. *Science* **217**, 1140–2.

Capone, D. G. and Taylor, B. F. (1977). Nitrogen fixation (acetylene reduction) in the phyllosphere of *Thallassia testudinum*. *Marine Biology* **40**, 19–28.

Carpenter, E. D. (1972). Nitrogen fixation by a blue-green epiphyte on pelagic *Sargassum*. *Science* **178**, 1207–9.

Carpenter, E. D. and Cox, J. L. (1974). Production of pelagic *Sargassum* and a blue-green epiphyte in the western Sargasso Sea. *Limnology and Oceanography* **19**, 429–36.

Dawes, C. J. (1974). *Marine Algae of the West Coast of Florida*, University of Miami Press, Miami, pp. 52–116.

Delwiche, C. C. (1981). The nitrogen cycle and nitrous oxide. In: *Denitrification, Nitrification and Atmospheric Nitrous oxide*, Delwiche, C. C. (ed.), pp. 1–15. Wiley-Interscience, New York.

Dugdale, R. C., Goering, J. J. and Ryther, J. H. (1964). High nitrogen fixation rates in the Sargasso Sea and the Arabian Sea. *Limnology and Oceanography* **9**, 507–10.

Gallon, J. R. (1980). Nitrogen fixation by photoautotrophs. In: *Nitrogen Fixation*, Stewart, W. D. P. and Gallon, J. R. (eds.), pp. 197–238. Academic Press, New York.

Hanson, R. B. (1977). Pelagic *Sargassum* community metabolism: carbon and nitrogen. *Journal of Experimental Marine Biology and Ecology* **29**, 107–18.

Kapulnik, Y., Kigel, J., Okon, Y., Nur, I. and Henis, Y. (1981). Effect of *Azospirillum* inoculation on some growth parameters and nitrogen content of wheat, sorghum and panicum. *Plant and Soil* **61**, 65–70.

Phlips, E. J. and Hansen, P. (1987). Growth, photosynthesis, carbohydrate production by two unicellular, marine blue-green algae, *Synechococcus* sp. (cyanophyta). *Journal of Phycology* (in press).

Phlips, E. J., Willis, M. and Verchick, A. (1986). Aspects of nitrogen fixation in *Sargassum* communities off the coast of Florida. *Journal of Experimental Marine Biology and Ecology* **102**, 99–119.

Rippka, R., Deruelles, J., Waterbury, J. B., Herdman, M. and Stanier, R. Y. (1979). Genetic assignments, strain histories, and properties of pure cultures of cyanobacteria. *Journal of General Microbiology* **111**, 1–161.

Rogers, P. A. and Reynaud, P. S. (1982). Free-living blue-green algae in tropical soils. In: *Microbiology of Tropical Soils and Plant Productivity*, Dommergues, Y. R. and Diem, H. G. (eds.). M. Nijhoff/Dr W. Junk Publishers, The Hague, The Netherlands.

Ryther, J. H. and Dunstan, W. M. (1971). Nitrogen, phosphorus, and eutrophication in the coastal marine environment. *Science* **171**, 1008–13.

Smith, R. L., Schank, S. C., Milam, J. R. and Baltensperger, A. A. (1984). Responses of *Sorghum* and *Pennisetum* species to the N_2-fixing bacterium *Azospirillum brasilense*. *Applied and Environmental Microbiology* **47**, 1331–6.

Stewart, W. D. P. (1965). Nitrogen turnover in marine brackish habitats. I. Nitrogen fixation. *Annals of Botany* **29**, 229–39.

Van Liere, L. and Walsby, A. E. (1982). Interaction of cyanobacteria with light. In: *The Biology of Cyanobacteria*, Carr, N. G. and Whitton, B. A. (eds.), pp. 10–45. University of California Press, New York and Los Angeles.

13
Manipulation of *Sargassum* for Biomass Production

J. F. PRESTON III, T. ROMEO
Microbiology and Cell Science Department, University of Florida—IFAS, Gainesville, Florida, USA

A. GIBOR and M. POLNE-FULLER
Biology Department, University of California, Santa Barbara, California, USA

INTRODUCTION

Marine macro-algae (seaweeds) have received attention throughout recorded history, mainly for their food and fertilizer value. A number of species have proved to be of significant commercial value for the production of chemical commodities. Several of the red seaweeds (Rhodophyta) have provided agar and carageenan, while several of the browns (Phaeophyta) have provided alginate, mannitol and iodine. The commercial exploitation of marine seaweeds has been the subject of several publications (Woodward, 1966; Hoppe and Schmid, 1969; Doty, 1977; Neushul, 1977).

Recent concern for the limited supply of fossil fuels has focused consideration on marine seaweeds as a renewable biomass resource for conversion to methane. Species of *Macrocystis*, the giant kelp, which occur as large stands along the Pacific coasts of North and South America, and have served as a significant commercial source of alginate in the past, have received particular attention. Several groups have studied the macro-algae in areas of growth dynamics, physiology, and genetics (Clendenning, 1964; North, 1971; Neushul, 1977; North, 1977; Wilson et al., 1977; Lobban, 1978; Kuga and Neushul, 1980; Kuwabara and North, 1980; Gerard, 1982). Conditions have been defined for the efficient fermentation of *Macrocystis* to methane (Chynoweth et al., 1980; Sowers and Ferry, 1985).

*Florida Agricultural Experiment Station Journal Series No. 5967

and the future development of an economically competitive system is anticipated. Kelp farms have been considered (Neushul, 1977; North, 1977), for providing sufficient material for the commercial production of methane. Similarly the potential of *Laminaria* species as additional marine sources of biomass has been researched (Gerard and Mann, 1979; Brinkhuis *et al.*, 1984).

In considering marine algae that might compete with various land-based crops, we should evaluate our present understanding of their genetics and mechanisms of reproduction. While there have been significant efforts over the years that have defined the sexual cycles of many species (Neushul, 1977; Van der Meer, 1981), a breeding program comparable to any one of those developed for agronomically important land-based species has not been implemented. Recent advances have been made with the application of somatic cell genetics to cloning and propagation of plant lines containing desired phenotypes, and to introduction of new phenotypes through somatic cell hybridization (Wenzel, 1980; Schieder and Vasil, 1980; Chaleff, 1981; Gleba and Evans, 1984). Such techniques have provided promising approaches for improving the genetic stock of the marine algae as well. The application of these approaches to the genetic manipulation of both the Rhodophyta (Polne-Fuller and Gibor, 1986c) and Phaeophyta (Tsung-Ci *et al.*, 1978) has occurred with some success and promise for the future.

To apply emerging technologies to genetic manipulation of marine algae, the phenotypic properties that would be desirable for the growth, harvesting, handling and processing of the algal biomass must be identified. A question in the land plant field might be posed: What are the important genes that should receive attention? Genes that control photosynthesis, cell division, composition (with respect to palatability as well as feed conversion), herbicide resistance, drought tolerance and temperature tolerance (and optima) would all be important. The research efforts directed at land plants are concerned with the identification of specific genes pertaining to each of these categories. Considering the relatively formative stage of both research interest and effort for the marine algae, the genetic basis for chemical composition and growth physiology might be given a high priority at this time.

Much of the previous relevant work on the physiology and genetics of the Phaeophyta has dealt with the cold water species. Warm water macro-algal species that might qualify as candidates for biomass production include several species of *Sargassum*. We have identified some of the compositional properties of species of the genus *Sargassum*, and have sought to define conditions for the isolation of axenic cultures of candidate species of this genus. We have sought also to prepare viable protoplasts for the development of methods involving protoplast fusion to select lines with improved quality and yields for the production of biomass for conversion to methane.

CHEMICAL COMPOSITION OF THE BROWN ALGAE

The marine algae in the family Phaeophyceae possess significant quantities of the structural acidic polysaccharide alginate which comprises as much as 23% of the annual tissue weight (dry) of *S. filipendula* (Davis, 1950). Lesser quantities of cellulose and the sulfated poly-L-fucan, fucoidin, also comprise the structural polymer fraction. The monomeric polyol, D-mannitol, and a poly-D-β(1-3)-glucan, laminarin, are also present in significant quantities as storage carbohydrates with the apparent respective roles of sucrose and starch common to many of the higher plants. Mannitol and alginate are in general the organic constituents present in the highest quantities, and represent a portion of the biomass that is readily fermented to methane (Chynoweth *et al.*, 1980; Sowers and Ferry, 1985). Relative levels of these components have been established for several different species (Table 1). Production of elevated levels of these would be desirable phenotypes to be introduced through breeding programs. Genes that code for enzymes which regulate levels of mannitol and alginate would then be candidates for manipulation.

Alginate is the predominant structural polymer in the brown algae. It comprises most of the structure of the intercellular matrices and surrounds each cell. The polyuronide structure of alginate suggests a role analogous to that of pectin, which comprises the outermost structural component of the cell walls of higher plants. As in the case of pectin, the ability of the alginate to complex with divalent metal ions is important for its ability to confer structure. The interspersion of D-mannuronan (poly(ManA)) linked β-(1-4) and poly-L-guluronan (poly(GulA)) linked α-(1-4) in alginate provides a differential ability to complex with divalent metal cations, particularly calcium. L-Guluronate residues may be derived directly from D-mannuronate residues in the intact alginate polymer through enzyme catalyzed epimerization at carbon 5, and this reaction has been observed in the brown algae, *Pelvetia* (Madgwick *et al.*, 1973), and in species of *Azotobacter* capable of synthesizing bacterial alginate (Skjak-Braek and Larsen, 1985; Larsen *et al.*, 1986). The ratio of the guluronate to

TABLE 1
Alginate and Mannitol Levels in Phaeophyta Species

Species	Alginate % dry wt	Mannitol % dry wt	References
S. fluitans	18·8	4·6– 9·5	Aponte et al (1983); Preston and Jiminez, unpublished data, University of FL
S. natans	19·9	4·4– 6·9	Aponte et al. (1983); Preston and Jiminez, unpublished data, University of FL
S. filipendula	13·3–23·5	13·0–16·0	Davis (1950); Preston and Jiminez, unpublished data, University of FL
S. wightii	21·3–31·7	1·2– 5·5	Rao (1969)
L. cloustoni	14·0–22·0	5·0–18·0	Black (1948)
M. pyrifera	14·1	18·7	Chynoweth et al. (1980)
F. vesiculosus	13·8–17·2	5·5–16·1	Black (1949), Reed et al. (1985)

mannuronate in alginate increases with the age of the tissue in several species of the brown algae (Haug et al., 1974). This indicates that the different structures of alginate define the structure of a given tissue upon interaction with divalent metal cations. Spectroscopic studies have led to a postulated ribbon-like structure for the poly(ManA) regions, and an 'eggbox' structure for the calcium-poly(GulA) complexes within the portions alginate (Grant et al., 1973). A schematic representation of the structural changes in the polymer that result from the epimerase reaction is shown in Fig. 1. A cooperative process results from the complexing of Ca^{2+} with adjacent dimers of GulA on one polymer and adjacent GulA dimers on a second polymer strand. A sequence of 20 GulA residues on a single strand is probably necessary before stable structures are formed as a result of Ca^{2+} complexes between two alginate polymers.

Cellulose has been identified as a structural component in *Fucus* zygotes (Quatrano and Stevens, 1976), where it is found as 20% of the cell wall several hours after embryo fertilization. Alginate levels were greater than 60% of the cell wall under the same conditions. Cellulose comprises 8·8% of the volatile solids of *Macrocystis pyrifera*, which corresponds to 4·8% of

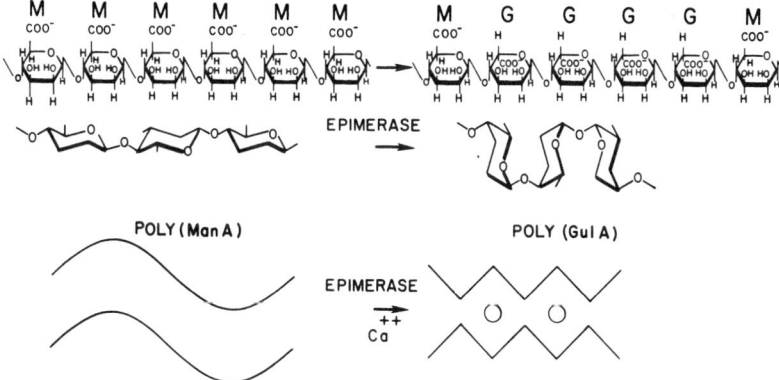

FIG. 1. Schematic representations of alginate structure, with the linear array of Haworth structures given at the top, the conformational structures given next, and the effect of calcium on the formation of complexes between two polymeric strands of alginate given at the bottom. The epimerase catalyzed conversion of β-linked D-mannuronate to α-linked L-guluronate residues in the intact polymer is shown, with the conversion of the ribbon structure of poly(ManA) to the catenated structure of poly(GulA) and the formation of the 'eggbox' structure upon the complexing of two polymer strands with Ca^{2+}.

the dry weight (Chynoweth et al., 1980). Cellulose has ranged from 1 to 10% of the dry weight of marine algae (Percival and McDowell, 1967). It is presumed that the cellulose serves the same structural role it plays in the higher plants, possibly providing a rigid layer surrounding the plasma membrane.

The sulfated polymer fucoidin is present in varying quantities, with levels found from less than 1% in *Macrocystis pyrifera* to 24% for *Fucus spirilis* and *Pelvetia canaliculata* (Percival and McDowell, 1967). After fertilization of *Fucus vesiculosus* embryos, the levels of fucoidin increase until it represents 20% of the cell wall (Quatrano and Stevens, 1976). While its function is not established, it has been suggested as a possible analog of xyloglucan of higher plant cells (Quatrano, 1982) which contributes to the structural rigidity through its strong association mediated via hydrogen bonding with cellulose (Valent and Albersheim, 1974). Fucoidin has also been proposed to comprise an extracellular mucilaginous layer, and due to its hygroscopic properties, it may serve to protect the brown algae against dehydration (McCully, 1966; Evans et al., 1973).

STRATEGIES FOR THE IMPROVEMENT OF QUALITY AND QUANTITY OF *SARGASSUM* AS A CULTIVATED CROP

The pelagic species of *Sargassum*, i.e. *S. fluitans* and *S. natans*, have been reported to comprise a significant amount of biomass, with estimates from 3·7 to 40 million Mg as an annual yield in the Sargasso Sea alone (Parr, 1939). The absence of a sexual reproductive cycle in either of these species has precluded the application of breeding approaches to improve the quality that would contribute to the efficiency and speed with which their biomass could be fermented to methane. The genetic manipulation of these species is therefore dependent on the use of the techniques of somatic cell genetics for the introduction of new phenotypes.

The genes that might be deliberately affected to improve the quality of the biomass are presently unidentified, but they may well be those that are involved in the synthesis of alginate and/or mannitol. Genes that affect the growth rate or yield of a candidate species should also be identified, although the biochemical complexity expected for the regulation of these properties suggests that the identification of many of these important genes may best be left to the research efforts that are currently applied to the improvement of agronomically important land plants.

The continued research on the development of procedures for preparing viable protoplasts followed by their regeneration into plants is thus critical for improvement of these species. With this aim in mind, two experimental strategies were adopted. One was to search for appropriate enzymes from bacteria which are associated with *Sargassum*, the other was to utilize available enzyme preparations, especially those derived from the digestive system of animals which graze on seaweeds.

BACTERIAL FLORA ASSOCIATED WITH PELAGIC SPECIES OF *SARGASSUM*

As a prelude to the isolation of viable unialgal and/or axenic cultures of *Sargassum*, tissues of actively growing pelagic species, *S. fluitans* and *S. natans*, both of which are common to the eastern coast of Florida, were examined by scanning and transmission electron microscopy. A large number of bacteria have been found on the surface of the tissues of actively growing *S. fluitans* (Preston *et al.*, 1985a). Sonication and antibiotic treatments designed to clean tissues for the successful isolation of axenic cultures of *Porphyra* (Polne *et al.*, 1980b) were only partially successful with respect to the removal of these bacteria from *Sargassum* tissues. The strong association of these bacteria with the alginate comprising the surface matrix of the algal tissues is presumably responsible for the difficulty in their removal. The removal of this outer layer appears to be required for the isolation of axenic cultures from vegetative tissues of the Phaeophyta species.

The enumeration and preliminary characterization of the bacteria associated with different species of *Sargassum* was undertaken by plating dilutions of sea-water samples containing organisms released by the sonication treatment described above. Approximately the same number of bacteria were detected upon growth on a minimal medium containing alginate as the sole carbon source as on a rich medium containing glucose and yeast extract (Table 2). While older tissues from a given specimen contain larger numbers of bacteria than the younger tissues, high numbers were obtained in all cases. The transfer of selected isolates onto a sea-water calcium alginate medium led to the demonstration that these were capable of secreting alginate depolymerizing enzymes which led to cleared zones around the individual colonies. The intensities and sizes of these zones were interpreted as a measure of the alginase activities secreted and were used as the basis for selection of strains for further study (Preston *et al.*, 1985b). The further characterization of seven alginase-secreting isolates has provided a tentative assignment of four isolates to the oxidative genus

TABLE 2
Comparison of Total and Alginate Metabolizing Bacteria Associated with *Sargassum* Species (g^{-1} of fresh tissue)

Species	Young Tissue		Old tissue	
	Rich medium[a]	Alginate medium[b]	Rich medium	Alginate medium
S. fluitans	$2 \cdot 9 \times 10^7$	$1 \cdot 8 \times 10^7$	$5 \cdot 5 \times 10^7$	$5 \cdot 5 \times 10^7$
S. natans	$2 \cdot 6 \times 10^7$	$6 \cdot 1 \times 10^7$	$7 \cdot 6 \times 10^7$	$1 \cdot 3 \times 10^8$
S. hystrix[c]	—	—	$6 \cdot 0 \times 10^6$	$1 \cdot 2 \times 10^7$
S. filipendula[d]	$4 \cdot 25 \times 10^6$	$1 \cdot 94 \times 10^8$	$1 \cdot 25 \times 10^2$	$2 \cdot 82 \times 10^8$

[a] Rich medium composition: 1% yeast extract; 1% glucose; 0·8% nutrient broth; 2% agar.
[b] Alginate medium composition: 1% Na alginate; 2% agar.
[c] Only old leaf tissue was used in the isolation of bacteria from *S. hystrix* tissue; algal specimens were not attached to substratum.
[d] Algal specimens were attached to substratum.

Alteromonas and three isolates to one or more facultative genera with a %GC inclusive in the range of 45·4–47·4.

The observation that most, if not all, of the bacteria associated with the surface of the *Sargassum* tissues utilized alginate as a sole carbon source was not surprising in view of the structural composition of these tissues as noted above. The intracellular and extracellular alginolytic enzymes of those isolates most active with respect to growth on and degradation of alginate were examined as a step toward the removal of these bacteria with the objective of obtaining axenic cultures and protoplasts. In all of the systems studied, lyase activities which effect the depolymerization of alginate through the nonhydrolytic eliminative cleavage of glycosidic linkages were quantified and compared with the activities determined by measuring the decrease in the viscosity of alginate solutions. There is a wide range in the substrate preferences of the intracellular (or cell bound) and the extracellular preparations when comparing one organism to another, or when comparing the different compartments for a given isolate (Table 3). The extracellular activities secreted by the bacteria were more active than the intracellular activities with respect to their ability to decrease viscosity (given as the reciprocal of the specific viscosity, or the specific fluidity value ϕ_{sp}). The higher the ratio of the increase in ϕ_{sp} to the formation of unsaturated nonreducing termini (measured as A^{548} with the thiobarbituric acid assay (Romeo *et al.*, 1986)), the greater the decrease in viscosity relative to the cleavage of glycosidic bonds and therefore the more random, i.e. endolytic, the mechanism of depolymerization.

Each isolate appears to secrete one or more enzymes that are unique with respect to substrate specificity. Both of the fermentative isolates, A and D, secrete enzymes, that are almost if not completely specific for poly(ManA) and the fermentative isolate G shows a marked preference for poly(GulA) over poly(ManA). The A isolate has been shown to secrete a single poly(ManA) specific endolyase which forms unsaturated dimers and trimers of mannuronate as true limit products; an unsaturated tetramer of mannuronate is also rapidly formed, although this is slowly degraded to the trimer and monomer (Romeo and Preston, 1986a,b,c). The extent of growth of this isolate on the alginate provided as carbon source and the consumption of carbohydrate, at least through the exponential phase of a batch culture, is consistent with the conclusion that the A isolate is capable of degrading alginate through the endolytic cleavage of poly(ManA) regions with the selective uptake and utilization of the mannuronate oligomers (Preston *et al.*, 1986). The intracellular alginate lyases of this organism degrade poly(GulA) and poly(ManA), and include exolyases.

TABLE 3
Substrate Specificities and Modes of Cleavage of Alginate by Intracellular and Extracellular Alginate Lyases

Isolate	Metabolism[a]	Intracellular[b]			Extracellular[c]			
		$\Delta\mu mol\ min^{-1}$		ϕ_{sp}[d]	$\Delta\mu mol\ min^{-1}$		ϕ_{sp}	
		Poly G	Poly M	Alginate	ΔA^{548}	Poly G	Poly M	Alginate	ΔA^{548}
FM	O	0·19	0·53	0·24	1·2	0·30	0·60	0·65	20·0
A	F	0·69	1·14	0·76	8·6	0·00	0·37	0·23	12·0
B	O	0·026	0·091	0·032	—	0·28	0·29	0·37	—
C	O	0·015	0·042	0·018	—	0·42	0·37	0·65	—
D	F	0·83	0·92	0·88	—	0·087	1·06	0·98	—
G	F	1·29	2·45	1·20	7·3	2·23	0·35	3·14	10·5

[a] Isolates A–G were obtained from actively growing *S. fluitans* tissues. Isolate FM was obtained from decaying *Sargassum* tissues. The characterization of these organisms has been described (Preston et al., 1985b), with the assignment of either oxidative (O) or fermentative (F) mode of metabolism.
[b] Intracellular activities were obtained after disrupting bacteria followed by partial purification to remove anionic polymers (Romeo, 1986). This fraction may include activities bound to the cell surface as well as those which are intracellular. Values given are μmoles of unsaturated product (Δ) formed min^{-1} g^{-1} (wet weight) of cells. All activities presented were calculated after subtracting activities observed in the absence of added substrate.
[c] Extracellular activities were measured in the medium after concentration and dialysis but without further purification. Values were determined as for the intracellular preparations.
[d] Slopes of straight lines obtained by plotting specific fluidity, ϕsp, against the results of TBA assays, A^{548}, measured during depolymerization of alginate, indicate relative levels of endolytic versus exolytic cleavage.

Therefore, this bacterium should have the ability to utilize GulA and/or (GulA, ManA) oligomers that might be provided by enzymes secreted by other organisms.

The G isolate, with its apparent preference for the poly(GulA) portions of alginate, has been examined further. This organism secretes one or more alginate lyases which degrade poly(ManA), poly(GulA), and native alginate; with the alginate as substrate, the unsaturated dimers and trimers are formed which correspond to some of the products which accumulate in the medium of batch cultures of this organism grown with alginate as the sole carbon source (Preston *et al.*, 1986). The G isolate appears to be an example of a fermentative bacterium which functions in a commensual relationship with other associated bacteria and which may be considered to function synergistically with respect to the complete utilization of alginate.

The ability of the extracellular alginases secreted by the A and G isolates to remove alginate from the tissues of *Sargassum* species was evaluated. The exposure of tissues of *S. filipendula* to the poly(ManA) lyase from the A isolate for either 4, 8 or 24 h released approximately half the amount of alginate (estimated from the amount of unsaturated residues generated upon treatment of the solubilized polymers with the G enzymes) that was released upon initial treatment with the G enzymes (Table 4). Pretreatment of the tissues with EDTA did not appreciably affect the amount of alginate solubilized by either enzyme preparation. In contrast, the treatment of the tissues from *S. fluitans* indicated that either enzyme preparation released the same amount of alginate, and that this release was enhanced to a significant extent, i.e. 25%, by pretreatment of the tissues with EDTA. On the assumption that the treated tissues were 90% water and that the treatment with the G enzyme preparation, either as the first or second enzyme, produces an approximately equimolar mixture of the dimeric and trimeric uronide limit products each containing a single unsaturated residue at the nonreducing terminus, the maximum amounts of alginate released were greater than 90% of the dry weight for both the *S. filipendula* (24 h exposure to the G enzyme, minus EDTA pretreatment, minus post treatment with the A enzyme), and for the *S. fluitans* (8 h exposure to the G enzyme, plus EDTA pretreatment, plus post treatment with the A enzyme). The apparent dramatic decrease in the amount of alginate detected for the 24 h exposure of the *S. fluitans* tissues may be the result of the consumption of the products by proliferating bacterial contaminants. The absence of such a decrease for the 24 h exposure of the *S. filipendula* suggests that either these tissues contained few bacterial contaminants capable of consuming the alginate degradation products, or

TABLE 4
Degradation of *S. filipendula* and *S. fluitans* Tissues by Extracellular Alginate Lyases from Bacterial Isolates A and G

			μmol nonreducing termini, $\pm$ second enzyme[d]					
			4 h		8 h		24 h	
Organism[a]	Enzyme[b]	EDTA	−	+	−	+	−	+
S. filipendula	A	−	0·036	0·017	0·066	0·155	0·132	0·237
	A	+	0·032	0·087	0·061	0·135	0·129	0·233
	G	−	0·144	0·220	0·229	0·324	0·529	0·517
	G	+	0·133	0·172	0·248	0·250	0·489	0·473
	none[c]	−	0·000	0·008	0·000	0·012	0·000	0·019
	none	+	0·000	0·010	0·002	0·018	0·009	0·006
S. fluitans	A	−	0·111	0·318	0·157	0·429	0·013	0·006
	A	+	0·146	0·424	0·207	0·587	0·019	0·008
	G	−	0·220	0·307	0·282	0·405	0·018	0·009
	G	+	0·303	0·422	0·490	0·542	0·112	0·007
	none	−	0·000	0·009	0·002	0·015	0·002	0·011
	none	+	0·002	0·019	0·002	0·021	0·016	0·015

[a] Apical foliar tissues from *S. filipendula* and *S. fluitans* were sonicated and 12·5 mg portions were finely chopped, placed in wells of a microtiter plate, and incubated 15 min in PESI lacking added Ca^{++} and Mg^{++} in the presence or absence of 1·0 mM EDTA.

[b] The PESI solutions were removed and replaced with enzyme solutions prepared from isolate A or G containing 0·0018 units of activity, as assayed with alginate as the substrate. The enzyme solutions were buffered at pH 7·8 with 0·1 M sodium hydrogen phosphate and contained 0·5 M NaCl. One unit of enzyme activity is defined as that which will catalyze the formation of one μmole of unsaturated uronate residue in 1 min at room temperature.

[c] These tissue samples were incubated in buffer with no added enzyme.

[d] At 4, 8 and 24 h the contents of wells were sampled, excluding pieces of tissue, and the content of unsaturated nonreducing terminal residues generated by the action of alginate lyase on the tissues was determined by the TBA assay. Additional samples from tissues incubated with alginate lyase activity from isolate A or with buffer solution were added to equal amounts of alginate lyase activity from isolate G, and samples from wells containing tissues incubated with enzyme from isolate G were added to equal volumes of enzyme from isolate A. These mixtures were incubated for 14 h before measuring unsaturated terminal residues.

that these tissues contained and/or secreted compounds that inhibited the growth of such contaminants. In any case, tissues from different *Sargassum* species contain alginate that is differentially susceptible to endolytic alginate lyases with different substrate specificities.

Previous studies were reported on the release of alginate from *S. filipendula* upon treatment with the extracellular poly(ManA) specific endolyase and/or the intracellular enzyme mixture containing lyase activities toward poly(GulA) and poly(ManA) structures, and with exolytic mechanisms (Romeo et al., 1986). The endolytic activity was most effective in releasing the alginate from the tissues. Considering these studies collectively, the bacterial flora associated with the actively growing *Sargassum* tissues may best be described as a consortium of oxidative and fermentative organisms which secrete predominantly endolytic alginate lyases with varying specificities for the poly(ManA) and poly(GulA) structures that comprise alginate. The different specificities of these enzymes, their ability to produce discrete dimeric and trimeric limit products, and the inability of at least some of the bacteria to assimilate all of these products, suggest this consortium to be a collection of unique organisms which evolved in a commensual configuration based upon their ability to degrade alginate and may have developed a mutual dependence upon different members. The dimeric and trimeric products would then provide carbon and energy sources for growth after assimilation and conversion to the monomer, 4-deoxy-L-erythro-5-hexoseulose uronic acid (Preiss and Ashwell, 1962a,b). A scheme for the degradation and assimilation of the alginate by these organisms is given in Fig. 2.

PREPARATION OF PROTOPLASTS FROM *SARGASSUM* SPECIES

The treatment of specimens of *S. fluitans* and *S. filipendula* with individual bacterial alginases or mixtures failed to release protoplasts or even single cells under conditions in which a significant portion of the alginate was released (Romeo et al., 1986). In none of the treatments with bacterial alginases was there significant tissue maceration observed. Experiments in which partially purified alginase was used in conjunction with the fungal cellulase complex from *Trichoderma reesei* failed to release protoplasts or single cells from *S. fluitans* (Romeo, 1986). Pretreatment with EDTA often assisted in the softening of tissues, but did not complement the alginases to the extent that single cells or protoplasts were released. It is

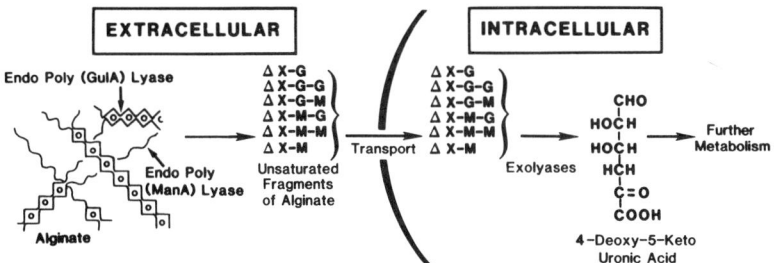

FIG. 2. Scheme for the proposed metabolism of alginate by bacteria, specifically the A and G isolates, associated with actively growing tissues of *Sargassum*. Alginate polymers may be released from the tissue by endolytic alginate lyases secreted by the bacteria and are further depolymerized by these enzymes to form trimeric and dimeric limit products, each containing the same unsaturated residue at the nonreducing terminus. These oligomers might then be assimilated by the bacteria and degraded further by exolytic alginate lyases to form 4-deoxy-5-keto-uronate, which would then be dissimilated to provide the energy requirements of the bacterial cells.

probable that other enzymes are needed for the effective removal of the cell walls from the vegetative tissues of the *Sargassum* species, and the enzymes that degrade fucoidins may be the most likely candidates for complementing the other enzymes that have been used.

TISSUE DISSOCIATION AND PROTOPLAST PRODUCTION WITH COMMERCIAL ENZYMES

Tissues of *Sargassum* free of contaminating organisms (axenic) were used according to procedures described in detail by Polne-Fuller and Gibor (1986a). In brief, the seaweed surfaces were cleaned by repeated washing in an ultrasound cleaning bath, followed by treatment with a mild antiseptic (usually a 1% solution of Betadine, an iodinated polyvinyl pyrolidone preparation), and then the treated and washed tissues were immersed in a sterile antibiotics mixture for 1 week. Subsequently the tissues were washed and subjected to sterility tests. For these the tissues were placed on sea water–agar medium enriched with nutrient broth (0·5%), yeast extract (0·8%) and glucose (0·1%) for 2 weeks at 22°C. Tissues which were free from contamination were then immersed in a liquid medium of the same composition for an additional week. It is best to test each piece of tissue individually in a separate test tube. In some seasons troublesome fungi were associated with the *Sargassum* tissues; these were controlled by the use of the fungicides Kytoconozol and Imazalil, at 10–50 mg ml^{-1} from Jannses Pharmaceutica.

Some tropical species of *Sargassum* harbored endophytic fungi which were not eliminated by the surface cleaning treatments. From such tissues the apical few millimeters of actively growing tips were often found to be free of fungi and could serve as good starting material. The claim to sterility of the material is defined by the sterility tests which were used; these were adequate for the physiological studies for which the tissues were intended.

TISSUE DISSOCIATION

A variety of commercially available crude enzyme preparations, some from the entrails of herbivorous marine animals, were used. The tissues were submerged in the enzyme solutions at room temperature, on a shaker. After treatment for about 18 h the tissues were gently dispersed in a loose-fitting glass tissue homogenizer. The effectiveness of the enzyme treatment was evaluated by the release of protoplasts and general softening of the tissues. The following enzyme preparations were tried (concentrations mg ml^{-1} are indicated in brackets):

1. Abalone powder, Sigma (30 mg ml^{-1})
2. Abalone powder, freshly prepared (10)
3. *Patella vulgata* (limpet), Sigma (10)
4. Sea urchins' entrail powder, freshly prepared (20)
5. Laminarinase, Sigma (5)
6. Cellulase, Onozuka R-10 (5)
7. Cellulase, Onozuka R-S (5)
8. Macerozyme, Calbiochem (2)
9. Pectinase Nutritional-Biochem. (5)
10. An amoeba enzyme preparation, freshly prepared (10^7 cells ml^{-1}). Various combinations of these enzymes were also tried. The enzyme solutions were sterilized through 0·22 μm Milipore filters.

Highest yields of protoplasts (3 × 10^9 g^{-1} of fresh tissue) were obtained with mixtures containing enzyme preparations No. 2 + No. 7 (Polne-Fuller, Fig. 3A, unpublished data), No. 3 + No. 7 (Fisher, unpublished data, University of California, Santa Barbara), both prepared in sea water containing 0·6M sorbitol, and No. 10 which was a crude extract of a cultured amoeba (isolated from *Sargassum muticum*), and yielded proto-

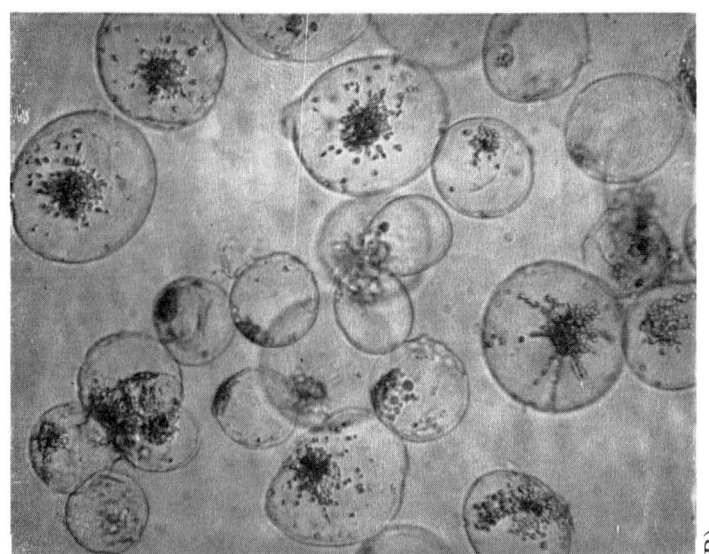
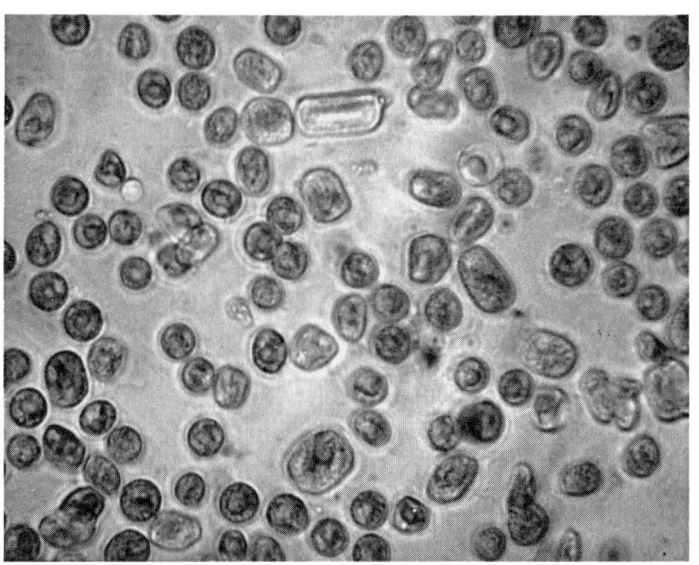

FIG. 3. Isolated protoplasts from *Sargassum* tissues. (A) dense, pigmented protoplasts from stipe epidermal (surface) tissues of *S. muticum* treated in enzyme mixture No. 2 + No. 7 for 16 h. Cell size is between 30–50 µm. (B) large, clear protoplasts from stipe cortical tissues treated in amoeba extract for 10 h. Cell size between 120 and 250 µm.

plasts from cortical tissues (Fig. 3B). Success in tissue dissociation depended on the nature of the plant material. The composition of the extracellular matrices varied with the species, the season, and the different tissues of the same plant.

Best yields of protoplasts were obtained by treating young meristematic tissues (Fisher, unpublished data, University of California, Santa Barbara). Blades, stipes, vesicles and holdfast tissues released lower yields as they got older and more differentiated. The separated protoplasts were rinsed from the enzymes by sinking through a column of PES + 0·6% sorbitol medium in 15 cm tall test tubes. Four rinses were carried out, each time removing most of the supernatant. The cells were resuspended in PES + sorbitol medium (PES, Provasoli, 1968).

WALL REGENERATION

Protoplast preparations were examined for the presence of cell walls by two criteria: (1) Exposure to low osmotic pressure (diluted sea water) caused the bursting of protoplasts, while walled cells survived. (2) Calcofluore-white stained walled cells but not protoplasts. For wall regeneration, the isolated and washed protoplasts were suspended in the above-mentioned medium composed of PES medium and sorbitol. Calcofluore stainable thin walls were observed after an overnight incubation at 18–22°C. The osmolarity of the culture medium was then gradually reduced by daily 20% dilution with PES. By the end of 1 week the cells were transferred to PES medium.

CALLUS CULTURES

Another approach to vegetative cultivation of *Sargassum* was by culturing small tissue fragments on agar solidified PES medium. Callus-type growth was obtained from such tissues (Polne-Fuller and Gibor, 1986b). Light-beige or 'clear' calli, and dark brown or 'pigmented' calli were obtained (Figs. 4A, B). Both type calli grew rather slowly and were maintained on agar medium for over 2 years.

The calli were easy to dissociate and viable cells and protoplasts were obtained from them. Gentle dissociation of calli in a loose-fitting homogenizer or by other mechanical agitation produced viable cell suspensions. Treatment of calli with abalone enzymes for 2–10 h produced viable

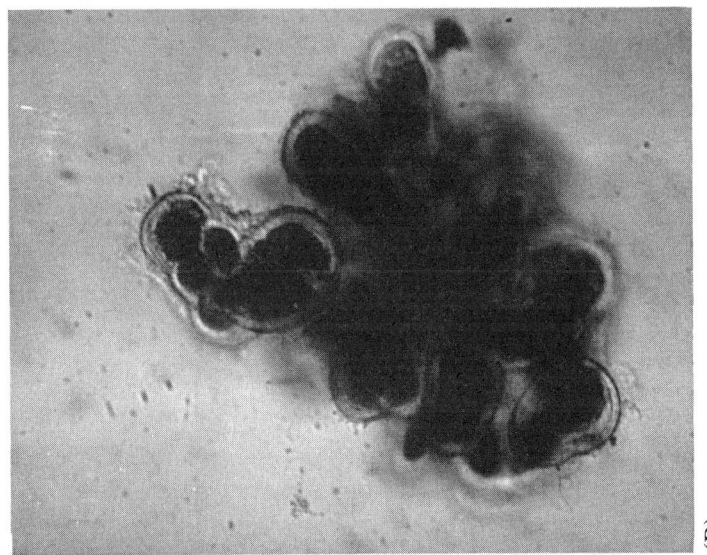
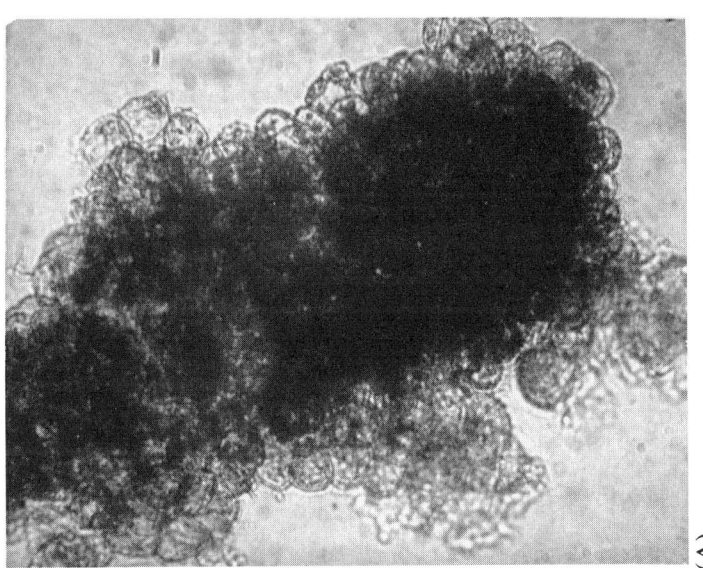

FIG. 4. Calluses of *Sargassum muticum*. (A) clear callus of *S. muticum*, originated from cortical tissue and grown on agar medium. Cell size between 80 and 150 μm. (B) Fragments of dark callus, originated from epidermal (surface) cells. The densely pigmented cell in the clusters have thick walls. Cell size between 30 and 50 μm.

FIG. 5. Regenerated plantlets, from cells similar to those in Fig. 3(B) (×1).

protoplasts. Such cells which were isolated from calli grew well in liquid cultures and developed into normal-looking plantlets after two weeks (Fig. 5).

Callus growth is also seen sometimes in nature, especially on holdfasts probably in response to abrasion or grazing. In the laboratory, calli developed from a variety of wounded tissues. Under the conditions tested, about one third of the tissue pieces derived from holdfasts or stipes produced calli. One sixth of the tissue fragments of vesicles or young sporlings produced calli under these culture conditions.

From their ability to grow and regenerate, callus cells were in a modified physiological state as compared to cells and protoplasts which were isolated directly from vegetative tissues. The results obtained on isolation of protoplasts from *Sargassum* tissues are in Table 5. Under the tested conditions, calli grew very slowly (2–4 weeks generation time) and colonies reached only 2–3 mm in diameter when divisions stopped, although cell viability persisted. Better culture conditions which may enhance division rates should be developed.

Vegetative propagation of *Sargassum* can now be accomplished by maintaining desired plant material as a callus culture and when needed, dissociating the calli into cell suspension and regenerating plantlets from them.

TABLE 5
Protoplasts from Sargassum

Tissue source	Treatment	Wall regeneration	Divisions	Differentiation
From tissues:				
Stipes	Mixed enzymes	Calcofluor	Few forming	No difference
Vesicles	for 20 hours:	positive in	clumps of	
Holdfast	No. 2 + No. 7;	15 h	2–3 mm in	
	No. 3 + No. 7;		diameter	
Meristems	No. 10			
From calluses:				
Pigmented	Mixed enzymes	Yes	Regularly	Plantlets
calluses	No. 2 + No. 7			in 1 month
	2–15 h			
Clear	Mixed enzymes	Yes	Frequent	No difference,
calluses	No. 2 + No. 7			few rhizoids
	2–4 h			

However, yet to be learned is how to induce active growth and regeneration from isolated cells and protoplasts obtained directly from vegetative tissues of *Sargassum*, and eliminating the extra step of callus preparation. Regenerated plantlets from isolated cells and protoplasts from calli are certainly encouraging.

The large number of alginase secreting bacteria associated with the actively growing tissues of the pelagic as well as the benthic species of *Sargassum* may prove to be a fortuitous discovery. Clearly, they secrete a number of endolytic alginate lyases with different substrate specificities. These may yet prove to be useful reagents in conjunction with other enzyme activities that appear to be included in the abalone preparation. Some of these bacteria may also prove important as producers of growth factors to enhance cell divisions in calli (Provasoli *et al.*, 1977).

SUMMARY

Marine Phaeophyta are considered to be a promising source of renewable biomass for methane production. The chemical composition of several species of *Sargassum* was compared to other species of Phaeophyta. Clonal breeding of dissociated cells and somatic hybridization of protoplasts have been considered for improving the seaweeds' properties as sources of readily fermentable biomass.

A large number of alginate degrading epiphytic bacteria have been found associated with actively growing *Sargassum* plants. Endolytic lyase enzymes were obtained from such bacteria. These enzymes have specificities for different portions of alginate molecules. Treatment of *Sargassum* tissues by these enzymes caused effective release of much of the alginate wall content, however, no significant release of protoplasts resulted. Protoplasts were obtained from *Sargassum* tissues by treatments with crude enzyme preparations obtained from limpets and abalone entrails, mixed with cellulase. Protoplasts from calli developed new calli and callus cells successfully regenerated plantlets. Obtaining protoplasts from calli and regenerating *Sargassum* plantlets from them is certainly within the realm of feasibility.

ACKNOWLEDGEMENTS

We are grateful for the assistance of Ms Donna Huseman for the preparation of figures and of Ms Martina Champion for the preparation of the text.

REFERENCES

Aponte de Otaola, N. E., Diaz-Piferrer, M. and Graham, H. (1983). Seasonal variation and anatomical distribution of alginic acid in *Sargassum* sp. found along the coasts of Puerto Rico. University of Puerto Rico. *Agronomy* **67**, 469–75.
Black, W. A. P. (1948). Seasonal variation in chemical constitution of some of the sublittoral seaweeds common to Scotland. Part I. *Laminaria cloustoni. Journal of the Society of Chemistry and Industry* **67**, 165–8.
Black, W. A. P. (1949). Seasonal variation in chemical composition of the littoral seaweeds common to Scotland. Part II *Fucus serratus, Fucus vesiculosus, Fucus spiralis* and *Pelvetia canaliculata. Journal of the Society of Chemistry and Industry* **68**, 183–9.
Brinkhuis, B. H., Bariani, E. C., Breda, V. A. and Brady-Campbell, M. M. (1984). Cultivation of *Laminaria saccharina* in the New York marine biomass program. *Hydrobiologia* **116/117**, 266–71.
Chaleff, R. S. (1981). *Genetics of Higher Plants*, Cambridge University Press, New York, 184 pp.
Chen, L. and Taylor, A. (1978). Medullary tissue culture of the red alga *Chondrus crispus. Canadian Journal of Botany* **56**, 883–6.
Chynoweth, D. P., Sambhurath, G. and Klass, D. L. (1980). Anaerobic digestion

of kelp. In: *Biomass Conservation Processes for Energy and Fuels*, Zaborsky,O. R. and Soter, S. (eds)., pp. 315–38. Plenum Publishing Corporation, New York.
Clendenning, K. A. (1964). Photosynthesis and growth in *Macrocystis pyrifera*. *Proceedings of the 4th International Seaweed Symposium*, deVirville, A. D. and J. Feldmann (eds.). Macmillan, New York.
Davis, F. W. (1950). Algin from *Sargassum*. *Science* **111**, 150.
Doty, M. (1977). Eucheuma-current marine agronomy. In: *The Marine Plant Biomass of the Pacific Northwest Coast*, Krauss, R. (ed.), pp. 203–14. Oregon State University Press, Corvallis, Oregon.
Evans, D. A. (1984). Genetic techniques. In: *Handbook of Plant Cell Culture*, Vol. 1, Evans, D. A., Sharp, W. R., Ammirato, P. V. and Yamada, Y. (eds.), pp. 291–321. Macmillan, New York.
Evans, D. A. and Bravo, J. E. (1984). Protoplast isolation and culture. In: *Handbook of Plant Cell Culture*, Vol. 1, Evans, D. A., Sharp, W. R., Ammirato, P. V. and Yamada, Y. (eds.), pp. 124–78. Macmillan, New York.
Evans, L. V., Simpson, M. and Callow, M. E. (1973). Sulfated polysaccharide synthesis in brown algae. *Planta, Berlin* **110**, 237–52.
Gerard, V. A. (1982). Growth and utilization of internal nitrogen reserves by the giant kelp *Macrocystis pyrifera* in a low-nitrogen environment. *Marine Biology* **66**, 27–35.
Gerard, V. A. and Mann, K. H. (1979). Growth and production of *Laminaria longicruris* (Phaeophyta) populations exposed to different intensities of water movement. *Phycologia* **15**, 33–41.
Gleba, Y. and Evans, D. A. (1984). Genetic analysis of somatic hybrid plants. In: *Handbook of Plant Cell Culture*, Vol. 1, Evans, D. A., Sharp, W. R., Ammirato, P. V. and Yamada, Y. (eds.). Macmillan, New York.
Grant, G. T., Morris, E. R., Rees, D. A., Smith, P. J. C. and Thom, D. (1973). Biological interactions between polysaccharides and divalent cations. The eggbox model. *FEBS Letters* **32**, 195–8.
Haug, A., Larsen, B. and Smidsrod, O. (1974). Uronic acid sequence in alginate from different sources. *Carbohydrate Research* **32**, 217–25.
Hoppe, H. A. and Schmid, O. J. (1969). Commercial products. In: *Marine Algae*, Levring, T., Hoppe, H. A. and Schmid, O. J. (eds.), pp. 286–367. Cram, DeGruyter and Co., Hamburg.
Kuga, V. A. and Neushul, M. (1980). The ontogeny of an abnormal specimen of *Macrocystis angustifolia* Bory (Phaeophyta). *Phycologia* **19**, 334–7.
Kuwabara, J. S. and North, W. J. (1980). Culturing microscopic stages of *Macrocystis pyrifera* (Phaeophyta) in aquil, a chemically defined medium. *Journal of Phycology* **16**, 546–9.
Larsen, B., Skjak-Braek, G. and Painter, T. (1986). Action pattern of mannuronan C-5-epimerase: generation of block copolymeric structures in alginates by a multiple-attack mechanism. *Carbohydrate Research* **146**, 342–5.
Lobban, C. S. (1978). The growth and death of the *Macrocystis* sporophytes (*Phaeophyceae, Laminariales*). *Phycologia* **17**, 196–212.
Madgwick, J., Haug, A. and Larsen, B. (1973). Polymannuronic acic C-5-epimerase from the marine alga *Pelvetia canaliculata*. *Acta Chemica Scandinavica* **27**, 3593–4.
McCully, M. E. (1966). Histological studies on the genus *Fucus*. I. Light

microscopy of the mature vegetative plant. *Protoplasma* **62**, 205–30.

Neushul, M. (1977). The domestication of the giant kelp, *Macrocystis* as a marine plant biomass producer. In: *The Marine Plant Biomass of the Pacific Northwest Coast*, Drauss, R. W. (ed.), pp. 163–81. Oregon State University Press, Corvallis, Oregon.

North, W. J. (1971). Growth of individual fronds of the mature giant kelp, *Macrocystis*. *Beihwege Nova Hedwigia* **32**, 123–68.

North, W. J. (1977). Algal nutrition in the sea—management possibilities. In: *The Marine Plant Biomass of the Pacific Northwest Coast*, Krauss, R. W. (ed.). pp. 215–30. Oregon State University Press, Corvallis, Oregon.

Parr, E. (1939). Quantitative observations on the pelagic *Sargassum* vegetation of the western North Atlantic. *Bulletin of the Bingham Oceanographic Collection* **VI**, 1–94.

Percival, E. and McDowell, R. H. (1967). *Chemistry and Enzymology of Marine Algal Polysaccharides*, Academic Press, New York, 260 pp.

Polne-Fuller, M. and Gibor, A. (1984). Developmental studies of *Porphyra* species. I. Cell differentiation and protoplast regeneration in *Porphyra perforata*. *Journal of Phycology* **20**, 609–16.

Polne-Fuller, M. and Gibor, A. (1986a). Tissue culture of seaweeds. In: *Seaweed Cultivation for Renewable Resources*, Bird, K. and Benson, P. H. (eds.). Gas Research Institute, Chicago, Illinois (in press).

Polne-Fuller, M. and Gibor, A. (1986b). Callus growth in seaweeds, induction and culture. XII International Seaweed Symposium, Sao Paulo, Brazil (in press).

Polne-Fuller, M. and Gibor, A. (1986c). Calluses, cells, and protoplasts in studies towards genetic improvement of seaweeds. *2nd International Symposium of Genetics in Aquaculture*. Davis, California.

Polne-Fuller, M., Biniaminov, M. and Gibor, A. (1984). Vegetative propagation of *Porphyra perforata*. *Hydrobiologia* **116/117**, 308–13.

Polne-Fuller, M., Saga, N. and Gibor, A. (1986). Algal cell, callus, and tissue cultures, and selection of algal strains. *Nova Hedwigia* **83**, 30–6.

Preiss, J. and Ashwell, G. (1962a). Alginic acid metabolism in bacteria. I. Enzymatic formation of unsaturated oligosaccharides and 4-deoxy-L-erythro-5-hexoseulose uronic acid. *Journal of Biological Chemistry* **237**, 309–16.

Preiss, J. and Ashwell, G. (1962b). Alginic acid metabolism in bacteria. II. The enzymatic reduction of 4-deoxy-L-erythro-5-hexoseulose uronic acid to α-keto-3-deoxy-D-gluconic acid. *Journal of Biological Chemistry* **237**, 317–21.

Preston, J. F., Romeo, T., Gibor, A. and Polne-Fuller, M. (1985a). Investigations of *Sargassum* species for bioconversion to methane: mannitol levels, temperature requirements, and protoplast formation. 1984 International Gas Research Conference, pp. 567–79. Government Institute, Rockville, Maryland.

Preston, J. F., Romeo, T., Bromley, J. C., Robinson, R. W. and Aldrich, H. C. (1985b). Alginate lyase-secreting bacteria associated with the algal genus *Sargassum*. *Developments in Industrial Microbiology* **26**, 727–40.

Preston, J. F., Romeo, T. and Bromley, J. C. (1986). Selective alginate degradation by marine bacteria associated with the algal genus *Sargassum*. *Journal of Industrial Microbiology* **1**, 235–44.

Provasoli, L. (1964). Symbiotic relationship between micro-organisms and seaweeds. *American Journal of Botany* **51**, 681.
Provasoli, L. (1968). Media and prospects for the cultivation of marine algae. In: *Cultures and Collections of Algae*, Watanabe, A. and Hattori, A. (eds.). pp. 63–75. Japan Society of Plant Physiology, Tokyo.
Provasoli, L., Pintner, I. J. and Sampathkumar, S. (1977). Morphogenetic substances for *Monostroma oxyspermum* from marine bacteria. *Journal of Phycology* **13**, 56. (Supplement.)
Quatrano, R. (1982). Cell-wall formation in *Fucus* zygotes: a model system to study the assembly and localization of wall polymers. In: *Cellulose and Other Natural Polymer Systems. Biogenesis, Structure, and Degradation*, Brown, R. M., Jr (ed.), pp. 45–59. Plenum Press, New York.
Quatrano, R. S. and Stevens, P. T. (1976). Cell wall assembly in *Fucus* zygotes. *Plant Physiology* **58**, 224–31.
Rao, M. U. (1969). Seasonal variations in growth, alginic acid and mannitol contents of *Sargassum wightii* and *Turbinaria conoides* from the Gulf of Mannar, India. In: *Proceedings of the Sixth International Seaweed Symposium*, Margalef, R. (ed.), pp. 580–4, Pergamon Press, New York.
Reed, R. H., Davison, F. R., Foster, J. A. and Foster, R. (1985). The osmotic role of mannitol in the Phaeophytaian appraisal. *Phycologia* **24**, 35–47.
Rees, D. A., Morris, E. R., Thom, D. and Madden, J. K. (1982). Shapes and interactions of carbohydrate chains. In: *The Polysaccharides*, Vol. 1, Aspinall, G. O. (ed.). pp. 194–290, Academic Press, New York.
Romeo, T. (1986). Catalytic and structural properties of alginate lyases from bacterial epiphytes of *Sargassum*. Ph.D. dissertation, University of Florida, Gainesville.
Romeo, T. and Preston, J. F. (1986a). L.C. analysis of the depolymerization of (1+4)-β-D-mannuronan by an extracellular alginate lyase from a marine bacterium. *Carbohydrate Research* **153**, 181–93.
Romeo, T. and Preston, J. F. (1986b). Purification and structural properties of an extracellular (1-4)-β-D-mannuronan specific alginate lyase from a marine bacterium. *Biochemistry* **25**, 8385–91.
Romeo, T. and Preston, J. F. (1986c). Depolymerization of alginate by an extracellular alginate lyase from a marine bacterium: substrate specificity and accumulation of reaction products. *Biochemistry* **25**, 8391–6.
Romeo, T., Bromley, J. C. and Preston, J. F. (1986). Alginate lyases of varying substrate specificity from marine bacteria. In: *Biomass Energy Development*, Smith, W. H. (ed.), pp. 303–20. Plenum Press, New York.
Saga, N., Uchida, T. and Sakai, Y. (1978a). Clone *Laminaria* from Single Isolated Cells. *Bulletin of the Japanese Society of Fisheries* **44**, 87.
Saga, N., Motomura, T. and Sakai, Y. (1978b). Induction of callus from the marine brown Alga *Dictyosiphon foeniculaceus*. *Plant and Cell Physiology* **23**, 727–30.
Saga, N., Polne-Fuller, M. and Gibor, A. (1986). Protoplasts from seaweeds: production and fusion. In: *Proceedings of Biotechnology of Algal Biomass Production*. University of Colorado Press, Boulder, Colorado (in press).
Schieder, O. and Vasil, I. K. (1980). Protoplast fusion and somatic hybridization. *International Review of Cytology*, Supplement **11B**, 21–46.

Skjak-Braek, G. and Larsen, B. (1985). Biosynthesis of alginate: purification and characterization of mannuronan C-5-epimerase from *Azotobacter vinelandii*. *Carbohydrate Research* **139**, 273–83.

Smidsrod, O. and Haug, A. (1965). The effect of the divalent metals on the properties of alginate solutions. I. Calcium ions. *Acta Chemica Scandinaviea* **19**, 329–40.

Sowers, K. R. and Ferry, J. G. (1985). Characterization of a marine methanogenic consortium. 1984 International Gas Research Conference, pp. 527–36. Government Institutes, Rockville, Maryland.

Tsung-Ci, F., Chi-Hsun, T., Yu-Lin, O., Chin-Chin, T. and Ten-Chin, C. (1978). Some genetic observations on the monoploid breeding program of *Laminaria japonica*. *Scientia Sinica* **21**: 401–10

Valent, B. S. and Albersheim, P. (1974). The structure of plant cell walls. V. On the binding of xyloglucan to cellulose fibers. *Plant Physiology* **4**, 105–8.

Van der Meer, J. (1981). Genetics of *Gracilaria tikvahiae* (Rhodophyceae). VII. Further observations of mitotic recombination and the construction of polyploids. *Canadian Journal of Botany* **59**: 787–92.

Wenzel, G. (1980). The potential and limits of classical genetics in plant breeding. In: *Plant Cell Cultures: Results and Perspectives*, Sala, F., Parisi, B., Cella, R. and Ciferri, O. (eds.). Elsevier/North-Holland Biomedical Press, Amsterdam.

Wilson, K. C., Haaker, P. and Hanan, D. (1977). Kelp restoration in southern California. In: *The Marine Plant Biomass of the Pacific Northwest Coast*, Krauss, R. W. (ed.), pp. 183–202. Oregon State University Press, Corvallis, Oregon.

Woodward, F. N. (1966). The seaweed industry of the future. 5th International Seaweed Symposium, Young, E. G. and McLachlen, J. L. (eds.). pp. 55–71. Pergamon Press, Oxford.

14

Alternative Production Systems: Root and Other Herbaceous Crops

S. K. O'HAIR
Tropical Research and Education Center, University of Florida—IFAS, Homestead, Florida, USA

G. H. SNYDER
Everglades Research and Education Center, University of Florida—IFAS, Belle Glade, Florida, USA

J. M. WHITE
Central Florida Research and Education Center, University of Florida—IFAS, Sanford, Florida, USA

S. M. OLSON
North Florida Research and Education Center, University of Florida—IFAS, Quincy, Florida, USA

and

L. S. DUNAVIN
Agricultural Research and Education Center, University of Florida—IFAS, Jay, Florida, USA

INTRODUCTION

Root and herbaceous crops have many qualities which make them of potential value for biomass production. The root crops are valued for their ability to concentrate starches or sugars in storage organs at or below ground level, providing for a flexible harvest schedule. The others are known for their ability to produce an abundance of biomass in relatively short time periods, or are most productive during periods of cool weather when other crops are dormant or not grown. Thus, they can be utilized to

TABLE 1
Root and Herbacous Crops Evaluated for Biomass Potential in Florida

Genus	Species	Cultivars (Locations)[a] [Potential][b]
Alocasia	macrorrhiza	Giant taro (1)[L]
Amorphophallus	rivieri	Elephant yam (1)[L]
Beta	vulgaris	Korsroe (4)[L], Hugin (4)[G], Krake (4)[L], Kyros (4)[G], Meka (4)[L], Monara (4)[L], Giant half sugar mammoth long red (5)[G], Giant half sugar rose (4)[G], US H-11 (5)[G], Peroba (5)[L], Monobamba (2)[G], Monoblane (1)[L], Monorosa (5)[G], Capax (1)[L], Bario (1)[L], Barwi (1)[L], Barsien (1)[L]
Brassica	alba (mustard)	Florida broadleaf (3)[L], Savannah (2)[L], Tendergreen (3)[L], Giant curled (1)[L],
	napobrassica (rutabaga)	Purple top (4)[L], Bangholm fama (2)[L], Magnificent (1)[G]
	napus (rape)	Winfred forage (3)[E], Emerald (3)[G], Dwarf essex (1)[G]
	oleraceae (kale)	Midas (3)[E], Siberian (1)[G], Vates (2)[L], Maris kestrel (2)[L], Debrian (2)[L], Marrow stem (2)[L]
	rapa (campestris) (turnip)	Civasto (3)[E], White knight (4)[E], Purple top (4)[L], Tyfon (3)[L], Just right (1)[G], Jobandi (1)[G], Cyclone (2)[G], Jobe (1)[G], Sirus (1)[L], Samson (1)[L], Marko (1)[L], Shogoin (3)[L], Royal crown (3)[L], White egg (1)[L], G > W > 21 (1)[L], Kowai kabu (1)[E], Frisia (1)[L]
Cichorium	intybus	Rouge de Verone (1)[L], Asparagus (1)[E]
Colocasia	esculenta	Lehua maoli (4)[L], Malanga islena (3)[G], Bun long (3)[E], Sawa bastora (3)[L], Senorita (3)[L], Magaulusima (3)[L], Sawa puetata (3)[L], Keoni (3)[G], Blanc (3)[L], Souf (3)[L], Common (3)[L],

maintain a year round supply of fresh biomass. Many were selected in breeding programs for their food qualities; therefore, yield maximization may have been a low priority or negative selection factor. Others have been selected for their value as animal feed, being highly productive in minimal input situations. Thus, these selections are more likely to have potential in biomass production systems.

Trials evaluating the biomass production of potential bioenergy crops

TABLE 1—contd.

Genus	Species	Cultivars (Locations)[a] [Potential][b]
Cucurbita	foetidissima	Noir (3)[L], Chi-sen (1)[G], Costa Rica (1)[L], Aquatilis (4)[L] Buffalo gourd (1)[L]
Cyrtosperma	chamissonis	Swamp taro (1)[L]
Dausus	carota	Egmont gold (4)[G], Holmes improved (4)[L], Taranaki strong top (4)[G], 25-9X QM (1)[L], Waltham hicolor (4)[L], Danvers 126 (1)[G], Blanche a collect vert hors-terre (3)[L], Juane obtuse du doubs (3)[L], Labbericher (3)[L]
Helianthus	tuberosus	Sunchoke (9)[L]
Ipomoea	batatas	Georgia Jet (5)[G], Centennial (5)[E], Jewel (5)[G], Red Jewel (4)[G], Nugget (2)[L], Georgia red (2)[L], Rose centennial (1)[G], Jasper (1)[G], Julian (1)[L], Gold rush (1)[L], White star (1)[L], Acadian (1)[L], Porto Rico (1)[L], Morado (4)[L], Ga. TG-3 (2)[G], Rojo blanco (2)[L], Travis (1)[L], Picadita (1)[L], Canary (1)[L], White (1)[L], W-119 (2)[L], W-115 (2)[L], W-125 (2)[L], W-149 (2)[L]
Manihot	esculenta	CMC-40 (8)[E], M. Col-1684 (8)[L], HMC-2 (8)[L], CMC-92 (8)[L], Senorita (1)[L]
Pachyrrhizus	erosus	Yam bean (1)[L]
Raphanus	sativus	Silentina (1)[L], White celestial (1)[L]
Tragopogon	porrifolius	Mammoth sandwich island (1)[L]
Xanthosoma	sagitifolium	South Dade white (1)[L]

[a] (Locations) = number of locations where the crop was evaluated.
[b] [Potential] = potential of the crop in biomass production L = low, G = good, E = excellent.

covering 16 genera, 20 species and over 116 cultivars (Table 1) were established among 15 research centers of the University of Florida, Institute of Food and Agricultural Science, representing widely varying ecosystems (O'Hair, 1982; O'Hair et al., 1983a; Snyder and O'Hair, 1986). When possible, several cultivars were evaluated within each crop for assessing potential for genetic improvement. Since plantings were established with bioenergy production as a goal, energy inputs were kept to a

TABLE 2
Yields, Season of Growth, Time to Harvest and Planting Density of Herbaceous and Root Crops Considered for Biomass Production in Florida

Crop	Planting Density ($1000\ ha^{-1}$)	Season	Days to harvest	Plant biomass yield $Mg\ ha^{-1}\ crop^{-1}$	$kg\ ha^{-1}\ day^{-1}$
Cassava	10	Spring	300	15·3	51
Chicory		Fall/winter	120	5·6	133
Cruciferous					
Kale	48	Fall/winter	80–100	11	142
Radish	800	Fall/winter	100	4–10	38
Rutabaga	300	Fall/winter	80–100	2·9	29
Turnip	300	Fall/winter	100–130	10	120
Fodder beet	300	Fall/winter	160–220	6–13	59
Jerusalem artichoke		Spring	160	10	62
Sweet potato	26	Spring/summer	130–160	22–28·5	180
Taro		Spring	180–240	15–35	145

minimum. Harvesting of the most promising crops was sequential during seasonal development to identify the peak in yield potential.

Yields, season of growth, time to harvest and planting densities varied among the crops evaluated (Table 2). In addition, culture and advantages vary. Therefore, with the exception of the cruciferous crops, which are similar in production characteristics, each will be discussed separately.

AROIDS

Plants in this group are members of the Araceae family and include taro (*Colocasia esculenta* (L.) Schott.), cocoyam (*Xanthosoma* spp.), and elephant yam (*Amorphophallus rivieri* Durieu). Taro, exemplary of these crops, is the only crop in this family which has promise for bioenergy production in the US. Taro, also known as malanga islena, dasheen and eddoe, grows best during warm weather. The preferred time to plant is in early spring and to harvest is before freezing temperatures kill leaves. Taro can be grown as a perennial; however, yields per unit area and time are highest during the first year of growth. The cultivars are divided into two basic varietal groups: (1) *C. esculenta* var. *esculenta* which produces a large main corm (enlarged stem) and a few (4–8) cormels (outgrowths from the corm), and (2) *C. esculenta* var. *antiquorum* which produces many small corms or cormels as enlarged side shoots from the parent corm. Some cultivars produce long stolons on the soil surface during certain stages of plant growth. In general, however, stoloniferous types have less corm and cormel production than nonstoloniferous types, and probably have received less attention because they are less suitable for food production. A stoloniferous type (*C. esculenta* var. *aquatilis* Hassk.) is naturalized in central and southern Florida and along freshwater bodies in Louisiana. Large biomass yield differences have been recorded in Florida among cultivars, suggesting cultivar collection and evaluation is a promising means for improving biomass yields. For example, in cultivar trials conducted in 1984 and 1985, total biomass production by 'Bun long' was approximately 2–4 times that of 'Lehua maoli' (Snyder and O'Hair, 1986; Snyder, G., unpublished data). With a plant spacing of 20 cm, total 'Bun long' biomass production exceeded 35 Mg ha^{-1}.

Tolerance to flooding is one of the important attributes of taro; although flooding is not an absolute requirement. However, they do grow best in soil that is intermittently or continuously flooded (Fig. 1). Although plants respond to fertilization, they can produce a crop on infertile soils that are not productive for other crops. In Florida, evapotranspiration from flooded taro plots totaled approximately 1200 mm during the 300 day cropping period (Shih and Snyder, 1985). Evapotranspiration from flooded taro plots decreased in relation to open pan evaporation as leaf area index increased. For most nonflooded crops the reverse occurs.

Planting can be mechanized, utilizing sets comprised of a short section of the main stem and 1–2 cm of the attached petiole. Pieces of corms or cormels can also be used as propagules; however, planting schemes utilizing machinery have not been developed.

Taro plants grow up to 2 m high and form a starch-filled corm at or below the soil surface. The stoloniferous types may be preferred in a biomass production system, since replanting steps might be eliminated once the crop is established. High density plantings (20–30 cm row-spacings) produced higher biomass yields than plants spaced 60 cm (Fig. 2). This presents a problem for cultivars that do not readily produce stolons, since large numbers of propagules are needed for crop establishment. Close spacing also suppressed corm enlargement and decreased the number of cormels, but biomass on an area basis was greater and maximum yield occurred earlier with close spacing (Snyder and O'Hair, 1986; Snyder, G., unpublished data).

FIG. 1. Flooded field of taro in the Everglades.

A method that uses cumulative leaf area index to determine the time of maximum biomass yield, regardless of plant spacing, has been described for 'Lehua maoli' taro (Shih and Snyder, 1984). The first phase of taro growth largely involves production of leaves. Later (July in south Florida) corms begin to enlarge and cormels form. As the underground parts develop, leaf growth is reduced. By November, in southern Florida, the leaf canopy of March-planted taro is noticeably reduced from that observed in June and July. In one study, total top growth declined from 4800 kg ha^{-1} in July to 1650 kg ha^{-1} in November, while total plant biomass remained fairly constant over the same time period because of corm and cormel development (Snyder and O'Hair, 1986). Corms and cormels contain more dry matter (20–25%) and more starch than other plant parts. The stoloniferous types that produce little in the way of below ground biomass nevertheless may be useful for leaf production. Leaves can be harvested several times during a growing season without seriously depleting the stand (Snyder and O'Hair, 1986), whereas, harvest of corms and cormels generally will necessitate crop replanting. Taro cultivars may exist that produce acceptable quantities of corms and cormels and stolons to

FIG. 2. Effect of plant spacing (20, 40, 60 cm) on relative size of 'Lehua maoli' taro corms (left) and on number of cormels (right) per six-plant plot.

facilitate reestablishment of the crop if some undisturbed material is left in the field for crop regeneration.

Taro culture is slightly different from the other crops since conditions of periodic flooding are preferred. Where possible, taro should be kept flooded to a few centimeters depth, although it appears to suffer little from intermittent periods without flooding provided that adequate soil moisture is maintained. Nevertheless, genetic variation in ability to tolerate continuous flooding has been observed, with some cultivars growing poorly under a regime of continuous flooding. Flooding also aids in weed control. Phosphorus and nitrogen, where needed, should be applied early in the growth cycle. Nitrogen applied late in the growth cycle can be expected to spur leaf production at the expense of corms and cormels. Taro has a fairly high requirement for potassium, and in a number of studies has responded well to lime, presumably because it has a high calcium requirement (Sunell and Arditti, 1983). In a south Florida organic soil, taro biomass yield was increased by 32 and 36% by pre-plant phosphorus fertilization at 50 and 127 kg ha^{-1}, respectively, but not by nitrogen or potassium (Snyder, G., unpublished data). Some cultivars of taro are fairly tolerant of salinity (de la Pena, 1983). Few pest and disease problems are known in the southern US. Weeds, probably the greatest problem, can be suppressed to some degree by early flooding and by close plant spacing. Harvest-timing is not critical as long as it is accomplished before or shortly after above-ground parts are killed by frost.

The ability of taro to produce a crop in a variety of soils including wetland sites unsuitable for many other crops is a major advantage. Other advantages include: both the leaves and stems are potentially convertible to methane, few pest problems are known, culture stages from planting to harvesting are rather flexible, and mechanized harvesting is possible.

Production of taro is limited to Florida and southern coastal areas where the growing season is long. Tolerance of low moisture soils and propagation by vegetative means are major restraints. The lack of breeding and selection programs for producing cultivars that are high in biomass production with minimal inputs is a problem needing research.

CASSAVA

Cassava (*Manihot esculenta* Crantz) is also known as yuca, tapioca and manioc in the tropics. There are a few closely related species that grow in the US. However, none has been thoroughly evaluated for biomass production.

Cassava is planted at the start of the warm season as rains begin and grows for 8 months in northern Florida or up to 10 months in the southern parts before harvesting. The most rapid growth occurs during the warmest months. Since it is a perennial shrub, it can remain unharvested for more than one season in the southern parts of Florida.

Well-drained soils are preferred for cassava; although it can tolerate occasional periods of flooding lasting several hours (O'Hair, 1982). Long periods of flooding can result in root rot and major yield loss. Nearly all soil types, including infertile sandy soils, support good growth of cassava.

Cassava is propagated by using portions of mature stem, usually 25 cm in length, placed in moist soil with at least the lower 2/3 covered. Planting densities greater than 10 000 cuttings ha^{-1} did not result in higher yields. Dense plantings add to costs by requiring additional planting material and complicating mechanization, a practice not yet perfected.

Plants usually grow to shrub, 2 m or more high, with enlarged starch-filled roots forming during the latter parts of the growing season and continuing to enlarge from one season to the next. On a dry weight basis, the starch content decreased from 96 to 45% while fiber content increased as the roots aged (O'Hair, 1984).

Weed control is required for the first 2 months or until a full leaf canopy has developed; afterward, little attention is needed. Fertilization at a rate of 90 kg N and 180 kg P and K ha^{-1} is adequate for the season and irrigation is usually not necessary. Pests and diseases common to the US can be controlled by utilizing resistant clones such as 'CMC-40'.

The main advantages of cassava are its high root starch content which averages 30% (O'Hair et al., 1983a) and its high yield potential of as much as 9·3 Mg of starch-filled roots ha^{-1} and stem and leaf production of 6 Mg ha^{-1} (O'Hair, 1982; O'Hair et al., 1983a). Cassava is easy to grow and reliable in its production, especially on marginal soils where it outperforms other crops. Untapped genetic material and breeding potential make this a crop worthy of continued attention.

One of the main disadvantages of cassava is the difficulty in maintaining planting material through the winter, since the plants are killed by freezing temperatures. In addition, the bulkiness and perishability of the propagules make transportation and storage a problem. The inability to compete with weeds during early growth stages, lack of mechanization, perishability of roots after harvest, and long growing seasons are also problems.

CHICORY

Chicory (*Cichorium intybus* L.) grows best during cool weather. The crop is planted from seed which should be tested for viability prior to planting. Plants grow and are managed in a fashion similar to that of leaf lettuce. The tap root enlarges as carbohydrates are stored during later growth stages. Weeds must be controlled during early plant growth stages. Plants respond to fertilization, and irrigation is essential during periods of unusually low rainfall.

The greatest advantage of chicory is its ability to produce large quantities of biomass during cool periods. Total plant yields at Homestead have been as high as 16 Mg ha^{-1} after 119 days of growth (O'Hair et al., 1983a). Uneven stands due to poor seed germination and maintaining good weed control during early growth stages are problems. Lack of extensive trials as a biomass crop limits recommendations for chicory.

CRUCIFEROUS CROPS

The plants in this group are in the family Cruciferae and include: kale, mustard, radish, rape, rutabaga and turnip. They grow best during the cooler months, with the older plants having varying degrees of cold hardiness and tolerance. Planting is usually done in late summer or early spring after danger of frost has passed. Seeds are planted at densities ranging from 48 000 to 800 000 plants ha^{-1} for kale and radish, respectively. Seedlings in moist soil grow quickly to mature plant heights of 0·5 m or more. Turnip, rutabaga, radish and related hybrids produce enlarged stems, while rape, mustard and kale produce an abundance of leaves on short stems (Janick, 1963). When a portion of the stem and roots are left intact at harvesting, plants resprout to produce a second crop.

The plants in this group are the highest yielding of the cool season crops with turnip and forage rape cultivars producing between 10 and 11 Mg biomass ha^{-1} within 80 days after planting (DAP) (O'Hair et al., 1986). Rapid growth in a relatively short time period during the cool season allows them to fill a production gap. Multiple cropping is possible in the southern US. Since interspecific hybrids are possible, the potential to develop high-yielding biomass hybrids appears promising.

Their use in the southern US is restricted to the fall and winter months and to locations where irrigation can be applied. Young plants do not tolerate freezing temperatures. Insects and foliage diseases can be a problem.

FODDER AND SUGARBEET

Fodder beet (*Beta vulgaris* L.) is also known as beetroot, mangel and mangel-wurzel. Sugarbeet is a selection within this species. Their growth requirements are similar to the crucifers. In most parts of the southern US, planting for a fall or early winter crop provides the highest yields since long cool growing seasons are preferred.

Fodder beets are planted as seeds which can be divided into mono-germ and poly-germ, whereby, the former will produce single plants and the latter may produce several plants at each seed drop, the preferred situation, since seed costs are lower.

Plants have been selected for high biomass production and storage of sugars in tap roots. However, the yields in Florida have not been favorable in comparison to the cruciferous crops. The highest yields of 13 Mg ha^{-1} by

216 DAP were recorded in northern Florida from 'US H 11' sugarbeet (O'Hair, 1982) while the highest yield in southern Florida was 7·4 Mg ha^{-1} (O'Hair et al., 1983a). Because of warm humid conditions, poor storage, rapid rot and deterioration of roots, and foliar diseases, fodder beets are not well adapted to production in the southern parts of the US.

FODDER CARROT

Fodder carrot (*Daucus carota* L.) production requirements are similar to chicory. High yields of roots and availability of high-yielding hybrids are advantages although seeds costs are high. Yields are not consistently good. Highest yields were at Sanford, in central Florida, with approximately 11 Mg ha^{-1} by 177 DAP (O'Hair, 1982); at other locations yields were less than 4 Mg ha^{-1} (O'Hair, 1982; O'Hair et al., 1983a). Susceptibility to leaf diseases and nematode damage reduces yields.

JERUSALEM ARTICHOKE

Jerusalem artichoke (*Helianthus tuberosus* L.) is propagated vegetatively with rhizomes in the spring after danger of frost. Plants grow best in well-drained soil of medium to high fertility. Moist soil or irrigation soon after planting prevents desiccation, a major problem in Florida (O'Hair, 1982).

Plants grow to heights of 2 m forming erect shrubs, storing inulin in rhizomes. Crop rotation is essential and it is not recommended to plant after a crop that is susceptible to southern blight (*Sclerotium rolfsii* Sacc.). The disease problems have reduced yields to no production at all in many trials in Florida. In one apparently disease-free trial at Homestead, yields were less than 10 Mg ha^{-1} (O'Hair, 1982). Because chemical control would not be cost effective and genetic resistance is not known, this crop cannot be recommended for biomass production in the southern US.

SWEET POTATO

Sweet potatoes (*Ipomoea batatas* (L.) Lam.) grow best during periods of warm weather, being propagated vegetatively after the danger of freezing has passed. Plants prefer well-drained soil and do not tolerate flooding.

Planting density can range up to 26 000 plants ha^{-1}; thus, the cost of planting material can be a major expense. Sweet potatoes grow as vines on the ground, producing enlarged roots filled with starch and sugar by 130 DAP (O'Hair et al., 1986).

Fertilizer regimes, similar to those recommended for cassava, and early weed control are needed. Harvesting can begin between 130 and 220 DAP, before freezing conditions arrive; but, for efficient use of land it is best to harvest between 130 and 160 DAP (O'Hair et al., 1986). In this case, yields can approach 22 Mg ha^{-1}. Yields have varied among the clones, with some performing better in cooler weather than others (O'Hair et al., 1983b). Clones also vary in maturity characteristics with early maturing clones providing the opportunity for double cropping in regions with a long growing season. 'Centennial' demonstrates early maturing characteristics in comparison with 'Morado' a late maturing sweet potato (O'Hair et al., 1986).

Sweet potato advantages include high concentrations of starches and sugars in the roots, and yields in the warm season crops approach the maximum expected biomass production for plants (O'Hair et al., 1983a,b). Unfortunately, crop establishment is labor intensive, since propagule production and collection is not mechanized. For that reason, tissue culture propagation and 'gel-seeding' are being explored by others (Liu and Cantliffe, 1984). Sweet potato weevil (*Cylas formicarius elegantus* Summers) is a major problem in the southern US causing considerable yield loss. Chemical control of this insect is possible, but probably not cost effective in a biomass production system. Plants do not tolerate flooding or freezing temperatures.

SUMMARY

Sweet potato and several of the cruciferous crops have consistently shown the most promise for biomass production among those we evaluated. When grown in succession, two crops of sweet potato and one of turnip can produce up to 41 Mg biomass ha^{-1} yr^{-1}. Cassava and taro are lower yielding; however, they can provide moderate yields in ecosystems that are not suited for other bioenergy crops. Research needs for the crops that are vegetatively propagated include breeding and selection for types that can be grown from botanical seed or development of cost-effective 'seeding' of somatically derived plants. Crucifers more tolerant of warm temperatures that produce higher yields under low fertility conditions are needed. The sweet potato weevil and the high cost of propagation are the major problems with sweet potato which otherwise is an attractive biomass crop. For the most part, these objectives are readily obtainable once research efforts are centered on these goals.

REFERENCES

de la Pena, R. S. (1983). Agronomy. In: *Taro*, Wang, J.-K. (ed.), pp. 167–79. University of Hawaii Press, Honolulu, Hawaii.
Janick, J. (1963). *Horticultural Science*, pp. 535–6. Freeman and Company, San Francisco, California.
Liu, J. R. and Cantliffe, D. J. (1984). Somatic embryogenesis and plant regeneration in tissue cultures of sweetpotato (*Ipomoea batatas* Poir.). *Plant Cell Rept.* **3**, 112–15.
O'Hair, S. K. (1982). Root crop evaluation, selection and improvement in Florida for energy applications. In: *Energy from Biomass and Wastes Symposium*, Vol. 6, pp. 135–65. Institute of Gas Technology, Chicago, Illinois.
O'Hair, S. K. (1984). Cassava root qualities and yield in a subtropical environment. In: *Root Crops in the Caribbean*, Dolly, D. (ed.), pp. 161–6. Faculty of Agriculture, University of the West Indies, St. Augustine, Trinidad.
O'Hair, S. K. (1986). Edible aroids: taxonomy and horticulture. *Horticultural Reviews* (in press).
O'Hair, S. K., Snyder, G. H. and Morton, J. F. (1982). Wetland taro: a neglected crop for food, feed and fuel. *Proceedings of the Florida State Horticultural Society* **95**, 367–74.
O'Hair, S. K., Locascio, S. J., Forbes, R. B., White, J. M., Hensel, D. R., Shumaker, J. R. and Dangler, J. M. (1983a). Root crops and their biomass potential in Florida. *Proceedings of the Florida Soil Crop Science Society* **42**, 13–17.
O'Hair, S. K., McSorley, R., Parrado, J. and Matthews, R. F. (1983b). The production and qualities of Cuban sweetpotato cultivars in Florida. *Proceedings of the American Society of Horticultural Science Tropical Region* **27B**, 35–41.
O'Hair, S. K., Dangler, J. M., Everett, P., Forbes, R. B., Locascio, S. J., Olson, S. M., Shumaker, J. R. and White, J. M. (1986). Cruciferous and root crops for year-round biomass production. In: *Biomass Energy Development*, Smith, W. H. (ed.), pp. 173–84. Plenum Publishing Company, New York.
Plucknett, D. L. (1983). Taxonomy of the genus *Colocasia*. In: *Taro*, Wang, J.-K. (ed.), pp. 14–19. University of Hawaii Press, Honolulu, Hawaii.
Shih, S. F. and Snyder, G. H. (1984). Leaf area index and dry biomass of taro. *Agronomy Journal* **76**, 750–3.
Shih, S. F. and Snyder, G. H. (1985). Leaf area index and evapotranspiration of taro. *Agronomy Journal* **77**, 554–6.
Snyder, G. H. and O'Hair, S. K. (1986). Biomass production from taro (*Colocasia esculenta*) in subtropical wetlands. In: *Biomass Energy Development*, Smith, W. H. (ed.), pp. 185–96. Plenum Publishing Company, New York.
Sunell, L. A. and Arditti, J. (1983). Physiology and phytochemistry. In: *Taro*, Wang, J.-K. (ed.), pp. 34–140. University of Hawaii Press, Honolulu, Hawaii.

15
Alternative Production Systems: Marine Crops

K. BIRD, B. LAPOINTE, D. HANISAK, J. RYTHER
Harbor Branch Oceanographic Institution, Fort Pierce, Florida, USA

and

C. DAWES
Biology Department, University of South Florida, Tampa, Florida, USA

INTRODUCTION

Marine crops may represent significant biomass resources for areas with extensive coasts. Florida, a peninsula, has a wide continental shelf where water depths decline very gradually, decreasing 1 m for every 5 km from the coast. When the Florida coast between the cities of Pensacola and Tarpon Springs was examined for potential areas suitable for seaweed cultivation, 190 000 ha were determined to be between water depths of 0·5 and 1·5 m, and 1 900 000 between 2·5 and 20 m. This analysis did not include areas of shipping lanes, national parks and refuges, competing resources (oyster and scallop beds), or barren, sandy areas. This protected and extensive marine area, which supports a rich tropical and subtropical marine flora (Dawes, 1974), is well suited for seaweed cultivation to provide a major energy resource for the state of Florida and the adjacent southeastern US.

Marine biomass research was initiated in Florida in 1974 and concentrated on maximizing seaweed biomass yields. Some of the important contributions were: establishing yields of the red alga (*Gracilaria tikvahiae* McLaughlin), at 150 dry ash free Mg ha^{-1} yr^{-1} in small intensive culture systems (Lapointe and Ryther, 1978); determining effects of stocking density on seaweed biomass yields (Ryther, 1982a); making efficient use of digester effluents as recycled fertilizer for marine and aquatic plants (Hanisak, 1981); and developing a nutrient pulse fertilization strategy where large amounts of nutrients are supplied for short periods, rather than continual fertilization (Ryther *et al.*, 1981). Over 20 clones of

Gracilaria have been isolated, of which 3–4 have high year round biomass yields (>80 Mg ha^{-1} yr^{-1}), including ash in intensive outdoor culture systems (these systems receive 2–4 exchanges day^{-1} and are water turnover limited as discussed below, hence these data do not represent maximum yields). In terms of large scale systems, the most limiting factor to high Gracilaria yield was water turnover rate in the ponds, vaults, or raceways (Ryther, 1982b). Twenty or more daily water exchanges were required for high sustained biomass yields of 25 a.f.g. m^{-2} day^{-1}, while 0·25–0·5 turnovers in 0·10 ha ponds resulted in yields of 3–5 a.f.g. m^{-2} day^{-1}. This turnover effect occurred even when nutrient additions were adjusted to allow for equal total available nitrogen and phosphorus at different water turnovers. Photosynthesis apparently causes shifts of the CO_2 concentration in these enclosed systems through a rise in pH and a shifting of the dissolved inorganic carbon species concentrations away from high CO and bicarbonate to high concentrations of carbonates, creating C limitation (Ryther and DeBusk, 1982).

Cultivation of Gracilaria spp. for energy feedstocks could take place on the shallow, natural tidal flats (0·5–1·5 m depths) in simple enclosures of net fences supported by pilings. The natural tidal flushing could provide 1–2 water exchanges per day, possibly resulting in ash free biomass yields as high as 25 Mg ha^{-1} yr^{-1}, based on turnover experiments in raceways and ponds (Ryther, 1982b). Preliminary economics for this system indicate feedstock costs in the $2–3 GJ^{-1} range, depending on biomass yields (Bird, 1987).

A floating marine plant, Sargassum, was first proposed by Ryther (1982b) as economic analyses of offshore Macrocystis (kelp) systems indicated the kelp supporting system (the offshore farm structure) was the greatest component of the capital costs (Sullivan et al., 1981). A floating seaweed would only have to be contained, and such marine containment systems have already been developed for pelagic fish culture (e.g. Mitsui, Inc. and Bridgestone, Inc.). Given the potential economic benefits of floating seaweeds for marine biomass, the focus was placed on Sargassum as a means to use the larger offshore Florida areas. In addition, agricultural engineers have been developing containment and harvesting equipment for water hyacinths, a floating freshwater aquatic weed (see Chapters 7 and 8). It may be possible to eventually adapt this system to Sargassum cultivation. Since little was known about Sargassum, the emphasis of the research reported here has been on determining short-term high yields, long-term sustained yields, nutrient requirements, management practices and seasonal changes in growth and composition.

MARINE BIOMASS PRODUCTION RESEARCH

Laboratory Growth Studies

While outdoor cultivation research was the major emphasis of the marine biomass program, laboratory culture studies were used to isolate the effects of different environmental factors on seaweed growth. *Gracilaria tikvahiae* can grow over a temperature range of 12–36°C (maximal growth at 24–30°C) and a salinity range of 6–42⁰/oo, indicating that *Gracilaria* could be cultivated under any combination of temperature and salinity conditions found nearshore to Florida.

In contrast to *Gracilaria*, pelagic species of *Sargassum* have a much narrower tolerance for salinity, with growth observed from 18 to 42⁰/oo. Oceanic salinity (ca. 36⁰/oo) is required for maximal growth; significant reduction of growth occurs even at 30⁰/oo. However, as far as temperature is concerned, *Sargassum* grows well over a broad temperature range (18–30°C), a range which encompasses that normally found for surface ocean water in Florida.

Growth of both *Gracilaria* and *Sargassum* apical tips is saturated at relatively low light levels (ca. 100 μE m^{-2} s^{-1}). The saturation intensity is both clone specific and influenced by temperature, with the light requirement decreasing with temperature. No evidence is apparent of photoinhibition of growth in pelagic *Sargassum* at light levels which inhibit *Gracilaria* spp. (ca. 400 μE m^{-2} s^{-1}), indicating that pelagic *Sargassum* is physiologically well adapted for utilizing light in the high light environment associated with its pelagic habitat.

Nutrients

Historically, the macronutrients nitrogen (N) and phosphorus (P) have been considered primary limiting nutrients to aquatic plant growth. Most important in freshwater systems is P, whereas N is considered most important in marine systems (Stumm and Morgan, 1981). Clearly trace metals, particularly iron (Fe), manganese (Mn), and copper (Cu), can also severely limit algal growth in marine waters (Huntsman and Sunda, 1980). Possible trace metal inhibition or limitation of kelp growth by artificially upwelled deep seawater was an early concern of the offshore *Macrocystis* program (Manley, 1983). Single nutrient enrichments providing Fe, Mn and Cu to *Gracilaria* and *Sargassum* typically enhanced growth over that of the controls in most cases. The enhanced growth of *Gracilaria* and *Sargassum* receiving trace metal enrichments demonstrates their importance to maintaining maximum growth and physiological state of these marine biomass species (Lapointe and Hanisak, 1985). Potential N or P limitation to *Gracilaria* or *Sargassum* cultivation in tropical waters was investigated by exposing seaweeds to single enrichments of these individual nutrients (NH_4^+ for *Gracilaria* and NO_3^- for *Sargassum*) in molar ratios of 17:1 of N:P, approximately the same as the ratio of these elements in natural plants. Phosphorus appears as important as N for *Gracilaria* cultivation, and was more important than N for *Sargassum* growth in tropical waters (Lapointe and Hanisak, 1985).

Yields

Yields reported for *Gracilaria* have been from energy-intensive culture systems, i.e. tanks and/or ponds that receive vigorous aeration and rapid exchange of pumped seawater, culture modes that are clearly not economical for production of biomass for energy. Virtually no previous research with pelagic *Sargassum* documents possible yields with this plant. Therefore, to determine potential yields of *Gracilaria* and pelagic *Sargassum* under conditions representative of an *in situ* energy farm, they were cage cultured during both winter and summer in Pine Channel in the Florida Keys. These cultures received nutrient enrichment and were maintained at algal densities that maximize yields (e.g. 2–5 kg wet wt m^{-2}). Short-term maximal yields of *Gracilaria* (ca. 35 g m^{-2} day^{-1}, including ash) were similar during both summer and winter (Table 1). Similarly, maximal yields of *Sargassum natans* (L.) J. Meyer were similar during summer and winter (ca. 15 g m^{-2} day^{-1}) although the maximal yield obtained with *S. natans* (33 g m^{-2} day^{-1}) occurred during the winter (Table 1). In contrast to *S. natans*, yields obtained with *Sargassum fluitans* Borgeson were maximal during summer months (ca. 20 g dry wt m^{-2} day^{-1}) and fell to lower levels during winter.

The following year, long-term *in situ Sargassum* growth experiments were conducted to determine annual yields. Stocking densities were reduced to those capable of sustained yields (ca. 2 kg m^{-2}). Experiments conducted in two locations, a nearshore area with high nutrient and $CaCO_3$ sediment loading ('greenwater') and an offshore area with low nutrients and sediment loadings ('bluewater') showed higher sustained yields which ranged from 5 to 12 g m^{-2} day^{-1} in bluewater conditions (Fig. 1). Gellenbeck (1986) has carried out a small research effort on *Sargassum muticum* Fensholt yields using the *Gracilaria* intensive culture system, and obtained low sustained yields (ca. 4–5 g m^{-2} day^{-1}) compared to those reported here. Even under these conditions, the initial yields decreased, indicating that these data probably do not reflect potential sustained yields.

TABLE 1
Yields of *Gracilaria tikvahiae*, *Sargassum natans*, and *Sargassum fluitans*, Grown in Cage Culture Over a Two-week Period during Summer and Winter 1983–1984. Values Represent Means ± 1 Standard Deviation

Species and date	Algal density (kg wet wt m^{-2})	Temperature (°C)	Yield (g m^{-2} day^{-1})
S. natans 6/2–6/17/84	2·4	29 ± 3	15·4 ± 0·7
S. natans 1/20–2/2/84	5·0	20 ± 3	33·6 ± 3·2
S. fluitans 6/2–6/17/84	2·4	29 ± 3	19·1 ± 2·8
S. fluitans 1/20–2/2/84	2·4	20 ± 3	4·2 ± 1·2
G. tikvahiae 2/14–3/2/83	5·0	22 ± 3	37·1 ± 5·4
G. tikvahiae 6/2–6/17/84	5·0	29 ± 3	35·5 ± 7·0

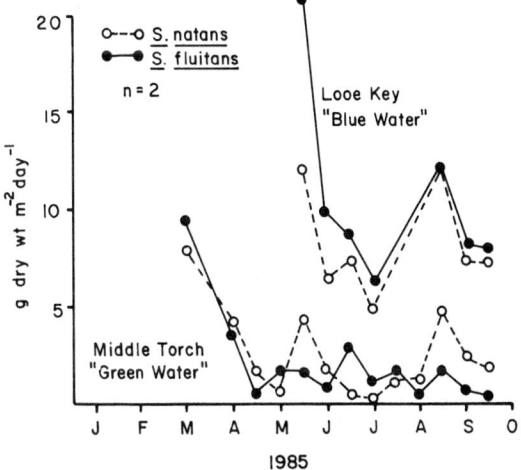

FIG. 1. Yields of *Sargassum fluitans* and *S. natans* in the Florida Keys in a 'greenwater' and 'bluewater' habitat.

Perhaps the only way to really determine the yields of these species would be to conduct an offshore test farm experiment. Such an experiment is not warranted, however, until more is learned about the cultivation requirements and management of these organisms.

Benthic *Sargassum* species have been compared to the pelagic species in laboratory cultures. Typically, benthic species show greater growth rates over a wider range of salinities and temperatures than do pelagic species. Benthic species can be grown in suspended cultures rather than attached, and the new growth is a morphology more suitable for a floating habit. It is not known, however, if the higher yields of these benthic species are sustained in floating cultures or whether significant compositional changes occur.

Population Studies—Florida West Coast

Growth patterns and proximate composition of *Sargassum filipendula* C. Agardh and *S. pteropleuron* Grunow, both common to the west coast of

TABLE 2
Levels of Proximate Constituents (Percent of Dry Weight) in the Upper 5 cm Tips of *Sargassum filipendula* N = 5, ± 1 SD

Month	Ash	Protein	Acid soluble carbohydrate	Lipid
1983				
May	25·8 ± 4·0	5·6 ± 1·1	11·2 ± 4·0	1·6
June	33·4 ± 1·5	4·8 ± 0·8	9·8 ± 2·0	1·3
July	30·2 ± 2·4	4·2 ± 0·4	17·2 ± 1·2	2·8
Sept.	34·0	4·3 ± 0·5	14·5 ± 1·8	3·3
Nov.[a]	32·4 ± 2·7	4·2 ± 0·8	17·2 ± 1·9	2·8
Dec.[a]	29·4 ± 3·4	5·0 ± 0·2	19·2 ± 1·1	3·1
1984				
Jan.[a]	40·0	3·0 ± 0·0	10·0 ± 2·5	2·6
Mar.	32·0 ± 2·1	5·5 ± 0·5	25·5 ± 1·0	3·3
Apr.	28·3 ± 2·9	5·7 ± 0·8	25·8 ± 1·7	1·8
May	35·6 ± 2·5	6·6 ± 0·5	19·8 ± 1·2	3·0
June	36·1 ± 0·5	6·7 ± 0·5	20·0 ± 0·8	1·6
July	36·4 ± 5·6	6·1 ± 0·4	20·5 ± 0·8	1·0
Aug.	24·4 ± 0·6	8·0 ± 0·0	22·0 ± 1·2	1·8
Sept.	31·4 ± 2·3	8·2 ± 0·5	22·0 ± 1·0	1·7
Nov.[a]	26·0	6·0 ± 0·0	20·2 ± 1·1	1·9
Mean	31·7 ± 4·4	5·6 ± 1·4	18·3 ± 5·0	2·2 ± 0·8

[a] Reproductive plants.

Florida, were examined over an 18 month period in 1983 and 1984. These two species, growing in estuarine conditions, showed similar patterns of rapid growth from March to June, maturity in July and August, and are reproductive by October to December. After release of their zygotes, the parent plants die back to perennial basal stipes in January and February (Dawes, 1985). During the periods of rapid growth, *S. pteropleuron* had growth rates of $0.3 \, \text{cm day}^{-1}$, while *S. filipendula* had growth rates of $0.8 \, \text{cm day}^{-1}$. Ash content was higher in tips of *S. filipendula* than *S. pteropleuron*, showed no seasonal patterns, and ranged from 25 to 36%. Alkaline soluble protein content was highest in the spring for both species, ranging from 3 to 8%, while acid soluble carbohydrate was highest in the summer and early fall and ranged from 10 to 25%. Lipid levels are low and show no seasonal pattern (Tables 2 and 3). Acid insoluble carbohydrate would be comprised of mainly alginates and fiber. Alginate concentration can range from 10 to 20% in the benthic *S. pteropleuron* (Daly and Prince, 1981).

TABLE 3
Levels of Proximate Constituents (Percent of Dry Weight) in the Upper 5 cm Tips of *Sargassum pteropleuron* (Grun). N = 5, ± 1 SD

Month	Ash	Protein	Acid soluble carbohydrate	Lipid
1983				
July	28·7 ± 5·9	4·7 ± 0·8	15·8 ± 0·8	1·2
Aug.	27·7 ± 2·7	4·2 ± 0·8	16·2 ± 1·9	1·7
Sept.	31·8 ± 3·7	4·5 ± 0·5	15·0 ± 2·2	1·2
Oct.	30·5 ± 3·6	4·7 ± 0·5	16·3 ± 2·8	1·4
Nov.[a]	27·0 ± 1·3	4·5 ± 0·8	18·2 ± 1·3	1·4
Dec.[a]	26·0 ± 2·2	4·5 ± 0·5	18·2 ± 1·6	0·6
1984				
Feb.[a]	30·0	4·0	14·0	0·8
Mar.	27·0	6·0	23·0	0·9
April	29·5 ± 0·5	6·0 ± 0·0	22·0 ± 0·7	0·8
May	34·7 ± 3·5	7·0 ± 0·0	20·3 ± 1·2	2·5
June	35·5 ± 3·5	6·5 ± 0·5	20·0 ± 0·0	2·7
July	30·0 ± 2·0	8·5 ± 0·5	22·0 ± 0·0	1·3
Aug.	27·0 ± 1·7	7·9 ± 0·4	22·6 ± 0·6	1·3
Sept.	28·8 ± 0·8	5·6 ± 0·6	20·8 ± 0·8	1·7
Oct.[a]	28·0	6·2 ± 0·5	20·8 ± 1·3	1·3
Mean	29·5 ± 2·8	5·7 ± 1·4	19·0 ± 3·0	1·4 ± 0·6

[a] Reproductive plants.

Light intensity-photosynthetic rate studies showed that the compensation points for these two benthic species were similar, and quite low. The maximum photosynthetic rates were achieved at about 50% full sunlight, and photoinhibition was infrequent but occurred occasionally in young plants (Fig. 2; Dawes, 1985). Photosynthetic and respiratory responses to different combinations of salinity and temperature showed both species had broad tolerances and seasonal photosynthetic patterns (Dawes, 1985). Growing tips taken from rapidly growing plants of both species exhibited high photosynthetic and respiration rates in March through May, a slight decrease in July, and then a resumption of higher photosynthetic rates in August and September prior to dropping sharply in November, the period of reproductive onset. During this period, the plants appeared healthy yet showed little responses to changes in temperature or salinity. This may reflect a metabolic shift from a growth mode to a reproductive mode, as the plants shift photosynthate into new reproductive structures.

Reproductive phenologies vary among the west coast populations of these two species which are found in stenohaline ($30-34^0/oo$) as well as estuarine conditions ($5-35^0/oo$), and in deep water populations (20 m) as well as in shallow water. Each population has a distinctive reproductive cycle. For example, the Filman Bayou population of *S. filipendula* did not die back after it became reproductive as do the more estuarine populations. The deep water population is in a rapid growth phase in December through February, and only becomes reproductive later, probably dying back in April; vegetative bases are present only in November. The reproductive phenologies of Florida *Sargassum* spp. are still poorly understood, yet they indicate that a large genetic base is available for plant breeding research.

SUMMARY

Prior to this marine biomass research program on *Sargassum*, the only information available on this genus was from biological oceanography research investigating the role of pelagic species in oceanic productivity and nutrient cycling, *S. muticum* as a noxious marine weed, or its ecological roles or utilization. Lack of experience with *Sargassum* hindered the initiation of biomass research with these plants; nonetheless, important information has been gained for nutrient and trace metal requirements of *Sargassum*. While yield studies have not been conducted on a scale compared to *Macrocystis* and *Gracilaria*, the sustained *Sargassum* yields of

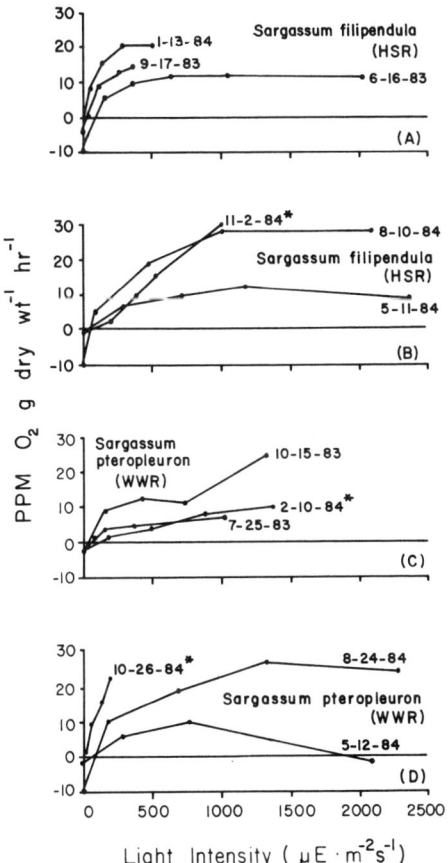

FIG. 2. Photosynthesis–photon flux density curves (PI curves) for a population of *Sargassum filipendula* (HSR = Homosassa River) and another population of *S. pteropleuron* (WWR = Weeki Wachee River, at Gomez Rock). Asterisk indicates use of reproductive plants; all points are a mean of three replicates.

3–9 a.f.g. m^{-2} day^{-1} are within the range of *Macrocystis*, although lower than *Gracilaria*. The high, short term *Sargassum* yields suggest that improved cultivation techniques may increase these preliminary sustained yields. Perhaps more importantly, the benthic *Sargassum* spp. represent a wide species and population gene bank which could be tapped to breed *Sargassum* cultivars specific for energy farms. Taxonomists (Taylor, 1960) have commented on the great amount of hybridization within this genus, which may indicate that seaweed breeders may be able to cross different species and populations. Priorities lie in determining whether the faster growing benthic species will continue their higher growth rates when grown in a floating mode, better outdoor culture system designs for small scale yield experiments and germling grow out, better culture techniques, and factors influencing plant composition, as well as methods for germplasm preservation and maintenance of cultivars. Genetic engineering techniques could prove valuable. Long term efforts must focus on the actual development of *Sargassum* cultivars through selection of inbred lines.

ACKNOWLEDGEMENTS

This is Harbor Branch Contribution No. 511. The US Department of Energy, the University of South Florida, and the Harbor Branch Foundation provided additional financial support.

REFERENCES

Bird, K. T. (1987). Tropical macroalgal culture for bioconversion to 4 methane. In: *Energy From Biomass and Waste X*, Klass, D. (ed.), pp. 1283–92, Institute of Gas Technology, Chicago, Illinois.

Daly, E. L. and Prince, J. S. (1981). The ecology of *Sargassum pteropleuron* Grunow (Phaeophyceae, Fucales) in the waters of South Florida III. Seasonal variation in alginic acid content. *Phycologia* **20**, 352–7.

Dawes, C. J. (1974). *Marine Algae of the West Coast of Florida*, University of Miami Press, Coral Gables, 201 pp.

Dawes, C. J. (1985). Studies on two species of *Sargassum* on the west coast of Florida. A topical report to the Gas Research Institute. Chicago, Illinois, 31 pp.

Gellenbeck, K. (1986). Biomass yields, growth, and bioconversion of *Sargassum muticum*. *Beihefte zur Nova Hedwigia* **83**, 107–15.

Hanisak, M. D. (1981). Recycling the residues from anaerobic digesters as a nutrient source for seaweed growth. *Botanica Marina* **24**, 57–61.

Huntsman, S. A. and Sunda, W. G. (1980). The role of trace metals in regulating phytoplankton growth. In: *The Physiological Ecology of Phytoplankton*, Morris, I. (ed.), pp. 285–328. University of California Press, Berkeley.

Lapointe, B. E. and Hanisak, M. D. (1985). Productivity and nutrition of marine biomass systems in Florida. In: *Energy from Biomass and Wastes IX*, Klass, D. (ed.), pp. 111–26. Institute of Gas Technology, Chicago, Illinois.

Lapointe, B. E. and Ryther, J. H. (1978). Some aspects of the growth and yield of *Gracilaria tikvahiae* in culture. *Aquaculture* **15**, 185–93.

Manley, S. (1983). Micronutrient uptake and translocation by *Macrocystis pyrifera* (Phaeophyta). *Phycology* **20**, 192–201.

Ryther, J. H., Corwin, N., DeBusk, T. A. and William, L. D. (1981). Nitrogen uptake and storage by the red alga *Gracilaria tikvahiae*. *Aquaculture* 26, 107–15.
Ryther, J. R. (1982a). Cultivation of macroscopic marine algae and freshwater aquatic weeds. Solar Energy Research Institute, SERI/TR-98133-1A, Golden, Colorado.
Ryther, J. R. (1982b). Biomass production, anaerobic digestion, and nutrient recycling of small benthic or floating seaweeds. Solar Energy Research Institute, SERI/TR-98133-1B, Golden, Colorado.
Ryther, J. R. and DeBusk, T. A. (1982). Significance of carbon dioxide and bicarbonate-carbon uptake in marine biomass production. In: *Energy from Biomass and Waste VI*, Klass, D. (ed.), pp. 221–36. Institute of Gas Technology, Chicago, Illinois.
Stumm, W. and Morgan, J. (1981). *Aquatic Chemistry*, John Wiley, New York.
Sullivan, R. J., McGinn, R. J., Jain, K. and Engel, M. (1981). *Systems Analysis Studies on Marine Biomass Commercial Application*, General Electric Company, Philadelphia, Pennsylvania.
Taylor, W. R. (1960). *Marine Algae of the Eastern Tropical and Subtropical Coasts of the Americas*, University of Michigan Press, Ann Arbor, Michigan.

16
Alternative Production Systems: Nonconventional Herbaceous Species*

P. MISLEVY
*Agricultural Research and Education Center, University of Florida—
IFAS, Ona, Florida, USA*

J. P. GILREATH
*Gulf Coast Research and Education Center, University of Florida—
IFAS, Bradenton, Florida, USA*

G. M. PRINE
*Agronomy Department, University of Florida—
IFAS, Gainesville, Florida, USA*

and

L. S. DUNAVIN
*Agricultural Research and Education Center, University of Florida—
IFAS, Jay, Florida, USA*

INTRODUCTION

Conventional food, feed and fiber crops have often been considered as potential renewable energy sources while the use of nonconventional plants is a rather recent concept (Adamson, 1983, 1984; Foster and Karpiscak, 1983; Gilreath et al., 1983; Coppola and Brunori, 1984; Vasudevan et al., 1984). Management of crops, such as corn and sorghum, requires large inputs of energy-dependent materials, such as fuel, fertilizer, pesticides, and often requires irrigation and labor-intensive management. This results in unfavorable energetic/economic conditions, as well as supply uncertainties when conventional uses result in high commodity demands. However, intensified management can yield significantly more units of lignocellulose per input unit per hectare, if botanic

*Florida Agricultural Experiment Station Journal Series No. 7107

capacity is correctly tailored to warm-season opportunity (Alexander, 1985).

Utilization of nonconventional plants may be an attractive alternative since many can grow under a wider range of cultural and climatic conditions than domestic crops. Lawson et al. (1983) indicated natural vegetation to be an excellent source of nonconventional plants because it does not compete for land with agriculture and forestry. Additionally, many nonconventional plants have few known pests, are highly competitive and are capable of growing on soils generally unsuitable for domestic crop production. Utilized in a rotation scheme, annual nonconventional plants could be produced by production practices with lower inputs of fertilizer, or by utilizing residual fertilizer and other inputs remaining from production of a previous crop (Gilreath et al., 1983). The use of perennial nonconventional plants is also desirable since they would eliminate the expense of annual establishment. Some of these plants could possibly utilize marginal agricultural soils that have poor drainage, low fertility, cold spots, pest infestations, or other factors which limit production of conventional crops.

Research reported here was designed to assess the potential yield and crop production problems of various nonconventional grass and broadleaf herbaceous species for biomass. Approximately 20 nonconventional herbaceous genera consisting of some 50 entries were screened at various locations throughout Florida (Table 1). Those entries that established slowly, had poor seedling vigor, problems with disease, insects or nematodes, low yield, or those that required excessive cultural inputs were dropped from further testing. Also, those entries for which inadequate data are available (e.g. alemangrass, *Arundo phragmites*) are not discussed further. Four grasses and six broadleaves with adequate databases and showing promise as biomass crops were selected for discussion here.

GRASS SPECIES

Erianthus arundinaceum (Retz.) Jesw. (Erianthus) is a tall-growing, perennial bunch grass which can attain a height of 6–7 m. This subtropical plant initiates growth in early spring (February) and reaches a height of 2–2·5 m by mid-May, but much biomass accumulation takes place during the hot, humid summer. This grass develops a deep root system which can tolerate both drought and excessive moisture. When established from culms with 2 or 3 nodes, vegetative establishment was slow, with many culms failing to grow, producing poor stands. Excellent stands were established by planting rooted culms on 1 m centers. Once established, plants grew rapidly, with many culms occupying a semi-erect position (20–45°). This lodging may make harvesting more difficult. The percentage of sugar solids (Brix) of the single entry tested was low (+6%). The species (Table 2) required very short days before flowering and has a large stem diameter (about 3 cm).

Annual dry biomass (DB) yields of *Erianthus* grown over a two-year period at Ona averaged 46·4 Mg ha^{-1}. This species did not show a significant response to fertilization or application of a nematocide (Table 3). Nematode injury appeared minimal with only low numbers of *Trichodorus minor* Colbran (stubby root) recorded.

Saccharum spontaneum L. (energy cane) is a tall-growing, C_4 pathway species which can attain heights of 6 m. Accessions of this species range from bunch to rhizomatous growth habit, small to large stem, early to late flowering, and can have a Brix content of 9–19% (Table 2). These tropical to subtropical entries initiate spring growth slightly later than Napiergrass (*Pennisetum purpureum* L.), producing maximum DB accumulation under moist, hot, summer conditions with long days and high solar radiation. Energy cane tolerates a wide range of soil textures and organic matter contents and droughty to saturated soil conditions. However, plants can be seriously retarded when exposed to standing water during semi-dormant cool winter conditions. Energy cane can be easily established from 'seed' pieces (stem cuttings) containing 2–3 nodes. Placing seed pieces in cultivated soil with adequate moisture, at 9–13 cm below the soil surface in a horizontal position (using a tree planter), followed by packing with the tractor wheel as successive rows are planted, resulted in excellent plant stands. This method conserves moisture while providing good soil contact with the seed piece and protecting it from freezing. This seeding method requires no post-seeding irrigation, and results in excellent plant stands. The pre-emergence herbicides atrazine (2-chloro-4-ethylamino-6-isopropylamino-S-triazine) and alachlor (2-chloro-2'-6'-diethyl-N-(methoxymethyl)-acetanilide) were each applied at 2·2 kg ha^{-1} and provided excellent broadleaf and annual grass control. Skips (section of row where plants did not develop) sometimes occur within a land area seeded to *Saccharum* entries. However, the amount of skips depends on the plant entry. Accession L 79-1002, established from seed pieces placed in the row every 0·3 m, resulted in few nongerminating seed pieces.

Dry biomass yields of the better *Saccharum* selections ranged from

TABLE 1
Grass and Broadleaf Entries Screened for Biomass Production from 1981 to 1985 at Various Locations in Florida

Scientific name	Common name	Biomass potential
	Grasses and reeds	
Arundo donax L.	Giant reed	Inadequate data available
Echinochloa polystachya (H.B.K.) Hitchc.	Aleman	Inadequate data available
Erianthus arundinaceum (Retz.) Jesw.	None	High yield, tolerance to wide range of cultural and soil conditions
Phragmites communis (Cav.) Trin. ex. Steud.	Common reed	Inadequate data available
Saccharum spontaneum L.	Energy cane	High yield, tolerance to wide range of cultural and soil conditions
Setaria magna Griseb.	Giant foxtail	High yield and tolerance to wet soil
Tripsacum dactyloides L.	Eastern gamagrass	Low-medium yield, tolerant to wide range of cultural and soil conditions
	Broadleaves	
Aeschynomene americana L.	Aeschynomene	Low yield, could become subsequent weed
Alysicarpus vaginalis	Alyce clover	Low yield, could become subsequent weed
Amaranthus australis (Gray) J. D. Sauer	Giant pigweed	High yield, tolerance to wet soil
Amaranthus hybridus L.	Smooth pigweed	Moderate yield, rapid regrowth
Amaranthus viridis L.	Slender pigweed	Low yield
Ambrosia artemisiifolia L.	Ragweed	High yield, tolerant to wide range of cultural and soil conditions

Baccharis sp. L.	Salt bush	Low yield, poor seedling vigor
Cajanus cajan (L.)	Norman pigeonpea	[a]Medium yield
Chenopodium album L.	Lambsquarters	Low yield, high fertility requirement
Crotalaria lanceolata E. Mey	Slenderleaf crotalaria	Medium yield
Crotalaria spectabilis Roth.	Showy crotalaria	Medium yield, could become subsequent weed
Cyperus rotundus L.	Purple nutsedge	Low yield
Desmodium cinerascens	None	Medium yield, poor seedling vigor, not persistent
Eclipta alba (L.) Hassk.	Eclipta	Low yield
Eupatorium capillifolium (Lam.) Small	Dogfennel	High yield, tolerant of wide range of cultural conditions, perennial
Glycine max Merr.	Jupiter soybean	Low yield
Indigofera hirsuta L.	Hairy indigo	[a]Medium yield, could become subsequent weed
Macuna deeringiana (Bort.) Merr.	Velvetbean	[a]Medium yield
Ricinus communis L.	Castor bean	Low yield, no tolerance to wet soil
Sida acuta Burm. F.	Prickly sida	Low yield, seed dormancy problems
Sida rhombifolia L.	Arrowleaf sida	High yield, tolerance to wet soils

[a] Reddy et al., 1986

TABLE 2
Effect of Nitrogen Rate on Dry Biomass Yield and Agronomic Characteristics of Saccharum spp. (Energy Cane Hybrids) and Erianthus arundinaceum Grown for Three Years, 1983–85 at the AREC, Ona

Energy cane hybrids	Brix (% sugar solids)	Nitrogen rate ($kg\,ha^{-1}$)			Physiological stage[a]	Lodging	Morphology[b]	Stem size[c]	1985 flowering date
		56	($Mg\,ha^{-1}$)	168					
US 79-1002	18	52·7		43·7	V	No	B	M	[e]
L 79-1002[d]	11	53·6		79·1	H	No	WR	M	10-23
US 72-1153	10	71·9		75·5	H	No	B	L	11-13
US 72-1289	17	47·1		31·4	H	No	B	M	11-13
US 78-1012	13	41·4		33·5	H	Yes	B	M	11-11
US 72-1183	11	34·9		32·8	H	No	WR	M	[e]
US 78-1014	13	26·5		55·4	H	Yes	B	M	11-16
US 78-1011	13	9·3		7·8	H	No	B	M	11-11
CP 70-1133	19	26·6		21·5	H	No	B	L	11-21
US 72-1089	19	45·3		30·5	H	No	B	M	10-28
US 78-1015	13	20·9		32·8	H	No	B	L	11-21
US 78-1013	16	35·5		25·7	H	No	B	M	11-21
US 74-19	10	40·4		40·8	H	Yes	B	S	10-28
US 79-1006	9	27·3		13·6	H	No	SR	S	11-5
IK 76-110[f]	6	2·3		[g]	V	No	B	L	[e]

[a] H = headed, V = vegetative, by early December.
[b] B = bunch, SR = strongly rhizomatous, WR = weakly rhizomatous.
[c] S = small, M = medium, L = large (equal to normal sugarcane).
[d] Average dry biomass yield of this entry grown in northwest Florida (Jay) over two years was 30·8 $Mg\,ha^{-1}$.
[e] Plants did not flower by early December.
[f] This entry is not a Saccharum but Erianthus arundinaceum.
[g] Plants died in this treatment.

Planted November, 1982 from stem pieces containing 2–3 nodes in Ona fine sand (sandy, siliceous, hyperthermic Typic Haplaquod) harvested in early December each year.

TABLE 3
Effect of Fertilizer and Nematocide Rates on Dry Biomass Yields of Perennial Grasses Harvested from 1982 to 1984 at Ona and Jay, FL

Treatment (kg ha^{-1})		Grass entry (Mg ha^{-1})			
Fertilization	Nematocide	L79-1002 energy cane (Ona)[b]	Erianthus (Ona)[b]	Eastern gamagrass (Ona)[c]	Eastern gamagrass (Jay)[c]
Normal	0	85.2	46.4	4.3 a[a]	23.0
Half normal	0	79.5	54.0	2.2 b	—
Normal	6.7	—	53.8	4.3 a	—
		NS	NS		

[a] Means within the column followed by the same letter are not different ($P < 0.05$) according to Duncan's Multiple Range Test. NS = nonsignificant.
[b] Data represent a 2 year average.
[c] Data represent a 3 year average.
Normal fertilizer rate at Ona was 168–25–93 kg ha^{-1} N, P_2O_5, K_2O except for eastern gamagrass which received 84–13–47 kg ha^{-1}.
Normal fertilizer rate at Jay was 260 kg ha^{-1} N applied in 3 applications, 57 and 108 kg ha^{-1} P and K respectively.

$26 \cdot 2$ Mg ha^{-1} at Jay (northwest Florida) to $79 \cdot 1$ Mg ha^{-1} at Ona (south-central Florida) (Tables 2 and 4). Percent sugar solids in the juice (Brix) of *Saccharum* accessions tested at Ona ranged from 9% for US 79-1006 to 19% for US 79-1022 and US 72-1089 (Table 2). Winter survival plays an important role in northwest Florida. Minimum temperatures of $-12°C$ caused mortality of 0–100% among accessions of this species (Table 4).

Energy cane L 79-1002 tested at Ona gave comparable yields at normal and half normal (for forage crops) fertility levels, $85 \cdot 2$ and $79 \cdot 5$ Mg ha^{-1} DB, respectively (Table 3). These DB yields for energy cane are similar to those reported by Giamalva *et al.* (1984). This entry supported less than half the nematode population recorded on Napiergrass PI 300086.

Tripsacum dactyloides L. (eastern gamagrass) is an upright, perennial, warm season, rhizomatous grass which can attain a height of $1 \cdot 5$–$3 \cdot 0$ m.

TABLE 4
Percentage Plant Stand and Dry Biomass Yield of Energy Cane (*Saccharum*) Hybrids Grown in Northwest Florida, 1983–85

Energy cane hybrid	Plant stand[a] (%)	Yield (Mg ha^{-1})
US 78-1012	100	$26 \cdot 2^b$
US 72-1289	83	$18 \cdot 7^b$
US 79-1017	83	$16 \cdot 9^b$
US 72-1137	67	$20 \cdot 7^c$
US 78-1015	67	$18 \cdot 5^c$
US 78-1014	33	$15 \cdot 0^b$
US 79-1022	67	$19 \cdot 1^c$
CP 65-357	33	$9 \cdot 0^b$
B2-23-5 × CP 65-357	17	$8 \cdot 3^b$
US 79-1018	0	$25 \cdot 7^a$
CP 63-588	0	$22 \cdot 1^a$
B5-34-7 × CP 63-588	0	$12 \cdot 9^a$

[a] Three months following freeze (-12 to $-9°C$) of Dec., 1983.
[b] Yields represent 3 years' data. Plants were exposed to 2 severe freezes in 2 years.
[c] Means represent 2 years' data.
[d] Means represent 1 year's data, plants died after Dec. 1983 freeze.
Plants transplanted to the field (Tifton sandy loam soil) (Plinthic Paleudult) April 1, 1983 from green house, and harvested in early November.
Fertilization: 50 days after transplanting 1983,
   27–35–67 kg ha^{-1} N–P–K; 31 May 1984, 36–47–89 kg ha^{-1} followed by 74 kg ha^{-1} N on 19 June 1984; 1 April 1985, 36–47–89 kg ha^{-1} followed by 74 kg ha^{-1} N on 29 April 1985.

Some 40–50% of the overall plant height consists of a flower stalk with leaves developing in the crown region and attaining a height of 1·3 m. Plants grow on moist, well-drained soils, from the northeastern US through the midwest to Nebraska and the southern states. They will not tolerate long periods of standing water. The major growth period for this species is from early spring through summer. Leaf tips can be killed by frost but many of the leaves remain green all winter in Florida. Seed production is from July to September; but, few seeds are viable which makes seedling establishment difficult.

Dry biomass yields at Jay averaged 23 Mg ha^{-1} or about 500% higher than those obtained at Ona (Table 3). Dry biomass yields at Ona averaged 4·3 Mg ha^{-1} under normal forage fertilization (84–13–47 kg ha^{-1} N–P–K) and about 2·2 Mg ha^{-1} when no fertilizer was applied. This entry is not recommended for biomass in south-central Florida because of low DB yield, however, it may prove useful on cooler, well-drained, loamy soils of northwest Florida and similar southeastern USA environments.

Setaria magna Griseb, (giant foxtail) is a tall annual bunch grass which grows from mid-spring through early fall. This species is widely distributed throughout Florida and especially on organic soils which have a moderate to high nitrogen status. Moderate tiller production occurs on sandy soils and intensifies with increasing nitrogen rate. Maximum growth rate occurs during warm, moist weather with DB yields as high as 14 Mg ha^{-1} when given 76 kg ha^{-1} nitrogen fertilization. Increasing the rate of nitrogen has produced little additional DB, but lower N rates resulted in very poor growth. Plants of this species produce large quantities of seed in an indeterminate manner that shatter from the plant as the inflorescence desiccates. Because seed of *S. magna* do not survive well in irrigated Spodosols, this grass has not become a weed problem in horticultural crops grown on these soils. Establishment can occur by direct seeding, either broadcast or drilled, with no appreciable seed dormancy observed. Attempts to propagate plants through rooted cuttings of stem tissue have not been successful.

A linear increase in yield was obtained when plant population was increased from 27×10^3 to 108×10^3 plants ha^{-1} (Gilreath, 1986a). Production of giant foxtail on raised beds improved growth over that obtained when grown without beds, and polyethylene mulch was shown to further increase yields. When grown in rotation with horticultural crops using polyethylene mulch, the mulch and residual fertilizer (Csizinszky and Gilreath, 1985; Gilreath and Jones, 1985) are passed along to the subsequent crops. As much as 25% of the original fertilizer can be available to the subsequent crop (Marlow and Geraldson, 1976). This production scheme could be utilized by many of the annual species discussed in this chapter. Management of this grass as a multiple-harvest biomass crop has not been successful as the plants do not regrow after cutting.

The main advantage of *S. magna* is its tolerance to wet soils. This is countered by a requirement for nitrogen fertilizer, poor growth in the fall and during drought periods, and poor regrowth potential.

HERBACEOUS BROADLEAF SPECIES

Amaranthus australis (Gray) Sauer (giant pigweed), is a large, upright annual plant, with a hollow stem which can attain heights of 3 m or more with a basal diameter in excess of 20 cm. The species is dioecious with female plants being shorter, more branched and typically weighing considerably more than male plants. Plants develop an open branch structure and females produce large quantities of small, shiny black nondormant seeds. The germination of seed has been observed to increase with the application of nitrate nitrogen to the parent plant. Plants can be established from both seed and rooted cuttings (Gilreath, 1986b). Root initiation is quite rapid, resulting in rooted cuttings ready for field planting in less than 2 weeks. In some cases, plants have been produced from nonrooted cuttings which were planted directly in the field. Regrowth from crowns of previously harvested plants is generally poor.

A. australis can be produced from early spring through mid-fall with maximum growth occurring during the warmer months. Plants respond to nitrogen levels up to 67 kg ha^{-1}. Growth is generally rapid, with DB yields of 18 Mg ha^{-1} produced within 69 days (Gilreath, 1986a) (Table 5). Growth is generally slow during the rainy season due to low nitrogen levels. Varying plant populations, from 12×10^3 to 48×10^3 plants ha^{-1}, had no effect on yield due to compensation in individual plant size (Gilreath, 1986a). Weed control is less of a problem in giant pigweed than with many other broadleaf species because of its rapid growth rate and competition. This species is also extremely allelopathic to other species.

This species is very tolerant of wet soils and thrives in irrigation ditches and other low areas of standing water (Table 5). Growth of *A. australis* is very poor under dry soil moisture conditions. Cutting of this large-stemmed species is easy and not energy intensive.

Amaranthus hybridus L. (smooth pigweed) is an upright, multi-branched, herbaceous annual which is smaller (seldom over 1·5 m in

TABLE 5
Evaluation of Production Factors of Nonconventional Weedy Species Relative to their Potential as Feedstocks for Methane from Biomass

Species	Maximum biomass yield ($Mg\ ha^{-1}$)	Factor rating[a]								
		Response to low fertility	Flooding tolerance	Drought tolerance	Insect tolerance	Disease tolerance	Seed dormancy tolerance	Ease of vegetative propagation	Regrowth from crown	Overall potential[b]
Amaranthus australis	18	2·5	5	2	4	5	4	5	1	3
Amaranthus hybridus	16	3	3	2	3	2	4	3	3	4
Ambrosia artemisiifolia	14	4	3	3	5	5	3	3	3	4
Eupatorium capillifolium	44	5	5	4	5	5	3	5	5	5
Ricinus communis	11	2	1	2	5	4	4	1	2	1
Setaria magna	15	2	3	1	5	5	5	1	1	2
Sida rhombifolia	39	4·5	5	3	3·5	4	2	5	4	5

[a] Factors were evaluated using a 1–5 rating scale where 1 = unsatisfactory, 2 = below satisfactory, 3 = satisfactory, 4 = above satisfactory, and 5 = excellent.
[b] Overall potential was a subjective evaluation based on a wide range of observed parameters, including those presented here, which were weighted to reflect the relative importance of each. Thus, the overall rating may not equal the arithmetic average of the individual ratings reported here.

height) than its relative, *A. australis*. *A. hybridus* has produced dry biomass yields as high as 16 Mg ha^{-1} (Gilreath et al., 1983), with 56 kg ha^{-1} N on polyethylene mulch (Table 5). In addition to good yields, this species can tolerate multiple cutting if harvested before floral maturation (Gilreath et al., 1983). Establishment is rapid from seed or rooted cuttings during the warm season (late March until late November in southwest Florida (Gilreath, 1986b)). Growth is poor during the rainy season because of stress and the incidence of rust *Albugo bliti* which has been reported in epiphytotic proportions (Gilreath and Jones, 1985), however, the rust has not been observed on *A. australis*. *A. hybridus* has considerable tolerance of dry soil conditions but grows well in moist to wet sites, provided the plant is not exposed to standing water for long periods of time.

In population studies DB yield increased linearly from 1·0 to 2·4 Mg ha^{-1}, when seeded at 27×10^3 and 215×10^3 plants ha^{-1}, respectively (Gilreath, 1986a). Smooth pigweed produces large quantities of viable seed and easily reestablishes. Control of this species as a weed in most horticultural crops is easily done with many preemergence herbicides.

Ambrosia artemisiifolia L. (ragweed) is an erect, herbaceous annual which grows to approximately 2 m in height. Stems of large mature plants can become somewhat woody, especially near the base. Although generally regarded as a warm season plant, it can be found growing during the cool season under subtropical conditions.

Ragweed grows well in both slightly dry to moist soils (Gilreath, 1986a) and can be established from seed or rooted cuttings (Table 5). *A. artemisiifolia* does not produce large numbers of seed and some seed dormancy exists. However, ragweed can become a serious weed pest because it is difficult to control. Considerable diversity occurs between accessions of this species, both in growth form and in phenology. Allergic reactions are caused by pollen from blooms occurring over a considerable time, contact with the plant's vegetation and volatile compounds emitted from some plants during drying.

The highest yield obtained for *A. artemisiifolia* was 14 Mg ha^{-1} of dry biomass when planted at a density of 54×10^3 plants ha^{-1} (Gilreath, 1986a). Yield increased with increasing plant population, while the weight of individual plants decreased. Little information is available on the response of this species to fertilization or other cultural variables.

Eupatorium capillifolium (Lam.) Small (dogfennel) is a warm season perennial which is widely distributed throughout the southeastern United States. In southwest Florida, dogfennel can be grown from March through November. Plants have a bunch habit of growth and can attain heights greater than 2 m (Gilreath *et al.*, 1983). The number of stems arising from a crown is dependent upon soil fertility and plant spacing. Stem tissue contains chlorophyll and remains green and herbaceous until close to physiological maturity, at which time lignification occurs in the lower portion of stems. The leaves of *E. capillifolium* are small and appear as almost hair-like appendages on stems, but develop a plume-like appearance as they increase in size.

E. capillifolium is a promising biomass crop since it has a moderate-to-fast growth rate (Gilreath *et al.*, 1983) and produces dry biomass yields as high as 44 Mg ha^{-1} (Table 5), with plants averaging 40% DM (Gilreath, 1986a,b). Dogfennel is tolerant to a wide range of soil pH, fertility levels and soil water conditions. Dogfennel also has no observed insect or disease problems (Gilreath, 1986b). It can produce two crops per year from one planting in the spring (Table 5). Plants may be established from either seed or rooted stem cuttings (Gilreath, 1986b); however, seed are small, easily windblown, and sometimes difficult to collect since they shatter at maturity. Crowns can successfully overwinter and initiate new growth once warm weather returns (Gilreath, 1986b). Dry biomass yields increased when plant population increased from 27×10^3 to the maximum of 215×10^3 plants ha^{-1} (Gilreath, 1986a). Even higher populations may be necessary to achieve maximum yield. Dogfennel has been shown to respond to nitrogen fertilization and to supplemental potassium, when supplied in conjunction with nitrogen (Gilreath *et al.*, 1983).

Ricinus communis L. (castor bean) is a warm-season, annual/perennial plant which is cultivated as an oil crop in some parts of the world and occurs naturally throughout south Florida, especially on irrigation canal banks. It has an upright, multi-branched growth habit with most of its foliage near the branch tips. Castor bean can be grown in well-drained, moist soil during the spring and fall in south Florida. It does not grow well on dry soils. Poor growth has been observed on Spodosols during the hot, rainy summer months and is believed to be due to excessive moisture in the root zone (Gilreath *et al.*, 1983; Gilreath, 1986b). Although generally herbaceous, stem tissue may become somewhat woody with age. Plants produce large, viable seed in late fall. Spring seed production can also occur; however, seeds are usually hollow with no viable embryo. Seed size and rapid emergence of vigorous seedlings make this species ideal for direct seeding with precision planters. Since regrowth is poor and plants do not overwinter well from crowns (Gilreath, 1986b), castor bean is basically a single-harvest annual plant. Tip cuttings do not root well and are not a practical means of stand establishment (Gilreath, 1986b).

Castor bean has been shown to respond to both nitrogen and potassium, alone or in combination, with the highest yield obtained where the two were combined (Gilreath et al., 1983).

Sida rhombifolia L. (arrowleaf sida) is a warm-season upright annual which thrives in moist-to-wet soils. Optimum growth rate occurs when maximum ambient temperature exceeds 30°C. *S. rhombifolia* has been observed to exceed 2 m (Gilreath et al., 1983) and will grow from spring through fall; however, if planted too early in the spring, it shifts from vegetative to reproductive growth, as a photoperiodic response. Thus, it is not advisable to plant *S. rhombifolia* until early May when the days are sufficiently long to eliminate spring reproduction. Plants are herbaceous when young, but become woody as they approach maturity.

Arrowleaf sida produces large quantities of easily harvested seed which mature in late fall (Gilreath, 1986b). Establishment of plants from seed is easy, provided seed dormancy has been broken (Gilreath, 1986b). Freshly harvested seed will not germinate for at least 4 months after planting. Dormancy can be overcome by subjecting the seed to temperatures in excess of 35°C. Plants are produced easily from rooted cuttings of shoot tips (Adamson, 1984). Spider mites can reduce plant growth during hot, dry weather; however, once frequent rains occur, mites cease to be a problem.

S. rhombifolia responds quite dramatically to nitrogen fertilization and has produced 39 Mg ha^{-1} of dry biomass with 56 kg ha^{-1} of nitrogen (Gilreath et al., 1983) (Table 5). Potassium fertilization produced modest yield increases, and application of both nitrogen and potassium did not increase yield over that obtained with nitrogen alone. Plants tend to have high dry matter contents, averaging about 30% at maturity. Increasing the plant population from 27×10^3 to 215×10^3 plants ha^{-1} enhanced yield significantly, with highest gain associated with 108×10^3 plants ha^{-1} (Gilreath, 1986a). Increased growth has occurred with increases in nitrogen rate from 34 to 101 kg ha^{-1}. This species responds to good root environment. Highest yields were on raised, polyethylene-mulched beds.

Arrowleaf sida may hold considerable potential as a biomass crop due to its high yields of dry biomass, excellent growth during hot, wet weather, good response to nitrogen fertility, and ease of maintenance.

SUMMARY

E. arundinaceum, *S. spontaneum*, *S. magna*, *A. australis*, *E. capillifolium* and *S. rhombifolia* all are capable of producing high biomass yields ($> 15\,\mathrm{Mg\,ha^{-1}\,yr^{-1}}$) and are tolerant to a wide range of cultural and soil conditions. Further, serious insect or disease problems were not observed on these species. Since *E. arundinaceum*, *S. spontaneum* and *E. capillifolium* and perennials, the need for costly reestablishment is eliminated. Among annuals to be rotated with highly fertilized horticultural crops, *S. rhombifolia*, *A. artemisiifolia*, and *A. hybridus* show promise as biomass crops.

REFERENCES

Adamson, W. C. (1983). Weeds for oil, polyphenol, or hydrocarbon production in the southeast. In: *Proceedings of the 3rd Annual Solar and Biomass Energy Workshop, Atlanta, Georgia,* p. 155. US Dept of Agriculture, Washington, DC.

Adamson, W. C. (1984). Dog-fennels for soil and polyphenol production. In: *Proceedings of the 4th Annual Solar and Biomass Energy Workshop, Atlanta, Georgia,* p. 134. US Dept of Agriculture, Washington, DC.

Alexander, A. G. (1985). *The Energy Cane Alternative*, Elsevier, Amsterdam, 509 pp.

Coppola, F. and Brunori, A. (1984). The effect of the herbicide Gramixel on the heptane fraction of *Euphorbia* biomass. *Biomass* **4**, 59–68.

Csizinszky, A. A. and Gilreath, J. P. (1985). The utilization of residual nutrients in fallow vegetable fields for the production of herbaceous phytomass in west central Florida. In: *Proceedings of the Soil and Crop Science Society of Florida* **44**, 223–7.

Foster, K. E. and Karpiscak, M. M. (1983). Arid land plants for fuel. *Biomass* **3**, 269–85.

Giamalva, M. J., Clarke, S. J. and Stein, J. M. (1984). Sugarcane hybrids of biomass, *Biomass* **6**, 61–8.

Gilreath, J. P. (1986a). Effect of plant population on biomass production of six weed species. In: *Biomass Energy Development*, Smith, W. H. (ed.), pp. 207–16. Plenum Press, New York.

Gilreath, J. P. (1986b). Development of a production system for weeds as biomass crops. *Biomass* **9**, 135–44.

Gilreath, J. P. and Jones, J. B. (1985). White rust epiphytotic on *Amaranthus hybridus* in south Florida incited by *Albugo bliti*. *Plant Disease* **69**, 542.

Gilreath, J. P., Pitman, W. D. and Rockwood, D. L. (1983). Production of nonconventional crops for methane feedstock. In: *Proceedings of the 1983 International Gas Research Conference*, Hirsch, L. H. (ed.), pp. 340–50. Government Institute, Incorporated, Rockville, Maryland.

Lawson, G. J., Callaghan, T. V. and Scott, R. (1983). Biofuels from natural vegetation in the United Kingdom: The management of novel energy crops. In: *Energy from Biomass*, Strub, A., Chartier, P. and Schleser, G. (eds.), pp. 212–21. Elsevier Applied Science, London.

Marlow, G. A. and Geraldson, C. M. (1976). Results of a soluble salt survey of commercial tomato fields in southwest Florida. In: *Proceedings of the Florida State Horticulture Science Society* **89**, 132–5.

Reddy, K. C., Soffes, A. R. and Prine, G. M. (1986). Tropical legumes for green manure. I. Nitrogen production and the effects on succeeding crop yields. *Agronomy* **78**(1), 1–4.

Vasudevan, P., Gujral, G. S. and Madan, M. (1984). *Saccharum munja* Roxb., an underexploited weed. *Biomass* **4**, 143–9.

17

Alternative Production Systems: Woody Crops

D. L. ROCKWOOD[a] and G. M. PRINE[b]

[a] *Forestry Department,* [b] *Agronomy Department, University of Florida—IFAS, Gainesville, Florida, USA*

INTRODUCTION

Research on woody species for energy production in Florida began in 1978 (Rockwood *et al.*, 1983). Initially, 15 species were established in statewide experiments to assess the potential for achieving maximum biomass production in much shorter rotations and with more intensive culture than traditionally employed for timber crops. In 1981, this research effort was expanded to woody crops having potential as methane feedstocks, specifically focusing on (1) screening of species, (2) selection of superior genotypes within promising species, (3) establishing cultural options for these species and genotypes, and (4) evaluating system inputs and outputs in pilot-scale studies with three species.

MATERIALS AND METHODS

Species Screening and Improvement
Initial selection was within certain commercial timber species and from broad-scale screening of other promising species. Genetic materials, experimental design, and analytical procedures have been previously summarized for a *Eucalyptus grandis* Hill ex Maid. genetic base population (Reddy *et al.*, 1985), for an *E. grandis* clonal trial (Rockwood, 1986), and for 58 new species in screening trials (Rockwood and Devalerio, 1986). Five eucalypts in a source trial planted July, 1984 near LaBelle using three replications of 8-tree row plots in a randomized complete block

*Florida Agricultural Experiment Station Journal Series No. 7282

design were measured in October, 1985 for height and survival; species were compared by Duncan's multiple range test.

In spring 1979, 373 accessions of several *Leucaena* species were planted at 1×1 m spacing in duplicate 5-tree plots in Lakeland fine sandy soil at Gainesville. In spring 1982, a broadcast application of 1120 kg ha^{-1} of 0–10–20 (N–P$_2$O$_5$–K$_2$O) fertilizer was made to prevent nutrient limitations. In Spring 1983, 556 kg ha^{-1} of a similar fertilizer was applied. Biomass ratings of all accessions were made in fall 1982, and 62 high-ranking accessions (all *L. leucocephala* (Lam.) deWit) were harvested for biomass yield. Fifty-three of the same accessions were harvested again in winter 1983. Each season, the dry matter yields were determined after a killing freeze and most of the dead leaves and twig tips had fallen to the ground. Four or more stems were selected for dry matter determination and were machine chipped. The fresh chips were weighed and dry matter yields determined after oven drying to a constant weight at 60°C. Twelve of the highest dry biomass yielding accessions were analyzed for energy content with a bomb calorimeter.

Above-ground biomass samples from 12 species were provided to others for chemical analyses and methanogenesis bioassay.

Mineral content of the harvested biomass of the 12 highest-yielding *Leucaena* accessions was determined on the 1982 samples. In 1983, leaping plant lice (*Heteropsylla* spp.) attacked the young leaves and tips, precluding determination of mineral contents.

Culture Improvement

A *Pinus elliottii* var. *elliottii* Engelm. fertilization and spacing trial planted in January, 1980 near Trenton (Rockwood, 1986) was measured in October, 1985 for height, diameter breast height (DBH), and survival. Dry biomass ha^{-1} was estimated by predictive equations for five spacings and four site amendments and compared using Duncan's multiple range statistical test.

Production System Evaluation

Five pilot-scale plantings involving *E. grandis*, *Pinus clausa* var. *immuginata* Ward, and *P. elliottii* were installed from 1982 to 1984 to assess operational-level aspects of biomass plantations (Rockwood, 1986). All five were prepared using conventional forest site preparation equipment and procedures, except that bedding was necessary at two sites, which was performed using plows modified to form closely spaced beds. Tree planting done by machine employed conventional planters for *E. grandis* at Palmdale (PT1) and *P. clausa* at Hosford (PT5) and a double row planter (1·5 m between rows) for *P. clausa* and *P. elliottii* near Perry (PT3 and PT4). *P. elliottii* near Bradford (PT2) was hand-planted. Two planting densities were used for PT1 (12 ha each) and PT2 (4 ha each), three densities (20 ha each) for PT3 and PT4, and one density in the 4 ha PT5. Economic and energetic inputs were monitored, and permanent sample plots were installed and measured periodically.

RESULTS AND DISCUSSION

Species Screening and Genetic Improvement

Selection of species for biomass quality and production in the various climatic zones and sites in Florida has been emphasized (Table 1). Some 16 of 58 species previously untested for biomass production performed well in screening trials, and subsequent coppice growth of six species was exceptional (Rockwood and Devalerio, 1986). Specifically recommended for further evaluation on particular sites were *E. amplifolia, E. camaldulensis, L. leucocephala, Liquidambar styraciflua, Platanus occidentalis,* and *Sapium sebiferum.*

Additional genetic tests suggest that certain eucalypts warrant greater

TABLE 1
Planting Zones for Promising Woody Species for Methane Feedstocks in Florida

Species	Planting zone
Eucalyptus amplifolia Naud.	Fertile intermediate sites in the North and Central
Eucalyptus camaldulensis Dehnh.	Sandhills in the Central and South
Eucalyptus grandis	Organic soils and 'palmetto prairie' in the South
Leucaena leucocephala	Fertile intermediate sites in the North and Central
Liquidambar styraciflua L.	Wet and intermediate sites in the North
Pinus clausa	Sandhills in the North and Central
Pinus elliottii	'Flatwood' and intermediate sites in the North and Central
Platanus occidentalis L.	Wet and intermediate sites in the North
Sapium sebiferum (L.) Roxb.	Wet and intermediate sites Statewide

consideration. Several sources of *E. amplifolia*, and the species overall, did well in early height and frost-resilience near Gainesville (Rockwood, 1986). Four *E. amplifolia* sources survived acceptably at LaBelle but were second to *E. camaldulensis* for survival through one rigorous winter (Table 2). Overall, these unimproved *E. camaldulensis* were tallest and even exceeded the performance of a superior *E. grandis* progeny (42% survival, 0·6 m height). Few of the new *E. grandis* sources equalled the superior progeny, a finding in accordance with earlier comparisons of new introductions with succeeding generations of genetic selection (Reddy *et al.*, 1985). Most importantly, *E. camaldulensis* and *E. amplifolia* appear to have sufficient frost-resilience for culture in colder parts of Florida.

TABLE 2
Performance of *Eucalyptus* and *Leucaena* Sources in Biomass Trials

Species/source group-trait	Mean	Range
E. amplifolia		
Four sources—15-mo Survival (%)	60b[a]	54–63
Height (m)	0·3B	0·2–0·4
E. camaldulensis		
18 sources—15-mo Survival (%)	82a	0–100
Height (m)	1·0A	0–1·4
E. dunnii Maid.		
Three sources—15-mo Survival (%)	11c	0–29
Height (m)	0·4B	0–0·4
E. grandis		
31 sources—15-mo Survival (%)	16c	0–42
Height (m)	0·6AB	0–0·9
E. robusta Sm.		
One source—15-mo Survival (%)	54b	—
Height (m)	0·7AB	—
L. leucocephala		
62 sources harvested in		
1982—12-mo Yield (Mg ha^{-1})	—	5·9–37·8
12 best sources harvested in		
1982—12-mo Yield (Mg ha^{-1})	29·3	24·3–37·8
Heat Value (cal g^{-1})	4 700	4 620–4 780
1983—12-mo Yield (Mg ha^{-1})	24·7	19·9–30·7
Heat Value (cal g^{-1})	4 730	4 670–4 800

[a] *Eucalyptus* survival means with the same lower case letter and *Eucalyptus* height means with the same upper case letter are not significantly different at the 5% significance level.

TABLE 3
Genetic Resources, Representative Genetic Gains, and Maximum Achieved Dry Biomass Yields for Six Woody Species

Species	Genetic base	Genetic gain	Yield $(Mg\,ha^{-1}\,yr^{-1})$
Eucalyptus grandis	529 progenies and 400 clones	+90% in coppice volume per tree	24
E. amplifolia	12 sources	+19% in survival	23
Leucaena leucocephala	72 sources	+34% in biomass (ha^{-1})	38
Pinus clausa	150 progenies	+55% in biomass (ha^{-1})	9
P. elliottii	1 360 progenies	+85% in biomass (ha^{-1})	13
Sapium sebiferum	150 progenies	+25% in height	21

Leucaena top growth is killed annually by cold temperatures in subtropical Florida. Nevertheless, these coppice-formed shrubs, usually less than 6 m in height, could be productive if harvested annually during the winter. The above-ground coppice biomass of 62 Leucaena accessions varied from 5·9 to 37·8 Mg ha^{-1} in 1983 (Table 2). Biomass yield was not significantly different among the 12 best accessions and averaged 29·3 Mg ha^{-1} in 1982 and 24·7 Mg ha^{-1} in 1983. The lower biomass yields in 1983 were probably related to attack by leaping plant lice. Following the severe winters of 1982–83 and 1983–84, leaping lice were not evident in 1984 but did reappear in late September, 1985. The importance of this insect to Leucaena production is not clear at present.

The commercial timber species E. grandis, P. clausa and P. elliottii have produced well in short rotation intensive culture (Table 3). Productivity of E. grandis has exceeded all other species in frost-infrequent southern Florida (Rockwood et al., 1984). The pines offer modest but assured productivity on lands in northern and central Florida (Dippon, 1985).

Genetic base populations of various size are in place for six species with excellent short-term prospects (Table 3). Conventional genetic improvement strategies with the more extensive genetic materials of E. grandis and the pines can increase productivity from 55 to 90% (Frampton and Rockwood, 1983; Reddy et al., 1985). Use of the more frost-resilient and/or faster-growing sources of E. amplifolia, L. leucocephala, and S. sebiferum can provide meaningful gains.

Culture Improvement

Modest site amendments promote or accelerate growth of *E. grandis* and *P. elliottii* but may not be necessary for *P. clausa*. While phosphorus (P) (50 kg ha^{-1} applied as ground rock phosphate) is required for adequate stand development of *E. grandis* on 'palmetto prairie', nitrogen (N) application alone or in conjunction with P may not be economical (Rockwood *et al.*, 1984). Through five years, only high rates of sewage sludge have increased growth of *P. clausa* (Rockwood, 1986), possibly because of the effects of residual organic matter in these sandy soils. Response of *P. elliottii* is generally positive but dependent on site types. On less fertile, somewhat poorly drained 'flatwoods', fifth-year responses to N/P applications up to 200/100 kg ha^{-1} have been as high as 50% (Rockwood, 1986). Even greater changes were observed with sewage sludge amendments. Response of *P. elliottii* to similar fertilizers on more fertile sites is less but still significant (Table 4), as a combination of low P and intermediate N increased productivity over a control by 23% after six years.

The biomass of the 12 highest-yielding *Leucaena* accessions contained 0·073, 0·009, 0·061, 0·029 and 0·008 g kg^{-1} of N, P, K, Ca and Mg, respectively. The nutrients removed in top growth biomass were equivalent to 210, 30, 180, 80 and 20 kg ha^{-1}, correspondingly, of N, P, K, Ca and Mg. *Leucaena*, a legume, can fix N through symbiosis with *Rhizobium* bacteria. Continuous annual removal of the above quantities of nutrients probably will result in the need for fertilization on most soils in the southeastern United States. *Leucaena* harvesting during a period from mid-November to about March 1 would allow needed nutrients to be applied following harvest. *Leucaena* is adapted only to well-drained soils.

Planting density dramatically influences tree size, productivity (Fig. 1), and rotation length. A density of 43 300 trees ha^{-1} produced greatest *E. grandis* biomass at 8 months on 'palmetto prairie', but this relative advantage over lesser densities had decreased by 33 months (Rockwood *et al.*, 1982). On organic soil, individual trees were always larger in diameter at 1 600 than at 10 000 trees ha^{-1}, but biomass ha^{-1} at 1600 trees ha^{-1} was 40% of that at 10 000 trees ha^{-1} at 12 months and 70% at 36 months. Associated estimates of rotation length were 18 months for the 10 000 trees ha^{-1} density and 36 months for 1600 trees ha^{-1}.

Similar yield responses of *P. elliottii* to five densities have not yet been observed (Table 4). Through 6 years, the highest yield was achieved by the 43 300 trees ha^{-1} density, which had 14 more Mg ha^{-1} than the 25 100 density. However, imminent mortality of the smaller trees at the higher densities (Frampton and Rockwood, 1983) is likely to lead to greatest yield being achieved by the less dense plantings by age 10.

TABLE 4
Stem Dry Weight of 6-year-old *Pinus elliottii* in Response to Fertilization and Planting Density

Fertilizer level ($N/P\ kg\ ha^{-1}$)	Weight $Tree^{-1}$ (kg)	Weight Ha^{-1} (Mg)	Planting density ($trees\ ha^{-1}$)	Weight $Tree^{-1}$ (kg)	Weight Ha^{-1} (Mg)
0/0	5·8	54·1b[a]	4 800	10·6a	47·6
50/50	6·6	61·1ab	8 400	6·6b	53·7
150/50	7·8	66·8a	14 600	4·1c	57·6
200/100	6·2	57·4ab	25 100	2·6cd	61·3
			43 300	1·9d	75·5

[a] Values with the same letter are not significantly different at the 5% level of significance.

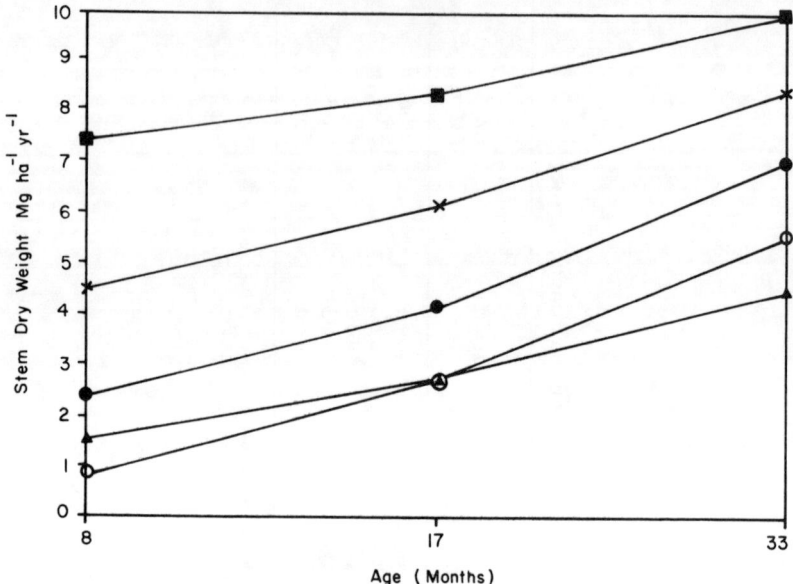

FIG. 1. Biomass productivity of *Eucalyptus grandis* on 'palmetto prairie' in response to planting densities of 4800 (○), 8400 (▲), 14 600 (●), 25 100 (X), and 43 300 (■) trees ha^{-1}.

Production System Evaluation

Preliminary economic and energetic analyses using inputs estimated from the pilot tests and productivities achieved in experiments were promising, especially for *E. grandis* (Rockwood et al., 1984). On organic soils, while more biomass was produced at 10 000 than 1600 trees ha^{-1} (e.g. 45·9 versus 28·1 Mg ha^{-1} after 24 months), associated energy output/input ratios (23·9 versus 23·4 were similar, and net cost per Mg ($14.05 versus $12.83) favored the lesser density because of variable costs related to plantation establishment. Coppicing of *E. grandis*, as implemented by two coppice rotations following the seedling stand, led to improved economics and energetics in comparison to repeated seedling rotations. While the 1600 trees ha^{-1} density with 2- to 3-year rotations was better, the optimal stand density was speculated to be between the two densities examined.

E. grandis does not grow as rapidly and is more susceptible to frost on the 'palmetto prairie' site represented by PT1. While early survival was excellent (Table 5), tree vigor was low due to P deficiency because of the loss of triple-superphosphate in the excessive rains that followed planting.

TABLE 5
Development of *Eucalyptus grandis*, *Pinus clausa*, and *Pinus elliottii* in Pilot Tests

Species—trait	Stand density		
	1	2	3
Eucalyptus grandis			
Target density (trees ha^{-1})	5 000	3 333	
Planted density (trees ha^{-1})	3 764	2 848	
Planting success (%)	75	85	
5-mo survival (%)	98	99	
height (m)	0·8	0·7	
38-mo survival (%)	98	98	
height (m)	2·1	2·0	
diameter (cm)	1·4	1·1	
Pinus clausa			
Target density (trees ha^{-1})	7 172	4 303	3 074
Planted density (trees ha^{-1})	5 722	3 589	2 463
Planting success (%)	80	83	80
2-mo survival (%)	91	78	80
12-mo survival (%)	85	74	78
34-mo survival (%)	63	65	64
height (m)	1·6	1·5	1·5
Pinus elliottii			
Target density (trees ha^{-1})	7 172	4 303	3 074
Planted density (trees ha^{-1})	4 870	3 444	3 025
Planting success (%)	68	80	98
2-mo survival (%)	99	98	98
12-mo survival (%)	69	71	65
34-mo survival (%)	50	40	40
height (m)	1·5	1·5	1·5

Tree growth and survival were further affected by exceptionally cold winters before the nutrient deficiency was corrected by the application of ground rock phosphate. To date, the productivity of *E. grandis* in PT1 is far below what has been achieved with typical culture and environmental conditions (Rockwood and Geary, 1982).

The *E. grandis* clonal trial adjacent to PT1 has demonstrated the productivity improvement possible through vegetative propagation (Rockwood, 1986). Through 20 months, rooted cuttings of 55 trees selected for superior vigor and hardiness averaged 2·4 m tall and 96% survival. The five best clones, by comparison, were 44% taller (3·1 m), considerably more frost-hardy, and had 100% survival. They are still markedly superior to the PT1 seedlings, which represented the best seedling planting stock available in 1982.

Biomass plantings of *P. clausa* received the lowest inputs of the three species in pilot tests, but projected growth has been too low to achieve favorable net returns (Dippon, 1985). Early stand development under operational culture, as illustrated by a lower than expected survival through 3 years in PT3 (Table 5), is critical. Similarly, with 60% survival and 1·0 m height 2 years after establishment at 4880 trees ha^{-1}, PT5 may also demonstrate operationally achievable stands of *P. clausa* on typical sandhills in Florida.

High-density plantings of *P. elliottii* appear to be feasible, and growth through 10 years may result in net costs of $24 Mg^{-1} and $1.19 GJ^{-1} at a stand density of 5000 trees ha^{-1} (Dippon, 1985). However, the machine planted PT4 indicates operational shortfalls that can occur (Table 5). Initial survival at each planting density was high, but extended drought, periodic excessive moisture, and heavy grass competition related to harrowing for site preparation have caused relatively low overall survival. Tree growth, although typical for the site, is below that observed in experiments. By comparison, second-year survival (99%) and height (1·6 m) in the hand-planted PT2 illustrate greater potential as a result of more intensive site preparation.

PT1, PT3, PT4 and PT5 reflect the difficulty of conventional tree planters in achieving high planting densities. Neither a 1 m within-row spacing in PT1 nor a 1·5 m distance in PT3 and PT4 were effectively maintained. For within-row spacings of 2 m or more, planting successes were between 80 and 98%. Given that between-row distances cannot be closer than 1·5 m using these planters, achievement of operational planting densities above 5000 trees ha^{-1} will require careful use of currently available equipment, development of special tree planters, or implementation of innovative tree establishment systems.

SUMMARY

Several perennial woody species, notably *E. grandis*, *E. amplifolia*, *L. leucocephala*, *P. elliottii* and *S. sebiferum*, have demonstrated sufficient biomass production to be considered for methane feedstocks. Such species may be grown under cultural options that appear economically and energetically feasible across a statewide forest and agricultural land base. Suitability of the biomass produced for fermentation to methane needs to be determined.

TABLE 6
Comparative Dry Biomass Yields under Various Cultural Regimes for Five Woody Species Recommended for Methane Production in Florida

Species	Culture	Annual Yield $(Mg\ ha^{-1})$
Eucalyptus grandis	1 500 trees ha^{-1}, 8-year pulpwood on 'palmetto prairie' (Geary et al., 1983)	7
	6 667 trees ha^{-1}, 2·7-year rotation (Suiter et al., 1980)	20
Leucaena leucocephala	Dense plantations at 3- to 8-year rotations (National Academy of Sciences, 1977)	20
Pinus clausa	12 400 trees ha^{-1}, 17·5-year experimental planting (Rockwood et al., 1980)	8
	1 700 trees ha^{-1}, 16-year-old pulpwood plantation (McNab and Carter, 1981)	6
P. elliottii	1 050 trees ha^{-1}, 9-year-old pulpwood plantation (Manis, 1977)	2
	1 200 trees ha^{-1}, 25-year-old pulpwood plantation (Bailey et al., 1982)	5
Sapium sebiferum	27 000 trees ha^{-1}, 4-year rotation (Scheld and Cowles, 1981)	10

In comparison with yields obtained elsewhere and/or with alternative culture (Table 6), the achieved yields of five candidates in closely spaced, short rotation systems in Florida are encouraging. Productivity of *E. grandis* at high planting density exceeded that of conventional pulpwood rotations in Florida and was similar to short rotation yields in Brazil. *Leucaena* yields from annual harvests were greater than estimated productivities for longer rotations in the Philippines. *Pinus clausa* and *P. elliottii* yields at close spacing were greater than those from typical culture, while the best productivity of *P. clausa* was comparable to a longer rotation yield observed at high planting density. *S. sebiferum* growth surpassed a figure derived from a denser planting on a poor site in Texas. As has been observed for one example of the 'wood grass' concept, namely, frequent harvesting over 2 years of poplars in California resulting in annual yields of 6–20 dry Mg ha^{-1} (Bryant et al., 1986), however, our data suggest that close spacings, e.g. above 10 000 trees ha^{-1}, under conventional tree-planting methods, are not economically feasible for energy feedstocks.

Achievement and documentation of the full potential of woody crops require research in the following areas:

1. Biomass properties—characterization of the variability in methane yield-related properties between and within species.
2. Genetic improvement—source/progeny/clonal testing and selection plus long-term expansion of genetic base populations and breeding.
3. Site matching and/or amelioration—identification of appropriate sites and fertilization/herbicide treatments.
4. Spacing and rotation—independent evaluation of planting density and harvesting factors and their relation to rotation length.
5. Biomass/methane yield modeling—prediction of productivity as a function of cultural practice, materials handling, and processing variables.

ACKNOWLEDGEMENTS

Additional support was provided by Subcontract No. 19X-09050C with Oak Ridge National Laboratory under Martin Marietta Energy Systems, Inc., contract DE-AC05-84OR21400 with the US Department of Energy. Experimental sites and/or materials were provided by Everglades Research and Education Center—IFAS, Belle Glade, FL; Consolidated Tomoka Land Company, Sebring, FL; Container Corporation of America, Fernandina Beach, FL; ITT-Rayonier, Inc., Fernandina Beach, FL; Gardinier-Big River, Inc., Bowling Green, FL; Lykes Brothers, Inc., Palmdale, FL; Buckeye Cellulose Corporation, Perry, FL; and St Joe Paper Company, Hosford, FL. The assistance of the USDA Forest Service, Southeastern Forest Experiment Station, Lehigh Acres, FL, was invaluable.

REFERENCES

Bailey, R. L., Pienaar, L. V., Shiver, B. D. and Rheney, J. W. (1982). Stand structure and yield of site-prepared slash pine plantations. Georgia Agricultural Experiment Station Bulletin 291, 83 pp.

Bryant, S. P., Bartikoski, R. G. and Scott, L. (1986). Commercializing high density annual rotation poplars: a status report. In: *Proceedings of Energy from Biomass and Waste X Conference*, Washington, D.C. (in press).

Dippon, D. R. (1985). Economic and energetic analyses of short-rotation culture of slash and sand pine. *Proceedings of the 7th Southern Forest Biomass Workshop*, Rockwood, D. L. (ed.), University of Florida, Gainesville, Florida, pp. 52–7.

Frampton, L. J., Jr and Rockwood, D. L. (1983). Genetic variation in traits important for energy utilization of sand and slash pines. *Silvae Genetica* **32**(1–2), 18–23.
Geary, T. F., Meskimen, G. F. and Franklin, E. C. (1983). Growing eucalypts in Florida for industrial wood production. USDA Forest Service General Technical Report SE-23, 43 pp.
Manis, S. V. (1977). Biomass and nutrient distribution within plantation and natural slash pine stands on wet coastal sites. M.S. thesis, University of Florida, Gainesville, 61 pp.
McNab, W. H. and Carter, A. R. (1981). Sand pine performance on South Carolina sandhills. *Southern Journal of Applied Forestry* **5**, 84–8.
National Academy of Sciences (1977). Leucaena: Promising forage and tree crop for the tropics. National Academy of Sciences, Washington, D.C., 115 pp.
Reddy, K. V., Rockwood, D. L., Comer, C. W. and Meskimen, G. F. (1985). Predicted genetic gains adjusted for inbreeding for an *Eucalyptus grandis* seed orchard. *Proceedings of the 18th Southern Forest Tree Improvement Conference*, Southern Forest Experiment Station, Gulfport, Mississippi, pp. 283–9.
Rockwood, D. L. (1986). Development of woody biomass cultural systems for Florida. In: *Biomass Energy Development*, Smith, W. H. (ed.), pp. 85–94. Plenum Publishing Corporation, New York.
Rockwood, D. L. and Devalerio, J. T. (1986). Promising species for woody biomass production in warm-humid environments. *Biomass* **11**, 1–17.
Rockwood, D. L. and Geary, T. F. (1982). Genetic variation in biomass productivity and coppicing of intensively grown *Eucalyptus grandis* in southern Florida. *Proceedings of the 7th North American Forest Biology Workshop*, University of Kentucky, Lexington, Kentucky, pp. 400–5.
Rockwood, D. L., Conde, L. F. and Brendemuehl, R. H. (1980). Biomass production of closely spaced Choctawhatchee sand pine. USDA Forest Service Research Note SE-293, 6 pp.
Rockwood, D. L., Geary, T. F. and Bourgeron, P. S. (1982). Planting density and genetic influences on seedling growth and coppicing of eucalypts in southern Florida, *Proceedings 1982 Southern Forest Biomass Workshop*, Southern Forest Experiment Station, New Orleans, Louisiana, pp. 95–102.
Rockwood, D. L., Comer, C. W., Dippon, D. R., Huffman, J. B., Riekerk, H. and Wang, S. (1983). Current status of woody biomass production research in Florida. *Proceedings of the Soil and Crop Science Society of Florida* **42**, 19–27.
Rockwood, D. L., Dippon, D. R. and Comer, C. W. (1984). Potential of *Eucalyptus grandis* for biomass production in Florida. *Proceedings Bioenergy* **84** (2), 86–93.
Scheld, H. W. and Cowles, J. R. (1981). Woody biomass potential of the Chinese tallow tree. *Economic Botany* **35**, 391–7.
Suiter, W., de Rezende, G. C. and Mendes, C. J. (1980). Production of short rotation *Eucalyptus* for energy sources. Sociedade de Investigacoes Florestals, Boletin Technico No. 3, 9 pp.

18
Biological Production of Methane from Biomass

P. H. SMITH, F. M. BORDEAUX, M. GOTO, A. SHIRALIPOUR
A. WILKIE
Microbiology and Cell Science Department, University of Florida—IFAS, Gainesville, Florida, USA

J. F. ANDREWS, S. IDE and M. W. BARNETT
Environmental Science and Engineering Department, Rice University, Houston, Texas, USA

INTRODUCTION

The methanogenic bacteria appear to be ancient forms of life (Woese and Fox, 1977). The hydrogen (H_2) and carbon dioxide (CO_2) outgassed from the earth's mantle (Walker et al., 1983) could well have supported the existence of these organisms long before the development of primitive plant species, hundreds of millions of years ago. Vast amounts of plant residue accumulated during the growth of the great coal forests in the latter part of the paleozoic period. This phenomenon can be explained by the assumptions that: (1) the range of microbial species prevalent at that time had inadequate biochemical capacity to efficiently catalyze complete anaerobic oxidation of the diversity of complex organic molecules produced by the many emerging plant species, and (2) aerobic processes were limited by ineffective oxygen transport within these vast deposits of plant material. Thus, ancient microbial conversions were insufficient to recycle the large amounts of biomass formed. Such a cycle would require anaerobic conversion of biomass to methane and carbon dioxide by microbial activities similar to those occurring in present-day anaerobic digesters. Incomplete anaerobic oxidations during this period could have been caused by organic acid accumulation, as occurs during digester failure, due to the absence of a sufficiently efficient hydrogenogenic microflora at that time.

*Florida Agricultural Experiment Station Journal Series No. 8018.

Anaerobic digestion of biomass has occurred throughout the existence of life on earth, with methanogenesis a most likely component of the process. Anaerobic digestion therefore is expected to be a highly reliable, stable and efficient process in those environments where its occurrence is essential to maintain the natural cycle of organic matter. The process may not be equally as effective in simulated environments such as anaerobic digesters. Indeed, the process is anticipated to be suboptimal in such environments since the forces of genetic selection, which have been in progress for millennia, have not had time to impact on such recently engineered environments. However, methanogenesis and other phases of the anaerobic digestion process should be amenable to improvement by the application of current biotechnological and genetic engineering techniques. Research is needed, therefore, to obtain a more complete understanding of the anaerobic digestion process and to identify potential areas for constructive improvement which will not produce undesirable or destructive environmental effects.

The first identifiable anaerobic biomass digesters were the ruminants. These animals may be described as motile, batch-fed, continuously mixed digesters (Hungate, 1966). Ruminant survival has depended on anaerobic digestion being a very reliable process when properly controlled. These living digesters possess many of the features needed for optimal digestion of biomass for the production of pipeline quantities of methane. They are highly efficient in materials handling, temperature control, mixing, mineral nutrition, pH control, *inter alia*.

Conceptually, a biomass conversion system may be considered from either an engineering or a biological perspective, or from a combination of the two. Engineering sensitivities in the fermentation industry have been analyzed by Cooney (1983). Since biomass has some similarities in chemical composition to domestic wastes, biomass conversion processes often resemble those used in waste treatment. The history of waste treatment has been reviewed by McCarty (1982). The processes employed and their biology have also been reviewed by Bryant (1979) and Wolfe (1983).

Conventional mixed digesters, which mimic natural biomass fermentations, have been utilized for many years in the treatment of wastes. Many design variations have been developed during the past 100 years (McCarty, 1982). However, large-scale production of methane using the anaerobic digestion process has been a recent research consideration, stimulated by the energy situation of the 1970s. One approach for improving methane-from-biomass systems is through improved digestion; another is to develop energy crops with improved convertibility. In anaerobic digestion of biomass for energy production, high conversion efficiency is required for economic feasibility. Biologically sensitive steps, which are manageable in waste treatment, become major stumbling-blocks, being negatively affected by high loading rates. Two readily identifiable, sensitive processes, which were addressed in this research, are the hydrolysis of large organic polymers and the conversion of organic acids.

Some of the large organic molecules found in plant material are highly resistant to biological degradation, while others may be toxic (Van Soest, 1983). This large and widely diverse group of molecules creates difficulties for the anaerobic digestion of biomass, limiting final yields of methane from plant feedstocks. Lignin, for example, is not appreciably degraded under anaerobic conditions (Hobson et al., 1974) and may also physically limit the hydrolysis of cellulose in plant materials (Van Soest, 1983). These problems are not amenable to major improvement by manipulation of the fermenter microflora using known techniques. An alternative approach would be to discover unique plant species which could serve as feedstocks for methane production with little residual organic substance.

A study of organic acid metabolism would provide new insight into the nature of the anaerobic oxidation of these acids and into factors which could affect the rates of these oxidations by species of bacteria present in digesters fed selected forms of biomass. The acids most significant in this regard are acetic acid, propionic acid and butyric acid (Smith, 1980).

Any improvement in the rates and yields of hydrolysis and acid conversion must not create a potential for perturbation of natural biological processes. These processes have been fine-tuned by evolution to guarantee successful continuity of the carbon cycle as it now exists, although the carbon cycle is being altered by combustion of fossil fuels. Biomass conversion represents an ecologically sensible way to maintain this continuity.

The process of anaerobic conversion of complex organic matter to methane may be viewed as a continuous, guided flow of electrons from electron donors to electron acceptors. In this regard the process resembles all life processes. This principle, which extends beyond such limited considerations as oxidation-reduction reactions, was first clearly elucidated by Kluyver and van Niel (1956) in their development of the concept of comparative biochemistry. The rate and stability of the process depend upon the continuity of a very large number of reactions, some of which cannot be quantitatively determined due to analytical limitations. For example, formate may be involved in the reaction sequence, but it is not currently possible to measure the reaction because of the rapid rate of exchange between formic acid and carbon dioxide (Smith, 1980).

A rational approach to the optimal regulation of the overall process requires an understanding of the interplay between the production and dissimilation of organic acids and the formation of acid precursors, leading to a perception of the critical reactions and their rates. However, the regulation and control of metabolic processes are difficult to define in specific terms, substantially because the regulatory processes occur at the cellular level, with strong interactions between different microbial species. In addition, the different control processes involved are not necessarily mutually exclusive and what appears to be an obvious explanation of results may, in fact, not be applicable (Conrad et al., 1976). The difficulty of obtaining a complete characterization of the reactions involved in biomass-to-methane conversion forces consideration of other approaches to an understanding of the roles of different biological mechanisms and of the possibilities for packaging appropriate biological material with optimal engineering designs.

There is a wide variety of currently functioning single-stage anaerobic digesters (Daniels, 1984). Viewed from an engineering perspective these include single-tank, stirred digesters with and without recycle, upflow and downflow anaerobic filters, anaerobic expanded bed and fluidized bed digesters, upflow anaerobic sludge blanket digesters and landfill digesters. There are almost limitless variations in the operational regimes for these different single-stage digesters.

The rumen fermentation suggests that the fermentation of biomass in more than one phase is biologically attractive. Detailed examination of ruminant physiology reveals the presence of at least two phases: (1) an anaerobic phase, in which organic acids are formed in the rumen, and (2) an aerobic phase, in which organic acids are metabolized in the animal tissue. The anaerobic phase is stratified into a high-solids subphase and a liquid subphase (Smith et al., 1956). Metabolic activity and concentrations of intermediates are greater in the high-solids than in the liquid subphase. Such a multiphase digestion system must have desirable lasting qualities since it has been maintained in Nature for millennia. Ghosh et al. (1985) compared two-phase and conventional single-stage anaerobic digestion of sewage sludge at mesophilic and thermophilic temperatures. Higher methane yields were obtained from the two-phase digesters than from the single-stage digesters. A two-phase digester was also compared to a novel three-phase digester and the three-phase digester's performance was superior.

The optimum number of phases for biomass conversion has not been established. In multiphase digestion of biomass, the number of phases could be quite large to allow regulation and control of such parameters as temperature, pH, microbial life cycles and population shifts, osmotic changes, oxidation–reduction potentials, added oxidants, introduction of unique microorganisms, production and removal of inhibitory levels of organic acids, *inter alia*. While the advantages of multiphase over single-stage anaerobic digestion have not been established at the commercial level, our research has focussed on factors important to the development of commercial multiphase digesters.

MULTIPHASE ANAEROBIC DIGESTION

In the context of a biomass fermentation system for the production of pipeline quantities of gas, interacting factors which impact directly on projected methane production costs include mass transfer, heat transfer and rate control.

Mass transfer involves solid biomass, water and dissolved material, and gases, with solid biomass being the most expensive to relocate, followed by water and then gas. Materials handling of feedstock biomass could be minimized in a multiphase digestion system in which the different phases are created by changing the microbiology of the fermentation, at time intervals dictated by the need to optimize different reactions for maximum fermentation productivity. The number of phases would be determined experimentally and could well vary, depending on the nature of the feedstock.

Heat transfer is another major cost factor. Heat transfer through the sides of a large digester can be minimized with appropriate insulation. However, a substantial amount of heat would still be required if large quantities of water were frequently added to the digester. Hence, water which has been previously heated in the digester should be recycled. The temperature of the digester is assumed to be elevated by the heat liberated during biological degradation of the feedstock. This assumption is based on the observation that temperatures in some experimental landfills may reach approximately 50°C (Halvadakis, 1983). Also, additional heat may be obtained by oxidizing some of the feedstock. Ideally, this would utilize digested biomass from which additional methane could not be economically obtained (in the absence of an added oxidant). This material could, conceivably, be oxidized aerobically or anaerobically by the addition of an oxidant such as nitrate, with the ultimate production of both heat and methane. In contrast to air, nitrate would not introduce nitrogen gas into the digester. Addition of nitrate may also enhance bioconversion by altering the microflora and the physical and chemical characteristics of the fermentation. In contrast to oxygen, nitrate is water soluble and could, therefore, be transported from one location to another in a liquid stream. Conversely, if purchased on a continuing basis, nitrate could be unattractive as an oxidant because of its cost. However, nitrate could be internally regenerated in the system by using established methods for the conversion of ammonia to nitrate.

Mass transfer is dealt with mainly by the digester design and other engineering requirements. Heat transfer is also largely an engineering problem. However, both are related in some ways to biological properties of the feedstock and microbial processes occurring in the digester. Rate control also affects a number of important aspects of the anaerobic digestion process. Rate control may be used to keep the various components of the fermentation in balance, to maximize the overall rate of the fermentation (thus reducing fermentation volume) and to alter the end-products, producing a gas with maximum methane content. Specific areas selected for consideration in this research, in terms of rate control, were the effects of biomass composition, salt concentration, oxidants, organic acid concentrations and rates of formation, and micronutrients.

The influence of biomass composition on fermentation is discussed in Chapter 20 (Bjorndal and Moore); the effect of salinity is evaluated in Chapter 19 (Smith *et al.*). In this chapter, we present a discussion of biomass species characteristics, the role of oxidants, acid dynamics, and micronutrient effects on biomass conversion. A mathematical model of a multiphase anaerobic digestion process is also presented. These, and other topics included in this program, should provide technical information for the development of the multiphase digestion system that we have proposed.

Evaluation of Plant Species as Feedstocks

Based on cost sensitivity analysis (see Chapter 3), methane yield is the most important economic parameter in the fermentation of a specific feedstock. Plant feedstocks must be selected based on economic considerations since the value of methane is low compared to other reduced forms of carbon. Two basic approaches to increasing yield are: (1) controlling plant composition, and (2) improving digestion systems. The amount of methane which can be obtained from a given amount of plant material by anaerobic fermentation is difficult to assess because it is a function of the many factors influencing the anaerobic digestion of organic matter. Plant composition (and methane production) is affected by variety, age, anatomy, agronomic practices, *inter alia* (see Chapter 20). Thus, a database ranking potential plant feedstocks, based on methane yield, could prove useful. No single procedure was identified which would classify different feedstocks in such quantitative terms. However, bioassay procedures, which have been developed to provide reasonable assessments of the amount of methane formed by the fermentation of organic wastes, appeared to be applicable to the determination of methane yields from plant species. Modifications were made in the procedure of Owen *et al.* (1979) to adapt it specifically to the analysis of plant materials. Analyses were conducted to determine methane yields from different plant species in order to select species which might be uniquely suited as feedstocks for methane production.

Plant samples were prepared for analysis by particle-size reduction, using methods compatible with tissue moisture content and texture. Volatile solids were determined and subsamples containing 200 mg volatile solids were placed in 250 ml serum bottles and 100 ml of diluent and inoculum was added. The amount of plant material added as substrate was empirically determined to give a measurable amount of gas without producing inhibition due to the build-up of metabolic end-products. The diluent was designed to include the anticipated nutritional requirements for the fermentation. The inoculum was obtained from a 4-liter laboratory digester which was maintained on a mixture of plant materials. This digester was initially established with an inoculum derived from a wide variety of microbial sources, including rumen fluid, sludge from a domestic wastewater treatment plant and material from other laboratory digesters. Material from a marine plant-fed digester was used to inoculate bioassays of marine species. After the addition of materials to the bioassay bottles, the gas phase was flushed and filled with an oxygen-free mixture of 80% nitrogen and 20% carbon dioxide. The bottles were sealed with butyl rubber stoppers, incubated at 35°C and assayed for methane biweekly, until measurable gas production ceased. Gas volumes were determined manometrically, and methane content by gas chromatography. These measurements were used to calculate methane production as standard (Gas Industry Standard, 15·5°C, 1 atm) cubic meters of methane per kilogram volatile solids added (m^3 kg^{-1} VS added).

The inoculum used was assumed to contain the microbial species needed for maximum degradation of the plant materials under anaerobic conditions. It was also assumed that these organisms remained viable during the course of the incubation. Since plant species vary considerably in composition, and since the possible number of different bacterial species present was enormous, this assumption could not be verified. Generally, many microbial species will maintain themselves at low population densities when the environmental requirements for growth are maintained at a low level. Conclusions derived from these assays were based on the assumption that microbial populations necessary for the degradation of complex biomass were maintained so that microflora active in the assay did not limit the total amount of methane formed.

Over 2000 samples were analyzed, in triplicate, covering a wide spectrum of potential plant feedstocks, e.g. different species, plants grown under different conditions of nutrition, plants of different ages and from such diverse environments as fresh water and the oceans. In addition, different parts (e.g. roots, tops) of some plants were compared with respect to methane yield (Table 1). The methane yields varied for the four major resource groups, freshwater aquatics, forage and grasses, roots and tubers, and marine species (Table 2). The yield from marine species was lower than from any other resource group. The highest yields were obtained from root crops, followed by forage and grasses, and freshwater aquatics, respectively. Treatment of plants with various nutrients, especially nitrogen, during the growth period (Table 3) and the age of plants at harvesting time (Table 4) also affected methane production.

These results identified a large reservoir of potential plant species for use in the production of methane from biomass. However, there was insufficient variation in the methane yields from different species to warrant selection of specific plant species, based solely on ultimate methane yield. Other factors, such as rate of digestion and cost of biomass

TABLE 1
Methane Production of Different Plant Parts of Water Hyacinth and Sweet Potato
(Shiralipour and Smith, 1984)

Species or plant part	Number of samples	Methane yield ($m^3 \, kg^{-1}$ VS added)	
		Average	Range
Water hyacinth shoots	9	0.32	0.26–0.43
Water hyacinth roots	9	0.18	0.13–0.24
Sweet potato shoots	6	0.23	0.19–0.26
Sweet potato tubers	6	0.35	0.31–0.43

TABLE 2
Methane Production of Different Biomass Resources
(Shiralipour and Smith, 1984)

Plant resource groups	Number of samples	Methane yield (m^3 kg^{-1} VS added)	
		Average	Range
Freshwater aquatics	203	0·22	0·07–0·43
Forage and grasses	153	0·24	0·07–0·41
Roots and tubers	86	0·33	0·19–0·43
Marine	57	0·21	0·08–0·38

TABLE 3
Effect of Nitrogen Treatment on Methane Production of Water Hyacinth after 11 Weeks Incubation
(Shiralipour and Smith, 1984)

N concentration (mg liter^{-1})	Methane yield (m^3 kg^{-1} VS added)		
	Root	Shoot	Entire plant
0	0·11	0·15	0·12
1	0·13	0·21	0·16
5	0·13	0·25	0·17
10	0·16	0·28	0·22

TABLE 4
Effect of Age on Methane Production of Napiergrass
(Shiralipour and Smith, 1984)

Age (days)	Number of samples	Methane yield (m^3 kg^{-1} VS added)	
		Average	Range
120	3	0·31	0·31–0·32
180	5	0·26	0·24–0·26
330	6	0·24	0·23–0·24

production, would also be important in the choice of a feedstock. A priority would be to select species which could be readily and inexpensively grown on available land and, subsequently, to improve these plants through breeding and the appropriate use of bioengineering techniques to enhance their convertibility to methane. Grasses with increased contents of high molecular weight carbohydrate would meet these criteria. They should also be easy to store and give high yields of methane.

Influence of Oxidants

Nitrate

Toxicity of added nitrate to methane formation is expressed almost immediately upon addition of nitrate to an unacclimated digester. The same is true of oxygen and air, although the anaerobic digestion of domestic waste is not particularly sensitive to air (Smith, 1980). However, this effect of nitrate may be one to which the digester can acclimatize as a result of microfloral changes.

Under anaerobic conditions, nitrate may be reduced via two different pathways. Reduction of nitrate to nitrite, nitric oxide, nitrous oxide and then to nitrogen gas is known as denitrification (Brock, 1974). If this pathway were to operate in our biomass-fed digesters, there would be no basis for the continuation of this research. Nitrogen in the gas phase usually reflects air entry due to engineering problems associated with either the containment structure or the pumping system. Alternatively, reduction of nitrate to nitrite and then to ammonium ion is a process known as dissimilatory nitrate reduction, which takes place in anoxic environments where high concentrations of substrates are available (Cole and Brown, 1980).

Laboratory size (4-liter) digesters were constructed and maintained at 55°C and fed a mixture of Bermudagrass and a commercial cattle-feed. The volumetric biogas productivity was approximately $1.5 \, m^3$ biogas $m^{-3} \, day^{-1}$. Additions of nitrate as low as 3 μM nitrate ml^{-1} mixed liquor per day ($\mu M \, ml^{-1} \, day^{-1}$) to a digester resulted in a marked decrease in gas production and elevated fatty acid concentrations; recovery was slow. Subsequently, instead of batch addition, nitrate was continuously infused (Rivard, 1983; Rivard and Smith, 1984). An empirical value for the maximum rate of nitrate infusion was determined. In these experiments, it was not possible to increase the rate of addition above 9.6 $\mu M \, ml^{-1} \, day^{-1}$ without disrupting the stability of the fermentation. The ammonia concentration in the digester increased to a level equal to that expected if all of the nitrate added were converted to ammonia. Nitrous oxide and nitrogen gas, which would indicate the occurrence of denitrification, were not detected in the gas phase. To further test for the possible occurrence of nitrogen gas production in a nitrate-infused digester, experimental and control digesters were tested for their ability to reduce nitrous oxide. The control digester did not reduce nitrous oxide, even after 40 days. The nitrate-infused digester reduced nitrous oxide, but only after four days. Isotopically labeled nitrate (^{15}N-nitrate) was incubated with a sample from the digester and 98·7% of the label was recovered in the ammonium ion and ammonia fractions. Denitrification was, therefore, not active under these experimental conditions (Rivard, 1983).

The amount of methane produced in the nitrate-infused digester was less than that in the control digesters. Possibly, the lower gas production rate observed in the nitrate-infused digesters could have been the result of increased ammonia concentration in the digesting sludge. This hypothesis was tested by adding ammonium chloride to a control digester at the same rate and to the same final equilibrium concentration as in the nitrate-infused digesters. No significant difference was observed between the ammonium-amended digester and other control digesters with respect to methane production or volatile fatty acid concentrations. The decrease in methane production in the nitrate-infused digester was equal to the amount of methane which would, theoretically, be produced from the amount of hydrogen required to reduce the added nitrate to ammonia. It was concluded that the lower methane production was the result of hydrogen utilization in the production of ammonia from nitrate.

Visual observation of the effluent solids showed a marked difference in appearance between that of a control digester and that of the nitrate-infused digester. Analysis of particle sizes from the sludge of the two digesters revealed that the average particle size of material from the nitrate-infused digester was approximately 10% smaller than that from control digesters (Rivard, 1983).

The microflora in the digesters to which nitrate had been continually added would be expected to have different characteristics than the microflora in the control digesters. Complete evaluation of the microflora of the two digesters at the species level would require many years of work. However, the relative importance of different physiological groups of organisms in the digesters should be expressed by differences in their population densities. The microbial populations of the digesters were examined and the number of hydrogen-utilizing methanogens in the nitrate-infused digester was less than observed in the control, while the number of nitrate-reducing bacteria was greater. The nitrate-infused digester also had a higher population of cellulolytic bacteria and an increased rate of cellulose degradation. The number of acetate-metabolizing methanogens was lower in the sludge with added nitrate, compared to the control digester sludge. This observation was not explained.

The data showed that nitrate was quantitatively converted to ammonia in a biomass-fed anaerobic digester and that, even at high rates of nitrate addition, it was possible to maintain an anaerobic fermentation of biomass to methane. Thus, operating a multiphase anaerobic digester system with a nitrate-amended phase included in the design may be possible without the development of operational problems due to nitrate toxicity. The data further showed that degradation of biomass in the presence of added nitrate is different, both qualitatively and quantitatively, from that obtained in the absence of added nitrate.

Sulfate

Reduction of sulfate to sulfide is anticipated to be a potential problem during digestion of marine biomass or in digester configurations where effluent recycle is practiced. To evaluate this possibility, two thermophilic digesters were continuously infused with sulfate. One digester had not received added nitrate; the other had been acclimated to nitrate and was receiving $9.6\ \mu M\ ml^{-1}\ day^{-1}$. No changes in VFA concentrations or methane production were observed in either digester at a sulfate addition rate of $0.2\ \mu M\ ml^{-1}\ day^{-1}$. After 18 days at this rate, the sulfate addition was increased to $2.4\ \mu M\ ml^{-1}\ day^{-1}$. A decrease in methane production occurred almost immediately in the digester receiving only sulfate. After eight days, its biogas production was 50% of that produced by control digesters. However, biogas production in the digester receiving sulfate in combination with nitrate did not drop to 50% of control rates until a total of 40 days had elapsed at the higher addition rate. Sulfate accumulated in the digester receiving nitrate, with a reduction to sulfide of approximately 30%. In the digester receiving only sulfate, measured sulfate concentrations never exceeded $0.1\ \mu M\ ml^{-1}$ (detection limit). In another experiment, a digester was infused with sulfate until methane production became inhibited. Addition of nitrate to this digester did not reverse the inhibition. These experiments support the hypothesis that nitrate is preferentially reduced in the presence of sulfate and nitrate.

The results suggest that a nitrate-amended digester could be advantageously utilized as part of a multiphase digestion system. Nitrate reduction by a carbohydrate substrate will, theoretically, produce more heat per electron pair than reduction of oxygen. This makes nitrate amendment potentially attractive for temperature regulation. The reported changes in biomass particle size suggest that nitrate amendment may also favorably affect biomass degradation. The demonstration of nitrate conversion to ammonia suggests that nitrate could be economically retained in a recycle process, while also enhancing the nitrogen nutrition of a multiphase digestion process. Further, since nitrate has been shown to be preferentially reduced in the presence of sulfate and nitrate, nitrate amendment may offer possibilities for the regulation and control of inhibitory sulfide concentrations.

Volatile Fatty Acid Dynamics
The formation of methane from biomass involves many intermediates and the interaction of many microbial species. Acids believed to be the predominant precursors of methane are carbonic acid, acetic acid, propionic acid and butyric acid. The formation of these acids has been extensively documented for the conversion of wastes to methane. However, the microflora involved in the metabolism of these acids are poorly understood. Acids larger than acetic acid are believed to be metabolized to acetic acid, hydrogen and carbon dioxide, which are then converted to methane (Boone and Bryant, 1980; McInerney et al., 1981). In addition to formation from other sources, hydrogen may also be produced from the oxidation of acetate (Zinder and Koch, 1984). The effects of these acids on the overall fermentation have not been elucidated, although they (with the exception of carbonic acid) have long been recognized as inhibitors of microbial growth. In the course of the fermentation of organic material, the rate of acid formation may exceed the rate of dissimilation, in which case the concentration of the acids will continue to increase and the fermentation will ultimately cease. This fact has led to the development of two-phase anaerobic digestion systems. Theoretical modeling has been used to characterize the inhibitory effects of acids on the digestion process. However, no data are available to evaluate the effects of the rates of formation or the concentration of these acids on the rate and stability of anaerobic digestion of biomass in a single-stage process. There has been no procedure available which permits the assessment of these phenomena, because batch addition of the acids does not take into account possible shifts in the microbial population. However, the question could be addressed if it were possible to alter the acid formation rates in isolation. This cannot be done by increasing the rate of substrate addition because this also increases the concentration of many other molecules.

Henson et al. (1986) assumed that the extracellular concentration of an individual acid was a function of its rates of formation and dissimilation. How these two factors interact to produce a specific concentration was not an immediate concern. It was further assumed that the microflora could not distinguish between acids produced biologically and acids continuously infused into the digester.

Based on these assumptions, stable thermophilic digesters were infused with organic acids at a constant rate (in addition to daily feeding with biomass), over time periods sufficient to permit the growth of new microflora capable of metabolizing both the infused acids and the products from the biomass feedstock. Following the establishment of new steady-state organic acid concentrations and methane production rates, the rates of infusion were increased. This was repeated until digester failure occurred.

The concentration of a compound infused into a digester that is diluted daily by batch feeding is a nonlinear function of time. We have therefore introduced a term, 'cumulative added organic acid concentration', to designate the concentration of the acid which would result from infusion (for a specific period of time) and washout, assuming that the acid was not metabolized. Measurements of methane production, calculation of cumulative added organic acid concentrations and measurement of actual organic acid concentrations in the digester permitted us to evaluate the ability of the system to adjust to increased rates of formation and higher concentrations of the acids.

Experiments were conducted using butyrate and propionate as the test acids. Butyrate could be infused at a rate of 10 mM day^{-1} (millimoles per liter of mixed liquor per day) with no perturbation of the digestion. However, when the infusion rate was doubled to 20 mM day^{-1}, there was a perturbation of the digestion (Henson et al., 1986). The infusion rate was then reduced to 15 mM day^{-1} and the digestion restabilized. This infusion rate was continued for a period of 200 days. The infusion rate was then increased again to 20 mM day^{-1}. On this second occasion the fermentation remained stable. After a period of 50 days of stable operation, the infusion rate was further increased to 25 mM day^{-1}. At this rate the digestion was unstable and a rapid, continuous increase in butyrate concentration was observed (Fig. 1). Methane production fell precipitously, with ultimate digestion failure. The similar experiment conducted with propionate yielded essentially the same results (Fig. 2).

The thermophilic digesters functioned stably at very high organic acid concentrations, with concentrations as high as 140 mM, 190 mM and 40 mM observed for acetate, propionate and butyrate, respectively. This suggests that, if other factors could be regulated to accommodate the digestion, it

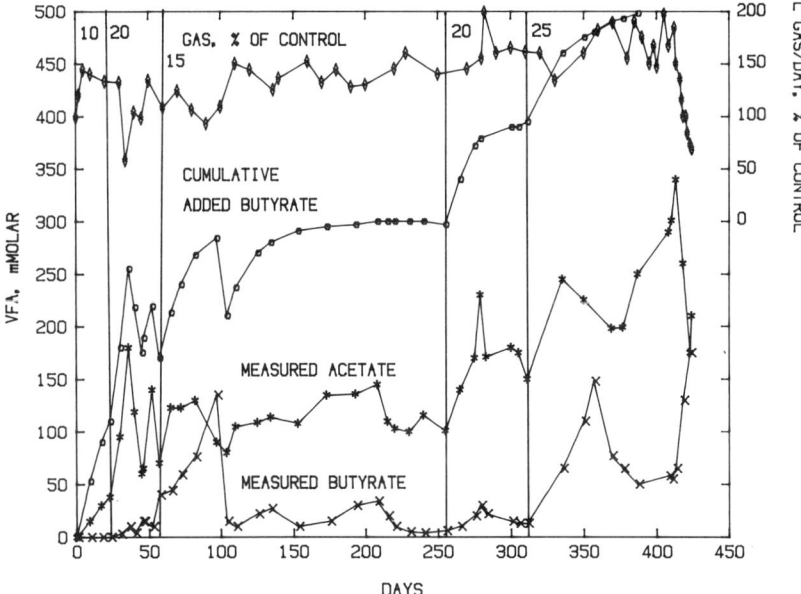

FIG. 1. Production of biogas and concentrations of acetate and butyrate in a single thermophilic digester (55°C) receiving continuous addition of sodium n-butyrate. The figure is subdivided into sections with the infusion rate (mM day^{-1}) given in the upper left corner for each section (Henson *et al.*, 1986).

may be possible to operate a phase of a multiphase digestion system at high concentrations of organic acids.

Effect of Micronutrients

Napiergrass (*Pennisetum purpureum* Schum.) was a feedstock selected for intensive study in this program. When standard mixed digesters were operated on this feedstock (supplemented with nitrogen and phosphorus) at a loading rate as low as 1·23 g VS liter^{-1} day^{-1}, at 35°C and a 20-day hydraulic retention time, the VFA concentrations were high and methane production was low (Wilkie *et al.*, 1986). An increase in the loading rate resulted in further elevation of the mixed liquor VFA concentrations. The reason for this condition was not apparent.

The carbon/nitrogen and carbon/phosphorus ratios were increased with no change in the fermentation profile. The concentrations of organic acids remained high and methane production was low, as expected in a fermentation where a substantial portion of the organic carbon would be

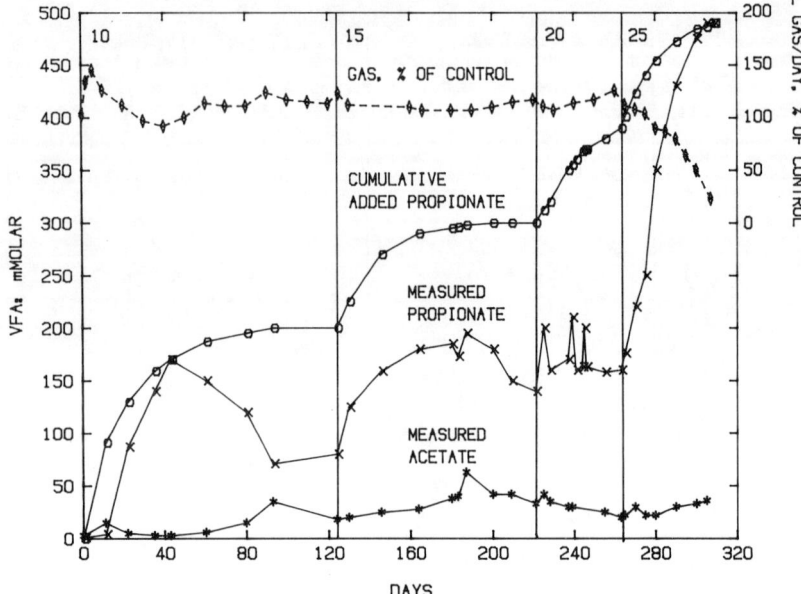

FIG. 2. Production of biogas and concentrations of acetate and propionate in a single thermophilic digester (55°C) receiving continuous addition of sodium propionate. The figure is subdivided as for Fig. 1 (Henson et al., 1986).

removed daily in the form of organic acids in the liquid effluent fraction. The response of the fermentation to increases in loading rates prevented experiments on specific perturbations of the fermentation, since any increase resulted in organic acid accumulation.

The importance of trace elements in anaerobic digestion, singly or in combination, has been reported (Speece and Parkin, 1985). Elemental analyses of the feedstock showed it to be devoid of nickel, cobalt, molybdenum and selenium (Wilkie et al., 1986). When a solution containing Ni, Co, Mo and Se was added with the daily feed, the concentrations of organic acids decreased rapidly (Fig. 3). Although the feedstock contained 0·41 g sulfur kg^{-1} dry matter, sulfate was also added since essential sulfides are lost in the product gas. In four separate experiments, the concentrations of acetic, propionic and butyric acids fell below the detection limits of 0·1 mM for all three acids (Fig. 3). Volumetric methane productivity (m^{3}CH$_4$ m^{-3} day^{-1}) increased to levels theoretically anticipated on the basis of complete metabolism of the organic acids present in the mixed liquor. The hydrogen concentrations in the gas phase remained low during the course of the investigation, varying from 17 to 24 ppm (Table 5). These results suggest that micronutrient addition substantially affected the organisms which metabolize volatile fatty acids, with relatively little effect on the organisms which metabolize molecular hydrogen, since no corresponding change was observed in the amount of end-product hydrogen.

The results obtained allow us to examine, quantitatively, the effects of a variety of parameters, such as organic acid concentrations, organic loading rates and pH, on the fermentation profile. The initial methane productivity of the digesters was limited since the loading rate was restricted by the high concentrations of volatile fatty acids in the mixed liquor. Subsequent to micronutrient supplementation, it has been possible through increased loading rates to further increase volumetric methane productivity from 0·5 to 2·6 $m^3 CH_4\ m^{-3}\ day^{-1}$, while still maintaining complete methanogenic conversion of the volatile fatty acids generated (Wilkie, unpublished). Work is in progress to determine an upper limit for the volumetric methane productivity of standard mixed, Napiergrass-fed digesters, through continued increases in loading rates.

The importance of micronutrients in anaerobic methanogenesis of energy crops may have been neglected due to the presence of seemingly adequate micronutrients and/or the practice of blending poor-quality feedstocks with nutrient-rich animal manures or sewage. The economic implication of these findings is tremendously important since trace elements would be a simple, low-cost, nutritional additive to biomass fermentations which are deficient in micronutrients. The micronutrient composition of energy crops should, therefore, be carefully determined relative to their utilization as feedstocks for methane production. Alteration of a fermentation profile by manipulation of micronutrients

TABLE 5
Average Methane Productivity and Average Evolved Biogas Composition from Four Reactors before and after Micronutrient Addition
(Wilkie et al., 1986)

Sample time (days):	Before					After				
	−29	−23	−16	−9	Average	+12	+19	+22	+28	Average
liter $CH_4\ day^{-1}$:	0·44	0·52	0·52	0·51	0·50	0·70	0·68	0·71	0·69	0·70
% CH_4:	69·7	69·3	65·9	69·7	68·7	62·4	63·7	65·7	61·2	63·3
% CO_2:	27·0	28·6	32·8	25·5	28·5	36·6	36·4	34·8	38·9	36·7
ppm H_2:	24	24	23	19	22·5	24	21	18	17	20

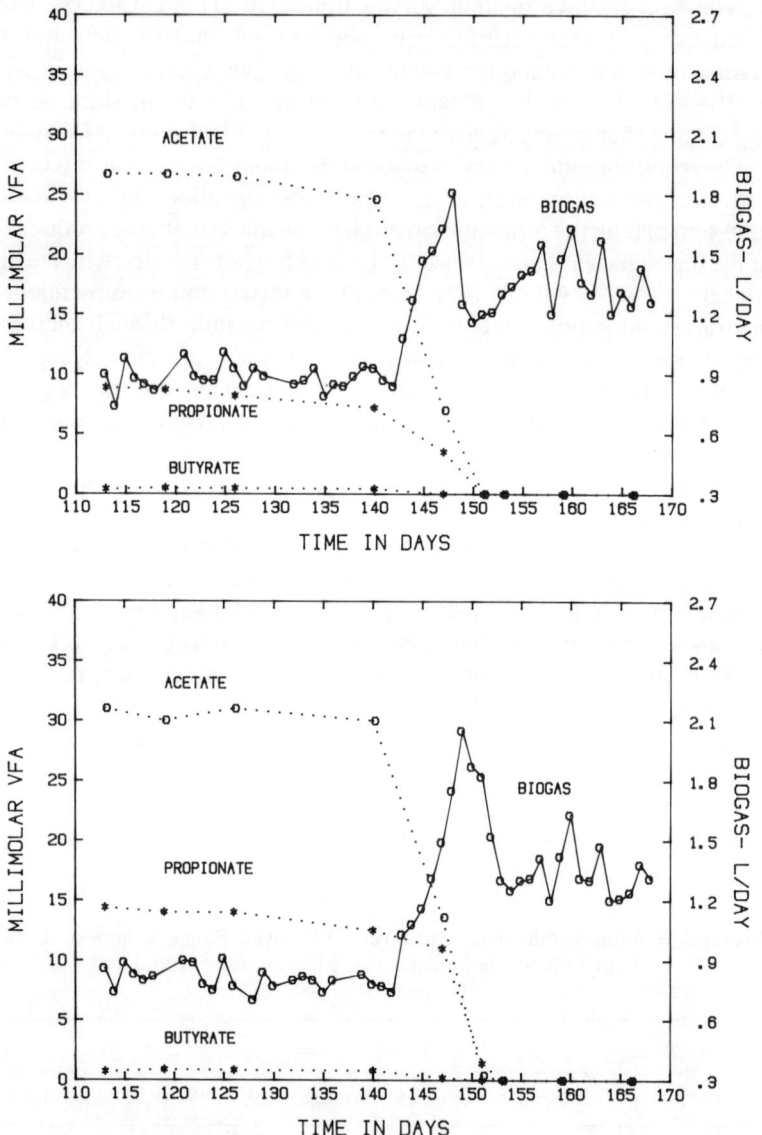

FIG. 3. VFA concentrations and biogas production in four Napiergrass-fed digesters, before and after micronutrient addition. Time axis is in days since start-up. Addition began on day 142. Last VFA measurements prior to addition were made on day 140 (Wilkie et al., 1986).

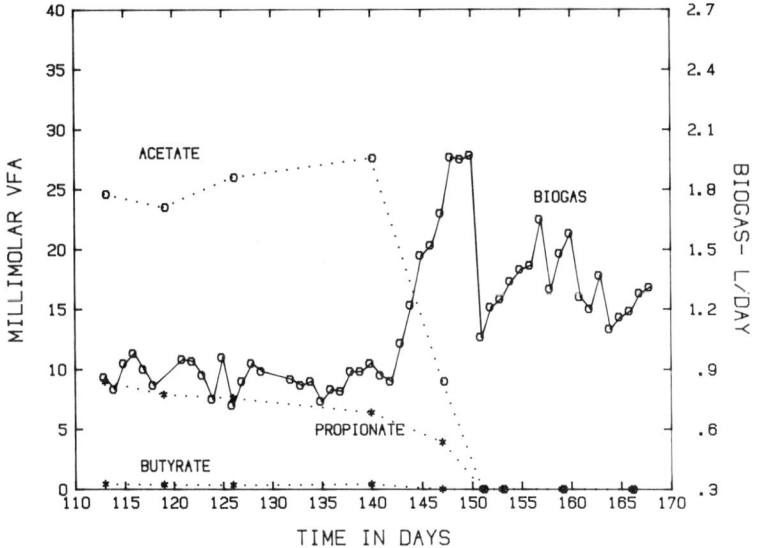
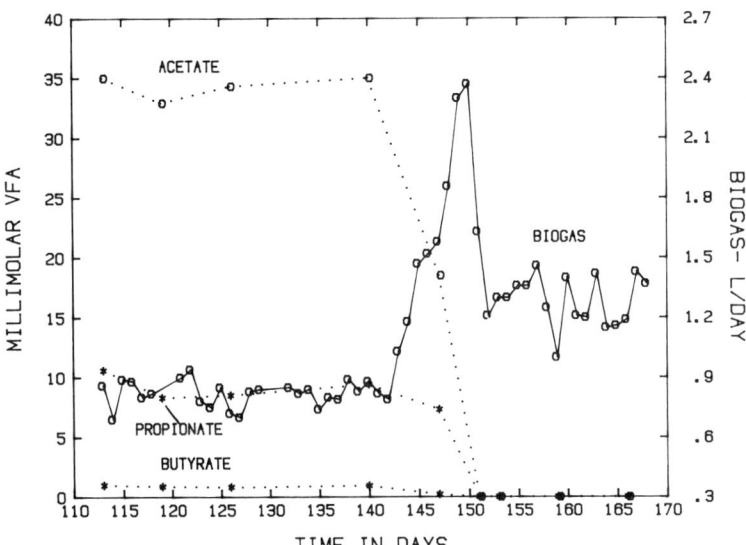

FIG. 3—contd.

may thus prove a useful tool in the regulation of specific aspects of multiphase digestion systems.

MATHEMATICAL MODELING

Mathematical modeling of an anaerobic digestion process can be helpful in the design, interpretation and evaluation of experimental research as well as in the design and operation of full-scale digesters. Errors in quantitative assessments become apparent when specific experimental data are found to be inconsistent with projections based on other accumulated data. Shortcomings in available information also become apparent when efforts are made to refine the model.

'First' models are normally developed using information from the literature and hypotheses formulated by the investigators. Such models can thus be considered as quantitative techniques for combining known facts and speculative information. These 'first' models should be kept as simple as possible, with the 'pieces' being put together carefully, and assumptions should be listed as each 'piece' is fitted into the model. The models should also be as mechanistic as possible, although empirical relationships may be used whenever mechanistic expressions are too complex or are not available.

In developing models to be used for guiding experimental research, data requirements must be carefully considered. Firstly, terms included in the model should be either measurable or easily calculated from other measurements. Secondly, it must be kept in mind that it is relatively easy to add additional terms to models, whereas experimental measurements can be both costly and time consuming.

As a point of departure, a simple model of a two-phase anaerobic digestion process was developed, based on existing information. The first phase, 'The Leaching Bed Phase', in which organic acids were formed, contained a high concentration of plant material. The methane-forming second phase, 'The Packed Bed Phase', received liquid effluent from the first phase. The complete process was termed the Leaching Bed–Packed Bed (LB–PB) process. A schematic of the overall reactor system is shown in Fig. 4. The number of components in such a system could be much larger, and would be determined by both engineering and biological considerations.

The objective of the modeling activities was to develop a mechanistic

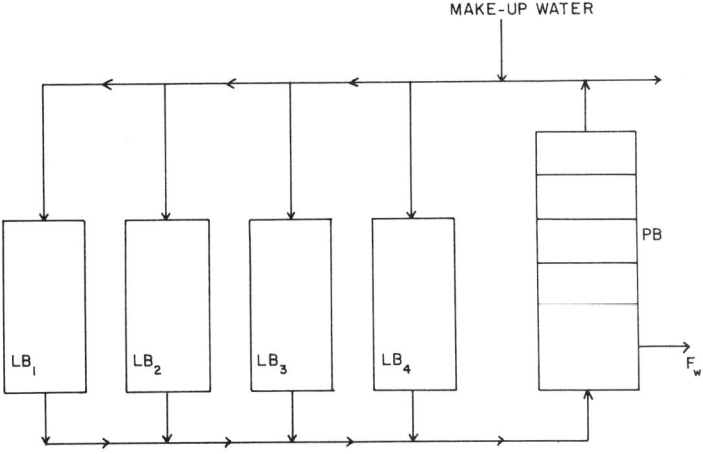

FIG. 4. Leaching Bed–Packed Bed (LB–PB) system.

mathematical model. The general approach used in the model development consisted of the five steps listed below:
1. Identification of the significant reactions.
2. Establishment of the stoichiometry.
3. Determination of the kinetics.
4. Characterization of the reactor.
5. Development of the model equations.

After the model was developed, a FORTRAN-based simulation package was prepared for the LB–PB process, on which experimental studies were being initiated. Simulations were conducted for planning experiments. Typical items of interest were 'first guesses' at loadings to be applied, types, location, accuracy and frequency of measurements to be made, and duration of the experiments. The modeling was iterative, in that the results of computer simulations, based on the model, could be tested experimentally and the results of the experiments could be used to refine the model. Hopefully, it would only be necessary to change the numerical values of the parameters in the model. However, for processes which are relatively unexplored, it is more likely that some changes in the basic structure of the model would be necessary. The model could then be subjected to field testing and, ultimately, could be used for process design and operation (including economic analysis).

All of the above steps should be repeated until the model is deemed satisfactory for its intended use. Final steps might include the conduct of sensitivity analyses for model simplification and the provision of suitable programming for linking the model into some overall larger model. Preparation of documentation and a user's guide for the model and simulation package are also important final steps.

The model developed for the LB–PB process and the simulations conducted to date should be considered as preliminary in nature and subject to change, since insufficient experimental data are available for quantitative model validation. Additional iterations between modeling, simulation and experimentation will be necessary before the model is used for purposes other than guiding model development and experimentation.

Leaching Bed Model

The amount of research which has been conducted on LB reactors is relatively small in comparison to that on PB reactors. Less theoretical knowledge is thus available and first models will, therefore, contain more uncertainty. Both the biological and the chemical reactions must be considered in modeling the process. A dynamic model is required since the LB reactor is operated in a batch mode.

Reaction Network

Almost any biomass (Napiergrass, etc.) used as a feedstock consists of many components, each of which will be biodegraded at a different rate (Eastman and Ferguson, 1981; Gujer and Zehnder, 1983; Bryers, 1985). The simplest structuring of the biomass is to assume that the biomass consists of two components, a rapidly biodegradable portion (BM_f) and a slowly biodegradable portion (BM_s). This structure is illustrated in eqn (1) which also shows the overall reaction network assumed for the model.

Reaction Network

$$BM_f \searrow \atop BM_s \nearrow \;\; SI \xrightarrow{X_a} \begin{array}{c} FA_t \\ + \\ CO_2 \\ + \\ Z \end{array} \xrightarrow{X_m} \begin{array}{c} CH_4 \\ + \\ CO_2 \end{array} \quad (1)$$

where, BM_f = rapidly biodegradable biomass
BM_s = slowly biodegradable biomass

SI = soluble intermediates
FA_t = total volatile fatty acids
CO_2 = carbon dioxide
CH_4 = methane
X_a = acid-producing bacteria
X_m = methane-producing bacteria
Z = net cations

The first step in the reaction network consists of parallel reactions in which the two solid biomass components are converted to soluble intermediates (SI). In the second step, the SI are converted to volatile fatty acids (FA_t) by acid-producing bacteria (X_a) with concurrent production of carbon dioxide, net cations (Z) and more acid-producing bacteria. The third step is the conversion of the volatile fatty acids by methane-producing bacteria (X_m) to form methane, carbon dioxide and more methane bacteria.

It was realized that the actual reaction network is more complicated than this, in that groups of microorganisms other than 'acid producers' and 'methane producers' are involved in the conversion of soluble intermediates to methane and carbon dioxide. Also, it is known that hydrogen is an important intermediate (Bryant, 1977; Hill, 1982; Mosey, 1983; Rozzi et al., 1985). Work is currently in progress to include some of these reactions in the model.

Stoichiometry
The stoichiometric relationships between the different components of eqn (1) were established by eight yield coefficients, as given in eqns (2) through (9):

$$Y_1 = \frac{\Delta SI}{\Delta BM_f} = \frac{\Delta SI}{\Delta BM_s} \tag{2}$$

$$Y_a = \frac{\Delta FA_t}{\Delta SI} \tag{3}$$

$$Y_b = \frac{\Delta X_a}{\Delta SI} \tag{4}$$

$$Y_c = \frac{\Delta CO_2}{\Delta SI} \tag{5}$$

$$Y_2 = \frac{\Delta X_m}{\Delta FA_t} \tag{6}$$

$$Y_d = \frac{\Delta CH_4}{\Delta FA_t} \tag{7}$$

$$Y_e = \frac{\Delta CO_2}{\Delta FA_t} \tag{8}$$

$$Y_z = \frac{\Delta Z}{\Delta SI} \tag{9}$$

In each case, the yield is defined as the ratio of the change in a product to the change in a reactant. All components, except for bacteria, methane, carbon dioxide and net cations, were expressed in chemical oxygen demand equivalents (COD_e). Common units of COD_e were used to facilitate the conversion of units as well as to permit establishment of the thermochemistry for later use in an energy balance.

Numerical values for the yield coefficients were based on assumptions as to how the components were transformed (Table 6). In the hydrolysis reactions, it was assumed that the COD_e of the soluble intermediates was equal to the COD_e of biomass destroyed, i.e. the yield coefficient (Y_1) was equal to 1·0. The value of the coefficient for internal generation of net cations (Y_z) was estimated from experimental data (Goto et al., 1986). The other coefficients were estimated based on theoretical considerations of free energy changes and ATP synthesis for typical volatile fatty acids, as described by Sawyer and McCarty (1978) and Mosey (1983), or they were empirically determined. Most of the values listed are in reasonable agreement with the tabulated values from experimental studies which were presented in the review by Henze and Harremoes (1983).

Reaction Kinetics

The expressions used in this preliminary model 'lump' the effects of both reaction and mass transfer. These can be separated at a later date if desired, but this will result in a more complicated model.

The rates of hydrolysis of BM_f and BM_s were assumed to be first order with respect to the concentration of BM_f and BM_s, respectively, as proposed by Eastman and Ferguson (1981) and Gujer and Zehnder (1983), among others. The kinetic expressions are presented in eqns (10) and (11):

$$r(BM_f) = -k_1(BM_f) \tag{10}$$

$$r(BM_s) = -k_1 p(BM_s) \tag{11}$$

TABLE 6
Summary of Parameter Values for the Leaching Bed and Packed Bed Models

Parameter	Value	Unit
Y_1	1·0	mg SI mg^{-1} BM
Y_2	0·035	mg X_m mg^{-1} FA_t
Y_a	0·75	mg FA_t mg^{-1} SI
Y_b	0·17	mg X_a mg^{-1} SI
Y_c	$0·13\,E^{-3}$	liter CO_2 mg^{-1} SI
Y_d	$0·33\,E^{-3}$	liter CH_4 mg^{-1} FA_t
Y_e	$0·33\,E^{-3}$	liter CO_2 mg^{-1} FA_t
Y_z	$0·80\,E^{-5}$	mol Z mg^{-1} SI
k_1	8·50	day^{-1}
$k_1 p$	0·45	day^{-1}
k_2	$0·16\,E^{-3}$	liter mg^{-1} day^{-1}
k_5	0·73	day^{-1}
k_6	0·024	day
k_{d1}	0·04	day^{-1}
k_{d2}	0·04	day^{-1}
K_a	$0·32\,E^{-4}$	mol $liter^{-1}$
K_{bc}	$1·00\,E^{-6}$	mol $liter^{-1}$
K_h	$3·23\,E^{-5}$	mol mm^{-1} Hg $liter^{-1}$
KI	3 000·0	mg $liter^{-1}$
KIp	4 000·0	mg $liter^{-1}$
KL_a	10·0	day^{-1}
K_s	20·0	mg $liter^{-1}$
D	25·5	liter mol^{-1}
MUhat	0·54	day^{-1}

where ( ) = concentration (mg COD_e $liter^{-1}$)

k_1 = first order rate constant for hydrolysis of BM_f (day^{-1})

$k_1 p$ = first order rate constant for hydrolysis of BM_s (day^{-1})

r = d/dt.

In the kinetics of hydrolysis, inhibition due to high volatile acid concentrations may occur. This possibility was not considered in the current model but should be explored in the future.

The growth of microorganisms is commonly modeled using a 'Monod function', which may be written as:

$$\text{Rate} = \text{MUhat} \frac{(\text{substrate})}{K_s + (\text{substrate})} \quad (12)$$

where MUhat = the maximum specific growth rate of the bacteria (day^{-1})

K_s = a saturation coefficient (mg COD_e $liter^{-1}$)

An inhibition term is sometimes added to the above to reflect the effect of high substrate concentration on growth rate. Such a term was incorporated in eqn (17).

The possible effects of kinetic limitation by mass transfer can be reflected in the numerical value of the saturation coefficient, K_s, with high values of K_s resulting from mass transfer limitation.

When substrate concentration is low as compared with the value for K_s, the Monod function can be simplified to a first order expression. This was assumed to be the case in the leaching bed so that the rate of growth of the acid-producing bacteria was proportional to the concentration of soluble intermediates, SI, and acid-producing bacteria, X_a, as shown in eqn (13):

$$\text{Rate of Growth of } X_a = k_2 \, (SI) \, (X_a) \left\{ 1 - \frac{(FA_t)}{(FA_t) + KI} \right\} \quad (13)$$

where k_2 = growth rate constant for X_a (liter mg^{-1} day^{-1})

KI = inhibition coefficient for X_a (mg COD_e $liter^{-1}$)

The last term in eqn (13) reflects the inhibitory effect of high concentrations of total volatile fatty acids on the growth of X_a. This function will be called the 'complementary Monod function'. Lower growth rates can result from either lower values of KI or higher values of FA_t.

Bacterial mass will decrease with time, and the rate of decay of the acid-producing bacteria was assumed to be proportional to their concentration.

$$\text{Rate of Decay of } X_a = k_{d1} \, (X_a) \quad (14)$$

where k_{d1} = decay rate constant for X_a (day^{-1}).

The net rate of growth for the acid-producing bacteria was obtained by combining eqns (13) and (14), as given below:

$$\text{Net Rate of Growth} = \text{Rate of Growth} - \text{Rate of Decay} \quad (15)$$

$$r(X_a) = k_2 \, (SI) \, (X_a) \left\{ 1 - \frac{(FA_t)}{(FA_t) + KI} \right\} - k_{d1} \, (X_a) \quad (16)$$

The rate of growth of the methane-producing bacteria was also assumed to be proportional to their concentration, with the relationship between volatile fatty acids concentration and growth rate following the inhibition function proposed by Andrews and Graef (1971).

$$\text{Rate of Growth of } X_m = \frac{\text{MUhat}(X_m)}{1 + K_s/(\text{FAH}) + (\text{FAH})/KIp} \quad (17)$$

where K_s = saturation coefficient (mg COD_e liter^{-1})
  MUhat = maximum specific growth rate for X_m (day^{-1})
  KIp = inhibition coefficient for X_m (mg COD_e liter^{-1})
  FAH = unionized volatile fatty acids (mg COD_e liter^{-1}).

In the above equation, it is assumed that the methane-producing bacteria can utilize only unionized volatile fatty acids as substrate. At high concentrations, the unionized volatile fatty acids can be inhibitory (Andrews and Graef, 1971). This has the effect of making the growth rate of the methane bacteria responsive to both total volatile fatty acid concentration and pH.

The rate of decay of the methane-producing bacteria was assumed to be proportional to their concentration, as given in eqn (18):

$$\text{Rate of Decay of } X_m = k_{d2}(X_m) \quad (18)$$

where k_{d2} = decay rate constant for X_m (day^{-1}).

Combining eqns (17) and (18) yields the expression for the net growth rate of X_m:

$$r(X_m) = \frac{\text{MUhat}(X_m)}{1 + K_s/(\text{FAH}) + (\text{FAH})/KIp} - k_{d2}(X_m) \quad (19)$$

Kinetic expressions for soluble intermediates (SI), volatile fatty acids (FA_t), net cations (Z), methane (CH_4) and carbon dioxide (CO_2) can be established using the above kinetic expressions and the yield coefficients, as shown below:

$$r(\text{SI}) = Y_1\{k_1(\text{BM}_f) + k_1p(\text{BM}_s)\}$$

$$- \frac{1}{Y_b}k_2(\text{SI})(X_a)\left\{1 - \frac{(\text{FA}_t)}{(\text{FA}_t) + KI}\right\} \quad (20)$$

$$r(FA_t) = \frac{Y_a}{Y_b} k_2(SI)(X_a) \left\{ 1 - \frac{(FA_t)}{(FA_t) + KI} \right\}$$

$$- \frac{1}{Y_2} \frac{MUhat(X_m)}{1 + K_s/(FAH) + (FAH/KIp)} \quad (21)$$

$$r(Z) = \frac{Y_z}{Y_b} k_2(SI)(X_a) \left\{ 1 - \frac{(FA_t)}{(FA_t) + KI} \right\} \quad (22)$$

$$rCH_4 = V \frac{Y_d}{Y_2} \left\{ \frac{MUhat(X_m)}{1 + K_s/(FAH) + (FAH)/KIp} \right\} \frac{1}{D} \quad (23)$$

$$rCO_2 = V \frac{Y_c}{Y_b} k_2(SI)(X_a) \left\{ 1 - \frac{(FA_t)}{(FA_t) + KI} \right\} \frac{1}{D}$$

$$+ V \frac{Y_e}{Y_2} \left\{ \frac{MUhat(X_m)}{1 + K_s/(FAH) + (FAH)/KIp} \right\} \frac{1}{D} \quad (24)$$

(note: $r(CO_2) = rCO_2/V$)

where V = reactor mixed liquor volume (liters)
D = gas constant (liter mol^{-1}).

It is expected that all reaction rates will be functions of temperature. Appropriate functions relating rates and temperature will need to be incorporated into the model if it is to be used to predict performance at different temperatures. The current model is for a temperature of 38°C.

Chemical Reactions

There are several chemical reactions which also influence process performance. The reactions shown below were used as a basis for modeling the variation of important chemical species in the liquid and gas phases of the LB reactor.

The total volatile fatty acids (FA_t) consists of both unionized volatile fatty acids (FAH) and ionized volatile fatty acids (FA^-), i.e.:

$$FA_t = FAH + FA^- \quad (25)$$

$$FAH \rightleftharpoons FA^- + H^+ \quad (26)$$

where FAH = unionized volatile fatty acids
FA$^-$ = ionized volatile fatty acids
H$^+$ = hydrogen ion

As previously mentioned, it was assumed that the unionized volatile fatty acids were both the growth-limiting substrate and the inhibiting agent (Andrews and Graef, 1971) for the methane bacteria. Ionization reactions are very fast in comparison to biological reactions so equilibrium, as given in eqn (26), may be assumed.

$$\frac{(FA^-)(H^+)}{(FAH)} = K_a \quad (27)$$

where ( ) = concentration (mol liter^{-1})
K_a = volatile fatty acids dissociation constant,
3·2E^{-5} mol liter^{-1} for acetic acid at 38°C, ionic strength of 0·02.

It was assumed that the system was buffered primarily by the carbon dioxide–bicarbonate system. The concentration of dissolved CO_2 in the liquid phase at equilibrium, $(CO_2)D_s$, can be obtained from Henry's law:

$$(CO_2)G \rightleftharpoons (CO_2)D \quad (28)$$

$$((CO_2)D_s) = K_h\, PCO_2 \quad (29)$$

where $(CO_2)G$ = gas phase CO_2
$(CO_2)D$ = liquid phase (dissolved) CO_2
$(CO_2)D_s$ = liquid phase CO_2 at equilibrium
K_h = Henry's law constant,
3·23E^{-5} mol mm^{-1} Hg liter^{-1} at 38°C, ionic strength of 0·02
PCO_2 = partial pressure of CO_2 in the LB gas phase, mm Hg.

Dissolved carbon dioxide, $(CO_2)D$, bicarbonate, HCO_3^-, and total dissolved inorganic carbon, $(CO_2)_t$, are related by the expressions given below. The concentration of carbonate ion, CO_3^{2-}, can be neglected if the range of validity of the model is restricted to a pH less than 8.

$$(CO_2)D + H_2O \rightleftharpoons H^+ + HCO_3^- \quad (30)$$

$$\frac{(H^+)(HCO_3^-)}{((CO_2)D)} = K_{bc} \quad (31)$$

$$((CO_2)_t) = ((CO_2)D) + (HCO_3^-) \quad (32)$$

where $(CO_2)_t$ = total dissolved inorganic carbon
HCO_3^- = bicarbonate
K_{bc} = dissociation constant, $1 \cdot 0E^{-6}$ mol liter^{-1} at 38°C, ionic strength of 0·02.

Finally, a charge balance equation can be used to relate the bicarbonate alkalinity (HCO_3^-), ionized volatile fatty acids (FA^-), total cations (C), total anions (A), hydroxyl ion (OH^-) and hydrogen ion (H^+):

$$(H^+) + (C) = (HCO_3^-) + (FA^-) + (OH^-) + (A) \tag{33}$$

where C = cations excluding hydrogen ion, H^+
A = anions excluding HCO_3^-, FA^-, and OH^-.

The hydrogen and hydroxyl ion concentrations are negligible compared to the concentrations of other ions, and, as previously discussed, the concentration of the carbonate ion can be neglected for pH values less than 8. These simplifications result in an equation relating the bicarbonate alkalinity to the ionized volatile fatty acids and net cation concentration, (Z):

$$(HCO_3^-) = (Z) - (FA^-) \tag{34}$$

$$(Z) = (C) - (A) \tag{35}$$

where Z = the net cation concentration.

The cations are assumed to be generated simultaneously with the conversion of soluble intermediates to volatile fatty acids by using the yield coefficient, Y_z, given in eqn (9).

Equations (25), (27), (31), (32) and (34) can be combined to give a quadratic expression for the bicarbonate alkalinity, HCO_3^-, in terms of FA_t, K_a, K_{bc}, $(CO_2)_t$ and Z. Equations (27) and (31) can be used to determine the concentration of hydrogen ion, H^+, and the concentration of unionized volatile fatty acids, FAH. These equations are summarized below:

$$(HCO_3^-) = \frac{-q + \{q^2 - 4pr\}^{1/2}}{2p} \tag{36}$$

where q = $((CO_2)_t) K_{bc} + (FA_t) K_a - (Z)\{K_a - K_{bc}\}$
p = $K_a - K_{bc}$
r = $-(Z)((CO_2)_t) K_{bc}$

$$(H^+) = K_{bc} \frac{((CO_2)_t) - (HCO_3^-)}{(HCO_3^-)} \qquad (37)$$

$$(FAH) = \frac{(H^+)(FA_t)}{K_a + (H^+)} \qquad (38)$$

An expression for the rate of transfer (T_g) of carbon dioxide from the liquid to the gas phase is required and the standard two-film gas transfer equation was used for this purpose. The dissolved CO_2 concentration at equilibrium, $(CO_2)D_s$, is calculated from Henry's Law (eqn (29)).

$$T_g = KL_a \{((CO_2)D_s) - ((CO_2)D)\} \qquad (39)$$

where T_g = dissolved CO_2 gas transfer rate
(mol liter^{-1} day^{-1})
KL_a = overall mass transfer coefficient for CO_2
(10·0 day^{-1}).

Reactor Characterization
The reactor used as a leaching bed must be characterized with respect to both the fluid and solid phases. This was accomplished on the basis of assumptions about mixing conditions and flow patterns into and out of the reactor. For the biomass, it was assumed that loading into the reactor was in an intermittent or periodic manner, i.e. 'batch' operation. The liquid phase was assumed to be a continuous flow, recycle stream from the PB, with the outflow being equal to the inflow. Complete mixing of the reactor contents was assumed for both the liquid and solid phases. If necessary, this assumption can be modified by using complete mixing reactors in series to approximate actual mixing conditions.

The effluent concentration of a component in a continuous flow, completely mixed reactor is equal to the concentration in the reactor. This is a characteristic of this type of reactor and is a reasonable assumption for soluble components in the liquid phase. However, it would be expected that the concentration of solid components, such as bacteria, would be more dependent on transfer between the solid and liquid phases which would be influenced by many factors such as particle size, intensity of mixing, etc. In the current model, it was assumed that the concentration of bacteria in the effluent from the LB was directly proportional to the bacterial concentration in the LB and to the effluent flow rate, and inversely proportional to the volume of the LB, as given in eqns (40) and (41):

$$(X_{a,out}) = k_6 \frac{F(X_a)}{V} \tag{40}$$

$$(X_{m,out}) = k_6 \frac{F(X_m)}{V} \tag{41}$$

where $(X_{a,out})$ = concentration of acid-producing bacteria in the effluent (mg liter^{-1})
$(X_{m,out})$ = concentration of methane-producing bacteria in the effluent (mg liter^{-1})
F = flow rate through the LB (liter day^{-1})
V = reactor volume (liters)
k_6 = empirical coefficient (day).

To summarize, each leaching bed was modeled as a completely mixed 'batch' reactor with respect to the solid phase and as a completely mixed, continuous flow reactor with respect to the liquid phase. To account for the transfer of bacteria from the solid to the liquid phase, the concentration of bacteria in the effluent from the LB was assumed to be directly proportional to both the flow rate through the LB and the concentration of bacteria in the LB, and inversely proportional to the reactor volume.

Material Balances

The appropriate reactions, stoichiometric relations, kinetics and type of reactor have been specified for the leaching bed. With this information, it was possible to prepare material balances with the resultant set of equations comprising the mathematical model for the process.

The material balance equation describes the time rate of change of a component within specified system boundaries. In the leaching bed, the boundaries are the same as the physical boundaries of the reactor. The general material balance may be written as:

$$\begin{bmatrix} \text{Rate of change} \\ \text{of a component} \end{bmatrix} = \begin{bmatrix} \text{Rate at which the} \\ \text{component enters} \\ \text{the reactor} \end{bmatrix} - \begin{bmatrix} \text{Rate at which the} \\ \text{component leaves} \\ \text{the reactor} \end{bmatrix}$$

$$\pm \begin{bmatrix} \text{Rate of generation} \\ \text{or utilization of} \\ \text{the component} \end{bmatrix} \tag{42}$$

The corresponding differential equation for each component and numerical values used in the mathematical model are summarized in a schematic diagram (Fig. 5) and Table 6.

Packed Bed Model
Different reactions predominate in the packed bed as compared to the leaching bed. During normal operation of the LB, the most significant reactions are the conversion of biomass to soluble intermediates and volatile fatty acids whereas, in the PB, the most significant reactions are the conversion of soluble intermediates to volatile fatty acids and thence to methane and carbon dioxide.

The same approach was used in developing a mathematical model for the packed bed reactor as for the leaching bed. However, since basically the same biological reactions, stoichiometry and kinetics are involved, the development here commences with a discussion of the reactor vessel and assumptions particular to the PB.

Reactor Characterization
Dahab and Young (1982), among others, have shown that the solids distribution in upflow PB reactors is such that a zone of concentrated solids exists in suspension in the lower portion of the reactor whereas, in the upper portion, the solids are primarily attached to the packing. It thus appears that the major function of the packing is to separate the solids from the liquid, permitting the accumulation of a high concentration of solids in the slurry portion of the PB which would result in a reduction of the reactor volume. Solids–liquid separation by this means is a complex process and cannot be mechanistically modeled at present. However, Guiot and van den Berg (1985) have taken advantage of this by including packing only in the upper portion of the reactor, thus reducing the amount of packing required and the overall cost of the reactor. The PB reactor modeled was assumed to be similar to the hybrid reactor proposed by Guiot and van den Berg.

The PB was modeled as two reactors in series: a slurry reactor, in which both the bacterial solids and the substrate were in suspension, and a fixed film reactor, in which the solids were attached to the packing. From a mixing point of view, the first reactor was considered to be completely mixed while the second reactor was considered to have plug flow characteristics.

To simplify simulation of the fixed film reactor, it was assumed that its mixing characteristics could be approximated by using a 'complete mixing

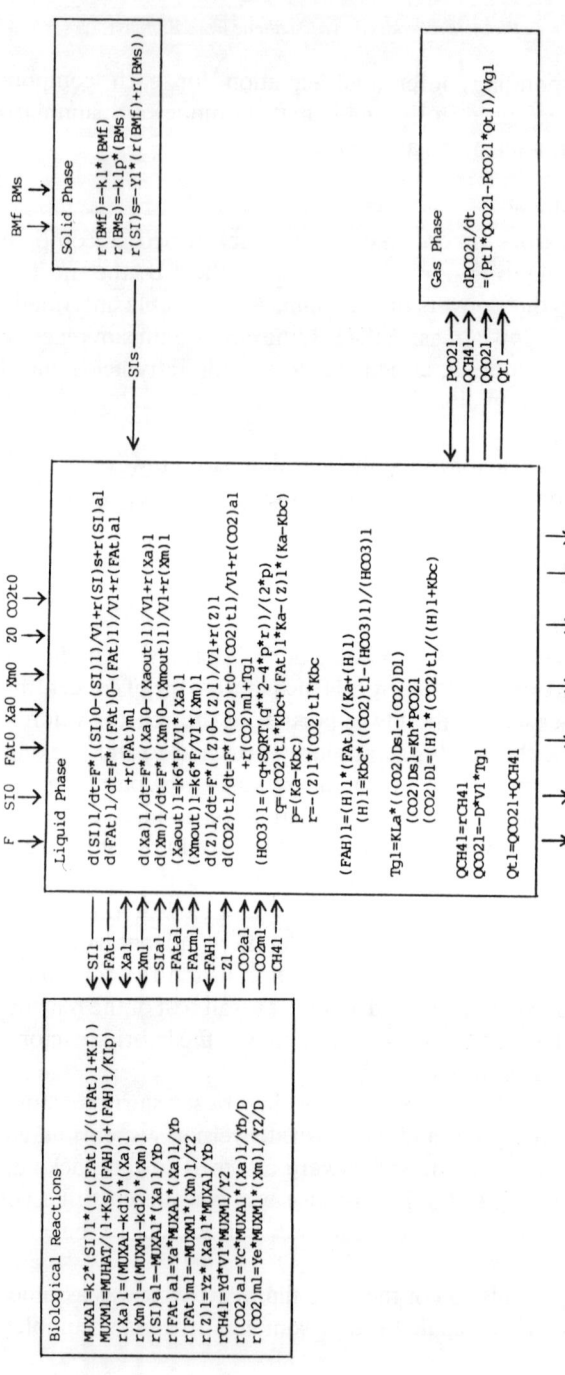

FIG. 5. Summary of Leaching Bed (LB) model.

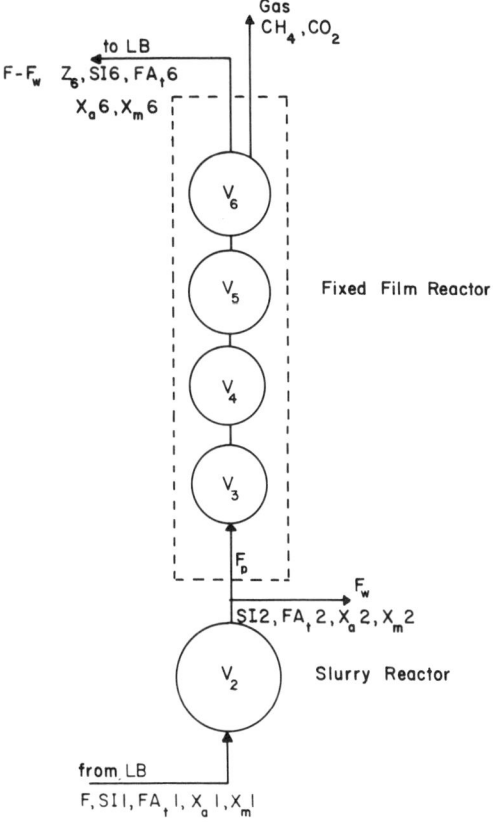

FIG. 6. Idealized reactor configuration for the packed bed.

tanks in series' approach. This is a common technique for simplifying solution of the equations involved and is justified since true plug flow is rarely attainable in practice. Four tanks in series were used as a first approximation.

The idealized reactor configuration for the packed bed is shown in Fig. 6. For the purpose of this preliminary model, it was assumed that the lower one-third of the total reactor volume was occupied by the slurry reactor and the upper two-thirds by the fixed film reactor, in accordance with the results of Dahab and Young (1982). In reality, the split between the volumes occupied by the two idealized reactors may be a relatively complex function depending upon several factors such as amount and type of packing, reactor design, gas evolution/unit volume, recycle flow rate, settling characteristics of the sludge, *inter alia*.

From a process dynamics and control point of view, an advantage of allowing the concentrated slurry to occupy only a portion of the reactor volume is that it provides a 'safety factor' (volume for the storage of excess sludge) if poor settling (bulking) of the sludge occurs. This would prevent the excessive loss of solids in the effluent from the PB and thus possible process failure. This also highlights the need to measure the volume occupied by the slurry portion, or level of slurry blanket in the reactor, for process operation and control.

Assumptions and Material Balances

Before preparing the material balances, the assumptions which govern the reactions, stoichiometry and kinetics, and how they fit into the material balances for the PB reactor are stated explicitly. These assumptions were:

1. Both the biological and chemical reactions taking place in the lower third of the PB (the slurry reactor) were the same as those in the LB, except there was no biomass (other than X_a and X_m) present.
2. There was a fixed thickness of biofilm on the packing in the fixed film reactor. This means that the concentrations of X_a and X_m in this reactor were assumed to be constant.
3. The only reaction occurring in the fixed film reactor was the conversion of volatile fatty acids to methane and carbon dioxide.
4. The rate of utilization of volatile fatty acids in the fixed film reactor was proportional to the concentration of the acids.
5. A limited range of pH values (6·5–7·5) was considered in the fixed film reactor. This was a reasonable assumption since most of the volatile fatty acids were utilized before reaching this portion of the PB. This assumption greatly simplified the gas–liquid equilibrium equations for the reactor, since concentrations of ionized volatile fatty acids and total volatile fatty acids were assumed to be equal (Andrews and Graef, 1971).
6. Concentrations of X_a and X_m in the effluent from the LB were assumed to be directly proportional to the flow rate and bacterial concentrations and inversely proportional to the volume of the LB, as previously discussed.
7. The mass of both the volatile fatty acid (X_a) and methane-producing bacteria (X_m), and thus their growth rate, was controlled not only by intentional wasting (F_w) of sludge from the slurry reactor but, also, by their concentration in the effluent. Concentrations of bacteria in the

effluent from the PB were assumed to be directly proportional to the flow rate and bacterial concentrations in the slurry reactor, and inversely proportional to the volume of the slurry reactor.
8. The growth of methane-producing bacteria (X_m) was assumed to be inhibited only in the slurry portion of the reactor. No inhibition was assumed in the fixed film reactor. This is reasonable since volatile fatty acid concentrations in the fixed film reactor were expected to be low under normal operational conditions.

Based on these assumptions, the function of the fixed film reactor was essentially that of 'polishing' the effluent of the slurry reactor by converting low concentrations of the acids to methane and carbon dioxide, and providing a 'buffer' zone to prevent the loss of bacteria during times of poor sludge settling or hydraulic shocks. The majority of the acids were converted to methane in the slurry reactor. The fixed film portion of the PB thus has a relatively minor influence on methane production so that a complex model of the reactions occurring in the fixed film reactor is not needed for estimating methane production rates.

Similar to the leaching bed reactor, material balance equations for each component in each portion of the PB reactor yielded mathematical expressions which comprised the model. A schematic diagram (Fig. 7) and Table 6 summarize the mathematical equations and numerical values of the coefficients used in the model. A subscripting convention was adopted to designate the origin of a given component relative to the idealized reactor configuration of Fig. 6. For this purpose, packed bed influent and the slurry reactor effluent were designated by the subscripts 1 and 2, respectively. In general, the completely mixed reactor with suspended growth and the reactors in series with fixed film growth were referred to as the 'slurry' reactor and the 'fixed film' reactor, respectively.

Leaching Bed/Packed Bed System

The two models presented for the LB and PB, respectively, were combined into a system by coupling the four leaching beds with the packed bed (Fig. 4). Each of the four LB reactors was operated as a batch reactor with respect to the solid phase and as a continuous flow reactor with respect to the fluid phase, except during times of charging with solids. The LB reactors were charged at different times so that the fluid flow to the PB reactor would be a mixture of streams from the LB reactors, each having a different composition and concentration. This dampens the concentration of volatile fatty acids in the influent to the packed bed, thus permitting it to operate in a more steady-state fashion.

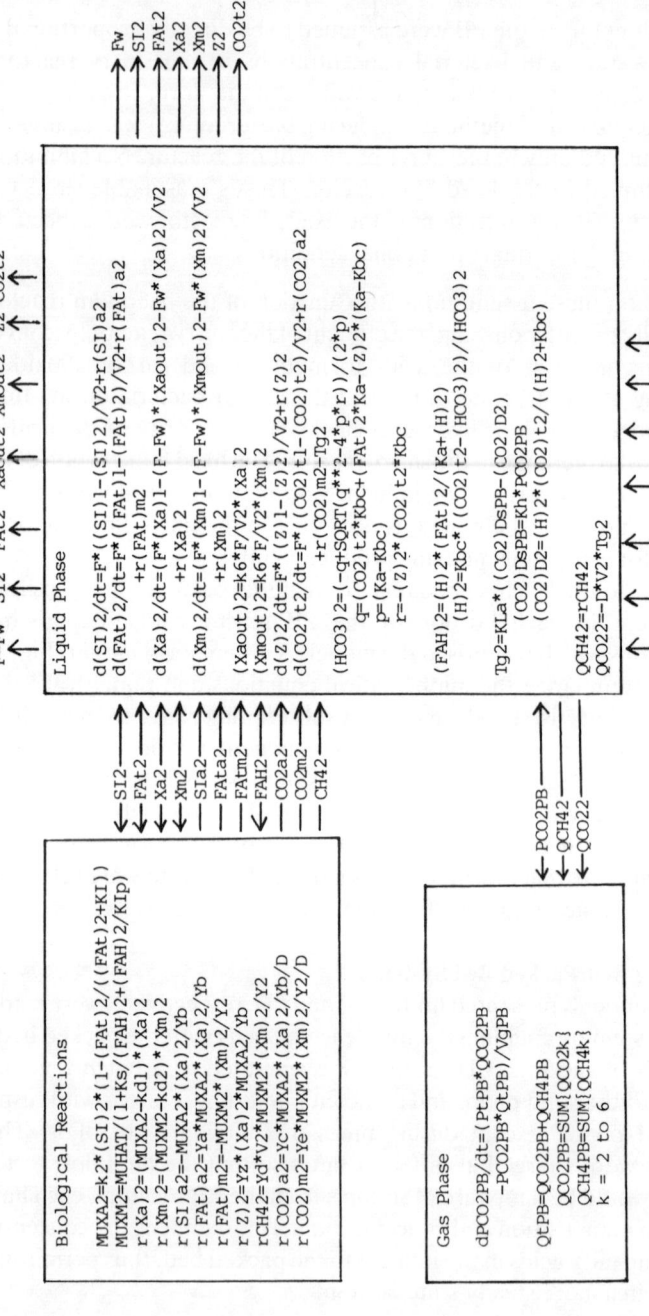

FIG. 7. Summary of Packed Bed (PB) model. (A) Slurry reactor portion.

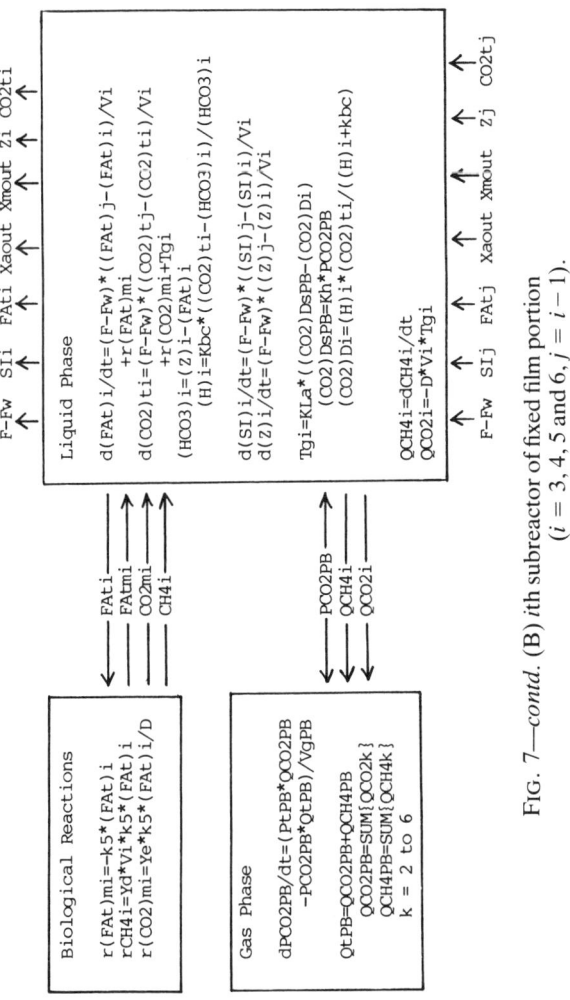

FIG. 7—*contd.* (B) *i*th subreactor of fixed film portion ($i = 3, 4, 5$ and $6, j = i - 1$).

There are several manipulatable variables for the system, such as: (a) make-up water, to replace losses and control salt content and alkalinity; (b) waste sludge from the PB reactor; (c) recycle around the PB reactor, and (d) recycle around the LB reactors and, perhaps, around each individual LB reactor. Selecting the proper operational strategy is therefore difficult.

Three possible reasons for recycling around the LB reactors are:

1. To bring the high volatile fatty acids (FA_t) effluent stream from the LB reactors into contact with the low FA_t effluent from the PB reactor, thus lowering the pH and releasing CO_2 with a possible reduction in gas scrubbing costs.
2. To maintain a high concentration of FA_t in the LBs for minimization of methane production in the LBs.
3. Inoculation of freshly charged LB reactors with appropriate microorganisms.

Proper adjustment of the ratio of recycle around the LB reactors to that around the reactor system, as well as perhaps recycle around each LB reactor, could also prove useful in avoiding inhibition of hydrolysis by high FA_t concentrations, if this proved to be a problem.

Preliminary simulations for partial model validation, sensitivity analyses and exploration of the effects of the recycle streams on overall system performance are given in the report by Andrews (1986). Additional research is in progress on comparing model predictions with experimental data, on simulations to explore operating conditions for minimizing the production of methane in the leaching beds, and on addition of a heat balance to the model. The heat balance will, ultimately, be expanded to an energy balance using COD values as estimates of heats of combustion.

SUMMARY

The evolutionary background of anaerobic methanogenesis in natural systems, the adaptation of these biological phenomena to commercial application and the basis for pursuing multiphase commercial systems for energy production were reviewed.

The evaluation of methane yield from a large number of plant species produced results consistent with general biological principles. High methane yields were obtained from plants having high contents of fermentable organic matter. Ultimate methane yields varied according to plant variety, anatomy, age and growth environment. Although methane yield is a major consideration in the selection of potential feedstocks, there exists a vast reservoir of plant species which give similar yields. Therefore, other factors such as crop productivity, transportation and storage requirements/costs may be more important.

Anaerobic digestion of biomass was studied under conditions of artificially elevated concentrations of nitrate, sulfate, butyric acid, propionic acid and micronutrients. Long-term, continuous-addition experiments showed that anaerobic digestion could occur in the presence of added nitrate without damage to the methanogenic process and without production of dinitrogen. Added nitrate appeared to provide a sparing action on the effect of added sulfate. Similar experiments with fatty acids showed that digesters were able to function in a stable manner with high addition rates of propionic or butyric acids. The addition of micronutrients to Napiergrass-fed anaerobic digesters produced a striking effect. Such additions enabled a five-fold increase in volumetric methane productivity, with concomitant decreases in volatile fatty acid concentrations.

A generalized mathematical model of a multiphase biomass digester was also developed, which can be expanded as specific data on selected digestion processes become available.

A general conclusion, which may be derived from these studies and from consideration of the modeling results, is that multiphase anaerobic digestion of biomass for methane production presents an attractive option for future research.

REFERENCES

Andrews, J. F. (1986). Models of the leaching and packed bed processes for converting biomass to methane. Progress Report. Department of Environmental Science and Engineering, Rice University, Houston, Texas.

Andrews, J. F. and Graef, S. P. (1971). Dynamic modeling and simulation of the anaerobic digestion process. In: *Anaerobic Biological Treatment Processes, Advances in Chemistry Series No. 105*, pp. 126–62. American Chemical Society, Washington, D.C.

Boone, D. R. and Bryant, M. P. (1980). Propionate-degrading bacterium, *Syntrophobacter wolinii* sp. nov. gen. nov., from methanogenic ecosystems. *Applied and Environmental Microbiology* **40**, 626–32.

Brock, T. D. (1974). *Biology of Microorganisms*, 2nd edn. Prentice-Hall, Inc., Englewood Cliffs, New Jersey. 852 pp.

Bryant, M. P. (1976). The microbiology of anaerobic degradation and methanogenesis with special reference to sewage. In: *Microbial Energy*

Conversion, Schlegel, H. G. and Barnea, J. (eds.), pp. 107–17. Pergamon Press, Frankfurt, West Germany.

Bryant, M. P. (1977). Microbial methane production—theoretical aspects. *Journal of Animal Science* **48**, 193–201.

Bryers, J. D. (1985). Structured modeling of the anaerobic digestion of biomass particulates. *Biotechnology and Bioengineering* **27**, 638–49.

Cole, J. A. and Brown, C. M. (1980). Nitrite reduction to ammonia by fermentative bacteria: a short circuit in the biological nitrogen cycle. *FEMS Microbiology Letters* **7**, 65–72.

Conrad, R. S., Sokatch, J. R. and Jensen, R. A. (1976). Relationship of metabolite inhibition of growth to flow-of-carbon patterns in nature. *Life Sciences* **19**, 299–320.

Cooney, C. L. (1983). Bioreactors: design and operation. *Science* **219**, 728–33.

Dahab, M. F. and Young, J. C. (1982). Retention and distribution of biological solids in fixed-bed anaerobic filters. In: *Proceedings: First International Conference on Fixed-Film Biological Processes*, Wu, Y. C., Smith, Ed. D., Miller, R. D. and Opatken, E. J. (eds.), pp. 1337–51. University of Pittsburgh, Pennsylvania.

Daniels, L. (1984). Biological methanogenesis: physiological and practical aspects. *Trends in Biotechnology* **2**, 91–8.

Eastman, J. A. and Ferguson, J. F. (1981). Solubilization of particulate organic carbon during the acid phase of anaerobic digestion. *Journal Water Pollution Control Federation* **53**, 352–66.

Ghosh, S., Henry, M. P. and Sajjad, A. (1985). Stabilization of sewage sludge by two-phase anaerobic digestion. Final Report, Cooperative Agreement No. CR 809982. US Environmental Protection Agency, Cincinnati, Ohio.

Goto, M., Wilkie, A. C., Bordeaux, F. M., Smith, P. H. and Andrews, J. F. (1986). A reaction model describing early events in anaerobic dissimilation of Napiergrass. In: *Energy from Biomass and Wastes*. Institute of Gas Technology, Chicago, Illinois. (In press.)

Guiot, S. R. and van den Berg, L. (1985). Performance of an upflow anaerobic reactor combining a sludge blanket and a filter treating sugar waste. *Biotechnology and Bioengineering* **27**, 800–6.

Gujer, W. and Zehnder, A. J. B. (1983). Conversion processes in anaerobic digestion. *Water Science and Technology* **15**(8/9), 127–67.

Halvadakis, C. P. (1983). Methanogenesis in solid-waste landfill bioreactors. Ph.D. Dissertation, Department of Civil Engineering, Stanford University, Stanford, California.

Henson, J. M., Bordeaux, F. M., Rivard, C. J. and Smith, P. H. (1986). Quantitative influences of butyrate or propionate on thermophilic production of methane from biomass. *Applied and Environmental Microbiology* **51**, 288–92.

Henze, M. and Harremoes, P. (1983). Anaerobic treatment of wastewater in fixed film reactors—a literature review. *Water Science and Technology* **15**(8/9), 1–101.

Hill, D. T. (1982). A comprehensive dynamic model for animal waste methanogenesis. *Transactions of the American Society of Agricultural Engineers* **25**, 1374–80.

Hobson, P. N., Bousfield, S. and Summers, R. (1974). Anaerobic digestion of

organic matter. *Critical Reviews in Environmental Control* **4**, 131–91.
Hungate, R. E. (1966). *The Rumen and Its Microbes*. Academic Press, New York. 533 pp.
Kluyver, A. J. and van Niel, C. B. (1956). *The Microbe's Contribution to Biology*, Harvard University Press, Cambridge, Massachusetts. 182 pp.
McCarty, P. L. (1982). One hundred years of anaerobic treatment. In: *Anaerobic Digestion 1981*, Hughes, D. E., Stafford, D. A., Wheatley, B. I., Baader, W., Lettinga, G., Nyns, E. J., Verstraete, W. and Wentworth, R. L. (eds.), pp. 3–22. Elsevier Biomedical Press, Amsterdam, The Netherlands.
McInerney, M. J., Bryant, M. P., Hespell, R. B. and Costerton, J. W. (1981). *Syntrophomonas wolfei* gen. nov. sp. nov., an anaerobic, syntrophic, fatty acid-oxidizing bacterium. *Applied and Environmental Microbiology* **41**, 1029 39.
Mosey, F. E. (1983). Mathematical modelling of the anaerobic digestion process: regulatory mechanisms for the formation of short-chain volatile acids from glucose. *Water Science and Technology* **15**(8/9), 209–32.
Owen, W. F., Stuckey, D. C., Healy, J. B., Jr., Young, L. Y. and McCarty, P. L. (1979). Bioassay for monitoring biochemical methane potential and anaerobic toxicity. *Water Research* **13**, 485–92.
Rivard, C. J. (1983). Effects of continuous addition of nitrate to a thermophilic anaerobic digestion system. Ph.D. Dissertation, University of Florida, Gainesville, Florida.
Rivard, C. J. and Smith, P. H. (1984). Effects of continuous addition of nitrate to a thermophilic anaerobic digestion system. *Abstracts of the Annual Meeting of the American Society for Microbiology*, **0-6** p. 190.
Rozzi, A. S., Merlinio, S. and Passino, R. (1984). Development of a four population model of the anaerobic degradation of carbohydrates. *Environmental Technology Letters*, **6**, 610–9.
Sawyer, C. N. and McCarty, P. L. (1978). *Chemistry for Environmental Engineering*, 3rd edn. McGraw-Hill, New York. 532 pp.
Shiralipour, A. and Smith, P. H. (1984). Conversion of biomass into methane gas. *Biomass* **6**, 85–92.
Smith, P. H. (1980). Studies of methanogenic bacteria in sludge. EPA-600/2-80-093. US Environmental Protection Agency, Cincinnati, Ohio.
Smith, P. H., Sweeney, H. C., Rooney, J. R., King, K. W. and Moore, W. E. C. (1956). Stratifications and kinetic changes in the ingesta of the bovine rumen. *Journal of Dairy Science* **39**, 598–609.
Speece, R. E. and Parkin, G. F. (1985). Nutrient requirements for anaerobic digestion. In: *Biotechnological Advances in Processing Municipal Wastes for Fuels and Chemicals*, Antonopoulos, A. A. (ed.), pp. 195–221. ANL/CNSV-TM-167, Argonne National Laboratory, Argonne, Illinois.
Van Soest, P. J. (1983). *Nutritional Ecology of the Ruminant*. O&B Books, Inc., Corvallis, Oregon. 374 pp.
Walker, J. C. G., Klein, C., Schidlowski, M., Schopf, J. W., Stevenson, D. J. and Walter, M. R. (1983). Environmental evolution of the archean-early proterozoic earth. In: *Earth's Earliest Biosphere: Its Origin and Evolution*, Schopf, J. W. (ed.), pp. 260–290. Princeton University Press, Princeton, NJ.
Wilkie, A., Goto, M., Bordeaux, F. M. and Smith, P. H. (1986). Enhancement of

anaerobic methanogenesis from Napiergrass by addition of micronutrients. *Biomass* **11**, 135–46.

Woese, C. R. and Fox, G. E. (1977). Phylogenetic structure of the prokaryotic domain: the primary kingdoms. *Proceedings of the National Academy of Sciences USA* **74**, 5088–90.

Wolfe, R. S. (1983). Fermentation and anaerobic respiration in anaerobic digestion. In: *Third International Symposium on Anaerobic Digestion Proceedings*, pp. 3–11. Third International Symposium on Anaerobic Digestion, 99 Erie Street. Cambridge, Massachusetts.

Zinder, S. H. and Koch, M. (1984). Non-aceticlastic methanogenesis from acetate: acetate oxidation by a thermophilic syntrophic coculture. *Archives of Microbiology* **138**, 263–72.

19
Microbial Aspects of Biogas Production[*]

P. H. SMITH, F. M. BORDEAUX, A. WILKIE, J. YANG

Microbiology and Cell Science Department, University of Florida — IFAS, Gainesville, Florida, USA

D. BOONE, R. A. MAH

Division of Environmental and Occupational Health Sciences, School of Public Health, University of California at Los Angeles, California, USA

D. CHYNOWETH and D. JERGER

Agricultural Engineering Department, University of Florida — IFAS, Gainesville, Florida, USA

INTRODUCTION

This research was undertaken for the conceptual development and provision of information for management of commercial digesters for the production of methane from biomass. The fundamental assumption underlying these efforts presumed that two major considerations, i.e. engineering and biological, should be addressed for the optimal (in economic terms) design and operation of such digesters. The biological factors involved are not easy to study by conventional methods used for pure cultures of bacteria, and may best be studied by examining the components of mixed cultures. Even though the relationship between engineering and biology may appear remote, engineering considerations greatly influence which biological phenomena should be selected for investigation. Thus, sampling sites were chosen which might yield bacterial isolates that could be useful in the study of methanogenesis under environmental conditions of specific interest to this program.

Anaerobic digestion consists of a complex series of reactions which are catalyzed by a mixed group of bacteria. In these reactions organic matter is

[*]Florida Agricultural Experiment Station Journal Series No. 8019.

converted in a stepwise fashion to methane and carbon dioxide. Polymers such as cellulose, hemicellulose, pectin and starch are hydrolyzed to oligomers or monomers which are then metabolized by fermentative bacteria with the production of hydrogen (H_2), carbon dioxide (CO_2), and volatile organic acids such as acetate, propionate and butyrate. The volatile organic acids other than acetate are converted to methanogenic precursors (H_2, CO_2 and acetate) by the hydrogen-producing organic acid oxidizers (HPOAO). Finally, the methanogenic bacteria produce methane from acetate or H_2–CO_2. Each biodegradable substrate which enters an anaerobic digester, and each extracellular intermediate produced from a substrate, creates a potential carbon and energy source to support growth of a mixture of bacteria. The more specialized the substrate, the more defined is the particular mixture of bacteria which develops. For less specialized substrates there may be a variety of bacteria capable of catalyzing their breakdown. Under steady-state conditions the dominant organism is normally the one whose characteristics are most favored by the environment (ability to use substrate at lower concentrations, ability to excrete products at higher concentration, more efficient energy production, faster growth rate, etc.). If the most-fit organism is not initially dominant, it soon becomes so. In mixed-culture systems such as digesters, high populations of an organism which is not normally dominant are sometimes desired. This could require continual additions of desired, nondominant organisms or some other control technique.

In stable anaerobic digesters, hydrolysis reactions are the only predominant reactions which are not carried to completion (Boone, 1982, 1984). The extent of hydrolysis depends largely on the chemical nature of the organic matter being degraded. Readily hydrolyzable substrates such as starch may be completely degraded in anaerobic digesters, whereas highly lignified cellulose is more recalcitrant. Essentially all of the oligosaccharides liberated from carbohydrate polymers during hydrolysis are fermented to methane, and so the extent of polymer hydrolysis determines the ultimate yield of methane. This fact, coupled with the small numbers of cellulolytic bacteria normally present in digesters (Iannotti *et al.*, 1982), may make it economical to control microbial populations such as the cellulolytic bacteria by repeated addition during steady-state operation. An alternative may be to manipulate these organisms genetically to enhance their activity. During the startup phase of anaerobic digesters, the use of an active inoculum is required for rapid initiation of a stable fermentation. Inoculation may be inadvertent, through the introduction of bacterial populations indigenous to the feedstock, or by design.

Often, part of the contents of one digester is used to inoculate another. Although the inoculum may not affect the fermentation which eventually develops (Varel et al., 1977), the use of efficient inocula may reduce the period of growth required before maximum methane productivity is attained. Inoculation with high numbers of slow-growing organisms is especially important. The slow growth-rate of some bacteria also dictates the solids retention time in anaerobic digesters. At short retention times these bacteria wash out of the digester more rapidly than they can reproduce themselves and, as the population is depleted, methane production decreases and the digester fails. Also, in laboratory studies of cattle-waste digesters the aceticlastic methanogenic reactions and propionate degradation take much longer to become established than other steps (Varel et al., 1977; Boone, 1982). Thus, development of effective methanogenic and HPOAO inocula may be especially important during the startup phase of digester operation. Research topics discussed here include mixed-culture enrichments and isolation of specific process-related species, namely cocultures of HPOAO bacteria, and pure cultures of cellulolytic and methanogenic bacteria obtained from a variety of environments.

DIGESTION OF WOODY BIOMASS

Wood is currently the most widely used biomass energy resource in the United States, accounting for over $3 \cdot 15 \times 10^9$ GJ energy yr^{-1}. Numerous types of combustion devices ranging from wood stoves to large industrial cogeneration boiler systems (Johnson et al., 1980) burn wood. Anaerobic digestion of wood has not been considered technically feasible without some type of pretreatment to increase its biodegradability. This poor anaerobic biodegradability has been attributed to its low moisture content, cellulose crystallinity and polymerization, and the degree of association between lignin and cellulose (Scheffer and Cowling, 1966; Cowling, 1975). In vitro rumen digestibility of wood species resulted in conversion efficiencies of only 5–20% for most hardwoods, while softwoods were essentially nondigestible (Baker et al., 1975). Poplar wood exhibited less than 10% in vitro dry matter digestibility in studies conducted by Wilson and Pigden (1964). The overall bioconversion efficiency of white fir was determined to be less than 5% using a batch bioassay procedure (Colberg et al., 1981). An inverse linear relationship between volatile solids reduction and lignin content was demonstrated in a study to evaluate the anaerobic biodegradability of several herbaceous and woody species (Chandler et al., 1980). Using a model developed by these investigators, an organic matter (volatile solids) reduction of only 15% would be predicted for hardwood feed with a lignin content of 25% dry weight.

Long-Term Batch Studies

Conventional semicontinuously-fed digestion studies (inoculated with digested sewage sludge) indicated that wood is highly refractory to anaerobic decomposition (Jerger et al., 1982). However, continued operation of these same digesters for 120 days in batch mode resulted in significant conversion with associated increases in methane production. These results were attributed to either enrichment and adaptation of a wood-degrading inoculum and/or increased retention of biological and feedstock solids resulting from the long-term batch incubation (Jerger et al., 1982, 1984).

Biochemical Methane Potential Studies

The anaerobic biodegradability of several woody species was examined using the biochemical methane potential assay (BMP). The highest methane yields of $0.32 \, m^3 \, kg^{-1}$ VS added were achieved from hybrid poplar and sycamore (Jerger et al., 1982; Chynoweth et al., 1985). Red and black alder and cottonwood exhibited methane yields of 0·28, 0·24 and $0.22 \, m^3 \, kg^{-1}$ VS added, respectively, whereas eucalyptus and loblolly pine exhibited poor anaerobic biodegradability with methane yields of 0·014 and $0.063 \, m^3 \, kg^{-1}$ VS added, respectively (Table 1). Volatile solids conversion efficiencies from these experiments confirmed the results obtained from gas production analyses. Volatile solids reductions of 56·7%, 53·8%, 32·5% and 32·3% were observed for sycamore, hybrid poplar, black alder and cottonwood, respectively. The validity of these unexpectedly high conversion data is supported by good material balances obtained from all reactors (Table 1).

The ultimate biodegradability of hybrid poplar was determined by the BMP assay in a 3 liter reactor with a 2 liter culture volume incubated at 35°C. The ultimate methane yield following a 105-day incubation period was $0.3 \, m^3 \, kg^{-1}$ VS added (Jerger et al., 1984). Volatile solids balances for the reactors sampled during the study typically ranged from 95% to 105%, confirming the methane yield data obtained from the long-term batch operation of conventional digesters (Jerger et al., 1982; Chynoweth et al., 1985). Reduction in volatile organic matter ranged from 28·6% at a 13-day incubation period to 58·9% for a 105-day incubation time. These values for

TABLE 1
Anaerobic Biodegradability of Woody Species[a]

Species	Methane yield, m^3 kg^{-1} VS added	Volatile solids reduction (%)	Recovery effluent volatile solids (%)	Recovery effluent carbon (%)
Black Alder	0·24	32·5	108	—
Cottonwood	0·22	32·3	106	107
Eucalyptus	0·014	1·00	107	—
Hybrid Poplar	0·32	53·8	106	101
Loblolly Pine	0·063	3·63	103	—
Red Alder	0·28	48·4	91·5	—
Sycamore	0·32	56·7	106	98·3

[a]Chynoweth et al. (1985).

volatile solids reduction are based on the amount of carbon recovered in the gas. Approximately 75% of the biodegradable fraction of the feed was converted within the first 30 days of digestion (Jerger et al., 1984).

Significant conversion of hardwood species can be accomplished by anaerobic digestion. The conversion efficiencies obtained were substantially higher than values reported elsewhere for woody biomass and indicate that the biogasification of wood may be a technically feasible process. However, fundamental information is lacking on the anaerobic digestibility of woody biomass as an energy source; thus, research is needed on the microbial ecology of the wood-to-methane fermentation. This research would lead to identification of factors limiting conversion and process rates, and could ultimately result in selection and isolation of organisms capable of improved performance.

SPECIFIC PROCESS-RELATED BACTERIAL SPECIES

Cellulolytic Bacteria

The anaerobic hydrolysis and fermentation of cellulose by rumen bacteria has been studied in detail (Hobson and Wallace, 1982a,b). Cellulolytic isolates have also been recovered from the lower gastrointestinal tracts of humans (Betian et al., 1977) and other animals (Davies, 1964; Dehority, 1977; Montgomery and Macy, 1982); from soils and decaying organic matter (Skinner, 1960; Madden, 1983; Petitdemange et al., 1984) and from sediments (Madden et al., 1982; Leschine and Canale-Parola, 1983). In contrast, little work has been done on cellulolytic bacteria from anaerobic digesters. Although isolates were not rigorously characterized, Iannotti *et al.* (1982) completed a study of bacteria from a swine manure digester which included quantification of cellulolytic bacteria. These researchers found (by using most-probable-number techniques) that less than 0·1% of the digester microflora was capable of fermenting cellulose or hemicellulose. Previously the predominant morphologies of known anaerobic cellulolytic bacteria were nonsporeforming rods (Hungate, 1950; Maki, 1954), but more recently cellulolytic clostridia have been isolated (Ng *et al.*, 1977; Patel *et al.*, 1980).

Two cellulolytic clostridia were isolated and characterized in the present study: *Clostridium cellulovorans* (Sleat *et al.*, 1984) and *Clostridium populeti* (Sleat and Mah, 1985). These isolates have higher cellulolytic rates than previously reported anaerobic species. Both new species were isolated from a methanogenic enrichment culture maintained on finely divided (*ca.* 0·8 mm diameter) hybrid poplar. The isolates have many similarities: they are anaerobic, mesophilic, sporeforming, cellulolytic bacteria. They stain Gram-negative and are rod-shaped. They ferment cellulose, cellobiose, xylan and pectin (as well as many monosaccharides and oligosaccharides) and produce H_2, CO_2, acetate, butyrate and lactate during fermentation of cellulose or cellobiose. Both species have pH optima near 7·0 and temperature optima near 35°C. The major differences between these two species are formate production (only *C. cellulovorans* produces it) and endospore morphology. The endospores of *C. populeti* are terminal and oval whereas those of *C. cellulovorans* are central to subterminal and oblong. Another distinguishing characteristic of *C. cellulovorans* is the cellular hydrolysis which produces clear-centered colonies. The centers are clear because all the cells there have apparently autolyzed. These cellulolytic bacteria maintain a very high rate of cellulose hydrolysis in culture. Measured rates of hydrolysis as fast as 29 g cellulose m^{-3} medium h^{-1} have been reached. These rates of cellulose hydrolysis match or exceed those reported for either thermophilic (Weimer and Zeikus, 1977) or mesophilic (Leschine and Canale-Parola, 1983) clostridia. Thus, these clostridia appear to be very desirable in anaerobic digesters.

Methanogenic Isolate from the Termite Gut

Termites are known to digest wood efficiently (Breznak, 1975) and also, in some cases, to evolve methane gas (Krishna, 1969). Cellulolytic bacteria have been isolated from the termite gut, but there had been no successful isolation of a methanogen reported from this environment when this research was initiated in 1980. Possibly unknown relationships exist between termite cellulolytic bacteria and methanogens unique to the termite gut. In addition, methanogens from the termite gut could have unknown metabolic features which could be useful in developing an understanding of the fermentation of woody biomass.

Termites were collected in the area of Gainesville, Florida, and examined for the presence of methanogens by microscopic examination of crushed termite guts for bacteria which would fluoresce following excitation at 420 nm (Doddema and Vogels, 1978). Many termites were examined, but no fluorescence was observed. Termites collected in San Juan, Puerto Rico, were examined using similar procedures and fluorescent cells were observed. Efforts were then made to isolate these bacteria.

Termites (*Nasutitermes nigriceps* or *Nasutitermes costalis*) were collected directly from nests. The termites were killed by exposure to nitrogen and carbon dioxide. Guts from the termites were used as an inoculum source for the development of liquid enrichments with H_2–CO_2 as the energy source. When methane formation occurred, the enrichment was used to inoculate agar roll tubes in a dilution series. Methanogenic bacteria grew in these dilutions. Three methanogens were isolated in pure culture and have been partially characterized. Of these, two were similar to known methanogens, *Methanobacterium bryantii* and *Methanobrevibacter arboriphilus*. The third organism was similar to *Methanobacterium formicicum* but formed a slime not observed in known strains of *M. formicicum*. The isolate had a significantly different DNA base ratio, i.e. 29·5 mol% G + C compared to 42·3 mol% G + C for *M. formicicum*. On the basis of these data, we concluded that the isolate was a member of a new methanogenic species (Yang *et al.*, 1985).

The isolate obtained does not appear to have features which would result in unique metabolic changes in an anaerobic digester. Other characteristics of this organism are still under investigation.

Halophilic Microflora

The production of pipeline quantities of methane gas could possibly require the development of very large digesters. The construction of such a digester by solution mining of salt domes is a possibility. The necessary engineering technology has been developed for the national strategic petroleum reserve (Davis, 1981). Liquid in such digesters would be saturated with salt, but little is known about hypersaline methanogenesis. Also, as biomass is converted to gas, salts accumulate. An effort was therefore undertaken to examine the biology of halophilic methanogenic bacteria and the biological production of methane from environments having high salt concentrations.

One objective was to isolate pure cultures of halophilic methanogenic bacteria which might provide insight into the nature of the biological catalysts which must be active in methane production when hypersaline conditions prevail. A second objective was to evaluate natural environments to determine possible limitations on the occurrence of methanogenesis at high salt concentrations.

To meet these objectives, several sampling sites were selected. Methane evolution has been reported from the north arm of the Great Salt Lake (Post, 1977). This location has a high salt concentration and receives a substantial amount of organic material as runoff from the surrounding land area. The lake is divided by a railroad causeway into north and south arms. Samples were collected from both arms. Salterns operated by the Leslie Salt Company, south of San Francisco Bay, were also attractive sites for the evaluation of natural environments because they exhibited a wide range of salinities and organic loads. Samples were collected from five salterns of different salinity, varying from that of near sea water to near saturation. All samples included sediment and the overlying brine. Samples were maintained under anaerobic conditions by replacing the head space air with oxygen-free gas. The samples were used either for enrichments or as a source of inoculum for immediate pure culture isolation attempts. Brine concentrations were measured, and the pH of each sample was determined by comparison with buffered solutions having the same salt concentration.

Numerous efforts to isolate pure cultures by conventional methods using H_2–CO_2, acetate, or formate as the energy sources were unsuccessful. The procedures were changed to include a more complex medium with the addition of methylated amines (Paterek and Smith, 1985). Colonies which fluoresced at 420 nm formed in this medium. Difficulties were encountered in isolation of pure cultures largely due to overgrowth by contaminating organisms. Further refinement of procedures to include the use of antibiotics led to the isolation in pure culture of a methanogenic bacterium from the Great Salt Lake (Paterek, 1983; Paterek and Smith, 1985). This isolate formed small circular colonies in agar roll tubes. Cells were irregular cocci with a diameter of $0·8$–$1·8\,\mu m$. Motility was not observed in wet mounts.

Growth of this halophilic organism was observed with methanol, monomethylamine, dimethylamine and trimethylamine, but not with H_2–CO_2, formate or acetate. Sensitivity to antibiotics was similar to that of other methanogenic bacteria. Both sodium and magnesium were required for growth. The maximum rate of methane formation occurred at sodium concentrations between 1·0 and 2·5 M, at a temperature of 35°C, and a pH of 7·5. The isolate was examined to determine phylogenetic relationships with other methanogenic bacteria. These cells were not closely related to known species (personal communication, E. Conway de Macario, NY State Department of Health, Albany, NY, 1985). The cells lysed readily in 0·001% Triton X-100 or 0·05% dodecyl sulfate. The cells also lysed in solutions of less than 0·3 M NaCl. The features of the isolate obtained differed from those of other known bacteria at both the genera and species level. It was therefore proposed to establish new genus and species names in the description of the isolate (Paterek and Smith, 1985). The name proposed was *Halomethanococcus mahi*. A similar isolate was obtained from the Leslie saltern in San Francisco Bay (Mathrani and Boone, 1985).

Samples obtained from five holding ponds of the Leslie saltern were examined in an effort to evaluate the salt tolerance of the methane fermentation in natural systems. These ponds all had high salt levels. Pond 1 was 20% saturated for sodium chloride; ponds 2, 3, 4 and 5 were 40%, 60%, 80% and 86% saturated, respectively. Sediments were obtained and incubated in the presence of added substrates. The added substrates were substances which would be anticipated to be essential intermediates in a fermentation of biomass in hypersaline conditions. They included hydrogen, acetate, propionate, butyrate, methanol, cellulose, trimethylamine, N-acetylglucosamine and chitin. The methanol and trimethylamine functioned in these experiments as controls to assure that viable methanogenic bacteria were present. Those samples incubated with hydrogen, acetate, propionate and butyrate only produced a few micromoles of methane from millimolar amounts of test substrate. These amounts were too small to be associated with a methanogenic fermentation of these latter substrates. Methane concentrations were higher in enrichments with chitin than with cellulose.

The pure culture obtained in these studies demonstrates the existence of a kind of bacterium which should be easy to manipulate using genetic engineering techniques. This organism, plus the information from the enrichment studies, suggests an area of genetic research which could ultimately lead to the development of an anaerobic fermentation system operable under hypersaline conditions. For this to occur, bacteria need to be modified to ferment large organic polymers when the suspending menstrum is hypersaline. Another biologically significant result of this isolation is that this organism may phylogenetically bridge the Halobacteriaceae and the Methanobacteriaceae.

Hydrogen-producing Organic Acid Oxidizing Microflora (HPOAO)
Intermediate pathways leading to the formation of methane gas in domestic sludge involve acetate as the major precursor of methane (Smith and Mah, 1966). Also, substantial amounts of acetate are formed from propionic and butyric acid as immediate precursors. Low molecular weight organic acids accumulate when digesters are induced to fail by increasing the loading rate or by decreasing the hydraulic retention time (Smith, 1980; McCarty, 1982). The characteristics of propionate-metabolizing bacteria have been reported by Boone and Bryant (1980) and a bacterium capable of metabolizing butyric acid has been described by McInerney et al. (1981). However, there had been no reported isolation of either thermophilic propionate-metabolizing bacteria or thermophilic butyrate-metabolizing bacteria in 1980.

It was assumed that large-scale anaerobic digestion of biomass would occur under thermophilic conditions, and that propionic and butyric acids would be key intermediates and play important roles in the control and regulation of the digestion process. Therefore, research was undertaken to isolate thermophilic bacteria capable of metabolizing these two acids. The techniques applied were similar to those previously reported (Boone and Bryant, 1980; McInerney et al., 1981). Attempts to culture a thermophilic propionate-utilizing organism failed repeatedly; however, a thermophilic butyrate-utilizing bacterium was obtained (Henson, 1983; Henson and Smith, 1985).

A thermophilic mixed digester fed a mixture of Bermudagrass and commercial cattle-feed was established. Samples from this digester were used to establish enrichments using butyrate as the major substrate. Inoculations of agar medium were made from these enrichments. Butyrate was used as the substrate in anaerobic roll-tube medium inoculated with dilutions from the enrichments. *Methanobacterium thermoautotrophicum* was used as the hydrogenotrophic partner. Tubes were incubated for an extended period of time after which two different colony types appeared. One colony type resembled a *Methanosarcina* colony. These colonies were irregular in shape and had a characteristic brownish-yellow color. Microscopic examination of this colony revealed the presence of three different morphological forms: a *Methanosarcina* sp., *M. thermoautotrophicum*, and a nonfluorescent curved rod. The second colony type was white, and contained bacteria similar to those observed in the brownish-yellow colonies, but without the *Methanosarcina* sp. Subcultures of the white colonies resulted in the development of similar colonies, with bacteria other than *M. thermoautotrophicum* and the curved rod not observed. Methane was produced by the two-member culture when it was incubated at 55°C but not at 45°C or 70°C. The addition of a mesophilic methanogen, *Methanospirillum hungatei*, to the coculture did not result in the formation of methane at mesophilic temperatures. It was concluded that the butyrate-utilizing member of the coculture was a thermophilic bacterium.

The addition of antibiotics known to interfere with cell wall synthesis of eubacteria, but not archaebacteria, resulted in the complete inhibition of methane production by the coculture, suggesting that the butyrate-oxidizer was a eubacterium.

The addition of hydrogen to the gas phase resulted in the total inhibition of butyrate metabolism. However, the observed inhibition did not result in death of the butyrate-utilizing bacteria. This was demonstrated by incubating the coculture for a period of 8 days under an atmosphere of 80% hydrogen, and then replacing the hydrogen with nitrogen. Following the removal of the hydrogen there was a lag period of approximately 8 days, after which butyrate was rapidly metabolized, demonstrating that the butyrate-utilizing coculture was still viable.

The significance of this work, in addition to the successful isolation of a unique microorganism, was the demonstration that an inhibitory hydrogen concentration did not kill the butyrate-utilizing bacteria. This suggests that, if hydrogen inhibition were to occur in an anaerobic digester, it could possibly be reversed by elimination of the hydrogen.

The previously described triculture from the brownish-yellow colony, consisting of a *Methanosarcina* sp., *M. thermoautotrophicum*, and a butyrate-utilizing curved rod, was incubated in the presence of butyrate and the rate and amount of methane formation was measured to determine if the presence of an acetate-utilizing organism would increase the amount of methane produced. The biculture of *M. thermoautotrophicum* and the butyrate-utilizing curved rod was incubated under identical conditions. As anticipated, more methane was produced by the triculture due to acetate metabolism. However, the increase was approximately 10% above that expected based on methane production in the absence of an acetate-utilizing organism, even when assuming no conversion of acetate into cell material.

This demonstration of the stimulation of butyrate catabolism by acetate removal suggests a potential use of acetate concentration in the control of anaerobic digesters.

Marine Microflora

Marine biomass is a potential feedstock for the production of pipeline quantities of methane gas. Microorganisms capable of producing methane under environmental conditions in which the suspending menstrum of the fermentation would be similar to sea water would be required. Considerable research has been conducted on the growth of marine plants and on the fermentation of the plant biomass to methane (Chynoweth et al., 1986). However, fundamental questions regarding potential organisms and their regulation in digesters designed and optimized for the fermentation of marine biomass have not been experimentally resolved. Thus, research efforts were directed toward the isolation of model organisms for the microbiological, biochemical and genetic study of marine methanogenesis. Research centered on two marine isolates with unique characteristics: a thermophilic methanogen and a mesophilic methanogen.

Thermophilic Isolate

An organism was isolated from a sampling site near the cooling water outfall of a coastal nuclear power plant (Florida Power Company, Crystal River, FL.). This site was selected because it was maintained at a relatively constant elevated temperature. Samples of sediment were enriched with hydrogen and carbon dioxide in liquid medium and then diluted in solid medium. The salt concentration of the medium was modified in order to resemble the natural environment from which the samples were obtained. Methanogenic colonies were identified by fluorescence upon exposure to light at 420 nm. Repeated subculture led to the isolation in pure culture of a thermophilic bacterium with characteristics unlike any known species. The cells were irregular cocci with an average diameter of $1 - 1\cdot3\,\mu$m. The isolate grew on either H_2–CO_2 or formate. Ethanol, acetate, propionate and pyruvate were not used. This isolate grew optimally at 55°C in 0·24 M sodium ion. Growth was slow at 37°C, and no growth occurred at 65°C. The characteristics of the organism are similar to those of the genus *Methanogenium*. However, since it differed from described species of that genus, the name *Methanogenium thermophilicum* was proposed (Rivard and Smith, 1982). This organism is easily grown and lysed, making it useful for genetic and biochemical research. The described isolate has features which could play an important role in the terminal step of methane formation during thermophilic digestion of marine biomass.

Mesophilic Isolate

The sampling site for the mesophilic isolate had no unique features which would suggest the presence of unusual methanogenic bacteria. Sediment cores were obtained from the edge of mosquito control canals in mangrove swamps located southwest of Georgetown, Grand Cayman, British West Indies. Research on the methanogenic microbial population in the cores led to the isolation of a new species of methanogenic bacterium with novel characteristics, *Methanomicrobium paynteri* (Rivard et al., 1983).

Subsamples from the cores were enriched in liquid medium with hydrogen as substrate and inoculated into agar roll tubes incubated at 37°C. The colonies which developed in the roll tubes were examined for fluorescence upon exposure to light at 420 nm. Fluorescent colonies were subcultured and a methanogenic bacterium was obtained which was morphologically unique relative to other marine methanogens. This isolate was extremely robust.

A pure culture was obtained by using dilution techniques. The organism was an irregular rod, $0 \cdot 6 \times 1 \cdot 5 - 2 \cdot 5 \mu m$ in size. It became coccoidal after lengthy incubation in the absence of substrate, yet remained viable.

Colonies were white and circular, with entire edges. H_2-CO_2 was the only tested substrate which supported growth. Ethanol, methanol, formate, acetate, propionate, monomethylamine, dimethylamine, trimethylamine, glutamate and glucose did not support growth. Antigenically, the organism was different from other known methanogens (personal communication, E. Conway de Macario, NY State Department of Health, Albany, NY, 1983). The organism had no identifiable requirements for organic nutrients other than acetate. Nucleotide cataloging placed it in the family Methanomicrobiaceae (personal communication, R. S. Wolfe and C. R. Woese, University of Illinois, Urbana, IL, 1983).

This organism is uniquely suited for biological investigations of methanogenesis by mesophilic marine methanogens since it is easy to grow in simple nutritional media, enriched only with acetate, and can be maintained for long periods of time in the absence of metabolizable substrate (Rivard et al., 1983).

Aceticlastic Microflora

About two-thirds of the methane produced in anaerobic digesters comes from the splitting of acetate (Smith and Mah, 1966; Varel et al., 1977; Boone, 1982) hence, part of the research focused on the investigation of aceticlastic methanogens. Prior to the start of this 5-year period, the only aceticlastic methanogen which had been isolated was *Methanosarcina barkeri* strain 227 (Mah et al., 1978) which was enriched and isolated from an anaerobic digester. The study of acetate degradation by this organism has continued. Throughout the 1970s it had been considered impossible for a methanogenic bacterium to derive energy by acetate decarboxylation (the aceticlastic reaction). This belief was based on the theoretical amount of energy released (insufficient for the production of one mole of ATP per mole of acetate degraded) and because of earlier reports of Zeikus et al. (1975) that H_2 was required for acetate catabolism. Smith and Mah (1980) demonstrated that acetate can support the growth of *M. barkeri* in the absence of any other energy source. Hydrogen was shown to be the preferred substrate and growth on H_2 was more rapid than on acetate (Smith and Mah, 1980; Ferguson and Mah, 1983). When H_2 was depleted, cultures immediately switched to acetate as substrate. During logarithmic growth, cultures doubled in about 34 h, suggesting that these organisms could maintain themselves in digesters with retention times of 2 days or less.

Methanosarcina mazei can catabolize acetate even more rapidly than *M. barkeri*. *M. mazei* was originally described by Barker (1936) and named *Methanococcus mazei*. Barker was unable to isolate it in pure culture. Mah recently duplicated Barker's enrichment techniques and obtained a mixed culture identical to that described by Barker. The organism was then isolated in pure culture (*Methanococcus mazei* S-6) and its life cycle was described (Mah, 1980). The organism was later renamed *Methanosarcina mazei* S-6 (Mah and Kuhn, 1984a,b) because it is more closely related to *Methanosarcina* than to *Methanococcus*. With acetate as substrate *M. mazei* grows with a doubling time of only 17 h, almost half that of *M. barkeri*. Individual *M. mazei* cells grow into large aggregates which often form 'cysts', which are large (up to 0·2 mm diameter) bodies with an apparently differentiated outer wall containing myriads of small (1–2 μm diameter) coccoid units. These coccoid units may be dispersed by physically rupturing the 'cysts'.

Another strain of *M. mazei*, strain LYC, which does not use acetate, was isolated (Liu *et al.*, 1985). Strain LYC produces a disaggregating enzyme which breaks up large aggregates of cells and also disaggregates the aceticlastic-type strain, *M. mazei* S-6 (see Chapter 23). This enzyme could be utilized for controlling the dispersion of *M. mazei* within and between multiphase digesters.

SUMMARY

Research on bacteria and the enzymes they produce has made a significant contribution to our understanding of the biological conversion of organic matter to methane. This has been due in part to recent scientific interest in the anaerobic bacteria as biological material useful in developing an understanding of basic life processes. One conclusion to be drawn from the mass of information currently being reported in the scientific literature is that the formation of methane from biomass is a much more complex process than had been previously recognized. Another conclusion is that there are unidentified bacteria with unique biochemical features which, if elucidated, could advance understanding of the regulation and control of methane formation from biomass. A third conclusion is that these bacteria are well adjusted biologically to their ecological roles. These three general conclusions lead to the assumption that isolation of selected new bacterial species could provide valuable material for future genetic and other biological investigations at the molecular level, in addition to providing insights into the microbial interactions occurring during anaerobic digestion.

In pursuit of unique microflora provided by natural selection, we have isolated a variety of previously unknown bacteria. The isolates include bacteria which hydrolyze large organic polymers, bacteria which produce methane under conditions of specific interest to our program and bacteria which oxidize organic acids. A general conclusion which can be drawn from the triculture data reported here is that methanogenesis from specific intermediates is enhanced by the interactions of a multiplicity of substrate linked microbial species.

Further investigation of not only individual microbial species but, equally important, their interactions, should provide new insight into the control and regulation of anaerobic digestion.

REFERENCES

Baker, A. J., Millett, M. A. and Satter, L. D. (1975). Wood and wood-based residues in animal feeds. In: *Cellulose Technology Research*, Turbak, A. F. (ed.), pp. 75–105, American Chemical Society, Washington, DC.

Barker, H. A. (1936). Studies upon the methane-producing bacteria. *Archiv fur Mikrobiologie* **7**, 420–38.

Betian, H. G., Linehan, B. A., Bryant, M. P. and Holdeman, L. V. (1977). Isolation of a cellulolytic *Bacteroides* sp. from human feces. *Applied and Environmental Microbiology* **33**, 1009–10.

Boone, D. R. (1982). Terminal reactions in the anaerobic digestion of animal waste. *Applied and Environmental Microbiology* **43**, 57–64.

Boone, D. R. (1984). Mixed-culture fermentor for simulating methanogenic digestors. *Applied and Environmental Microbiology* **48**, 122–6.

Boone, D. R. and Bryant, M. P. (1980). Propionate-degrading bacterium, *Syntro-*

phobacter wolinii sp. nov. gen. nov., from methanogenic ecosystems. *Applied and Environmental Microbiology* **40**, 626–32.

Breznak, J. A. (1975). Symbiotic relationships between termites and their intestinal microbiota. In: *Symbiosis*, Jennings, D. H. and Lee, D. L. (eds.), pp. 559–80, Cambridge University Press, Cambridge, UK.

Chandler, J. A., Jewell, W. J., Gossett, J. M., Van Soest, P. J. and Robertson, J. B. (1980). Predicting methane fermentation biodegradability. *Biotechnology and Bioengineering Symposium* **10**, 93–107.

Chynoweth, D. P., Jerger, D. E. and Srivastava, V. J. (1985). Biological gasification of woody biomass. In: *Proceedings of the 20th Intersociety Energy Conversion Engineering Conference*, Vol. 1, pp. 573–9, Society of Automotive Engineers, Inc., Warrendale, PA, USA.

Chynoweth, D. P., Fannin, K. F. and Srivastava, V. J. (1987). Biological gasification of marine algae. In: *Seaweed Cultivation for Renewable Resources*, Bird, K. and Benson, P. (eds.), Elsevier Science Publishers, New York, pp. 285–303.

Colberg, P. J., Baugh, K., Everhart, T., Bachmann, A., Harrison, D., Young, L. Y. and McCarty, P. L. (1981). Heat treatment of organics for increasing anaerobic biodegradability. SERI/TR-98174-1. Solar Energy Research Institute, Golden, Colorado, USA.

Cowling, E. B. (1975). Physical and chemical constraints in the hydrolysis of cellulose and lignocellulosic materials. *Biotechnology and Bioengineering Symposium* **5**, 163–81.

Davies, M. E. (1964). Cellulolytic bacteria isolated from the large intestine of the horse. *Journal of Applied Bacteriology* **27**, 373–8.

Davis, R. M. (1981). National strategic petroleum reserve. *Science* **213**, 618–22.

Dehority, B. A. (1977). Cellulolytic cocci isolated from the cecum of guinea pigs (*Cavia porcellus*). *Applied and Environmental Microbiology* **33**, 1278–83.

Doddema, H. J. and Vogels, G. D. (1978). Improved identification of methanogenic bacteria by fluorescence microscopy. *Applied and Environmental Microbiology* **36**, 752–4.

Ferguson, T. J. and Mah, R. A. (1983). Effect of H_2–CO_2 on methanogenesis from acetate or methanol in *Methanosarcina* spp. *Applied and Environmental Microbiology* **46**, 348–55.

Henson, J. M. (1983). Isolation of butyrate-utilizing bacteria from thermophilic and mesophilic methane-producing ecosystems. Ph.D. dissertation, University of Florida, Gainesville, Florida.

Henson, J. M. and Smith, P. H. (1985). Isolation of a butyrate-utilizing bacterium in coculture with *Methanobacterium thermoautotrophicum* from a thermophilic digester. *Applied and Environmental Microbiology* **49**, 1461–6.

Hobson, P. N. and Wallace, R. J. (1982a). Microbial ecology and activities in the rumen. Part I. *Critical Reviews in Microbiology* **9**, 165–225.

Hobson, P. N. and Wallace, R. J. (1982b). Microbial ecology and activities in the rumen. Part II. *Critical Reviews in Microbiology* **9**, 253–320.

Hungate, R. E. (1950). The anaerobic mesophilic cellulolytic bacteria. *Bacteriological Reviews* **14**, 1–49.

Iannotti, E. L., Fischer, J. R. and Sievers, D. M. (1982). Characterization of bacteria from a swine manure digester. *Applied and Environmental Microbiology* **43**, 136–43.

Jerger, D. E., Dolenc, D. A. and Chynoweth, D. P. (1982). Bioconversion of woody biomass as a renewable source of energy. *Biotechnology and Bioengineering Symposium* 12, 233–48.

Jerger, D. E., Razik, A. and Chynoweth, D. P. (1984). Biogasification of wood. GRI-84/0078. Gas Research Institute, Chicago, Illinois.

Johnson, R. C., Lay, R. K. and Newman, L. C. (1980). *Energy from Biomass: a Technology Assessment of Terrestrial Biomass Systems*, MTR-80W259. The MITRE Corporation, McLean, Virginia.

Krishna, K. (1969). Introduction. In: *Biology of Termites, Vol. I*, Krishna, K. and Weesner, F. M. (eds.), pp. 1–17, Academic Press, New York.

Leschine, S. B. and Canale-Parola, E. (1983). Mesophilic cellulolytic clostridia from freshwater environments. *Applied and Environmental Microbiology* 46, 728–37.

Liu, Y., Boone, D. R., Sleat, R. and Mah, R. A. (1985). *Methanosarcina mazei* LYC, a new methanogenic isolate which produces a disaggregating enzyme. *Applied and Environmental Microbiology* 49, 608–13.

Madden, R. H. (1983). Isolation and characterization of *Clostridium stercorarium* sp. nov., cellulolytic thermophile. *International Journal of Systematic Bacteriology* 33, 837–40.

Madden, R. H., Bryder, M. J. and Poole, N. J. (1982). Isolation and characterization of an anaerobic, cellulolytic bacterium, *Clostridium papyrosolvens* sp. nov. *International Journal of Systematic Bacteriology* 32, 87–91.

Mah, R. A. (1980). Isolation and characterization of *Methanococcus mazei*. *Current Microbiology* 3, 321–6.

Mah, R. A. and Kuhn, D. A. (1984a). Transfer of the type species of the genus *Methanococcus* to the genus *Methanosarcina*, naming it *Methanosarcina mazei* (Barker 1936) comb. nov. et emend. and conservation of the genus *Methanococcus* (Approved Lists 1980) with *Methanococcus vannielii* (Approved Lists 1980) as the type species. *International Journal of Systematic Bacteriology* 34, 263–5.

Mah, R. A. and Kuhn, D. A. (1984b). Rejection of the type species *Methanosarcina methanica* (Approved Lists 1980), conservation of the genus *Methanosarcina* with *Methanosarcina barkeri* (Approved Lists 1980) as the type species, and emendation of the genus *Methanosarcina*. *International Journal of Systematic Bacteriology* 34, 266–7.

Mah, R. A., Smith, M. R. and Baresi, L. (1978). Studies on an acetate-fermenting strain of *Methanosarcina*. *Applied and Environmental Microbiology* 35, 1174–84.

Maki, L. R. (1954). Experiments on the microbiology of cellulose decomposition in a municipal sewage plant. *Antonie van Leeuwenhoek* 20, 185–200.

Mathrani, I. M. and Boone, D. R. (1985). Isolation and characterization of a moderately halophilic methanogen from a solar saltern. *Applied and Environmental Microbiology* 50, 140–3.

McCarty, P. L. (1982). One hundred years of anaerobic treatment. In: *Anaerobic Digestion 1981*, Hughes, D. E., Stafford, D. A., Wheatley, B. I., Baader, W., Lettinga, G., Nyns, E. J., Verstraete, W. and Wentworth, R. L. (eds.), pp. 3–22, Elsevier Biomedical Press, Amsterdam.

McInerney, M. J., Bryant, M. P., Hespell, R. B. and Costerton, J. W. (1981). *Syntrophomonas wolfei* gen. nov. sp. nov., an anaerobic, syntrophic, fatty acid-oxidizing bacterium. *Applied and Environmental Microbiology* **41**, 1029–39.

Montgomery, L. and Macy, J. M. (1982). Characterization of rat cecum cellulolytic bacteria. *Applied and Environmental Microbiology* **44**, 1435–43.

Ng, T. K., Weimer, P. J. and Zeikus, J. G. (1977). Cellulolytic and physiological properties of *Clostridium thermocellum*. *Archives of Microbiology* **114**, 1–7.

Patel, G. B., Kahn, A. W., Agnew, B. J. and Colvin, J. R. (1980). Isolation and characterization of an anaerobic, cellulolytic microorganism, *Acetivibrio cellulolyticus* gen. nov., sp. nov. *International Journal of Systematic Bacteriology* **30**, 179–85.

Paterek, J. R. (1983). Ecology of methanogenesis in two hypersaline biocoenoses: Great Salt Lake and a San Francisco Bay saltern. Ph.D. dissertation, University of Florida, Gainesville, Florida.

Paterek, J. R. and Smith, P. H. (1985). Isolation and characterization of a halophilic methanogen from Great Salt Lake. *Applied and Environmental Microbiology* **50**, 877–81.

Petitdemange, E., Caillet, F., Giallo, J. and Gaudin, C. (1984). *Clostridium cellulolyticum* sp. nov., a cellulolytic, mesophilic species from decayed grass. *International Journal of Systematic Bacteriology* **34**, 155–9.

Post, F. J. (1977). The microbial ecology of the Great Salt Lake. *Microbial Ecology* **3**, 143–65.

Rivard, C. J. and Smith, P. H. (1982). Isolation and characterization of a thermophilic marine methanogenic bacterium, *Methanogenium thermophilicum* sp. nov. *International Journal of Systematic Bacteriology* **32**, 430–6.

Rivard, C. J., Henson, J. M., Thomas, M. V. and Smith, P. H. (1983). Isolation and characterization of *Methanomicrobium paynteri* sp. nov., a mesophilic methanogen isolated from marine sediments. *Applied and Environmental Microbiology* **46**, 484–90.

Scheffer, T. C. and Cowling, E. B. (1966). Natural resistance of wood to microbial deterioration. *Annual Review of Phytopathology* **4**, 147–70.

Skinner, F. A. (1960). The isolation of anaerobic cellulose-decomposing bacteria from soil. *Journal of General Microbiology* **22**, 539–54.

Sleat, R. and Mah, R. A. (1985). *Clostridium populeti* sp. nov., a cellulolytic species from a woody-biomass digestor. *International Journal of Systematic Bacteriology* **35**, 160–3.

Sleat, R., Mah, R. A. and Robinson, R. (1984). Isolation and characterization of an anaerobic, cellulolytic bacterium, *Clostridium cellulovorans* sp. nov. *Applied and Environmental Microbiology* **48**, 88–93.

Smith, M. R. and Mah, R. A. (1980). Acetate as sole carbon and energy source for growth of *Methanosarcina* strain 227. *Applied and Environmental Microbiology* **39**, 993–9.

Smith, P. H. (1980). Studies of methanogenic bacteria in sludge. EPA-600/2-80-093, U.S. Environmental Protection Agency, Cincinnati, Ohio.

Smith, P. H. and Mah, R. A. (1966). Kinetics of acetate metabolism during sludge digestion. *Applied Microbiology* **14**, 368–71.

Varel, V. H., Isaacson, H. R. and Bryant, M. P. (1977). Thermophilic methane production from cattle waste. *Applied and Environmental Microbiology* **33**, 298–307.

Weimer, P. J. and Zeikus, J. G. (1977). Fermentation of cellulose and cellobiose by *Clostridium thermocellum* in the absence and presence of *Methanobacterium thermoautotrophicum*. *Applied and Environmental Microbiology* **33**, 289–97.

Wilson, R. K. and Pigden, W. J. (1964). Effect of a sodium hydroxide treatment on the utilization of wheat straw and poplar wood by rumen microorganisms. *Canadian Journal of Animal Science* **44**, 122–3.

Yang, J., Bordeaux, F. M. and Smith, P. H. (1985). Isolation of methanogenic bacteria from termites. *Abstracts of the Annual Meeting of the American Society for Microbiology* **I-83**, 160.

Zeikus, J. G., Weimer, P. J., Nelson, D. R. and Daniels, L. (1975). Bacterial methanogenesis: acetate as a methane precursor in pure culture. *Archives of Microbiology* **104**, 129–34.

20
Chemical Characteristics and their Relation to Fermentability of Potential Biomass Feedstocks

K. A. BJORNDAL and J. E. MOORE

Animal Science Department, University of Florida—IFAS, Gainesville, Florida, USA

INTRODUCTION

Identification of the inherent variation in chemical constituents of potential feedstocks, and knowledge of the relation of chemical composition to anaerobic fermentability, should be of value in determining the suitability of feedstocks for biogasification to methane. In this study, techniques developed for the evaluation of forage quality for ruminants were used to determine chemical composition and potential fermentability of a wide range of feedstocks. An *in vitro* anaerobic fermentation with rumen microorganisms was used to measure fermentability because of the relative speed of the technique (1 week) and because of the many similarities between the rumen fermentation and the early stages of anaerobic digesters in which the end products are largely volatile fatty acids.

Feedstocks were analyzed for those chemical constituents that are determinants of fermentability—both rate and extent—in the rumen system. Samples were analyzed for cell wall constituents (CWC), an analysis which separates volatile solids into two fractions: cell contents, which are rapidly and completely fermented in anaerobic systems, and CWC, which vary widely in the rate and extent to which they are fermented in anaerobic systems. Because many potential feedstocks contain high percentages of cell walls, the extent of the fermentability of CWC may be an important determinant of overall fermentability. The percentage of lignin in CWC fraction was measured because lignin is known to be a major determinant of fermentability of CWC in ruminants (Van Soest, 1982).

Other plant components, such as tannins, adversely affect fermentability. Tannins are polyphenolic compounds that form insoluble complexes with protein (Swain, 1979). Some varieties of sorghum are known to contain high levels of tannin that interfere with fermentation (Harris et al., 1970). For this reason, the sorghum samples in this survey were analyzed for tannin.

More than 1400 samples, representing over 100 species, have been analyzed, yielding a broad foundation that can be used to select appropriate biomass feedstocks and to design further experiments to assess chemical constituents that may control the rates and ultimate yields of methane from anaerobic digesters. This report concerns only general relationships within broad plant subgroups.

METHODS

Samples of biomass feedstocks were collected from terrestrial, freshwater and marine environments by cooperating investigators. The feedstocks included natural vegetation, crops grown specifically as energy crops, and crop residues. Samples were dried at 60°C and ground in a Wiley mill to pass through a 1 mm screen.

The assay used to measure the fermentability of feedstocks, in vitro total volatile solids digestibility (IVTVSD), was modified from the Tilley and Terry (1963) anaerobic two-stage digestion technique (Moore and Mott, 1974). Rumen fluid was removed from a fistulated steer maintained on a constant diet of Bermudagrass hay and protein supplement, filtered through glass wool, and diluted 1:4 with a phosphate–bicarbonate buffer. Fermentation was carried out in 90 ml plastic centrifuge tubes containing 0·5 g of air-dry sample and 50 ml of rumen fluid:buffer. The tubes were stoppered with gas-release valves. After incubation at 39°C for 48 h, HCl and pepsin were added to stop fermentation and partially digest microbial and substrate protein. After an additional 48 h at 39°C, residues were recovered on glass wool, dried at 105°C, weighed, ashed at 500°C, and re-weighed. IVTVSD equals the percent of total volatile solids that disappears during the combined fermentation and acid–pepsin stages.

Total volatile solids (TVS) for the dried ground plant samples was determined as the percentage of dry matter that disappeared during ashing at 500°C for 3 h. Analyses for cell wall constituents (CWC) and permanganate lignin followed Goering and Van Soest (1970) with sodium sulfite and decalin omitted from the CWC procedure (Golding et al., 1985). The amylase-CWC technique was used to analyze samples with high starch contents (Robertson and Van Soest, 1977). Total nitrogen and phosphorus were measured using a block digester (Gallaher et al., 1975) and an automated analyzer (Hambleton, 1977). Two analyses were used to measure tannin content. The modified vanillin–HCl method (Price *et al.*, 1978) measured the concentration of condensed tannins in catechin equivalents. The protein precipitation method (Hagerman and Butler, 1978) determined the concentration of hydrolyzable and condensed tannins as well as phenolic compounds with similar protein-binding capacities, by measuring the amount of bovine serum albumin precipitated.

RESULTS AND DISCUSSION

There were wide ranges in IVTVSD and chemical composition (Table 1). Fermentability (IVTVSD) ranged from completely indigestible to almost completely digestible. Low and high values of IVTVSD were found among almost all plant subgroups, especially those with large numbers of samples. Ranges do not, however, accurately reflect the potential of a group or type of sample. Rather, the wide ranges in feedstock characteristics demonstrate clearly that each individual sample of feedstock must be examined if it is to be used in an appropriate manner as a substrate for methane production. Some of the variation among samples could be attributed to differences in plant parts. Composition and fermentability varied almost as greatly among parts of plants (Table 2) as they did among groups. Another source of variation may have been the maturity of plants at harvest.

The single component that best categorizes potential fermentability of feedstocks is percentage of CWC. We have divided the feedstocks into categories based on CWC percentages: low (CWC < 35%), medium (CWC 36–70%) and high (CWC > 70%) (Table 3). In general, plants in the low category—which include tubers, several vegetables, marine and some aquatic plants—are expected to have the highest fermentability. The middle category contains vegetable crops, grasses, most of the aquatic plants—both floating and emergent—and some of the terrestrial 'weed' species. The high category, which should have the lowest fermentability, includes wood samples, some of the grasses, a few of the emergent aquatic plants, and the remainder of the terrestrial 'weed' species.

The number of samples within a group affects interpretation because certain species, for which only a small number of samples were analyzed, may appear to have a much narrower range of cell wall percentages than others (Table 3). For example, eastern gamagrass, indiangrass and *Erianthus* are all listed only in the high category while Napiergrass, sorghum and sugarcane are all listed in both the medium and high

TABLE 1

Summary of Composition and Digestibility of all Biomass Samples and Subgroups. Data are Total Volatile Solids (TVS), *in vitro* TVS Digestibility (IVTVSD), Cell Wall Constituents (CWC), Lignin, Nitrogen and Phosphorus. All are Expressed as Percent of TVS, except TVS and Phosphorus are Percent of Dry Matter. Entries are the Mean ($\bar{x}$), Number of Samples (n) and the Range

		TVS	IVTVSD	CWC	Lignin	Nitrogen	Phosphorus
All samples	$\bar{x}$ (n)	89·9 (1 467)	54·2 (1 440)	62·4 (1 425)	10·3 (1 308)	1·5 (1 301)	0·26 (1 317)
	Range	3·5–99·6	0–98·7	7·2–96·5	0·6–68·0	0·1–7·0	0·01–1·12
Grasses	$\bar{x}$ (n)	93·8 (781)	52·7 (754)	70·8 (776)	7·6 (739)	1·0 (721)	0·21 (722)
	Range	43·3–99·3	15·0–87·0	23·5–94·8	1·0–33·2	0·1–5·4	0·02–0·96
Napiergrass	$\bar{x}$ (n)	93·0 (340)	49·3 (313)	78·2 (337)	8·3 (307)	1·0 (298)	0·23 (298)
	Range	73·8–99·3	15·0–79·4	41·2–91·8	3·5–20·3	0·1–3·0	0·06–0·96
Sorghum	$\bar{x}$ (n)	95·7 (285)	57·6 (285)	61·9 (285)	6·3 (284)	1·0 (284)	0·19 (284)
	Range	81·5–98·3	34·5–80·2	30·1–82·9	2·5–14·4	0·2–2·4	0·05–0·36
Brassicas	$\bar{x}$ (n)	79·0 (52)	80·9 (52)	32·9 (50)	11·5 (47)	3·2 (43)	0·50 (43)
	Range	54·6–89·8	33·5–93·8	23·4–59·8	5·9–35·0	1·5–6·2	0·24–0·83
Sweet potatoes	$\bar{x}$ (n)	91·3 (100)	75·8 (100)	36·0 (100)	9·8 (100)	1·1 (100)	0·25 (100)
	Range	69·0–97·7	41·0–98·7	10·9–70·4	1·3–28·9	0·4–3·0	0·12–0·43
Emergent aquatics	$\bar{x}$ (n)	90·1 (55)	58·8 (55)	48·2 (54)	8·9 (53)	2·0 (41)	0·24 (53)
	Range	76·7–97·8	11·9–85·2	14·2–86·5	2·5–26·9	0·8–5·7	0·07–0·54
Floating aquatics	$\bar{x}$ (n)	83·7 (146)	46·2 (146)	63·8 (123)	15·6 (99)	3·0 (108)	0·49 (108)
	Range	69·7–91·8	12·0–83·0	37·4–80·5	4·3–47·7	0·9–6·4	0·06–1·12
Water hyacinth	$\bar{x}$ (n)	83·0 (80)	43·7 (80)	67·0 (69)	10·2 (55)	3·0 (65)	0·52 (65)
	Range	69·7–88·1	12·0–80·1	43·8–80·5	4·3–19·7	0·9–5·9	0·06–1·12
Marine algae	$\bar{x}$ (n)	54·2 (30)	53·6 (30)	39·1 (29)	—	3·1 (20)	0·15 (20)
	Range	19·7–74·5	21·0–83·4	20·3–60·2	—	1·2–6·1	0·05–0·30
Red algae	$\bar{x}$ (n)	47·8 (11)	73·2 (11)	30·1 (11)	—	3·8 (4)	0·20 (4)
	Range	22·8–62·3	62·0–83·4	20·3–42·6	—	2·5–6·1	0·06–0·30
Brown algae	$\bar{x}$ (n)	59·3 (12)	35·5 (12)	43·1 (12)	—	2·3 (11)	0·13 (11)
	Range	51·8–66·8	21·0–49·8	34·6–46·7	—	1·2–2·9	0·07–0·22
Green algae	$\bar{x}$ (n)	57·8 (5)	60·3 (5)	45·6 (4)	—	4·3 (3)	0·12 (3)
	Range	19·7–74·5	34·8–80·4	36·8–50·6	—	2·6–5·3	0·05–0·22

TABLE 2
Composition and Fermentability of Plant Parts of Selected Biomass entries. Data as for Table 1

		TVS	IVTVSD	CWC	Lignin	Nitrogen	Phosphorus
Napiergrass							
Entire	$\bar{x}$ (n)	92·3 (230)	49·0 (230)	78·7 (227)	8·1 (222)	1·1 (217)	0·21 (217)
	Range	73·8–98·3	23·8–73·0	57·0–91·8	3·5–20·3	0·2–2·8	0·06–0·58
Leaves	$\bar{x}$ (n)	94·2 (31)	56·8 (29)	78·2 (31)	5·5 (30)	1·4 (30)	0·24 (30)
	Range	91·1–96·6	49·2–71·4	71·9–82·1	4·0–7·5	0·5–2·9	0·10–0·56
Stems	$\bar{x}$ (n)	94·3 (8)	54·4 (8)	80·0 (8)	8·7 (8)	1·5 (8)	0·29 (8)
	Range	88·0–97·1	39·4–70·3	74·0–84·9	5·4–12·0	0·7–3·0	0·17–0·46
Sheath	$\bar{x}$ (n)	94·8 (20)	43·4 (7)	86·0 (20)	8·8 (7)	0·5 (6)	0·21 (6)
	Range	91·6–97·1	36·4–53·4	82·4–88·5	7·7–9·8	0·4–0·6	0·13–0·38
Stem exterior	$\bar{x}$ (n)	96·9 (21)	28·0 (20)	85·8 (21)	15·7 (20)	0·3 (20)	0·21 (20)
	Range	91·0–99·3	15·0–55·2	79·2–90·3	8·5–17·8	0·1–1·2	0·06–0·32
Stem interior	$\bar{x}$ (n)	91·6 (19)	71·5 (13)	55·5 (19)	5·0 (14)	0·5 (12)	0·36 (12)
	Range	83·1–98·4	50·3–79·4	41·2–73·3	3·5–8·1	0·2–1·9	0·10–0·59
Inflorescence	$\bar{x}$ (n)	95·3 (11)	43·5 (6)	76·5 (11)	8·0 (6)	2·0 (5)	0·75 (5)
	Range	93·4–96·1	39·2–48·0	74·2–78·8	6·9–9·8	1·8–2·2	0·41–0·96
Sorghum							
Entire	$\bar{x}$ (n)	95·8 (132)	57·6 (132)	62·9 (132)	6·4 (132)	1·2 (131)	0·20 (131)
	Range	91·2–97·5	34·5–78·3	44·1–81·4	2·5–14·4	0·5–2·4	0·08–0·31
Leaves	$\bar{x}$ (n)	93·8 (23)	49·1 (23)	74·4 (23)	6·9 (23)	0·8 (23)	0·18 (23)
	Range	81·5–96·8	41·5–70·2	59·6–79·5	3·9–8·4	0·5–1·8	0·11–0·27
Stems	$\bar{x}$ (n)	95·1 (30)	55·2 (30)	64·9 (30)	6·2 (30)	0·4 (30)	0·16 (30)
	Range	92·2–98·3	43·1–69·5	48·6–82·9	3·2–8·8	0·2–0·7	0·09–0·27
Seed heads	$\bar{x}$ (n)	97·3 (27)	62·3 (27)	48·8 (27)	5·3 (26)	1·4 (27)	0·26 (27)
	Range	94·7–98·0	37·8–80·2	30·1–77·1	2·9–8·9	1·0–1·9	0·18–0·35
Sweet potatoes							
Tubers	$\bar{x}$ (n)	96·0 (50)	89·4 (50)	18·6 (50)	2·7 (50)	0·7 (50)	0·19 (50)
	Range	91·8–97·7	73·5–98·7	10·9–40·3	1·3–5·4	0·4–1·1	0·12–0·31
Foliage	$\bar{x}$ (n)	86·6 (50)	62·2 (50)	53·4 (50)	16·9 (50)	0·6 (50)	0·30 (50)
	Range	69·0–92·1	41·0–73·6	43·5–70·4	10·4–28·9	0·7–3·0	0·22–0·43

TABLE 3
Ranking of Species by Percent of Cell Wall Constituents (CWC) Expressed as Percent of Dry Matter

CWC Rank	Species
Low $CWC < 35\%$	Vegetable crops cabbage; radish; turnip; lettuce; celery; pepper fruits; cucumber; taro corms, cormels and foliage; fodder beet; chicory root; cassava tuber; sweet potato tuber; bean plants; carrot tubers and foliage; corn grain and roots Other: sorghum seed heads; *Pistia stratiotes*; seagrasses; marine algae

Medium $35\% < CWC < 70\%$	Vegetable crops
	  lettuce; bean plants; carrot foliage; taro non-storage roots; corn grain, roots, stalks and entire plant; pepper roots, plants and fruit; sweet potato foliage
	Grasses
	  Napiergrass entire plant, stem interior and leaves; sorghum entire plant, leaves, stalks and seed heads: rye; ryegrass; bermudagrass; millet; sorghum sudangrass hybrids; unimproved and domestic sugarcane
	Terrestrial plants
	  clover; vetch; *Amaranthus*; castorbean; pigweed; sida; saltbush; dogfennel; *Leucaena*; kenaf
	Aquatic plants
	  Eichhornia crassipes entire plant, roots and shoots: *Typha*; *Pontederia*; *Scirpus validus*; *Zizania*; *Polygonum*; *Iris*; *Sagittaria latifolia*; *Alternanthera philoxeroides*; *Lemna minor*; *Salvinia rotundifolia*; *Pistia stratiotes*; *Egeria densa*; *Hydrocotyle umbellata*; *Spirodela polyrhiza*; *Ipomoea aquatica*; *Azolla caroliniana*

High $CWC > 70\%$	Woody plants
	  pine; oak; sycamore; *Melaleuca*; *Eucalyptus*; poplar; *Casuarina*; *Mimosa*; *Acacia*; *Albizia*; *Spathodea*; *Sapium*; *Schinus*; *Leucaena*
	Grasses
	  Napiergrass entire plant, stems, leaves, sheaths, inflorescence, stem exterior; sorghum entire plant, stems, leaves, inflorescence; sorghum sudangrass hybrids; unimproved and domestic sugarcane; eastern gamagrass; indiangrass; *Erianthus*
	Terrestrial plants
	  cassava foliage; bean roots; sida; dogfennel; kenaf
	Aquatic plants
	  Phragmites australis; *Zizaniopsis miliaceae*

TABLE 4
Tannin Composition, both Vanillin–HCl Tannin (VTan) and Protein-precipitate Tannin (PTan) for all Sorghum Samples and Samples of Sorghum Plant Parts, Expressed as Percent of Total Volatile Solids. Each Entry is the Mean ($\bar{x}$), Number of Samples (n) and the Range. Entire Refers to Total of Above-ground Plant Parts

		VTan	PTan
All samples	$\bar{x}$ (n)	0·11 (134)	0·05 (134)
	Range	0–1·51	0–0·78
Entire	$\bar{x}$ (n)	0·09 (56)	0·05 (56)
	Range	0–0·70	0·01–0·23
Leaves	$\bar{x}$ (n)	0·06 (5)	0·01 (5)
	Range	0·01–0·09	0–0·02
Stems	$\bar{x}$ (n)	0·02 (10)	0·02 (10)
	Range	0–0·04	0·01–0·03
Seed heads	$\bar{x}$ (n)	0·24 (10)	0·09 (10)
	Range	0–1·51	0–0·78

categories. This differential distribution among categories is probably more a reflection of the low number of samples (three each) for eastern gamagrass, indiangrass and *Erianthus* and the very large number of samples (approximately 300 each) for Napiergrass, sorghum and sugarcane. The inherent variation of the former three grasses is not represented by the small number of samples, particularly because plant parts were not analyzed separately as they were in the latter three species.

Tannin values in the sorghum samples (Table 4) were quite low, perhaps because of oxidation of the tannin after the samples were ground (Price *et al.*, 1978). The purpose of the tannin analyses was not to measure the tannin content of fresh plant material, but of the feedstock when fermentability was measured. Therefore, tannin analyses were conducted on samples several weeks after they were ground in order to coincide with measurements of fermentability.

Two major determinants of the fermentability of a feed in ruminants is the percentage of CWC in the feed and the extent to which the CWC is lignified (lignin/CWC). Therefore, the initial model used to predict fermentability (IVTVSD) of biomass feedstocks included CWC and lignin/CWC. The results of the multiple regression analyses relating fermentability to CWC and lignin/CWC (Table 5) indicate that the percentage of variation accounted for by our model is low for some groups. For example, in emergent aquatic plants only 45% of the variation in fermentability is

TABLE 5
Predictions of Fermentability (IVTVSD) from Cell Wall Constituents (CWC), Lignin as a Percent of CWC (Lig), Nitrogen (N) and Phosphorus (P). Entries are Number of Samples (n), R^2 and $Sy.x$ values. Entire Refers to Total of Aboveground Plant Parts

	n	CWC, Lig		CWC, Lig, N, P	
		R^2	$Sy.x$	R^2	$Sy.x$
All samples	1 261	0·767	9·02	0·778	8·80
Grasses	718	0·793	5·88	0·833	5·29
Napiergrass	296	0·893	4·46	0·910	4·11
Entire	217	0·883	4·29	0·913	3·72
Leaves	29	0·522	3·30	0·662	2·89
Stems	8	0·979	1·80	0·997	0·95
Sheath	6	0·374	6·28	0·553	9·19
Stem exterior	20	0·805	3·89	0·828	3·89
Stem interior	11	0·750	2·42	0·885	1·89
Inflorescence	5	0·729	2·45	—	—
Sorghum	283	0·633	5·42	0·687	5·02
Entire	131	0·521	6·10	0·656	5·22
Leaves	23	0·899	2·67	0·953	1·91
Stems	30	0·892	2·23	0·904	2·20
Seed heads	26	0·704	6·16	0·761	5·79
Brassicas	43	0·912	4·68	0·944	3·82
Sweet potatoes	100	0·920	4·44	0·952	3·49
Tubers	50	0·512	4·47	0·833	2·68
Foliage	50	0·834	3·48	0·886	2·95
Emergent aquatics	41	0·451	13·90	0·675	10·98
Floating aquatics	96	0·879	6·56	0·884	6·48
Water hyacinth	55	0·843	6·45	0·873	5·92

explained by concentrations of CWC and lignin/CWC, and, for entire sorghum plants, only 52% of the variation is explained by our model. However, the percentage of variation accounted for by our model is much higher for other groups, such as brassicas (91%), floating aquatic plants (88%), sorghum leaves (90%) and sorghum stems (89%). A more useful statistical parameter is the standard error of the estimate $(Sy.x)$, which reflects the deviation that is unaccounted for by the regression. In some cases, low R^2 values are accompanied by low $Sy.x$ values which may be a reflection of the lack of variation in IVTVSD within a group of samples.

Addition of nitrogen and phosphorus data to the model had only a small effect on prediction of fermentability for most of the groups. However, a

TABLE 6
Predictions of Fermentability (IVTVSD) of Sorghum from Cell Wall Constituents (CWC), Lignin as a Percent of CWC (Lig), Vanillin-HCl Tannin (VTan), Protein-precipitate Tannin (PTan), Nitrogen (N) and Phosphorus (P). Entries are Number of Samples (n), R^2 and $Sy \cdot x$ values. Entire Refers to Total of Above-ground Plant Parts

	n	CWC, Lig		CWC, Lig, VTan		CWC, Lig, PTan		CWC, Lig, N, P		CWC, Lig, N, P, VTan		CWC, Lig, N, P, PTan	
		R^2	$Sy \cdot x$	R^2	$Sy \cdot x$	R^2	$Sy \cdot x$	R^2	$Sy \cdot x$	R^2	$Sy \cdot x$	R^2	$Sy \cdot x$
Entire	56	0·809	4·06	0·812	4·06	0·829	3·88	0·893	3·10	0·894	3·12	0·894	3·11
Leaves	5	0·984	2·10	0·990	2·39	0·999	0·64	—	—	—	—	—	—
Stems	10	0·951	2·18	0·960	2·14	0·951	2·35	0·977	1·75	0·980	1·82	0·990	1·29
Seed heads	10	0·423	9·27	0·818	5·62	0·924	3·63	0·512	10·09	0·880	5·59	0·951	3·58

few of the groups, e.g. Napiergrass leaves and sweet potato tubers, show a large increase in R^2 values, indicating that these two variables may have a significant effect on the fermentability of at least some feedstocks.

The effect of adding tannin to our model, both with and without nitrogen and phosphorus, varied among the sample groups (Table 6). There are two points from this table that should be stressed. First, adding protein-precipitate tannin yielded higher R^2 values and lower $Sy.x$ values than adding vanillin–HCl tannin for all regressions except for sorghum stems without nitrogen and phosphorus data. Second, tannin only gave a significant improvement to our models for sorghum seed heads. Tannin should have the greatest effect on the prediction for this group because the highest concentrations of tannin were measured in seed heads (Table 4) and because the fermentability of the other groups is already predicted quite well by CWC and lignin/CWC.

Fermentability can be predicted from the chemical constituents for which we analyzed. In a majority of cases, more than 80% of the variation in fermentability ($R^2 > 0.80$) could be accounted for from our model using four plant characteristics.

SUMMARY

Information presented here is intended to provide a general guide to plant composition of potential biomass feedstocks and its relation to an anaerobic fermentation system. We have shown that there is great variation, not only among species but also within species, in plant chemical composition. Within species, the variation is primarily due to stage of maturity; differential distribution of plant parts; and planting, harvesting and storage techniques. This great variation, and the factors involved, create a complex relationship between plant chemistry and fermentability that must be understood before the fermentability of a feedstock can be predicted from its chemical composition. Studies are underway to relate plant chemical composition to the potential for methane production. Our data demonstrate clearly that individual sources of biomass are different in terms of chemical composition and potential fermentability. Furthermore, there are differences among biomass sources in the relationships between composition and fermentability. Inappropriate generalizations about the relation of plant chemistry to fermentability may limit the development of practical bioconversion processes.

ACKNOWLEDGEMENTS

The authors appreciate the contributions of Mr Garry Foster in conducting the laboratory analyses.

REFERENCES

Gallaher, R. N., Weldon, C. O. and Futral, J. G. (1975). An aluminum block digester for plant and soil analysis. *Proceedings of Soil Science Society of America* **39**, 803–6.

Goering, H. K. and Van Soest, P. J. (1970). Forage fiber analyses (apparatus, reagents, procedures and some applications). Agriculture Handbook Number 379, United States Department of Agriculture, Washington, D.C.

Golding, E. J., Carter, M. F. and Moore, J. E. (1985). Modification of the neutral detergent fiber procedure for hays. *Dairy Science* **68**, 2732–6.

Hagerman, A. E. and Butler, L. G. (1978). Protein precipitation method for the quantitative determination of tannins. *Journal of Agriculture and Food Chemistry* **26**, 809–12.

Hambleton, L. G. (1977). Semiautomated method for simultaneous determination of phosphorus, calcium and crude protein in animal feeds. *Journal of the Association of Official Agricultural Chemists* **60**, 845–52

Harris, H. B., Cummins, D. G. and Burns, R. E. (1970). Tannin content and digestibility of sorghum grain as influenced by bagging. *Agronomy Journal* **62**, 633–5.

Moore, J. E. and Mott, G. O. (1974). Recovery of residual organic matter from in vitro digestion of forages. *Journal of Dairy Science* **57**, 1258–9.

Price, M. L., Van Scoyoc, S. and Butler, L. G. (1978). A critical evaluation of the vanillin reaction as an assay for tannin in sorghum grain. *Journal of Agriculture and Food Chemistry* **26**, 1214–18.

Robertson, J. B. and Van Soest, P. J. (1977). Dietary fiber estimation in concentrate feedstuffs. *Journal of Animal Science* **45** (Supplement 1), 254.

Swain, T. (1979). Phenolics in the environment. *Recent Advances in Phytochemistry* **12**, 617–40.

Tilley, J. M. A. and Terry, R. A. (1963). A two-stage technique for the in vitro digestion of forage crops. *Journal of the British Grassland Society* **18**, 104–11.

Van Soest, P. J. (1982). *Nutritional Ecology of the Ruminant*, O & B Books, Corvallis, Oregon.

21
Cellulase Enzymes for Enhancement of Methane Production from Biomass

M. GRITZALI,[a] A. SHIRALIPOUR[b] and R. D. BROWN Jr[a]

[a]*Food Science and Human Nutrition Department,* [b]*Microbiology and Cell Science Department, University of Florida—IFAS, Gainesville, Florida, USA*

INTRODUCTION

The conversion of biodegradable materials to methane in anaerobic wastewater treatment systems and laboratory scale reactors has focused attention on organisms and enzyme systems capable of degrading lignocellulose (Brown and Gritzali, 1984; Sahm, 1984). Many factors influence the success of methane production including reactor design, establishment of active, stable mixed cultures of microorganisms, identification and control of rate-limiting biochemical steps, and prospectively, genetic improvement of specific microorganisms to provide maximum rate and yield of methane from productive biomass species. The resistance of cellulose, the most abundant constituent of terrestrial biomass, to hydrolytic digestion has been noted and cocultures of anaerobic cellulolytic and methanogenic species have been used to explore the essential steps in the overall conversion process (Khan, 1977, 1980). Many limitations exist in anaerobic reactors supplied with complex biomass: solids loading levels are generally low (0·1–1·0%), residence times are long (3–5 weeks) and conversion is variable depending on the nature of the organic materials supplied (Sahm, 1984).

Cellulolytic enzymes from both aerobes and anaerobes have been investigated at both the fundamental (enzymological and genetic) level and in technological (enzyme or microbial) applications (Johnson *et al.*, 1982; Lamed *et al.*, 1983; Shoemaker *et al.*, 1983a,b; Brown and Gritzali, 1984; Béguin *et al.*, 1985; Grépinet and Béguin, 1986). The cellulolytic enzymes secreted by the aerobic fungus *Trichoderma reesei* form a system of enzymes which act synergistically to degrade cellulose to glucose

(Gritzali and Brown, 1979; Fägerstam and Pettersson, 1980; Henrissat et al., 1985). The high level of extracellular enzyme protein produced by this organism has led to intensive study of these enzymes and commercial production of the enzyme system. On the other hand, the level of cellulase enzyme activity observed in digesters or in ruminants is very low requiring sensitive techniques such as viscometric assays even to detect the endoglucanases which are characteristic of many anaerobic bacteria (Brown and Gritzali, 1984). The synergy displayed between endo- and exoglucanases from *T. reesei* and among exoglucanases from this organism has suggested the value of introducing such active enzymes into appropriate hydrolytic organisms by a series of genetic manipulations.

However, many preliminary questions regarding the structure, activity and synthesis of these enzymes needed to be answered to identify the specific properties and to insure the feasibility of such a project. Among such questions are the effects of these enzymes on methane production, the activity and persistence of these enzymes under bioreactor conditions, the interaction between fungal and bacterial enzymes, the gene and polypeptide sequence for each enzyme, and the essentiality of various domains within such enzymes.

Nature of Cellulase Enzymes

The cellulase system of *T. reesei* consists of three general types of enyme components: β-glucosidase (EC 3.2.1.21), endo-1,4-β-D-glucanase (EC 3.2.1.4) and exo-1,4-β-D-glucanase (EC 3.2.1.91) or exocellobiohydrolase (CBH). The optimal activities of each of these enzymes are observed at about pH 4·9–5·0 (Emert, 1973; Gritzali, 1979), values significantly less than neutrality which is commonly found to be optimal for methanogenesis. The stabilities of these enzymes under conditions of methane generation and in the presence of the mixed cultures which exist in such reactors had not been reported previous to this study. During the past 5 years the genetic basis for synthesis of the cellulases has been investigated in several laboratories, and gene sequences (and deduced polypeptide sequences) have been determined (Shoemaker et al., 1983b; Chen et al., 1987) for the two exocellobiohydrolases (CBH I and CBH II).

Berghem and Pettersson (1974) and Emert et al. (1974) reported the purification and properties of the β-glucosidase from *Trichoderma viride*, a species closely related to *T. reesei*. These workers found an enzyme protein of molecular weight 75–80 000 which was capable of cleaving aryl-glucosides and cellooligosaccharides to glucose. Further work by Chirico (1980) on the purification and characterization of the β-glucosidase from *T. reesei* revealed that the enzyme from this organism is indeed similar to that of *T. viride*. The enzyme is unstable at low pH values exhibiting a half-life of about 1 h at pH 3. Although only about 0·5% of the extracellular cellulase protein from *T. reesei* consists of the β-glucosidase, this very active enzyme is capable of hydrolyzing the oligosaccharides produced by the endo- and exoglucanases to glucose as rapidly as these soluble oligomers are formed. Thereby the inhibition by cellobiose of the depolymerizing enzymes is relieved, enhancing the overall rate of cellulose conversion to fermentable sugar (Gritzali and Brown, 1979). From anaerobes, such as *Clostridium thermocellum*, the low level of β-glucosidase activity seems to be predominantly cell-bound (Ait *et al.*, 1982) and is accompanied by phosphorolytic enzymes capable of cleaving cellobiose and cellooligosaccharides (Alexander, 1972a,b).

Endoglucanases (EG) are the most commonly encountered type of extracellular enzyme of the cellulase system; indeed, the trivial name 'cellulase' is often applied to endoglucanase activity. By cleaving bonds in the interior of a glucan (cellulose) molecule, an endoglucanase rapidly decreases the substrate molecular weight and provides chain ends which serve as substrates for the exocellobiohydrolases of those cellulase systems which possess both the exo- and endoglucanase activities. Most endoglucanases seem to have comparable specific activities on the most commonly used substrate carboxymethyl-cellulose (CMC); however, the first cellulase enzyme to be crystallized, endoglucanase D from *Clostridium thermocellum*, is reported to be some five times as active as those from other sources (Joliff *et al.*, 1986). The presence of endoglucanase activity alone is not enough to assure that an organism is capable of growth on cellulose as the sole carbon source. Several enzymes have been shown to be active in producing fermentable sugars including β-glucosidases, cellodextrin phosphorylases, and some endoglucanases (Brown and Gritzali, 1984). Thus, in evaluating the adequacy of endoglucanases to convert cellulose to glucose it is necessary to test not only growth of the organism, but also the nature of products arising from the enzymatic action and the presence and effect of related enzymes for conversion of the intermediate products to fermentable sugars (e.g. glucose).

Of the enzymic components of the *Trichoderma* cellulase system, the cellobiohydrolases I and II comprise some 80–85% of the enzyme protein (Gritzali and Brown, 1979). Of these, the highly acidic (pI = 3·9) CBH I, which contains about 6% carbohydrate, constitutes three quarters of the total CBH protein. The CBH II, containing nearly 20% carbohydrate, is the more active of the two enzymes. Not only do the two CBHs act synergistically together (Fägerstam and Pettersson, 1980) or with endoglucanases (Henrissat et al., 1985), but they are also capable of enhancing the degradation of crystalline cellulose when used together with all of the other components of the cellulase system (Brown and Gritzali, 1984). It is therefore interesting to evaluate the effect of these CBH enzymes on cellulose degradation by anaerobes which have extracellular cellulase enzymes consisting largely or entirely of endoglucanases. Since the gene and polypeptide sequences of the CBH proteins have been determined (Shoemaker et al., 1983b; Fägerstam et al., 1984; Chen et al., 1987), attention has been focused on the synthesis and activity of these enzymes.

In assessing the relevance of the fungal enzymes to cellulose degradation during methanogenic fermentations and the feasibility of using these cellulases to improve the rate of yield of methane, it was necessary to explore the effect of enzymatic pretreatment of substrate on subsequent methane generation, the persistence of active fungal enzymes during methane generation and the fate of individual enzymic components of the fungal system during this process. Preparation of messenger RNA and determination of the levels of message for the specific fungal cellulase enzymes formed part of the initial effort for cloning and expression of these enzymes in appropriate anaerobes from the methanogenic digesters. Use of purified enzyme components to test for synergism with isolated anaerobes provided another criterion for potential utility of specific fungal cellulase enzymes in the biomass conversion system. In a separate, but related, collaborative project the sequence of the gene (and protein) of the cellobiohydrolase II have been determined (Chen et al., 1987).

MATERIALS AND METHODS

Standard Biogasification Methodology

Prior to either the bioassay (i.e. biogasification) procedures or volatile solids (VS) determination, plant samples were dried at 60°C and then ground in a Christy–Norris hammer mill (1 mm mesh). For the VS determination samples were further dried at 105°C for 16 h and ashed at 550°C for 3 h. For bioassay procedures, to the dried plant (or other cellulosic) material equivalent to 0·20 g volatile solids in a 250 ml serum bottle were added defined medium (a mixture of vitamins, minerals, buffer, and reducing agents) and seed inoculum (liquor from a digester containing methanogenic bacteria) and the volume was brought to 100 ml with deionized water. The bottles then were transferred to a 35°C incubator. Preparation of the defined medium, seed inoculum and the procedures for gas measurements have been described by Shiralipour and Smith (1984).

Enzyme-treated Biogasification Samples
From a solution of 1 g of *Trichoderma* cellulase enzyme protein in 100 ml of deionized water (pH 4·5) 10 ml were added to a suspension of bioassay (cellulosic) material in 70 ml of deionized water. After adjustment to pH 5·0 for enzymic hydrolysis the suspension was incubated for 24 h; it was then adjusted to pH 7·0 for the methanogenic digestion and defined medium and seed inoculum were added to the enzyme and substrate suspension bringing the total volume to 100 ml. During subsequent incubation at 35°C aliquots of liquid and gas were analyzed periodically. For all samples at the 0·2% VS loading, values for samples without cellulosic substrates were determined and used to correct for methane evolved from cellulase enzyme protein.

Assays and Immunochemical Identification of Enzymes
Antisera to cellobiohydrolases were prepared by intramuscular injection of young adult rabbits with 1 mg of each antigen in complete Freund's adjuvant. A 1:3 ratio of aqueous antigen to adjuvant was employed. Six weeks after the primary injection, a single booster injection was administered using 0·5 mg of antigen in Freund's adjuvant. Antisera were collected 4 weeks following the booster injection, and further treated as described by Hurn and Chantler (1980). Titers and antiserum specificities were determined by the double diffusion method in agarose gels described by Ouchterlony (1968).

For polysome preparation, 5 g (wet weight) of *T. reesei* QM 9414 mycelia were broken in a Braun MSK homogenizer in the presence of 10 g of glass beads (0·17–0·18 mm). The cell homogenate was then processed and polysomes were prepared as described by Buell *et al.* (1978). Isolation of total polysomal RNA was accomplished by the method of Palmiter (1974), and mRNA was isolated from total polysomal RNA by two successive cycles of chromatography on oligo-dT cellulose (Buell *et al.*, 1978). Messenger RNA was translated using a rabbit reticulocyte lysate *in vitro* translation system, prepared in this laboratory by a modification of the method described by Maniatis *et al.* (1982) and individual proteins identified by immunoprecipitation and electrophoresis.

To assay components of the cellulase system specific techniques were used where available, whereas for mixtures of components or the entire cellulase system either microcrystalline cellulose (Avicel) or phosphoric acid–swollen cellulose (PSC) were used as substrates. Endoglucanase was assayed by the method of Shoemaker and Brown (1978) using carboxymethyl cellulose as substrate and the β-glucosidase was assayed using p-nitrophenyl β-D-glucoside as substrate (Emert, 1973). When either the intact cellulase system activity on Avicel or the synergistic activity of bacterial and fungal enzymes on PSC was determined, the production of reducing sugar was followed using the Nelson–Somogyi reagents (Gritzali, 1979).

RESULTS

Effects of Cellulase Enzyme Pretreatment on Subsequent Biogasification
Preliminary experiments by workers in the GRI/IFAS Biomass to Methane Program had identified several biomass types, such as sorghum, Napiergrass and water hyacinth, as possessing particular value for field productivity and thus as potential feedstock sources. These plants contain cellulose as the principal structural substance and would be expected to exhibit a limited convertibility based on the slow hydrolysis of this 1,4-β-glucan. Effects of cellulase enzyme added as a pretreatment to the cellulosic feedstock were evaluated after 14 weeks of methane generation in biogasification experiments (Table 1). Values shown are means determined for three replicate samples for each treatment and each control. The *Trichoderma* cellulase system (Gritzali and Brown, 1979) from cellulose-grown cultures of the fungus was added and allowed to act on the substrate for 24 h. Microcrystalline cellulose (Avicel PH-101) and pulped spruce fibers (Solka Floc) were used as positive controls for the effect of the enzymes on cellulose. Levels of volatile solids higher than the standard 0·2% were included to indicate whether the enzyme pretreatment alone would result in conversion of higher loadings of solids during methanogenesis. Enzyme additions increased in methane yields in each case at the 0·2% volatile solids loading with enhancements ranging from 1·7% to 14·7%. In five of the seven cases these increases were significant at $p \leq 0·05$ as determined by the Student's t-test (Table 1). Nearly theoretical levels of methane were produced in each case, except for Napiergrass and water hyacinth. For the standard cellulosic control materials loadings at the 1·0 and 2·0% volatile solids levels led to suppression of methane formation and copious production of fatty acids leading to pH values lower than 4·7 (data not shown). The plant materials were

TABLE 1
Effect of Added Cellulase Enzyme on Methane Yield

Biomass type	Biomass conc. (g VS/100 ml)	Enzyme added	CH_4 yield (Std m^3 kg^{-1} VS)	% Increase
Cellulose	0·2	–	0·408	
(crystalline)	0·2	+	0·415	1·7
	1·0	+	0·025	
	2·0	+	0·000	
Solka floc (B)	0·2	–	0·409	
	0·2	+	0·434	6·1[a]
	1·0	+	0·013	
	2·0	+	0·000	
Solka floc (M)	0·2	–	0·428	
	0·2	+	0·466	8·9[a]
	1·0	+	0·019	
	2·0	+	0·006	
Sweet sorghum	0·2	–	0·396	
	0·2	+	0·428	8·1[a]
	1·0	+	0·283	
	2·0	+	0·013	
Forage sorghum	0·2	–	0·384	
	0·2	+	0·422	9·9[a]
	1·0	+	0·264	
	2·0	+	0·050	
Napiergrass	0·2	–	0·358	
	0·2	+	0·371	3·6
	1·0	+	0·314	
	2·0	+	0·252	
Water hyacinth	0·2	–	0·258	
	0·2	+	0·296	14·7[a]
	1·0	+	0·220	
	2·0	+	0·119	

[a] Increased yield of methane significant at $p \leq 0.05$.

converted in fair yield to methane at the higher loadings, but did not reach the same level of conversion seen for the standard volatile solids concentration. These results encouraged a more detailed evaluation of exogenous enzyme stability during the methane fermentation in order that the feasibility for transformed organisms elaborating these, or other forms of genetically engineered enzymes, might be estimated.

The specific endoglucanase activities (Table 2) were determined for the

TABLE 2
Apparent Persistence of Cellulase Activity Added to Biogasification Samples

Biomass type	Volatile solids (g)	Cellulase enzyme addition	Cellulase specific activity after 14 weeks ($U\,mg^{-1}$)
Crystalline cellulose	0·2	+	<0·1
	1·0	+	11·5
	2·0	+	10·3
	0·2	−	<0·1
Solka floc (purified pulp fibers)	0·2	+	<0·1
	1·0	+	13·9
	2·0	+	9·5
	0·2	−	<0·1
Sweet sorghum	0·2	+	<0·1
	1·0	+	<0·1
	2·0	+	5·9
	0·2	−	<0·1
Forage sorghum	0·2	+	<0·1
	1·0	+	<0·1
	2·0	+	8·33
	0·2	−	<0·1

proteins in the supernatant liquid of biogasification samples incubated under the same conditions described for Table 1. These activities provided tentative evidence for residual *Trichoderma* cellulase enzyme since no detectable activity was observed in any case where the fungal enzyme had not been added. Although cellulolytic pretreatment had an effect at the 0·2% VS loading level, after 14 weeks of incubation no residual enzyme could be measured in those samples. In the samples where higher solids levels existed, some cellulase activity was detected. This might be due to protection of the enzyme by binding to the insoluble substrate or due to production of cellulase by anaerobes during the 14-week incubation. More specific means then were used to measure the disappearance of the cellulase system components at shorter incubation times during which residual activity would be expected to be in the process of inactivation or bioconversion.

Rate of Conversion of Fungal Cellulase to Methane
The production of methane from either (A) exogenous *Trichoderma* cellulase enzyme plus inoculum, (B) inoculum alone, or (C) from the added

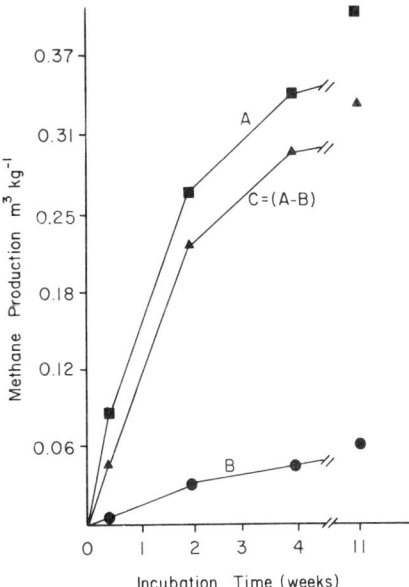

FIG. 1. Biogasification incubations showing methane production from the inoculum (with or without cellulase enzyme) as a function of incubation time at 35°C. Line (A) represents methane production from inoculum and added cellulase enzyme. Line (B) represents methane production from the inoculum alone, without added enzyme. Line (C) shows methane production apparently due to the added cellulase enzyme by difference. The value for methane formation which had become constant at 11 weeks in each case was taken to indicate 100% of potential conversion.

cellulase enzyme by difference is shown in Fig. 1. Most of the cellulase enzyme had disappeared after 4 weeks' incubation; however, the experiment was continued for 11 weeks to estimate the methane yield for complete bioconversion of the enzyme proteins. Using this information it was possible to plot (Fig. 2) an estimate of the enzyme remaining throughout the course of the incubation. This graphical representation allowed an estimate of 9·2 days for the half-life of the cellulase protein in the anaerobic digestion. These data are representative, at best, of the conversion of the enzyme protein and indicate neither the level of the residual exogenous cellulase enzyme *activity* nor the rate at which individual component enzymes are being inactivated or removed during the digestion.

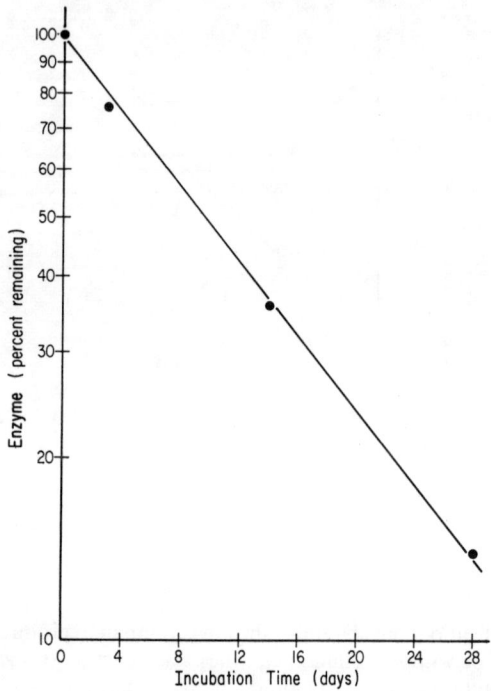

FIG. 2. Enzyme disappearance during biogasification. This is a replot of data from Fig. 1, showing a half-life for enzyme disappearance, under reactor conditions, of 9·2 days.

Kinetics of Component Cellulase Enzyme Disappearance from Biogasification Systems

In order to address these questions, it was necessary to assay the individual enzymes after a shorter exposure to the anaerobic fermentation. Since the cellobiohydrolases cannot be reliably assayed in the presence of the other enzymes of the cellulase system (i.e. endoglucanases and β-glucosidase), it was necessary to use specific antisera for the respective CBHs and quantify these results by comparison of precipitin lines obtained by known concentrations (i.e. dilutions) of the purified enzymes. When the capacity of the residual enzyme(s) to degrade microcrystalline cellulose (Avicel) was measured after 1 week exposure to the anaerobic digestion, it was found that only 14% of the initial activity remained (Table 3). This was, of course, much less than would be expected if the half-lives of the active enzymes were each 9–10 days. By use of immunochemical

TABLE 3
Cellulase Enzyme Activity Remaining after 1 Week in the Bioreactor

Enzyme	Immuno-chemical probe	Enzyme activity probe	Activity remaining (%)	Amount remaining $(mg\,ml^{-1})$
Cellobiohydrolase I	+		42	250
Cellobiohydrolase II	+		50	125
Endoglucanase		+	74	111
β-Glucosidase		+	35	6
Total (of cellulase enzyme components)				492
Cellulase (Avicelase)		+	14	

probes for the CBHs and by measurement of enzyme activities of the endoglucanase and β-glucosidase it was observed that the persistence of the individual enzymes of the cellulase system varied broadly. The CBH I and CBH II, respectively, were present at 42 and 50% (average = 46%) of their initial levels whereas the endoglucanase exhibited greater stability (74% remaining) and the β-glucosidase less stability (35% remaining) than the CBHs. When these values were used together with the known relative abundances of the enzymes in the initial exogenous cellulase system, it could be calculated that 492 mg ml^{-1} of active enzyme protein remained after 1 week. This represented 49·2% of the initial enzyme protein and, taking into account that some degraded or inactive enzyme protein might have remained unconverted to methane, is in fair agreement with the earlier estimate that the half-life for conversion of enzyme protein to methane was about 9·2 days.

The observation that only 14% of the Avicelase activity remained may be rationalized by considering how the cellulase enzymes act together synergistically. Whether they act sequentially or randomly, it is clear that for native cellulose each type of enzyme component is *necessary* for the maximum rate of degradation of cellulose to glucose. With the assumption that, during studies of initial rates of Avicel degradation, each enzyme is virtually saturated with respect to its substrate, the overall rate of hydrolysis (as measured by reducing sugars produced from Avicel) will be the product of the individual rates for each type of enzyme. Thus, as each rate is diminished by enzyme loss or inactivation it will decrease the overall (i.e Avicelase) rate to the same degree. From this model of synergistic action and by measurement of initial rates of Avicel degradation for the case illustrated in Table 3, it was predicted that the fraction of Avicelase activity initially present will, after 7 days, have decreased to 0·12 (0·46 × 0·74 × 0·35 = 0·12 which represents the product of the CBH × EG × BG residual fractions of initial enzyme). This prediction is in fair agreement with the observed value of 0·14 for residual Avicelase activity. The values observed for fractional activities remaining indicated that the most labile enzyme is the β-glucosidase, which contains little or no carbohydrate, and that the stability of the other glycoprotein enzymes increased with carbohydrate content. The observed half-lives of about 7 days for the cellulase enzyme components encouraged further investigation of a genetic basis for improvement of the hydrolytic anaerobes.

Synergism between Fungal Cellobiohydrolases and Cellulases of Digester Organisms
In order to select an enzyme for optimal impact on the digester system, the possibility of cooperative action of the pure fungal CBHs with bacterial cellulase was explored. For this purpose, three bacterial isolates (gifts from Dr P. H. Smith) which had been selected from biomass digesters for cellulolytic activity and for relative abundance in the digester were grown on cellulose and the secreted enzymes concentrated before use. The specific activities of the bacterial isolates for Avicel degradation were quite low (Table 4). However, when those bacterial enzymes were combined with fungal CBH I or CBH II, more than additive activity was observed and the percentage enhancement above additive values (synergism) varied from 10 to 40% with the highest values being found for combinations containing CBH II. This result targeted CBH II as the most favorable enzymic component for enhancement of the anaerobic digester system.

Molecular Biological Approach to Enhanced Methane Production
The cellobiohydrolases I and II each have molecular weights of about 54 000; however, the CBH II polypeptide is substantially smaller for the mature CBH II glycoprotein which contains some 18% by weight of carbohydrate. In preparation for construction of a cDNA library, isolation of the messenger RNA (mRNA) including those for the CBH enzymes, was required. Optimum conditions were established for isolation of polysomes present during active synthesis of the cellulase enzymes (Brown and Gritzali, 1984). An electron micrograph (Fig. 3) shows polysomes produced by cells of *Trichoderma reesei* induced to synthesize cellulase. From such polyribosomes was isolated mRNA which was used to direct

TABLE 4
Synergism between Fungal Cellobiohydrolases and Extracellular Cellulase Activity from Isolated Anaerobes

Enzyme	Specific activity ($\mu mol\, min^{-1}\, mg^{-2}$)		Synergism observed (%)
CBH I	0·19	—	—
CBH II	2·70	—	—
Isolate No. 1	0·07	—	—
Isolate No. 2	0·04	—	—
Isolate No. 6	0·15	—	—
CBH I + No. 1	0·29	0·26[a]	12
CBH II + No. 1	3·32	2·77[a]	20
CBH I + No. 2	0·26	0·23[a]	15
CBH II + No. 2	3·84	2·74[a]	40
CBH I + No. 6	0·37	0·34[a]	10
CBH II + No. 6	3·56	2·85[a]	25

[a]Specific activities expected if combined enzyme activities are additive.

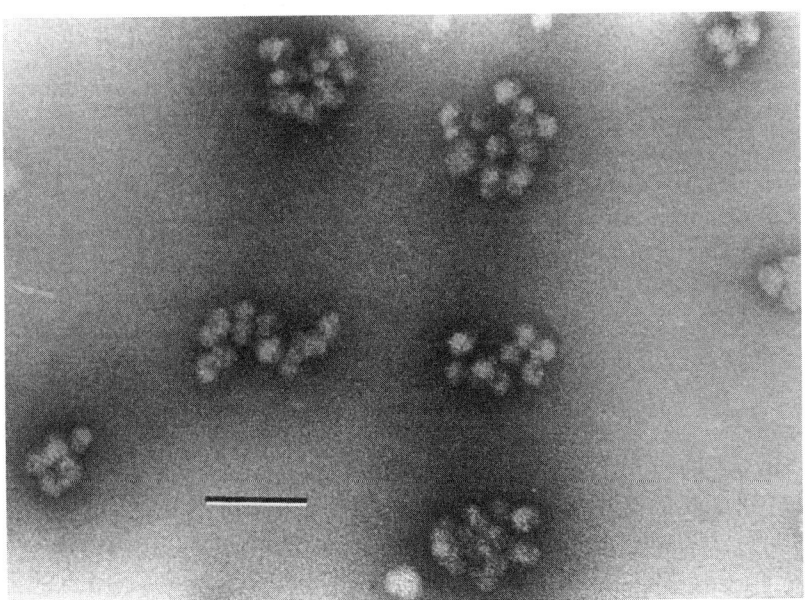

FIG. 3. Electron micrograph of large polysomes isolated from *T. reesei* QM 9414 (courtesy of Bonnie O'Brien).

TABLE 5
Cellulase Polypeptides Synthesized *in vitro* by Total mRNA Isolated from *T. reesei* QM 9414 Mycelia Cultured under Various Conditions

	Immunoprecipitable radioactivity (DPM)[a]			
	Growing cell		Resting cells	
Serum	Glucose	Cellulose	Induced	Control
Anti CBH I	289 ± 14	8 493 ± 1 240	29 278 ± 3 777	117 ± 23
Anti CBH II	264 ± 33	2 760 ± 238	17 566 ± 1 194	98 ± 15

[a] Immunoprecipitations were carried out using, in each case, an amount of *in vitro* translate which contained approximately 745 000 dpm of TCA precipitable radioactivity.

protein synthesis in a rabbit reticulocyte lysate system. By use of immunoprecipitation to identify the individual proteins (Table 5) the highest levels of cellobiohydrolase enzymes were observed to be synthesized in the presence of mRNA from induced resting cells. CBH I was the predominant protein, and the cellobiohydrolases accounted for about 6·2% of the proteins among the *in vitro* translation products. Although a cDNA library was then prepared, it was decided, in collaboration with Dr Darrel Stafford at the University of North Carolina at Chapel Hill, to proceed with a *Trichoderma* genomic library for selection and isolation of the CBH II gene. As a result of that project it is now known that the CBH II polypeptide has several regions of homology with CBH I but lacks key residues in the region which has been suggested to be the catalytic active site of CBH I (Chen *et al.*, 1987),

SUMMARY

This research established that fungal cellulase can enhance the anaerobic conversion of biomass to methane. In some cases the yield has approached the theoretical limits imposed by the volatile solids contents of the samples. Failure to convert higher solids loadings (>0·2% VS) was probably inherent in the fermentation system, the improvement of which was the subject of other research in this program. Although the fungal enzymes produced by the aerobic eukaryote *Trichoderma* lack several properties needed for activity in a conventional methane digester, these may be provided by genetic alteration prior to transformation of appropriate anaerobes. However, the fungal enzymes were found to be stable enough to persist for many days and should now be amenable to mutagenesis in preparation for useful transformation of hydrolytic anaerobes.

Among the properties of the CBH II to be modified are (1) the tentatively identified catalytic residues, for activity at pH values near neutrality; (2) the introns which must be removed from the gene for use in prokaryotes; (3) glycosylation which must be evaluated for essentially for stability; and (4) compatibility with the enzymes of the hydrolytic anaerobe specifically selected for transformation. Each of these approaches has a clear experimental basis and can be carried out with appropriate experience, skills and support.

ACKNOWLEDGEMENTS

The authors dedicate this chapter to Dr Elwyn T. Reese on the occasion of his 75th birthday with admiration and appreciation for his many scientific contributions and his devotion to the highest professional and personal values.

The authors gratefully acknowledge additional support by National Science Foundation Grant No. CPE-8314943.

REFERENCES

Ait, N., Creuzet, N and Cannanéo, J. (1982). Properties of β-glucosidase purified from *Clostridium thermocellum*. *Journal of General Microbiology* **128**, 569–77.

Alexander, J. K. (1972a). Cellobiose phosphorylase from *Clostridium thermocellum*. In: *Methods in Enzymology*, Vol. 28, Ginsburg, V. (ed.), pp. 944–8. Academic Press, New York.

Alexander, J. K. (1972b). Cellodextrin phosphorylase from *Clostridium thermocellum*. In: Methods in Enzymology, Vol. 28, Ginsburg, V. (ed.), pp. 948–53. Academic Press, New York.

Béguin, P., Cornet, P. and Aubert, J. P. (1985). Sequence of a cellulase gene of the thermophilic bacterium *Clostridium thermocellum*. *Journal of Bacteriology* **162**, 102–5.

Berghem, L. E. R. and Pettersson, L. G. (1974). The mechanism of enzymatic cellulose degradation. Isolation and some properties of a β-glucosidase from *Trichoderma viride*. *European Journal of Biochemistry* **46**, 295–305.

Brown, R. D., Jr and Gritzali, M. (1984). Microbial enzymes and lignocellulose utilization. In: *Genetic Control of Environmental Pollutants*, Omenn, G. S. and Hollaender, A. (eds.), pp. 239–65. Plenum Publishing Corporation, New York.

Buell, G. N., Wickens, M. P., Payvar, F. and Schimke, R. T. (1978). Synthesis of full length cDNAs from four partially purified oviduct mRNAs. *Journal of Biological Chemistry* **253**, 2471–82.

Chen, C. M., Gritzali, M. and Stafford, D. W. (1987). Nucleotide sequence and deduced primary structure of cellobiohydrolase II from *Trichoderma reesei* QM 9414. *Biotechnology* (in press).

Chirico, W. J. (1980). Purification and characterization of an extracellular β-glucosidase from *Trichoderma reesei* QM 9414. M.S. thesis, Virginia Polytechnic Institute and State University, Blacksburg, Virginia.

Emert, G. E. (1973). Purification and characterization of cellobiase from *Trichoderma viride*. Ph.D. dissertation, Virginia Polytechnic Institute and State University, Blacksburg, Virginia.

Emert, G. H., Jr, Gum, E. K., Lang, J. A., Liu, T. H. and Brown, R. D. Jr (1974). Cellulases. In: *Advances in Chemistry Series*, Vol. 136, Whitaker, J. R. (ed.), pp. 79–100. American Chemical Society, Washington, D.C.

Fägerstam, L. G. and Pettersson, L. G. (1980). The 1,4-β-glucan cellobiohydrolases of *Trichoderma reesei* QM 9414. *FEBS Letters* **119**, 97–100.

Fägerstam, L. G., Pettersson, L. G. and Engströni, J. A. (1984). The primary structure of a 1,4–β-glucan cellobiohydrolase from the fungus *Trichoderma reesei* QM 9414. *FEBS Letters* **167**, 309–15.

Grépinet, O. and Béguin, P. (1986). Sequence of the cellulase gene of *Clostridium thermocellum* coding for endoglucanase B. *Nucleic Acids Research* **14**, 1791–9.

Gritzali, M. (1979). Purification and characterization of an endo, 1,4-β-D-glucanase and two exo-1,4-β-D-glucanases from the cellulase system of *Trichoderma reesei*. Ph.D. dissertation, Virginia Polytechnic Institute and State University, Blacksburg, Virginia.

Gritzali, M. and Brown, R. D. Jr (1979). The cellulase system of *Trichoderma*. Relationships between extra-cellular enzymes from induced or cellulose-grown cells. In: *Advances in Chemistry Series*, Vol. 181, Brown, R. D., Jr and Jurasek, L. (eds.), pp. 237–60. American Chemical Society, Washington, D.C.

Henrissat, B., Driquez, H., Viet, C. and Schülein, M. (1985). Synergism of cellulases from *Trichoderma reesei* in the degradation of cellulose. *Biotechnology* **3**, 722–6.

Hurn, B. A. L. and Chantler, S. M. (1980). Production of reagent antibodies. In: *Methods in Enzymology*, Vol. 70, Van Vunakis, H. and Langone, J. J. (eds.), pp. 104–22. Academic Press, New York.

Johnson, E. A., Sakajoh, M., Halliwell, G., Madia, A. and Demain, A. (1982). Saccharification of complex cellulosic substrates by the cellulase system from *Clostridium thermocellum*. *Applied and Environmental Microbiology* **43**, 1125–32.

Joliff, G., Béguin, P., Juy, M., Millet, J., Ryter, A., Poljak, R. and Aubent, J. P. (1986). Isolation, crystallization and properties of a new cellulase of *Clostridium thermocellum* overproduced in *Escherichia coli*. *Biotechnology* **4**, 896–900.

Khan, A. W. (1977). Anaerobic degradation of cellulose by mixed culture. *Canadian Journal of Microbiology* **23**, 1700.

Khan, A. W. (1980). Degradation of cellulose to methane by a coculture of *Acetivibrio cellulolyticus* and *Methanosarcina barkeri*. *FEMS Microbiology Letters* **9**, 233–5.

Lamed, R., Setter, E. and Bayer, E. A. (1983). Characterization of a cellulose-binding, cellulase-containing complex in *Clostridium thermocellum*. *Journal of Bacteriology* **156**, 828–36.

Maniatis, T., Fritsch, E. F. and Sambrook, J. (1982). *Molecular Cloning, a Laboratory Manual*, Cold Spring Harbor Laboratory, Cold Spring Harbor, New York.

Ouchterlony, O. (1968). *Handbook of Immunodiffusion and Immunoelectrophoresis*, Ann Arbor Science Publishers, Ann Arbor, Michigan, 215 pp.

Palmiter, R. D. (1974). Magnesium precipitation of ribonucleoprotein complexes. Expedient techniques for the isolation of undergraded polysomers and messenger ribonucleic acid. *Biochemistry* **13**, 3606–15.

Sahm, H. (1984). Anaerobic wastewater treatment. In: *Advances in Biochemical Engineering and Biotechnology*, Vol. 29, Fiechter, A. (ed.), pp. 83–116. Springer-Verlag, Berlin.

Shiralipour, A. and Smith, P. H. (1984). Conversion of biomass resources into methane gas. *Biomass* **6**, 85–92.

Shoemaker, S. P. and Brown, R. D. Jr (1978). Enzymic activities of endo-1,4-β-D-glucanases purified from *Trichoderma viride*. *Biochemica et Biophysica Acta* **523**, 133–46.

Shoemaker, S. P., Watt, K., Tsitovsky, G. and Cox, R. (1983a). Characterization and properties of cellulases purified from *Trichoderma reesei* strain L27. *Biotechnology* **1**, 687–90.

Shoemaker, S. P., Schweickart, V., Ladner, M., Gelfand, D., Kwok, S., Myanibo, K. and Innis, M. (1983b). Molecular cloning of exo-cellobiohydrolase I derived from *Trichoderma reesei* strain L27. *Biotechnology* **1**, 691–6.

22
Ultrastructural Analyses of Methanogens

H. C. ALDRICH, D. S. WILLIAMS and R. W. ROBINSON
Microbiology and Cell Science Department, University of Florida — IFAS, Gainesville, Florida, USA

INTRODUCTION

The Archaebacteria, an assemblage of organisms which includes the methanogenic bacteria, are assuming increasing importance and now receiving more attention (Woese and Wolfe, 1985). The methanobacteria fall into three orders, the Methanobacteriales, the Methanococcales, and the Methanomicrobiales (Whitman, 1985). In an effort to establish a data base for the identification of bacterial species in mixed cultures such as anaerobic digesters, we examined, at the ultrastructural level, representative pure cultures of species from all three orders. Some, such as *Methanosarcina* spp., are morphologically so distinctive that they can be distinguished by shape alone. Others require additional sophisticated techniques such as gold-labeled antibody tagging (Smit and Todd, 1986). In addition to identification of methanogenic bacterial cells in natural populations, it is also important to understand in detail the structure of the cells, and especially how this structure relates to the biochemical events of methanogenesis. A long-range goal is to visualize and identify the cellular sites of methane formation, utilizing techniques of immunogold cytochemistry (Smit and Todd, 1986). Once this kind of basic information is in hand, one can attempt to optimize digester conditions to favor growth of species or even of cell cycle stages of a given species to enhance methane production rates. This chapter summarizes the techniques we have developed to study both pure cultures and digester samples, and then illustrates some of the more important genera and species of methanogens as examined with transmission electron microscopy.

ULTRASTRUCTURE TECHNIQUES

Suspension cultures of most species are adequately fixed for embedding and ultrathin sectioning by direct immersion of colony-forming species in a mixture of cacodylate-buffered (pH = 7·2) 2·5% glutaraldehyde and 2·5% formaldehyde (freshly made from paraformaldehyde) for 30 min; or by the addition of 5% buffered glutaraldehyde to an equal volume of the suspension medium before centrifugation and primary glutaraldehyde–formaldehyde fixation. After a buffer wash, cells are postfixed in 2% cacodylate-buffered osmium tetroxide. We use Spurr's low viscosity resin mixtures for embedment. Additional details may be found in Robinson et al. (1985) or Aldrich and Todd (1986).

Examination of biofilms presents some special problems, because the biofilm on digester surfaces is rather fragile and tends to slough away during processing in the various liquids used in preparation for EM. For scanning EM, we have had surprisingly good results by plunging digester packing material covered with biofilm directly into absolute ethanol, leaving it for 2–3 days in the refrigerator, and then air drying. This minimizes loss of components from the upper layers of the film and still results in acceptable preservation (see Fig. 1A in Chapter 23 in this volume). For plastic embedding and ultrathin sectioning, we drop razor-cut 1 mm thick sections of the film into warm 2·5% Bacto-agar or into 30% (w/v) bovine serum albumin, and then cover either type of sample with several ml of 2·5% cacodylate-buffered glutaraldehyde. Neither the agar nor the albumin prevents penetration of fixatives or plastics, and the albumin becomes a gel when exposed to the glutaraldehyde. The biofilm is thus held together in its original orientation and is protected from fragmentation during handling.

Methanosarcina (Methanococcus) mazei Mah and Kuhn

This species is a major component of the biofilm of packed-bed digesters we have examined (Robinson et al., 1984). It efficiently converts acetate to

---

FIG. 1. Transmission electron micrograph of a portion of a colony of *Methanosarcina mazei*. Light reticulum in cells is nucleoid region containing DNA. Dark ragged inclusions are polyphosphate. One larger, partly dissolved polyphosphate body is visible at arrow (×23 600).

FIG. 2. High magnification of cell membrane of *Methanosarcina mazei*. Typical tri-layered unit membrane (between large arrows) delimits cell. Outside that, a third dense layer is supported by periodic pillars (small arrows) (×280 000).

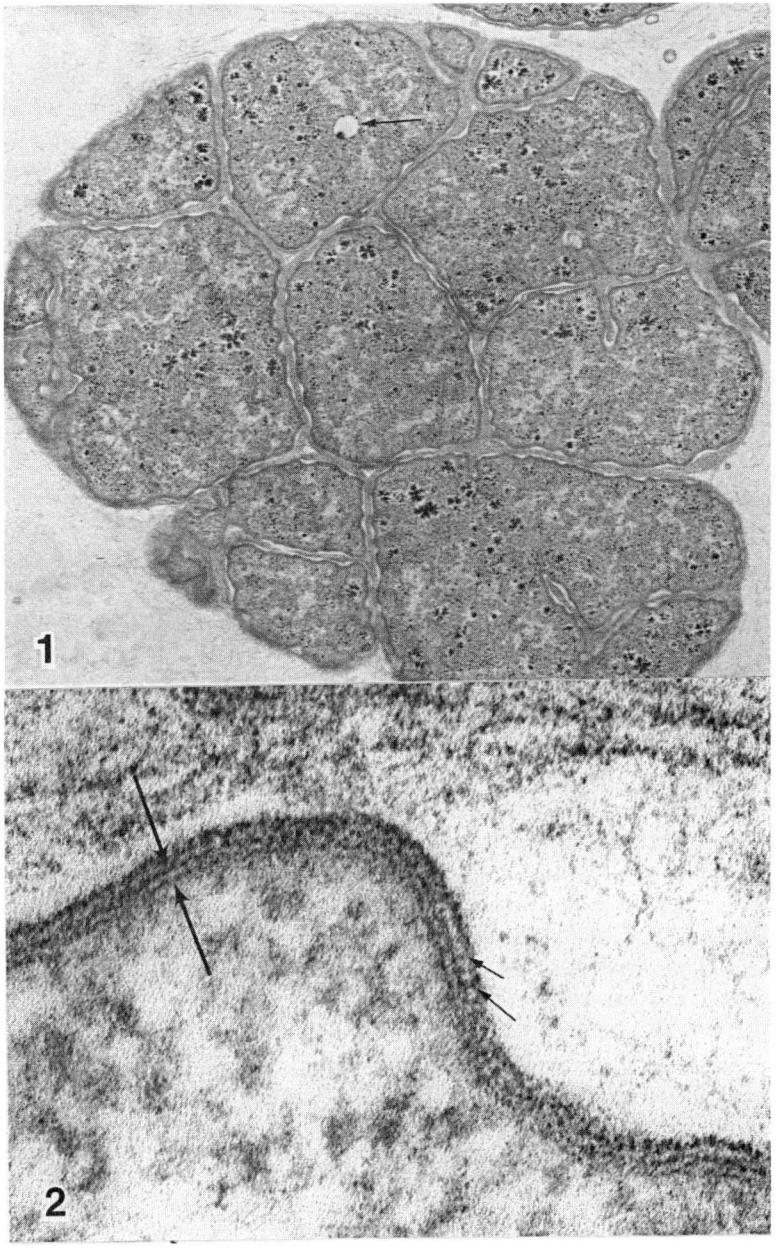

methane under anaerobic conditions (Mah, 1980). Hence, we have examined its ultrastructure in some detail (Aldrich et al., 1986).

In colonies which may reach 1 mm or more in diameter, individual cells of *M. mazei* are irregular and contain inclusions typical of prokaryotic cells (Fig. 1). The DNA is scattered, and stellate dark inclusions are common. These inclusions contain polyphosphate (Scherer and Bochem, 1983). We believe that they condense into the globular form more common in other prokaryotes (Fig. 1). The tripartite plasma membrane is coated with a single dark layer with repeating subunits (Fig. 2). In older (90 day) cultures (Fig. 3), many cells contain cylindrical intrusions of the plasma membrane; we call these intrusions 'exotubes' (Aldrich et al., 1986).

Other Species

The related species *Methanosarcina barkeri* Schnellen (Fig. 4) is indistinguishable ultrastructurally from *M. mazei*, but we are hopeful that tagging with specific gold-labeled antibodies (Smit and Todd, 1986) will enable us to differentiate the two at the EM level.

We have examined five species in the Order Methanobacteriales (Whitman, 1985). Cells of *Methanobrevibacter smithii* (Smith) Balch and Wolfe are short bacilli (Fig. 5) with some unusual internal inclusions which need additional study to characterize. External to the plasma membrane is a thin wall, more resembling the wall of a Gram-positive bacterium than the thinner layer outside *M. mazei*. *Methanobrevibacter arboriphilus* Zeikus and Henning is a straight bacillus with unusual tapered ends (Fig. 6). Again, a thin wall layer surrounds the plasma membrane. Cytoplasmic electron lucent inclusions are also present in some cells. Cells of *Methanobacterium bryantii* Balch and Wolfe are more convoluted than the other two (Fig. 7), with complete longitudinal sections rarely falling within a single section. Some cytoplasmic inclusions similar to those in *M. smithii* are also seen. *Methanobacterium formicicum* Schnellen (Fig. 8) and *M. thermoautotrophicum* Zeikus and Wolfe (Fig. 9) both resemble *M. arboriphilus* ultrastructurally.

Members of the Order Methanococcales examined thus far include *Methanococcus vannielii* Stadtman and Barker (Fig. 10) and *M. voltae* Balch and Wolfe (Fig. 11). These two species are virtually indistinguish-

---

FIG. 3. Portion of older colony of *M. mazei*. Aging cells are filled with cylindrical 'exotubes' (arrows) of unknown function ($\times 23\ 000$).

FIG. 4. Portion of young colony of *Methanosarcina barkeri*, showing same structures visible in Fig. 1. Fibrous matrix surrounding the cells is hardly stained in this preparation ($\times 29\ 000$).

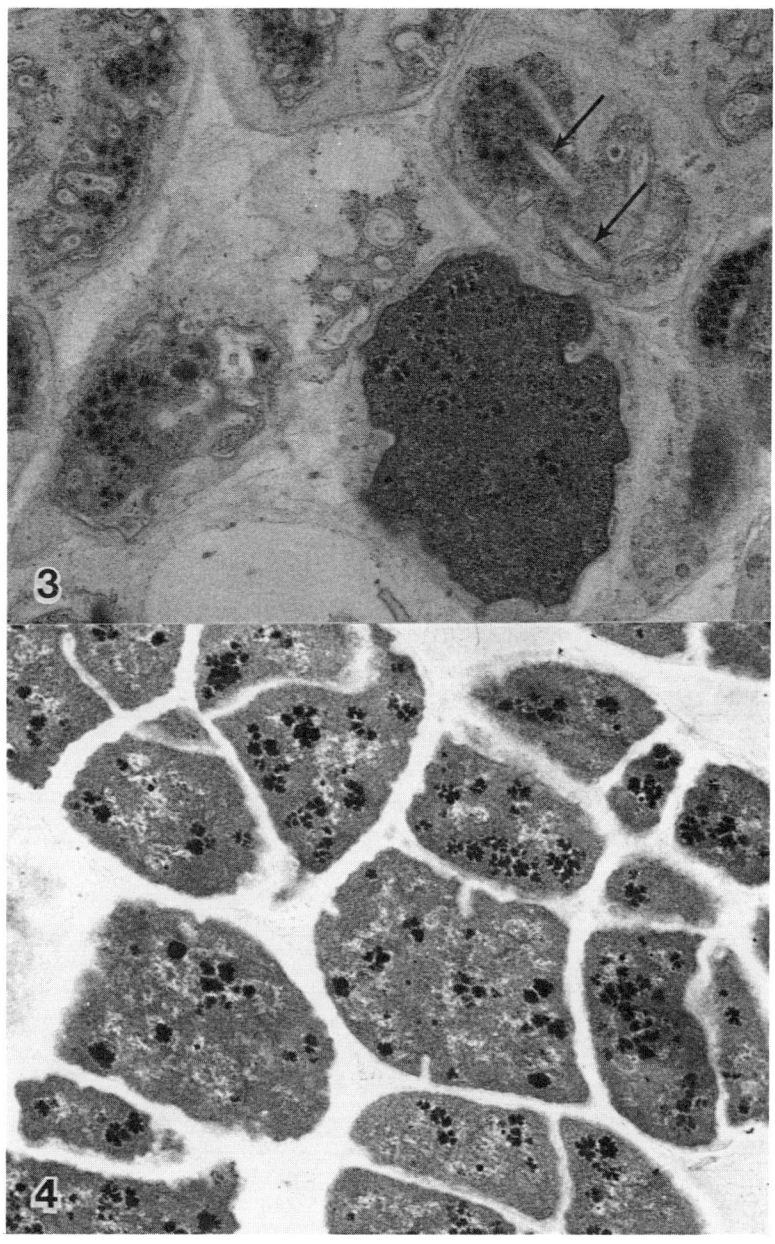

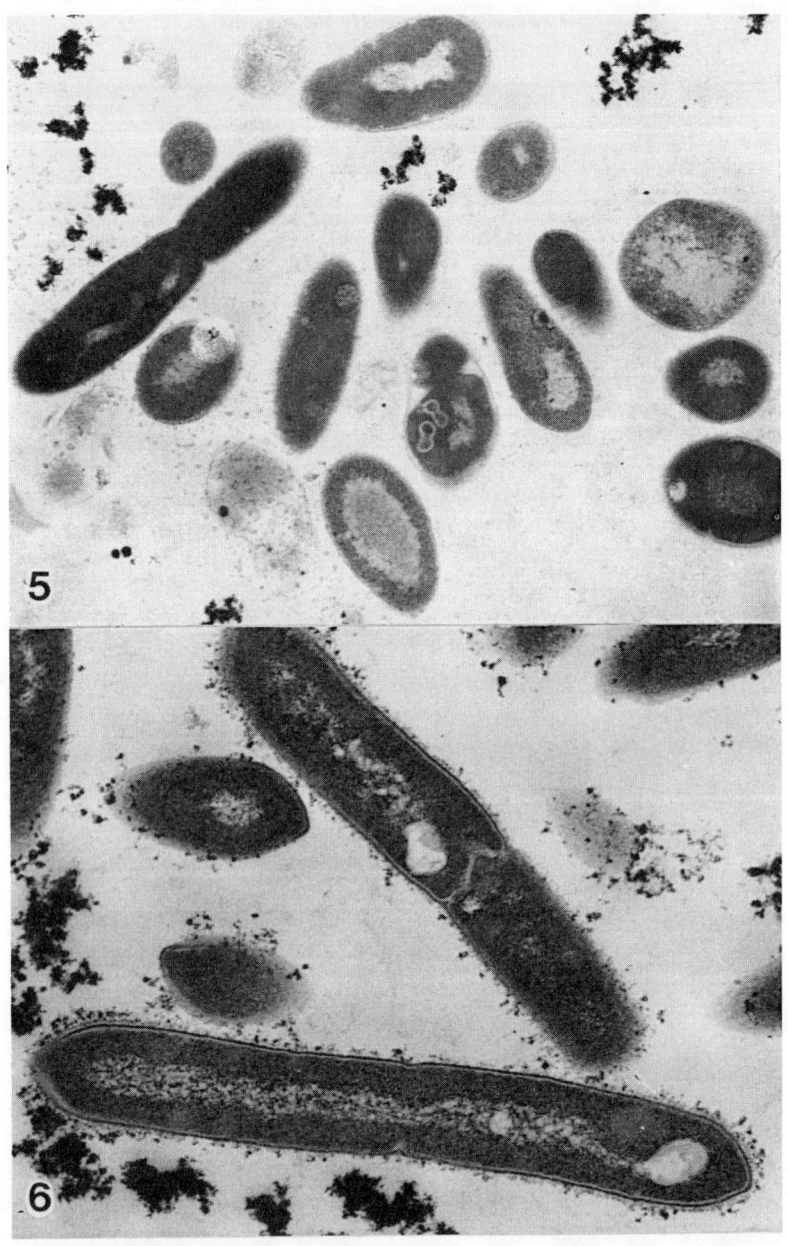

FIG. 5. Cells of *Methanobrevibacter smithii* (×25 000).
FIG. 6. Culture of *Methanobrevibacter arboriphilus* (×40 000).

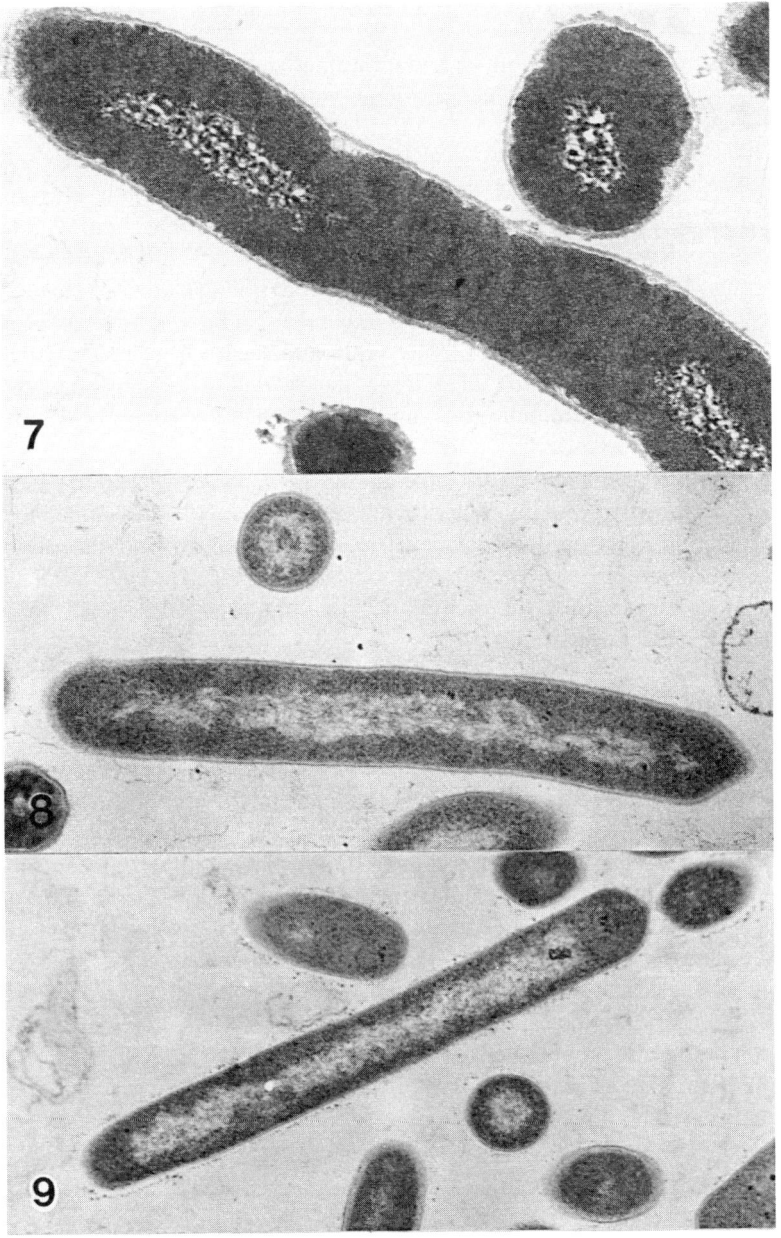

FIG. 7. *Methanobacterium bryantii*. Note irregular cell outline compared to bacteria in Figs 6, 8 and 9 (×60 000).
FIG. 8. *Methanobacterium formicicum*. Note sharply tapered ends (×38 300).
FIG. 9. *Methanobacterium thermoautotrophicum*. This cell exhibits same tapered ends as that in Fig. 8 (×36 300).

able ultrastructurally, with irregular outlines and a very thin layer outside the plasma membrane, which correlates with the observed fragility of both species.

In addition to the species of *Methanosarcina* discussed above, other members of the Order Methanomicrobiales examined include *Methanospirillum hungatei* Ferry, Smith and Wolfe, *Methanogenium thermophilicum* Rivard and Smith, and *Methanogenium marisnigri* Romesser and Wolfe. *M. hungatei* exhibits the typical curved profile in sections (Fig. 12). Its cells remain connected after division and are delimited by unusual and characteristic septa (Fig. 13). The septa and sheath have recently been studied in considerable detail by Beveridge *et al.* (1985) and by Shaw *et al.* (1985); both are proteinaceous, and the sheath is extremely resistant to chemical degradation.

M. thermophilicum (Fig. 14) resembles *Methanococcus voltae* and *M. vannielii* ultrastructurally. The cells are round to irregular in outline, and a single thin electron-dense layer surrounds the cell beyond the plasma membrane. Cells of *Methanogenium marisnigri* (Fig. 15) have a thicker, fuzzy coating outside the membrane, less organized than the *Methanosarcina* species described earlier.

SUMMARY

The methanogenic bacteria we have examined in ultrathin section with the transmission electron microscope exhibit a wide range of variation in structure. Features that appear most important for future concentrated examination include intracellular membranous inclusions, septa and wall/capsule/matrix layers beyond the plasma membrane. Similarity in morphology does not always agree with taxonomic position based on biochemical and physiological characters. For example, *M. thermophilicum* morphologically more closely resembles *Methanococcus* species than it does the other species of *Methanogenium* we examined. Clearly, an understanding of the structures of these unique organisms will require further study, including not only detailed examination of additional species, but also the use of sophisticated cytochemical techniques to

---

FIG. 10. *Methanococcus vannielii*. Cells are irregular, with central nucleoid (light areas). Dark precipitate came from growth medium ($\times 19\ 700$).

FIG. 11. *Methanococcus voltae*. Irregular cells, very similar to those of *M. vannielii*. DNA in nucleoid (arrows) has clumped during processing ($\times 42\ 700$).

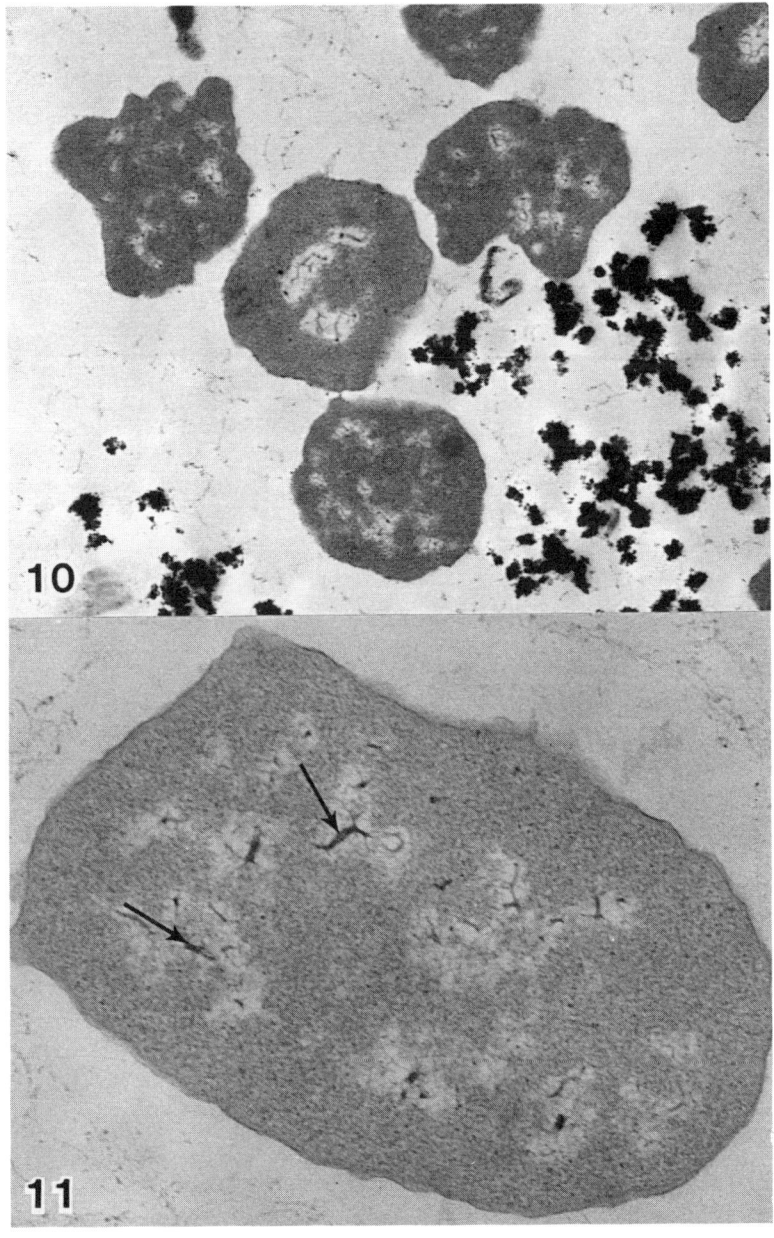

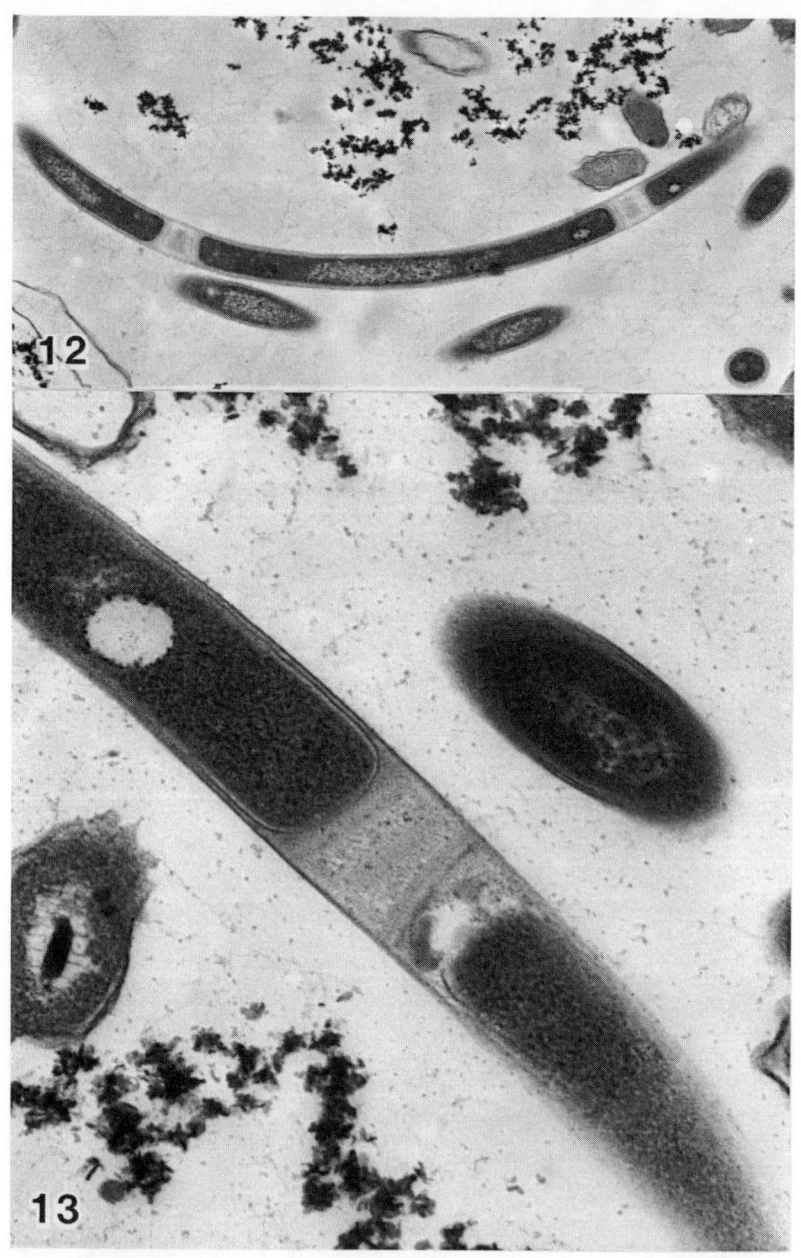

FIG. 12. Curved cells of *Methanospirillum hungatei* at low magnification (×15 400).

FIG. 13. Higher magnification of septum area of *Methanospirillum hungatei* (×57 100).

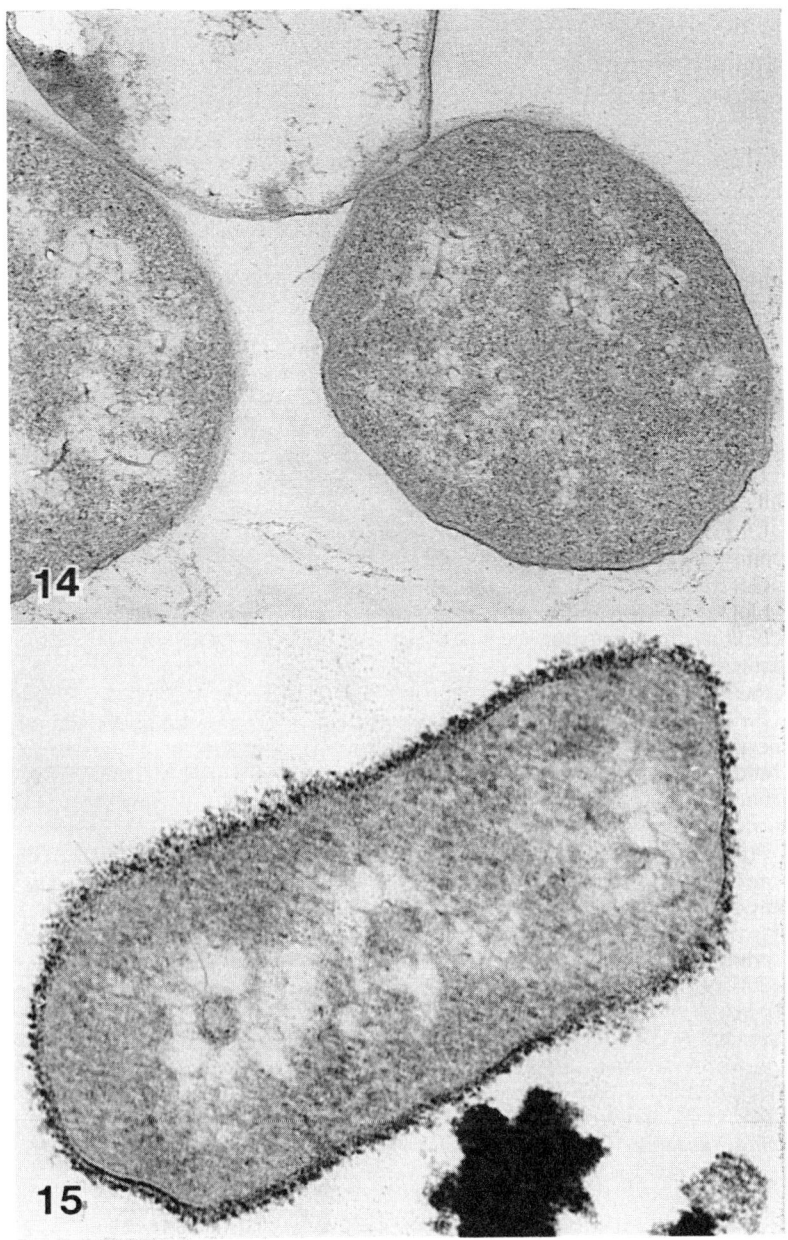

FIG. 14. *Methanogenium thermophilicum* (×42 000).

FIG. 15. *Methanogenium marisnigri* (×99 000).

determine the cellular sites of methane formation. Only when we fully understand these important cells can we really determine how to manipulate digester conditions and microbe populations to optimize methane production on a large scale, and this, after all, is the ultimate goal of these studies.

REFERENCES

Aldrich, H. C. and Todd, W. J. (1986). *Ultrastructure Techniques for Microorganisms*, Plenum Publishing Company, New York.

Aldrich, H. C., Robinson, R. W. and Williams, D. S. (1986). Ultrastructure of *Methanosarcina mazei*. *Systematic and Applied Microbiology* 7, 314–19.

Beveridge, T. J., Stewart, M., Doyle, R. J. and Sprott, G. D. (1985). Unusual stability of the *Methanospirillum hungatei* sheath. *Journal of Bacteriology* 162, 728–37.

Mah. R. A. (1980). Isolation and characterization of *Methanococcus mazei*. *Current Microbiology* 3, 321–6.

Robinson, R. W., Akin, D. E., Nordstedt, R. A., Thomas, M. V. and Aldrich, H. C. (1984). Light and electron microscopic examinations of methane-producing biofilms from anaerobic fixed-bed reactors. *Applied and Environmental Microbiology* 48, 127–36.

Robinson, R. W., Aldrich, H. C., Hurst, S. F. and Bleiweis, A. S. (1985). Role of the cell surface of *Methanosarcina mazei* in cell aggregation. *Applied and Environmental Microbiology* 49, 321–7.

Scherer, P. A. and Bochem, H. P. (1983). Ultrastructure of 12 Methanosarcinae and related species grown on methanol for occurrence of polyphosphate-like bodies. *Canadian Journal of Microbiology* 29, 1190–9.

Shaw, P. J., Hills, G. J., Henwood, J. A., Harris, J. E. and Archer, D. B. (1985). Three-dimensional architecture of the cell sheath and septa of *Methanospirillum hungatei*. *Journal of Bacteriology* 161, 750–7.

Smit, J. and Todd, W. J. (1986). Colloidal gold labels for immunocytochemical analysis of microbes. In: *Ultrastructure Techniques for Microorganisms*, Aldrich, H. C. and Todd, W. J. (eds.), Plenum Publishing Company, New York.

Whitman, W. B. (1985). Methanogenic bacteria. In: *The Bacteria, A Treatise on Structure and Function. Vol. VIII, Archaebacteria*, Woese, C. R. and Wolfe, R. S. (eds.), pp. 3–84. Academic Press, Orlando, Florida.

Woese, C. R. and Wolfe, R. S. (eds.) (1985). *The Bacteria, a Treatise on Structure and Function. Vol. VIII, Archaebacteria*, Academic Press, Orlando, Florida, 582 pp.

23
Properties that Affect Aggregation and Dispersion of Methanogens

A. S. BLEIWEIS, H. C. ALDRICH

Microbiology and Cell Science Department, University of Florida — IFAS, Gainesville, Florida, USA

and

R. A. MAH

Division of Environmental and Occupational Health Sciences, School of Public Health, University of California at Los Angeles, California, USA

INTRODUCTION

The microbiota of the anaerobic digester used for treatment of organic materials for commercial gas production is complex, yet has received little attention until recently. Studies reviewed in Chapter 22 have revealed that the most prevalent microbes recovered from the lower regions of anaerobic fixed-bed reactors, filled with various packing materials (e.g. wood chips) to increase biofilm formations, are sarcina-containing aggregates similar to *Methanosarcina* species. Figure 1A shows a scanning electron micrograph (SEM) and Fig. 1B is a transmission electron micrograph (TEM) of a typical adherent microcolony (Robinson *et al.*, 1984). Previously, Mah (1980) isolated from an anaerobic digester an acetotrophic methanogen that was initially named *Methanococcus mazei* S-6 before being renamed *Methanosarcina mazei* [Mah and Kuhn]. It became apparent that this and related species may be prominent gas-producing components of experimental and commercial anaerobic digesters. For this reason we initiated intensive studies of the physical and chemical ultrastructure of this methanogen, along with studies of the enzymatic disaggregation of mature cysts.

M. mazei was first described by Barker (1936). He discovered three morphotypes: isolated cocci, large irregular aggregates of these cocci, and sharply defined 'cysts' of various sizes. Zhilina and Zavarzin (1979) observed similar morphological types in mixed cultures which they too identified as *M. mazei*. These observations were confirmed with the isolation in pure culture of strain S-6 and the preliminary description of its life cycle by using light microscopy (Mah, 1980). Young cultures of S-6 resemble *Methanosarcina thermophila* TM1 with typical clusters (20 μm) of pleomorphic cells. These aggregates grow into uniformly dense spherical bodies ('cysts') of greater than 100 μm. Physical rupturing of cysts releases individual cocci (1–3 μm), thereby starting the life cycle anew.

In this section, we describe studies using SEM and TEM to determine the morphological changes that take place at the cell surface during microcolony disaggregation. Cytochemical techniques have been employed both at the electron microscopic level and on isolated matrix components to determine the nature of the cell surface of *M. mazei* (Robinson *et al.*, 1985). Also, we have isolated and characterized a new strain, called LYC, which spontaneously disaggregates to yield individual coccoid bodies (Liu *et al.*, 1985). This strain is morphologically and antigenically similar to strain S-6. We describe the enzymatic nature of this process and its relevance to the *M. mazei* life cycle.

RESULTS

Ultrastructure

SEM revealed microcolonies of up to 100 μm in size after 12 days of growth. These tended to aggregate in clusters of up to 5 mm at the bottoms of culture tubes. Multicellular lobes resembling heads of cauliflower (Fig. 2A) were prominent features of colonies, although individual cells were difficult to discern due to thick capsular (matrix) material. TEM of a typical lobe showed pleomorphic, irregular cells closely packed together (Fig. 2B). Individual cells were 1·0–1·5 μm in diameter and were surrounded by a 30–60 nm-thick matrix. Spheroidal, electron-dense granules (100 nm)

FIG. 1(A) Scanning electron micrograph of a biofilm surface from an anaerobic fixed-bed digester. A loose, open form of cyst is visible (arrow) ($\times$560). (B) Thin section micrograph of lower regions in the biofilm. Near the base of the film the cysts formed large, irregular masses (arrow). Internal granular material fills sections of many cells ($\times$5000) (1A and 1B from Robinson *et al.*, 1984, reproduced by permission of the American Society for Microbiology).

Properties that Affect Aggregation and Dispersion of Methanogens

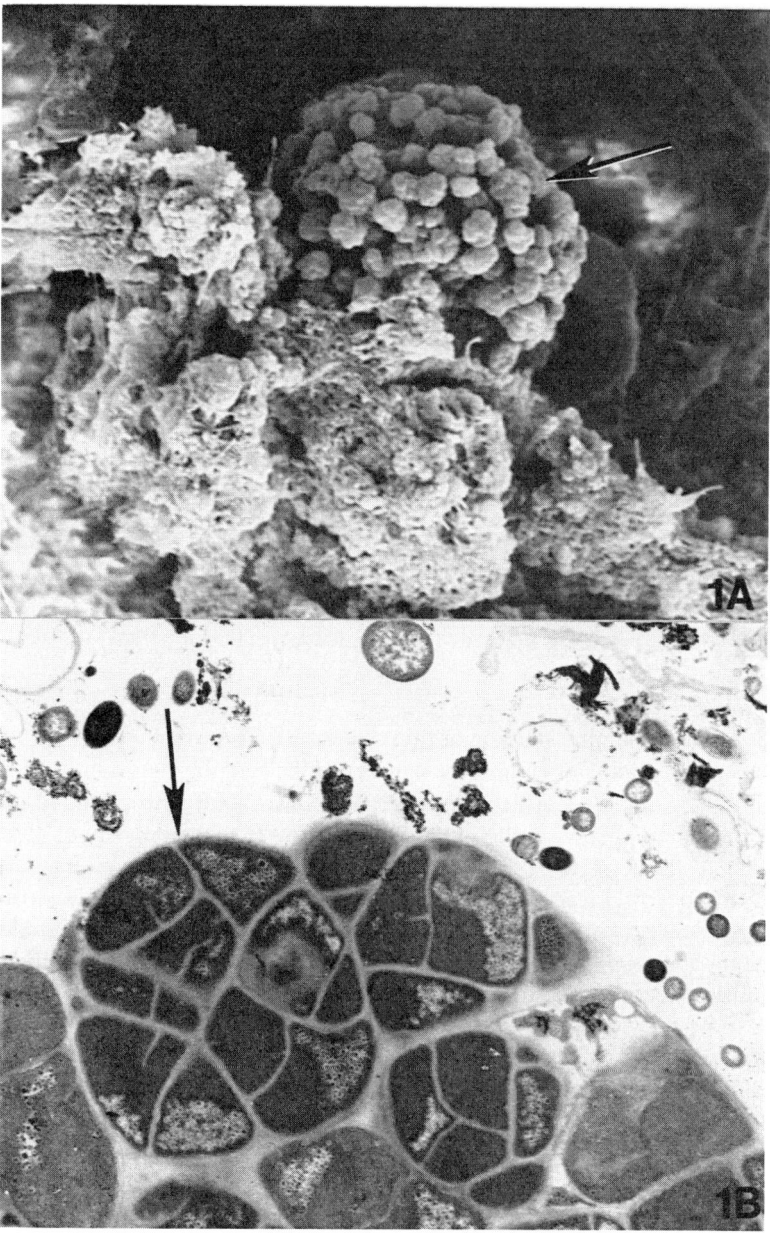

were scattered throughout the *M. mazei* cytoplasm. They were surrounded by an electron-transparent halo.

After growing for 2 weeks or longer, the colonies began to disaggregate, releasing single cocci into the medium. SEM now revealed individual cells on the surfaces of disaggregating colonies (Fig. 2C). Although most cells were still closely associated with neighboring cells, they did not form the tight multicellular lobes seen in younger cultures. Thin sections (Fig. 2D) revealed a loose arrangement of cells; disaggregating colonies appeared to be losing the cement-like matrix material that held the coccoid cells together.

The degree of cell–cell attachment appeared related to the compactness of the matrix. Figure 3A shows the matrix surrounding a cell in a disaggregating colony. It was a fibrous material that was shed in layers from the cells. Each individual layer was 17 nm thick and about 20–30 such layers were counted on each cell surface. Disaggregation was completed 48 h after onset and no sarcinal colonies remained. Free cells of strain S-6 were irregular and about 2 μm in size (Fig. 3B). Figure 3C shows a highly magnified view of the cell surface. Matrix material may be seen sloughing off the cell, and a typical unit membrane surrounded by cytoplasm. About 10 nm outside the plasma membrane was an additional electron-dense layer which is thought to be proteinaceous (Aldrich *et al.*, 1986).

Cytochemistry

Silver methenamine staining of thin sections of 12-day colonies (Fig. 4A) showed moderate labeling of the matrix and the cytoplasmic granules. This implies a polysaccharide composition, but we find phosphate in the granules and have suggested that it causes an artifactual deposition of the silver (Aldrich *et al.*, 1986). Thus, we doubt that the granules are polysaccharide. Fluorescein-lectin labeling (not shown) resulted in a strong fluorescence when whole colonies were treated with fluorescein-labeled peanut agglutinin, soybean agglutinin, or *Ricinus communis* [L.] agglutinin I. Other lectins did not react with *M. mazei* cysts; the reactive lectins

---

FIG. 2(A) SEM of 12-day sarcinal colonies of *M. mazei* growth in broth. The box delineates a multicellular lobe on the colony surface ($\times 1400$). (B) Thin section of a small 12-day colony showing the irregular cells enclosed by a fibrillar matrix (m) ($\times 10\,000$). (C) SEM of a disaggregating colony 2 weeks old. Single cells predominate on the surface ($\times 600$). (D) Thin section of a disaggregating colony. Cells are not tightly arranged and the matrix is very fibrous (arrows) ($\times 3000$) (2A–2D from Robinson *et al.*, 1985, reproduced by permission of the American Society for Microbiology).

Properties that Affect Aggregation and Dispersion of Methanogens 401

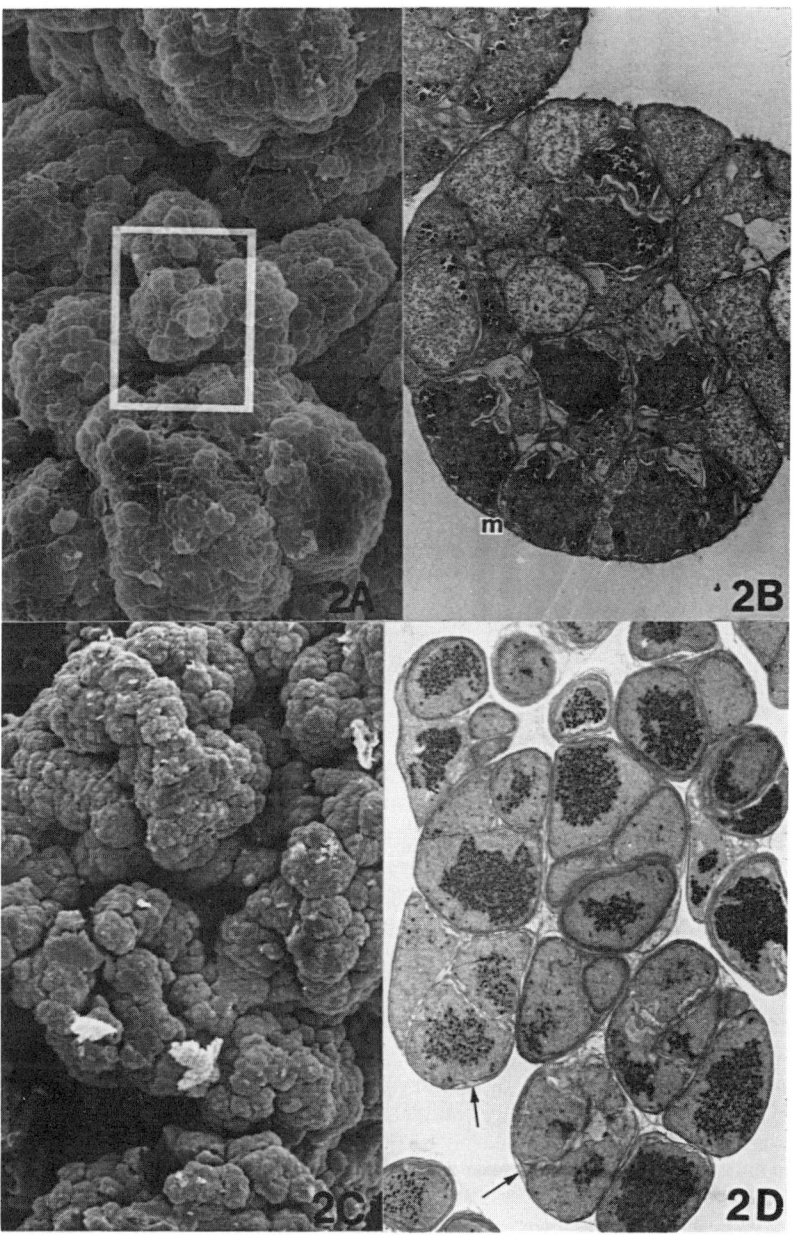

recognize N-acetylgalactosamine or galactosyl residues in polysaccharides. Figure 4B shows a distinct labeling of the matrix by the soybean agglutinin complexed with colloidal gold to allow visualization of the reaction.

Chemical Analysis of Isolated Matrix Material

Matrix material recovered from *M. mazei* colonies after glass bead disruption (Fig. 4C) retained the typical cellular contours seen within cysts (Fig. 2B). Thin-layer chromatography of acid hydrolysates revealed only two major products: galactosamine and galacturonic acid. Smaller amounts of glucosamine, glucose and glucuronic acid also were detected; no other sugars or amino sugars were found. Preliminary quantitative analyses indicate galactosamine and galacturonic acid are present in equimolar amounts. The qualitative nature of this polysaccharide, therefore, is consistent with the lectin reactivities discussed in the previous section. Studies to determine its structure and linkage patterns presently are underway. In all, polysaccharide constitutes 95% of the matrix, while the remaining 5% consists of only five amino acids: glycine, serine, lysine, ornithine and alanine (molar ratio: $1.0:1.8:0.9:0.7:0.2$). The possible presence of a cross-linking peptide has been considered by Robinson *et al.* (1985). These findings for *M. mazei* are in general agreement with those made for *M. barkeri* [Schnellen] by Kandler and Hippe (1977). These authors also failed to detect peptidoglycan but instead measured significant amounts of galactosamine.

Characterization of a Disaggregating Variant of *M. mazei*

During the course of methanogen isolations from samples of drilling-site sediments, an unusual variant of *M. mazei* was discovered. Although this strain, called LYC, forms colonies in roll-tube media that are similar to S-6 colonies and although strain LYC is antigenically related to other *Methanosarcina* species, this new isolate is far less acetotrophic than S-6 and grows better on H_2–CO_2, methanol or trimethylamine. More interest-

---

FIG. 3(A) High magnification of the matrix material of strain S-6 during colony disaggregation. The matrix is shedding from the cell in layers. Osmium–ferricyanide stain ($\times 27\,000$). (B) Free single cells of S-6 are irregular and contain cytoplasmic granules and polyphosphate-like bodies (P) ($\times 4\,000$). (C) At high magnification, a layer (L) is present that is closely associated with the plasma membrane (P) of single cells. Some matrix material was still visibly remote from the membrane ($\times 170\,000$). (3A–3C from Robinson *et al.*, 1985, reproduced by permission of the American Society for Microbiology).

Properties that Affect Aggregation and Dispersion of Methanogens 403

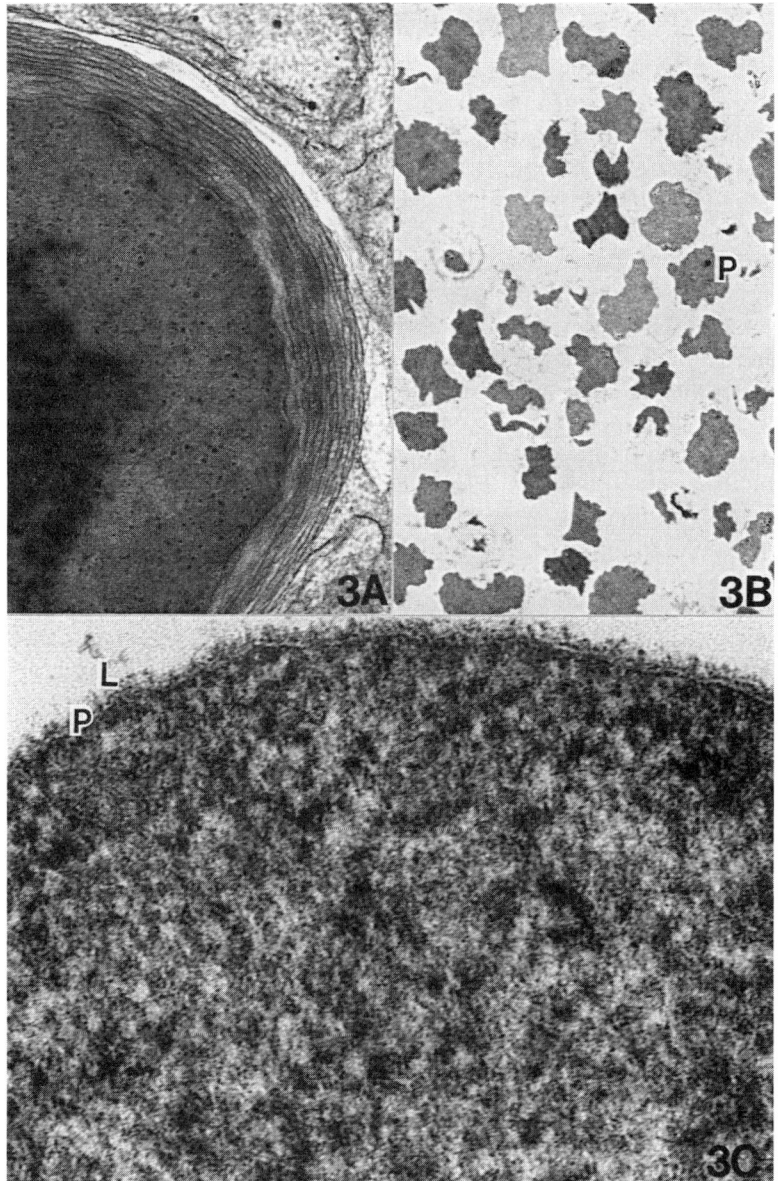

ing, however, is the fact that its cellular morphology varies with growth conditions.

In liquid medium at pH 7·0, cells grow first in small aggregates which soon (2 days) break down into small irregular cocci (1–3 μm) occurring singly or in pairs. The cocci appear to be devoid of capsular (matrix) material and are lysed by 1% sodium dodecyl sulfate. Individual cells as well as the aggregated forms exhibit UV fluorescence characteristic of methanogens. When grown at pH 6·0, however, strain LYC was indistinguishable from strain S-6 by both SEM and TEM (Figs. 2A and 2B). If during growth the pH was adjusted to 7·0, the cells began to disaggregate. Figure 5A shows the rupturing of the protective matrix and the emergence of a spheroplast during the earliest phases of this process. Samples taken later show the almost complete release of irregular cocci (Fig. 5B); the matrix material has been broken down to small fragments. At this point, the cocci have become susceptible to lysis by detergents. It is important to note, however, that disaggregation occurred only in the presence of a suitable growth substrate, a fact which distinguishes this process from the autolytic death phase of most other procaryotes.

It soon became apparent that disaggregation was an enzymatic phenomenon. Filter-sterilized supernatant fluids from pH 7·0-grown LYC cultures catalyzed a disaggregation of cellular clusters in the absence of growth substrates (Fig. 6). Disaggregation was accompanied by the release of individual cells which resulted in a marked increase in optical density of the culture. The 'disaggregatase' activity was destroyed by boiling for 10 min. Boiling the substrate (cell aggregates), however, increased the rate of cellular release by the enzyme (Fig. 6). The addition of ammonium sulfate (30% wt/vol) to crude supernatant fluids precipitated all enzymatic activity to effect a preliminary purification. This enzyme preparation was used to determine the optimal pH for activity. Disaggregation occurred between pH 5·0 and 7·4, with the greatest activity at pH 5·5 (Fig. 7). This low pH optimum was surprising because strain LYC appears to produce the enzyme at pH 6·3 or higher.

FIG. 4(A) Periodic acid–silver methenamine stain of thin sections of 12-day S-6 colonies. Stain over matrix material (m) suggests polysaccharide. Cytoplasmic phosphate granules (g) have also reacted (×23 000). (B) Staining with colloidal gold-labeled soybean lectin labels matrix material specifically (×21 000). (C) Isolated matrix material after colony disintegration of S-6 cysts with a Braun homogenizer (×3000). (4A–4C from Robinson et al., 1985, reproduced by permission of the American Society for Microbiology).

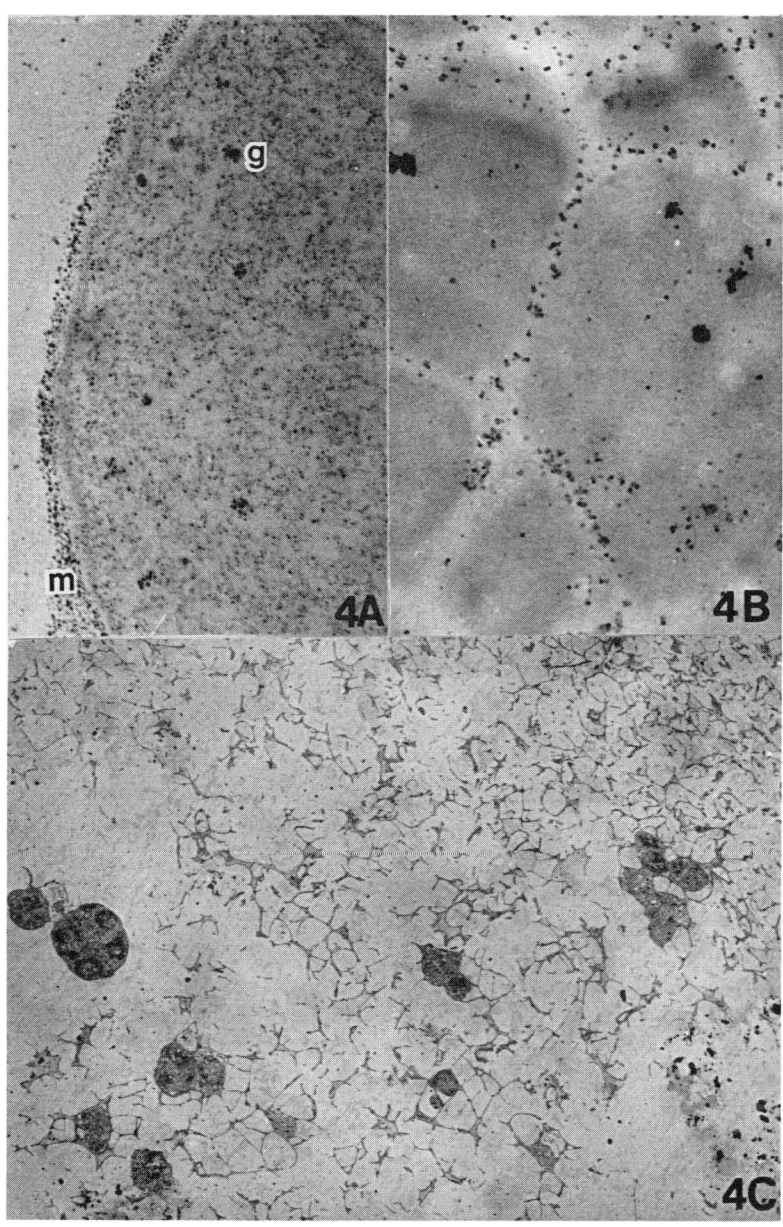

The substrate of the 'disaggregatase' is the galactosamine-galacturonic acid backbone of the matrix polysaccharide. When partially purified enzyme was incubated at 37°C for 24 h with the isolated heteropolysaccharide of strain S-6, there was a release of significant quantities of the uronic acid. Boiled-enzyme controls yielded no such release.

If the enzyme is released into the micro-environment of the digester, it is conceivable that closely related species could be caused to disaggregate. Indeed, enzyme preparations from LYC disaggregated *M. mazei* LYC, LYC-M (a mutant unable to produce this enzyme), S-6 and S-7, as well as *M. thermophila* TM1. *M. barkeri* 227, a common digester inhabitant, was not affected. It is conceivable, therefore, that such an enzyme may allow the dispersion of various methanosarcinae in anaerobic digesters by preventing the accumulation of static aggregates.

DISCUSSION

Increased demand for more efficient digester designs requires an understanding of basic operating principles. Various methanogenic species comprise the bulk of the microflora present in the biofilm and liquid phases of anaerobic fixed-bed digesters (Robinson *et al.*, 1984). Although no single methanogen predominated in any region of a digester, two fluorescent morphological types were most commonly found in the biofilms of nearly all regions. These were the long filaments commonly observed at the film surface and the pseudosarcina-containing cysts most commonly found in the lower regions. These were tentatively identified as belonging to the genera *Methanothrix* and *Methanosarcina*. These genera are known to utilize acetate for methane production and this was the predominant energy source in these experimental digesters (Robinson *et al.*, 1984).

The ultrastructure of the methanosarcina-like cysts found in digester biofilms (Figs. 1A and 1B) is similar to several strains of this genus, including *M. mazei*. Zhilina (1976) has described several biotypes of *Methanosarcina*; the biotype she designated as morphotype 1 most closely

---

FIG. 5(A) Disaggregation of *M. mazei* strain LYC colony at pH 7·0. The matrix material is shedding and a cell is seen protruding through the ruptured capsule (arrow) (×15 000). (B) A later stage of enzymatic disaggregation of strain LYC at pH 7·0. Matrix material is fragmented; and only naked, irregular cells are seen. These cells are osmotically sensitive (×23 000) (from Liu *et al.*, 1985, reproduced by permission of the American Society for Microbiology).

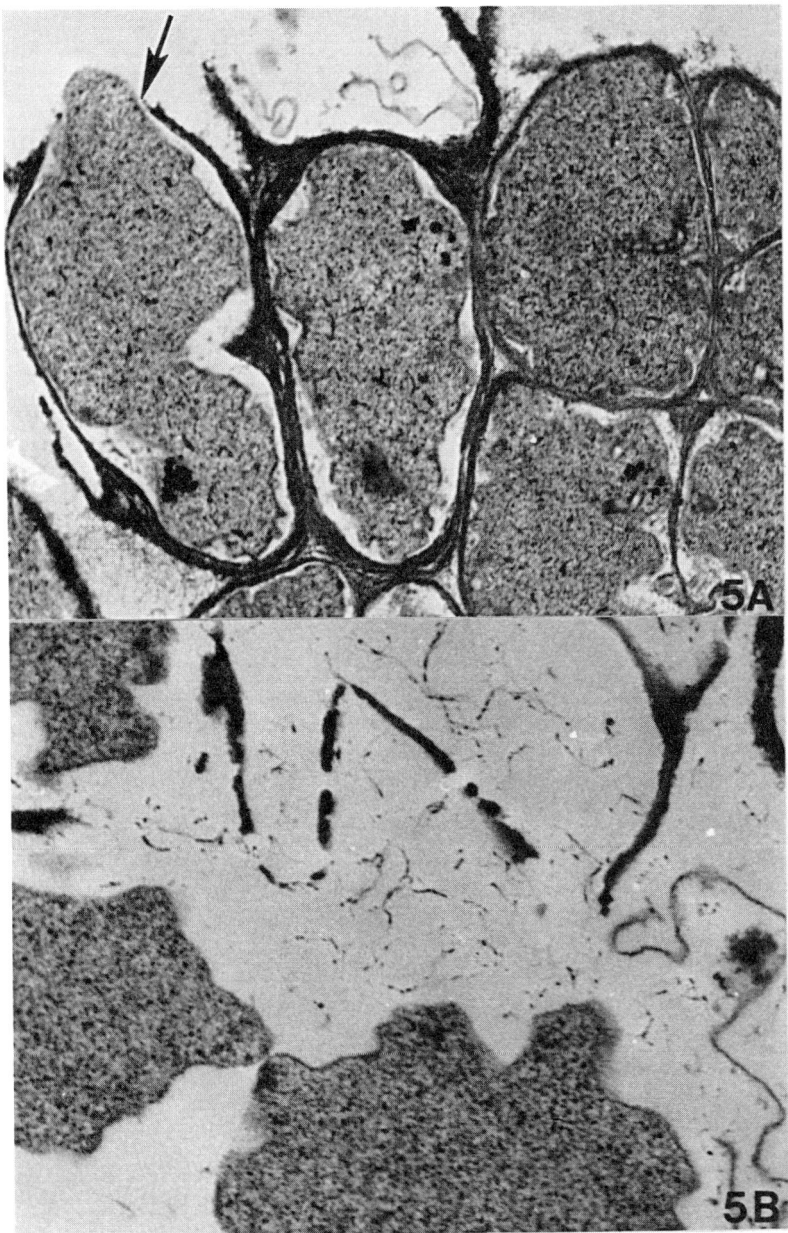

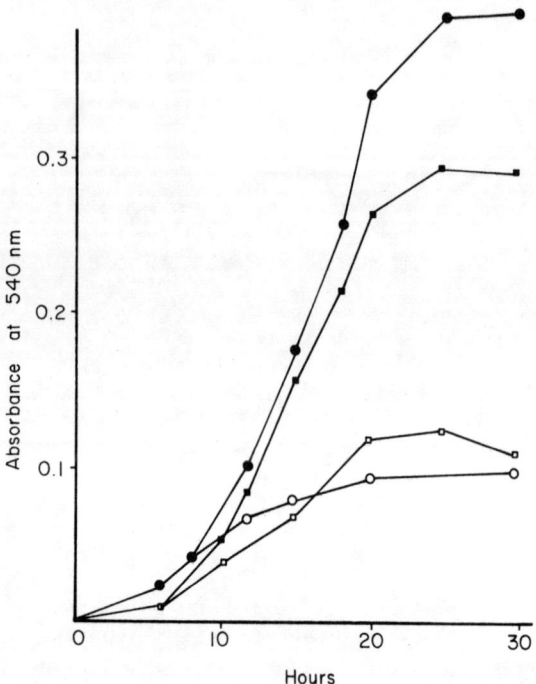

FIG. 6. Turbidimetric measurement of disaggregation of strain LYC. Solid symbols represent boiled cells used as the enzyme substrate; open symbols represent viable cells used as substrate. The source of enzyme was supernatant fluid (fresh = squares; reconstituted = circles) (from Liu et al., 1985, reproduced by permission of the American Society for Microbiology).

resembles the cysts in biofilm. Our studies of *M. mazei*, therefore, were based on the premise that this species, or others quite similar to it, are important components of digester biofilms.

The several investigators (e.g. Mah, 1980; Robinson et al., 1985) who have studied this species, or related strains, have concluded that 'life cycles' occur whereby single cells become aggregated, grow to form very large microcolonies ('cysts') and finally break down to release viable single cells again. The 'glue' that holds these cysts together, and probably also to the solid surfaces of digesters, is the so-called 'matrix' material. This is a heteropolysaccharide that is not really a cell wall but is an extended macrocapsule that allows daughter cells to remain adherent one to the other. This may be necessary for the survival of young clusters of cells since

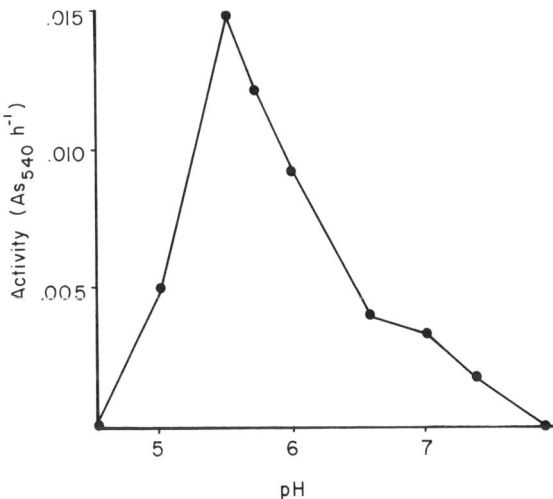

FIG. 7. Effect of pH on disaggregatase activity of ammonium sulfate-precipitated enzyme preparation from strain LYC. Activity was estimated by the linear increase in the optical density of boiled cells during incubation with the enzyme. As_{540} = absorbance at 540 nm (from Liu *et al.*, 1985, reproduced by permission of the American Society for Microbiology).

cross-feeding of growth-limiting nutrients may occur due to the close juxtaposition of cells in clusters. But there are inherent disadvantages to this arrangement as well. As the cyst grows in size it becomes more difficult for nutrients from the environment to reach cells located in the interior of cysts, and for toxic end-products of metabolism to escape quickly and completely from cell surfaces. A mechanism must exist therefore for the natural breakdown of aggregates while cells are still viable.

The isolation of *M. mazei* LYC reveals one such possible mechanism of disaggregation. Aggregates of strain LYC are large enough (1–2 mm) to see with the naked eye. But in the presence of methanogenic substrates at pH 6·3 or higher, there is an enzymatic breakdown of the matrix material allowing the release of individual cocci. The resulting large increase in the exposed surface-to-volume ratios of newly released cells should favor greatly increased rates of substrate permeability and should allow increased growth rates and methanogenesis. Since competition for available substrates (i.e. H_2 and acetate) is expected to be keen in the mixed-culture environment of the digester, a distinct metabolic advantage due to such increased surface-to-volume ratios is expected. Thus, cells that might have stagnated or died by being buried in cysts are able to survive and to start another putative life cycle.

This enzymatic disaggregation also would lead to the dispersal of many cell units to 'seed' the surrounding biofilm with viable *M. mazei* cells. When strain LYC was grown at pH 7·0, the viable count was about 3000-fold greater than that when strain LYC-M (the aggregated mutant) was grown under the same conditions, and the growth rate was 70% faster (Liu *et al.*, 1985). The ability, therefore, of this species to aggregate and then to disaggregate may assure its survival in the special environment of the anaerobic fixed-bed digester.

SUMMARY

Colonies of the methanogenic bacterium *Methanosarcina mazei* were examined with scanning and transmission electron microscopy. Irregular shaped cells surrounded by a loose, fibrillar matrix remain aggregated in colonies of 1 mm diameter or more. This matrix material is composed of galactosamine and galacturonic acid. Strain LYC is a recently isolated variant which spontaneously undergoes colony lysis under acid growth conditions, with protoplasts being released from the ruptured walls of the cells. A 'disaggregatase' enzyme responsible for this lytic activity has been isolated.

ACKNOWLEDGEMENTS

The results presented in this chapter represent the efforts of a large team of investigators at two universities. The University of California at Los Angeles group included Y. Liu, D. R. Boone and R. Sleat who worked on the LYC strain and enzymatic disaggregation. The University of Florida team included S. F. Hurst, who studied the chemistry of the matrix, as well as R. W. Robinson, D. E. Akin and D. S. Williams who conducted studies of the ultrastructure of *M. mazei*.

REFERENCES

Aldrich, H. C., Robinson, R. W. and Williams, D. S. (1986). Ultrastructure of *Methanosarcina mazei*. *Systematic and Applied Microbiology* **7**, 314–19.

Barker, H. A. (1936). Studies upon the methane-producing bacteria. *Archives Mikrobiologie* **7**, 420–38.

Kandler, O. and Hippe, H. (1977). Lack of peptidoglycan in the cell wall of *Methanosarcina barkeri*. *Archives of Microbiology* **113**, 47–60.

Liu, Y., Boone, D. R., Sleat, R. and Mah, R. A. (1985). *Methanosarcina mazei* LYC, a new methanogenic isolate which produces a disaggregating enzyme. *Applied Environmental Microbiology* **49**, 608–13.

Mah, R. A. (1980). Isolation and characterization of *Methanococcus mazei*. *Current Microbiology* **3**, 321–6.

Mah, R. A. and Kuhn, D. A. (1984). Transfer of the type species of the genus *Methanococcus* to the genus *Methanosarcina*, naming it *Methanosarcina mazei* (Barker 1936) comb. nov. et emend. and conservation of the genus *Methanococcus* (Approved Lists 1980) with *Methanococcus vannielii* (Approved Lists 1980) as the type species. *International Journal of Systematic Bacteriology* **34**, 263–5.

Robinson, R. W., Akin, D. E., Nordstedt, R. A., Thomas, M. V. and Aldrich, H. C. (1984). Light and electron microscopic examinations of methane-producing biofilms from anaerobic fixed-bed reactors. *Applied Environmental Microbiology* **48**, 127–36.

Robinson, R. W., Aldrich, H. C., Hurst, S. F. and Bleiweis, A. S. (1985). Role of the cell surface of *Methanosarcina mazei* in cell aggregation. *Applied Environmental Microbiology* **49**, 321–7.

Zhilina, T. N. (1976). Biotypes of *Methanosarcina*. *Mikrobiologiya* **45**, 481–9.

Zhilina, T. and Zavarzin, G. A. (1979). Cyst formation by *Methanosarcina*. *Microbiology (USSR)* **48**, 349–54.

24
Phospholipids to Monitor Microbial Ecology in Anaerobic Digesters

A. T. MIKELL Jr, T. J. PHELPS and D. C. WHITE

Institute for Applied Microbiology, Microbiology/Ecology Department, University of Tennessee, Knoxville, Tennessee, USA

INTRODUCTION

Analysis of the biomass community structure and nutritional status of the anaerobic digester microbial community was attempted using methods based on analysis of the lipids derived from the cells. Phospholipids, which are components of all cellular membranes, can be used as biomass estimators (Henson *et al.*, 1985). The ester-linked fatty acids of the phospholipids (PLFA) are often sufficiently unusual that they can serve as 'signature' biomarkers or for subsets of the community. Analysis of the PLFA can provide insight into the community structure of the digester microbial community (Henson *et al.*, 1985). These chemical techniques can be supplemented with a nondestructive analysis of the biofilm by Fourier transforming infrared spectroscopy (FT/IR). Biofilm analysis of FT/IR may lead to an on-line capability to monitor the microbes in a fermenter.

Microbes found in anaerobic digesters present a complex problem for analysis. It has been estimated, for example, that over 200 species of bacteria and at least 13 genera of protozoa can be isolated from the bovine rumen (Russell and Allen, 1984). Many of these organisms are attached to the surfaces of the substrates in microcolonies with organisms of diverse physiological activities. These anaerobic consortia can consist of organisms that ferment complex plant polymers into simpler sugars, those that form volatile fatty acids from these simpler saccharides, those that form acetate from the volatile fatty acids, and those that form the terminal products such as methane and hydrogen sulfide (Wolin, 1979). Microcolonies of mixed bacterial types bound together with extracellular polymers are readily detectable in the elegant transmission electron micrographs presented in this volume by Aldrich and his colleagues (Chapter 22). These complex assemblies have proved difficult to measure by the classical techniques of microscopy. Even in the water column where the microbiota exist in much simpler associations than in the digesters, the classical methods of microbiology that involve the isolation and subsequent culturing of organisms on petri plates can lead to gross underestimations of the numbers of organisms detectable in direct counts of the same waters (Jannasch and Jones, 1959). Kirsh (1969), Mah and Sussman (1968), and Varel (1984) all reported the recovery of about 10% of the nonmethanogenic bacteria present in digesters. In addition to the problems of providing a universal growth medium in the culture tube, the organisms must be quantitatively removed from the surfaces and from each other. Direct microscopic methods that require quantitative release of the bacteria from the biofilm can have the problem of inconsistent removal from some surfaces. Direct microscopy performed on biofilms can lead to underestimations of organisms rendered invisible by overlapping organisms in biofilms.

Organisms in complex consortia often differ markedly in size. The problems of calculations of biomass with microbes of different sizes can be minimized by the application of computer-based image enhancing software (Caldwell and Germida, 1984). This methodology works best when the density of organisms in the biofilms is low and overlapping is minimal. Even with computer-enhanced image processing in direct microscopy, we are left with the problem that the *in situ* methods often fail because the morphology of the microbe often offers little insight into the metabolic function or activity of the cells. Methane-forming bacteria, for example, come in all sizes and shapes (Zeikus, 1977). The problem is further complicated by the fact that in many environments only a tiny fraction of the organisms is active at any one time and aside from the observation of bacterial doubling (Hagstrom *et al.*, 1979), the morphology gives little evidence of the activity of the cells. The most direct method of determining the proportion of active cells in a given biofilm involves a combination of autoradiography and electron or epifluorescence microscopy. All these methods require metabolic activity in the presence of the substrates and are subject to the limitations of density of organisms and thickness of the biofilm in the field of view. With the necessity for inducing metabolic activity there is a danger of inducing artifically high levels of activity with the addition of the substrates. Application of the extremely sensitive chemical methods has shown the induction of activity in the sedimentary microbiota when subjected to minimal disturbances (Findlay *et al.*, 1985).

Isolation of the microbes in these consortia for viable counting or direct microscopic examination can provide insight into the details of the interactions that take place in mixed microcolonies if these isolates can be recombined into a system with similar properties. Since these consortia have much more versatile metabolic propensities than single species, it is important in understanding the interactions in anaerobic fermentation to preserve, as much as possible, the anatomy and metabolic interactions of these microcolonies. For that, a new type of analysis that does not involve quantitative removal of the microbes from surfaces or stimulation of new and possibly artifactual metabolic activities was developed.

Assays for Biomass and Community Structure Based on Phospholipid Analysis

Our laboratory has been involved in the development of assays to define microbial consortia in which the bias of cultural selection of the classical plate count is eliminated. Since the total community is examined in these procedures without the necessity of removing the microbes from surfaces, the microstructure of multi-species consortia is preserved. The method involves the measurement of biochemical properties of the cells and their extracellular products. Those components, generally distributed in cells, are utilized as measures of biomass. Components restricted to subsets of the microbial communities can be utilized to define the community structure. The concept of 'signatures' for subsets of the community based on the limited distribution of specific components has been validated by using antibiotics and cultural conditions to manipulate the community structure. The resulting changes agreed both morphologically and biochemically with the expected results (White *et al.*, 1980). Other validation experiments that involved isolation and analysis of specific organisms and finding them in appropriate mixtures, utilization of specific inhibitors and noting the response, and changes in the local environment such as the light intensity are summarized in reviews (White 1983, 1985).

Biomass Estimations

Phospholipids are found in the membranes of all cells. Under the conditions expected in natural communities the eubacteria contain a relatively constant proportion of their biomass as phospholipids (approximately 50 μ moles phospholipid g^{-1} dry wt) (White *et al.*, 1979b). Phospholipids are not found in storage lipids and have a relatively rapid turnover in some sediments, so the assay of these lipids gives a measure of the 'viable' cellular biomass (White *et al.*, 1979a). The phosphate of the phospholipids or the glycerol-phosphate and acid-labile glycerol from phosphatidyl glycerol-like lipids that are indicators of bacterial lipids can be assayed to increase the specificity and sensitivity of the phospholipid assay (Gehron and White, 1983).

Phospholipid analyses give a biomass that corresponds to that derived from analysis of other components of microbial cells. Estimations of microbial biomass based on the content of muramic acid, total recoverable intracellular adenosine nucleotides, lipopolysaccharides (LPS)-Lipid A of Gram-negative bacteria, teichoic acid glycerol and ribitol from Gram-positive bacteria, and total extractable phospholipid fatty acids from surface and subsurface sediments gave reasonably consistent estimates of the microbial biomass (White et al., 1979b; Gehron and White, 1983). Phospholipids, adenosine nucleotides, muramic acid, and the lipopolysaccharide of dead bacteria are rapidly lost from marine sediments (King and White, 1977; Moriarty, 1977; White et al., 1979a,c; Davis and White, 1980; Saddler and Wardlaw, 1980). This indicates that the chemical markers provide good estimates for the standing viable or potentially viable microbiota.

Analysis of Community Structure by Phospholipid Fatty Acids
The ester-linked fatty acids in the phospholipids (PLFA) are presently both the most sensitive and the most useful chemical measures of microbial biomass and community structure thus far developed (Bobbie and White, 1980; Guckert et al., 1985; White et al., 1984). The specification of fatty acids that are ester-linked in the phospholipid fraction of the total lipid extract greatly increases the selectivity of this assay as most of the anthropogenic contaminants as well as the endogenous storage lipids are found in the neutral or glycolipid fractions of the lipids. By isolating the phospholipid fraction for fatty acid analysis it proved possible to show bacteria in the sludge of crude oil tanks. The specificity and sensitivity of this assay has been greatly increased by the determination of the configuration and position of double bonds in monoenoic fatty acids (Nichols et al., 1985; Edlund et al., 1985) and by the formation of electron-capturing derivatives which, after separation by capillary GLC, can be detected after chemical ionization mass spectrometry as negative ions at femtomolar sensitivities (Odham et al., 1985). This makes possible the detection of specific bacteria in the range of 10–100 organisms. Since many environments, such as anaerobic digester substrates, often yield 150 ester-linked fatty acids derived from the phospholipids, a single assay provides a large amount of information. Combining a second derivatization of the fatty acid methyl esters to provide information on the configuration and localization of the double bonds in mono-unsaturated components provides even deeper

insight. By utilizing fatty acid patterns of bacterial monocultures, Myron Sasser of the University of Delaware in collaboration with Hewlett Packard has been able to distinguish between over 8000 strains of bacteria (Sasser, 1985). Thus, analysis of the fatty acids can provide not only the microbial biomass but insight into the community structure of microbial consortia as long as a significant number of organisms have been isolated from the consortia for the determination of 'signature' PLFA patterns.

Despite the fact that the analysis of PLFA cannot provide an exact description of each species or physiologic type of microbes in a given environment, the analysis provides a quantitative description of the microbiota in the particular environment sampled. With the technique of statistical pattern recognition analysis it is possible to provide a quantitative estimate of the differences between samples with PLFA analysis.

Potential problems with defining community structure by analysis of PLFA come with the shifts in fatty acid composition of some monocultures with changes in media composition or temperature (Lechevalier, 1977) some of which were defined in this laboratory (Frerman and White, 1967; Joyce *et al.*, 1970; Ray *et al.*, 1971). There is as yet little published evidence for such shifts in PLFA in nature where the growth conditions that allow survival in the highly competitive microbial consortia would be expected to severely restrict the survival of specific microbial strains to much narrower conditions of growth.

Analysis of other components of the phospholipid fraction gives insight into the microbial community structure. 'Signatures' for some of the microbial groups involved in anaerobic fermentations have been developed. The rate-limiting step in fermentations is the degradation of polymers. A second tier of microbes converts the carbohydrates and amino acids released from the biopolymers into organic acids, alcohols, hydrogen, and carbon dioxide. These are the anaerobic fermenters and some of these organisms contain plasmalogen phospholipids that are limited to this physiological class of anaerobes in the microbial world (Goldfine and Hagen, 1972). Plasmalogens can be assayed by their resistance to alkaline methanolysis and extreme sensitivity to mild acid (White *et al.*, 1979c). Other groups of anaerobic fermenters contain phosphosphingolipids with unusual sphingosine bases. These were detected in *Bacteroides* (Rizza *et al.*, 1970). Sphingosines are readily assayed in acid hydrolysates of the polar lipids by their amino groups or by GLC of the long chain bases (White *et al.*, 1969).

Phytanyl glycerol diethers found in the Archaebacteria can be assayed by high pressure liquid chromatography (HPLC) after appropriate derivatization (Martz et al., 1983). C. Mancuso in this laboratory has improved the sensitivity and resolution of the analysis of the diphytanyl glycerol and for the first time the bidiphytanyl glycerol ether lipids of the methanogenic bacteria (Mancuso et al., 1986). In the course of this work she has also been able to show the presence of isoprenologs of the aliphatic side chains of the diether lipids using highly sensitive GC/MS techniques (Mancuso et al., 1985).

The sulfate-reducing bacteria contain PLFA which can be utilized to identify at least a portion of these organisms. Some contain a unique profile of branched saturated and monounsaturated PLFA (Edlund et al., 1985; Parkes and Taylor, 1983; Taylor and Parkes, 1983; Dowling et al., 1986) that allows differentiation between those utilizing lactate and those using acetate and higher fatty acids. Detailed analysis of sulfate-reducing bacteria by N. Dowling of this laboratory strongly suggests that the majority of sulfate-reducing bacteria found in marine sediments and in waters used in the secondary recovery of oil are the acetate-utilizing strains (Dowling et al., 1986). The *Desulfobacter* type of sulfate-reducing bacteria can be induced in anaerobic digesters by amending the feedstock with sulfate (see below). These organisms can be active even in fermentations in which there is no added sulfate as they can recycle organic sulfur in the feedstock (Smith and Klug, 1981).

Nutritional Status
Certain bacteria form the endogenous lipid poly β-hydroxy butyrate (PHB) under conditions when the organisms can accumulate carbon but have insufficient total nutrients to allow growth with cell division (Nickels et al., 1979). A more sensitive assay based on gas liquid chromatography (GLC) of the components of the PHB polymer has been developed (Findlay and White, 1983). The sensitive assay of PHB has proved a useful means of defining the nutritional status of microbes in various environmental habitats.

Metabolic Activity
The analyses described above all involve the isolation of components of microbial consortia. Since each of the components is isolated, the incorporation of labeled isotopes from precursors can be utilized to provide rates of synthesis or turnover in properly designed experiments. Incorporation of 35-S-sulfate into sulfolipid can be utilized to measure activity in the microeukaryotes (White et al., 1980; Moriarty et al., 1985). Incorporation of 32-phosphate into phospholipids can be utilized as a measure of the activity of the total microbiota. The inhibition of phospholipid synthesis in

the presence of cycloheximide represents the microeukaryote portion of the lipid synthesis (White et al., 1980; Moriarty et al., 1985). Analysis of signatures by gas chromatography/mass spectrometry (GC/MS) makes possible the utilization of mass labeled precursors that are nonradioactive, have specific activities approaching 100%, include an isotopic marker for nitrogen, and can be efficiently detected using the selective ion mode in mass spectroscopy. The high specific activity makes possible the assay of critical reactions using substrate concentrations in the biofilms that are just above the natural levels. This is not possible with radioactive precursors. Improvements in analytical techniques have increased the sensitivity of this analysis. Utilizing a chiral derivative and fused silica capillary GLC with chemical ionization and negative ion detection of selected ions, it proved possible to detect 8 pg (90 femtomoles) of D-alanine from the bacterial cell wall (the equivalent of 1000 bacteria the size of *Escherichia coli* (Tunlid et al., 1985). In this analysis it proved possible to reproducibly detect a 1% enrichment of 15-N-D-alanine in the 14-N-D-alanine. The turnover of substrates utilized as electron donors can also be measured wth mass labeled compounds thereby avoiding radioactivity in fermenters.

Potential of On-line Monitoring of Digester Community Structure
The analysis of biofilms based on the isolation of chemical signatures is a destructive analysis and cannot be readily automated or utilized to give real-time monitoring of biofilms. To truly understand the interactions of microbial consortia the analysis should be nondestructive, sensitive and continuous, as well as have the resolution on the scale of micrometers—the sizes of microbial consortia themselves. The possibility of utilizing a nondestructive technique to monitor the chemistry of living biofilms is now possible with the Fourier transform infrared spectrometer (FT/IR).

The infrared portion of the spectrum is extraordinarily rich in information regarding the vibrational and rotational motions of atoms in molecules. Not only can specific infrared absorption be assigned to particular types of covalent bonds but the modifications of these bonds by the local electronic environment can be detected in the details of the spectra (Bellamy, 1958; Parker, 1981). The infrared spectrum of a compound has long been accepted as one of the best nondestructive identification techniques. One of the problems restricting the applications of infrared spectroscopy has been that the atomic interactions sensed in the infrared portion of the spectrum are at relatively low energies and the detection is relatively inefficient. This has precluded the full usage of the power of the analysis using complex materials isolated from the environment.

The advent of fast computers has made possible a new type of infrared spectral analysis. This has provided the technology to utilize the far infrared portions of the spectrum, to follow rapid reaction rates with changes in spectral intensity, and to utilize different types of sample exposures such as photoacoustic spectroscopy. The secret lies in the array processor computers that can perform Fourier transformations so rapidly that interference spectrscopy can be possible.

The attenuated total reflectance (ATR) cell makes possible the examination of living biofilms. The FT/IR can detect the vibration–rotation interactions of biofilms about 300 nm outside the surface of the germanium crystal used in the ATR cell. With this system it has proved possible to show that the carbohydrate rich initial fouling polymer coats the germanium surface exposed to sterile seawater in about 13 h (J. Guckert of this laboratory). A similar system has been utilized to follow the clotting sequence on various plastics inserted into the blood stream of living sheep (Gendreau and Jakobsen, 1978). This is clearly the way to follow biofilm formation and to potentially monitor fermentations continuously. Not only is the FT/IR nondestructive, rapid and sensitive, but it is possible to decrease the beam size to diameters approaching 10 μm which is the scale of the microbial interactions that must be monitored. Decreasing the area for analysis requires that longer analysis times be utilized to achieve the same sensitivity.

A summary of the use of FT/IR in microbial ecology has been published (Nichols *et al.*, 1985). The FT/IR examination by diffuse reflectance (DRIFT) of freeze-dried, powdered bacterial monocultures shows major differences. These findings together with the powerful technique of subtraction of one spectrum from another suggest that DRIFT could be utilized to recognize differences in community structure. Preliminary experiments indicate that examination of planktonic microbiota on pre-extracted filters by DRIFT can be correlated with a detailed examination of the lipid content.

Two measures have been identified as markers for the microbial nutritional status. The formation of PHB and the uronic acid-containing exopolysaccharide glycocalyx are responses to nutritional stress by some bacteria. Both polymers can be detected with the FT/IR. The polymers, such as gum arabic like the glycocalyx produced by *Pseudomonas atlantica*, show a prominent absorbance at ~1150 cm (−1) for C–0 stretch. The logarithm of the ratio absorbance at C–0 stretch to amide I gives an excellent correlation with mixtures of *E. coli* and gum arabic (Nichols *et al.*, 1985). This analysis replaces a three-week chemical tour-de-force involving GC/MS in the analysis of bacterial glycocalyx. The DRIFT spectrum of *E. coli* plus gum arabic and of *P. atlantica* induced to form polysaccharide glycocalyx are similar in appearance. Accumulations of PHA in bacteria or artificial mixtures of bacteria plus purified PHA show a linear correlation with the ratio of the carbonyl stretch at ~ 1750 cm (-1) to amide I. Using these recombination experiments as models, it proved possible to show DRIFT shifts in PHA and glycocalyx in the biofilms formed in anaerobic fermenters that were supplemented with various amendments (Henson et al., 1985). For example, amendments with propionate or butyrate showed similar biofilms compared with the unsupplemented or the biofilm of the digester amended with nitrate.

Development of the FT/IR, particularly in combination with ATR, offers a potentially rapid and nondestructive method to examine biofilms on the scale of the microbial consortia. The continued development of destructive analytical methods can be the essential validation for IR signatures.

APPLICATION TO ANAEROBIC FERMENTER COMMUNITIES

Microbial Biomass and Community Structure
Experiments performed in collaboration with P. H. Smith (University of Florida) and P. D. Brooks (this laboratory) utilizing digesters constructed from 4-liter aspirator bottles which were fed once daily with a mixture of Bermudagrass and cattle feed in a 3:1 mix were utilized to show shifts in microbial community structure (Henson et al., 1985). Analysis of PLFA showed a 3-fold increase after inoculation with sewage sludge and a 20-day incubation. The feed stock showed a simple PLFA pattern of saturated and unsaturated normal fatty acids. The loss of PLFA polyenoic fatty acids can be utilized to monitor degradation of the plant biomass feedstock. After inoculation and incubation the PLFA showed a markedly different pattern with increases in PLFA typical of microbes from anaerobic digesters. In the PLFA of the inoculated digesters, palmitic and stearic acids comprised only 35 and 45% of the total. PLFA typical of the phospholipids of anaerobic bacteria such as the saturated iso and anteiso branched 15, 16 and 17 carbon acids accounted for 30% of the PLFA. The monoenoic fatty acids showing the unsaturation in the omega 7 position (seven carbons from the aliphatic end of the molecule) comprised more than 10% of the total PLFA. Monoenoic PLFA with unsaturation at the omega (ω) 7 position are formed by the bacterial anaerobic desaturase pathway. PLFA analysis of digester slurries gives a quantitative means to determine the microbial biomass in an anaerobic digester.

The increase in microbial biomass with bacterial growth corresponded to a viable biomass of about 2–3 × 10 cells g^{-1} of solids calculated utilizing conversion factors based on monocultures (Gehron and White, 1983). This is roughly equivalent to that reported for direct counts from digesters (Iannotti et al., 1978; Kirsh, 1969).

The continuous addition of volatile fatty acids such as propionate or butyrate, or the terminal electron acceptors nitrate or sulfate at concentrations of 10 μmole ml^{-1} day^{-1} induced marked shifts in the community structure as indicated in the PLFA analysis. The butyrate amended digester slurry had significantly higher iso branched and monounsaturated branched 17 carbon fatty acids whereas the propionate-amended digester had higher proportions of 15 and 17 carbon normal saturated PLFA as well as 16 carbon monounsaturated trans configuration PLFA (Henson et al., 1985). The nitrate amended slurry showed highest proportions of iso branched 14 and 16 carbon fatty acids and the highest total biomass of all (Henson et al., 1985).

Amending a high solids digester with sulfate continuously over a 5-day period resulted in a loss of methane production (Table 1). The 'crash' was associated with great increases in the acetate but little change in the propionate concentrations. The total eubacterial biomass estimated from the ester-linked phospholipid fatty acids decreased 60%. The microbial community structure as indicated in the proportions of PLFA showed a decrease in the rate of methane production that paralleled increases in the 10 methyl branched 16 carbon PLFA characteristic of *Desulfobacter* sulfate-reducing bacteria (Dowling et al., 1986). Preliminary evidence from studies with J. M. Henson, M. J. McInerney and D. C. White suggests that the acetogenic bacterium *Syntrophomonas wolfei* contains an unusual (in eubacteria) polyenoic fatty acid in the PLFA. The activity of the acetogenic bacteria may account for part of the increased acetate. Growth of syntrophic acetogens is stimulated as the hydrogen they produce is more rapidly utilized by the sulfate-reducing bacteria. The data of Table 1 show an increase in the proportions of 18:3ω3 as the acetate concentration and 10 methyl 16:0 increases. These preliminary experiments illustrate the power of the PLFA analysis in resolving the interactions in the complex anaerobic consortia. To continue deepening this understanding, many more specific organisms with important metabolic functions must be isolated. This will allow the development of a larger catalog of 'signatures'.

Community Structure and Activity within a Water Hyacinth Digester

In experiments performed in collaboration with P. D. Brooks, C. A. Mancuso (of this laboratory) and D. P. Chynoweth (University of Florida)

TABLE 1
Microbial Activity and Community Structure Shifts in a Digester during a Crash Induced by Sulfate Infusion

	Days of sulfate treatment		
	0	2	4
Methane[a]	60	54	18
Acetate[b]	29	35	>50
Propionate[b]	13	12	12
Total diacyl-PLFA[c]	1·75 (0·8)	1·15 (0·6)	0·6 (0·2)
Mole % PLFA[d]			
i14:0[e]	0·3 (0)	<0·1	<0·1
i15:0	4·2 (0·2)	1·0 (0·2)	0·7 (0·1)
a16:0	1·9 (0·1)	0·4 (0·2)	<0·01
10me15:0	0·3 (0·1)	0·2 (0·1)	1·1 (1·0)
i16:0	9·0 (2·0)	7·0 (0·2)	4·3 (0·2)
16:1ω7c	1·0 (0·1)	0·6 (0·1)	<0·01
16:0	51·0 (6·0)	46·0 (0·5)	38·0 (4·0)
10me16:0	0·8 (0)	0·7 (0)	2·0 (1·0)
i17:0	6·0 (0·9)	6·8 (0·07)	4·3 (0·07)
a17:0	4·0 (0·5)	4·3 (0·07)	3·1 (0·1)
18:3ω3	6·7 (7·2)	19·0 (15·0)	32·0 (0·7)
18:1ω7c	1·6 (0·2)	2·5 (0·07)	2·0 (0·3)
18:0	4·3 (0·5)	5·3 (0·3)	7·8 (0·6)
20:0	0·6 (0·2)	0·9 (0·1)	1·2 (0·2)
22:0	0·9 (0·2)	1·7 (0·2)	2·7 (0)

[a] % of total yield.
[b] μmoles ml^{-1}.
[c] μmoles diacyl phospholipid fatty acids (PLFA)/g dry weight, mean (SD), $n = 4$.
[d] Mole % of total PLFA mean (SD), $n = 4$.
[e] PLFA indicated by prefix i, a, and 10me- indicate iso anteiso and mid chain (10 carbons from the carbonyl group) methyl branching, followed by the number of carbon atoms in the chain, followed by the : and the number of double bonds, followed by the number of carbons from the omega (ω) (alkyl) end of the molecule for the terminal double bond followed by the configuration (c = cis, t = trans) of the double bond in monoenoic PLFA.

at the Disney World water hyacinth digester, the eubacterial community structure was markedly different in samples recovered from different parts of the digester (Fig. 1). Comparing the feedstock, chopped hyacinth material, the stored feedstock, the bottom material, the top of the bottom material, the bottom of the floating mat, and the top of the floating mat in the digester, there was a progressive decrease in the proportions of poly-

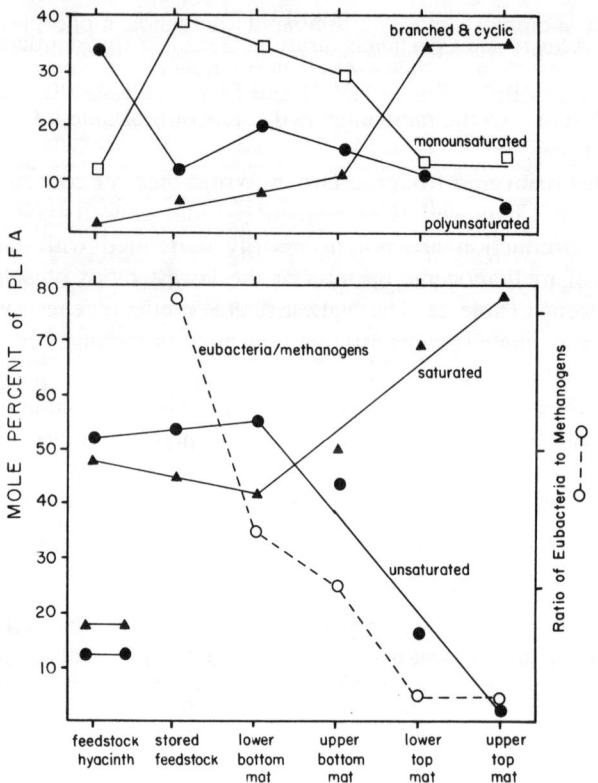

FIG. 1. Proportions of PLFA and ratio of eubacterial to methanogenic bacterial biomass in the lipids derived from the hyacinth feedstock and digester suspensions from the Disney World digester. Eubacterial and methanogenic biomasses were determined as the ratio of the diacyl PLFA to the archaebacterial ether phospholipids. Shifts in PLFA show shifts in eubacterial community structure.

unsaturated and monounsaturated PLFA. This decrease in proportions of unsaturated PLFA correlated with an increase in the branched and cyclic as well as the saturated PLFA. The proportion of methanogenic archaebacteria measured as the di- and tetraphytanylglycerol ether phospholipids (Mancuso et al., 1986) increased from none in the feedstock, to traces in the stored and partially fermented feedstock with more in the material at the bottom of the digester to the highest levels at the top of the digester. These data are reflected in the decrease in the ratio of diacyl phospholipid [eubacterial biomass]/ether phospholipid [methanogen phospholipid] (Fig. 1, lower panel). It will be important to compare these techniques with F-420 counts (direct microscopic counts of cells that fluoresce at 420 nm and correspond to the methanogens) and most probable number (MPN) enumerations.

Samples recovered from the Disney World digester collected with the help of D. P. Chynoweth (University of Florida) showed maximal rates of methane production are not necessarily correlated with the greatest biomass of methanogenic bacteria or the lowest ratios of eubacteria to methanogens (Table 2). The highest rate of methanogenesis was associated with the highest concentration of acetate in the digester. Maximum rates of methanogenesis occurred just above the bottom material of the digester. The maximum rate of methanogenesis was comparable to that found in the Tallahassee sewage treatment anaerobic digester. These data show that it is now possible to determine the relationship between the eubacterial nutritional status by PHA/PLFA ratio, the eubacterial community structure, the methanogenic bacterial biomass, and the yield of

TABLE 2
Relationship between Digester Community Structure and Methane Production

Sample	% solids	Diacyl PL[a]	Diether PL[b]	Acetate[c]	Methane[d]
Hyacinth	5.7	29.0	<0.1	—	—
Stored hyacinth feed	4.25	6.5	0.4	25.5	—
Digester top (8)[e]	6.88	8.3	3.4	1.9	14.0
(5)	6.58	12.5	5.3	0.3	10.0
(3)	0.75	9.8	2.0	0.9	0.7
(1s)	3.11	34.0	5.6	5.2	29.0
Digester bottom (B)	2.68	5.7	0.8	0.89	9.0
Sewage sludge	3.0	57.0	8.7	—	36.0

[a]Total nonmethanogenic microbial biomass based on diacylphospholipid (PL) fatty acids in μmoles liter^{-1}.
[b]Total methanogen biomass based on di- and tetra-ether phospholipids in μmoles liter^{-1}.
[c]Acetate concentration in mmoles liter^{-1}.
[d]Rate of methane production mmoles methane liter^{-1} day^{-1}.
[e]Samples 8 through 5 represent the top and lower portions of the floating mat in the digester, samples 1s and B the top and bottom of the material at the bottom of the digester, and sample 3 the suspension in the middle of the digester.

methane. The methanogenic efficiency in terms of the methanogen biomass of various zones in the digester were clearly different indicating that certain zones must be operating suboptimally.

J. J. Olie of this laboratory, utilizing high performance liquid chromatographic (HPLC) separation with detection by diode array spectrometry, has been able to show decreases in chlorophyll a and b as well as phaeophytin a, increases in phaeophorbide, and increases in several bacterial carotenoids between the feedstock hyacinths and the digester samples in the Disney World system. Possibly the most surprising finding of this analysis was the detection of large amounts of bacteriochlorophyll in the digester samples. None was detected in the feedstock. The role of these photosynthetic organisms that are known to utilize hydrogen in the dark to fix nitrogen is currently under investigation.

The ratio of the eubacterial endogenous storage product, PHB, to the total phospholipid fatty acids is a measure of the microbial nutritional status (Findlay et al., 1985). Low ratios of PHB/PLFA were found in digesters supplemented with butyrate, propionate, or in the unsupplemented control digesters compared to the feedstock. High ratios of PHB/PLFA were found in digesters supplemented with nitrate or in the feedstock Bermudagrass indicating that these microbial communities were not in balanced growth (Henson et al., 1985).

Preliminary experiments conducted by T. J. Phelps (this laboratory) and D. P. Chynoweth (University of Florida) at the Disney World hyacinth digester showed that estimates of methanogenesis before feeding were 9·8 (0·1) mmoles liter^{-1} day^{-1} and rose to 14 (1·3) mmole liter^{-1} day^{-1} 3 h after feeding. Analysis of the after-feeding sample showed 50% of the carbon in the methane came from 14-C-2-acetate, 37% from 14-C-carbon dioxide, and 13% from unknown precursors. In the digester before feeding the C-2-acetate accounted for 14 and 20% of the methane formation and carbon dioxide accounted for a surprising 79 and 86% on two separate occasions. The estimated total methane production based on the radiotracer transformation rates times respective pool sizes agreed with the rates of accumulation. The 12% of the 14-C-2-acetate oxidized to carbon dioxide after feeding was compared with 17 and 22% prior to feeding the digester. This suggests that a greater proportion of the acetate is syntrophically oxidized during the deprivation period. Clearly, further experiments following the fate of 14-C-acetate by HPLC must be performed to verify these findings. The utilization of 13-C-labeled precursors and analysis by HPLC/mass spectroscopy for the glycerol ethers (Mancuso et al., 1986) or GC/MS for the 'signature' PLFA of the syntrophs can allow nonradioactive measures of metabolic activity that can be compared to shifts in microbial community structure.

SUMMARY

The initial experiments utilizing the chemical methods for defining biomass, community structure, nutritional status and metabolic activities developed in this laboratory suggested they provide a quantitative description of the microbiota in anaerobic digesters. The PLFA provide a means to assess the viable microbial biomass and to follow the shifts in community structure as the feedstock changes. The methods for the ether lipids of the methanogenic archaebacteria provide a means for the first time to follow the relationship between the growth kinetics and the methane-forming activities of these most important organisms. Insight into the activities, nutritional status, and community structure of the eubacteria that provide the substrates for the methanogens and the activity of different groups of methanogens may provide means to predict conditions that lead to inefficiencies or digester crashes. The correlations between these chemical parameters and the shifts in the infrared spectra, as monitored by the FT/IR, may eventually provide a nondestructive, on-line monitoring capability. To achieve this goal will require careful monitoring of digester yields, and substrate kinetics as well as the isolation of specific monocultures for the generation of 'signature' catalogs.

REFERENCES

Bellamy, L. T. (1958). *The Infrared Spectra of Complex Molecules*, John Wiley, New York.

Bobbie, R. J. and White, D. C. (1980). Characterization of benthic microbial community structure by high resolution gas chromatography of fatty acid methyl esters. *Applied Environmental Microbiology* **39**, 1212–22.

Caldwell, D. E. and Germida, J. J. (1984). Evaluation of difference imagery for visualizing and quantitating microbial growth. *Canadian Journal of Microbiology* **31**, 35–44.

Davis, W. M. and White, D. C. (1980). Fluorometric determination of adenosine nucleotide derivatives as measures of the microfouling, detrital and sedimentary microbial biomass and physiological status. *Applied Environmental Microbiology* **40**, 539–48.

Dowling, N. J. E., Widdel, F. and White, D. C. (1986). Analysis of phospholipid ester-linked fatty acid biomarkers of acetate-oxidizing sulfate reducers and other sulfide forming bacteria. *General Microbiology* **132**, 1815–25.

Edlund, A., Nichols, P. D., Roffey, R. and White, D. C. (1985). Extractable and lipopolysaccharide fatty acid and hydroxy acid profiles from *Desulfovibrio* species. *Lipid Research* 26, 982–8.

Findlay, R. H. and White, D. C. (1983). Polymeric beta-hydroxy-alkanoates from environmental samples and *Bacillus megaterium*. *Applied Environmental Microbiology* 45, 71–8.

Findlay, R. H., Pollard, P. C., Moriarty, D. J. W. and White, D. C. (1985). Quantitative determination of microbial activity and community nutritional status in estuarine sediments: evidence for a disturbance artifact. *Canadian Journal of Microbiology* 31, 493–8.

Frerman, F. F. and White, D. C. (1967). Membrane lipid changes during formation of a functional electron transport system in *Staphylococcus aureus*. *Bacteriology* 94, 1854–67.

Gehron, M. J. and White, D. C. (1983). Sensitive assay of phospholipid glycerol in environmental samples. *Journal of Microbiology Methods* 1, 23–32.

Gendreau, R. M. and Jakobsen, R. J. (1978). Fourier transform infrared techniques for studying complex biological systems. *Applied Spectroscopy* 32, 326–8.

Goldfine, H. and Hagen, P. O. (1972). Bacterial plasmalogens. In: *Ether Lipids: Chemistry and Biology*, Snyder, E. (ed.), pp. 329–50. Academic Press, New York.

Guckert, J. B., Antworth, C. P., Nichols, P. D. and White, D. C. (1985). Phospholipid, ester-linked fatty acid profiles as reproducible assays for changes in prokaryotic community structure of estuarine sediments. *FEMS Microbiology Ecology* 31, 147–58.

Hagstrom, A., Larsson, U., Horstedt, P. and Normark, S. (1979). Frequency of dividing cells, a new approach to the determination of bacterial growth rates in aquatic environments. *Applied Environmental Microbiology* 37, 805–12.

Henson, J. M., Smith, P. H. and White, D. C. (1985). Examination of thermophilic methane-producing digesters by analysis of bacterial lipids. *Applied Environmental Microbiology* 50, 1428–33.

Iannotti, E. L., Fisher, J. R. and Sievers, D. M. (1978). Medium for the enumeration and isolation of bacteria from a swine waste digester. *Applied Environmental Microbiology* 36, 555–66.

Jannasch, H. W. and Jones, G. E. (1959). Bacterial populations in seawater as determined by different methods of enumeration. *Limnology and Oceanography* 4, 128–39.

Joyce, G. H., Hammond, R. K. and White, D. C. (1970). Changes in membrane lipid composition in exponentially growing *Staphylococcus aureus* during the shift from 37 to 25°C. *Bacteriology* 104, 323–30.

King, J. D. and White, D. C. (1977). Muramic acid as a measure of microbial biomass in estuarine and marine samples. *Applied Environmental Microbiology* 33, 777–83.

Kirsh, E. J. (1969). Studies on the enumeration and isolation of obligate anaerobic bacteria from digesting sewage sludge. *Developments in Industrial Microbiology* 10, 170–6.

Lechevalier, M. P. (1977). Lipids in bacterial taxonomy—a taxonomist's view. *Critical Review of Microbiology* 7, 109–210.

Mah, R. A. and Sussman, C. (1968). Microbiology of anaerobic sludge fermentation. I. Enumeration of the nonmethanogenic anaerobic bacteria. *Applied Environmental Microbiology* **16**, 358–61.

Mancuso, C. A., Odham, G., Westerdahl, G., Reeve, J. N. and White, D. C. (1985). C15, C20, and C25 isoprenoid homologues in glycol diether phospholipids of methanogenic archaebacteria. *Journal of Lipid Research* **26**, 1120–5.

Mancuso, C. A., Nichols, P. D. and White, D. C. (1986). A method for the separation and characterization of archaebacterial signature ether lipids. *Journal of Lipid Research* **27**, 49–56.

Martz, R. F., Sebacher, D. L. and White, D. C. (1983). Biomass measurement of methane-forming bacteria in environmental samples. *Journal of Microbiology Methods* **1**, 53–61.

Moriarty, D. J. W. (1977). Improved method using muramic acid to estimate biomass of bacteria in sediments. *Oecologia (Berlin)* **26**, 317–23.

Moriarty, D. J. W., White, D. C. and Wassenberg, T. J. (1985). A convenient method for measuring rates of phospholipid synthesis in seawater and sediments: Its relevance to the determination of bacterial productivity and the disturbance artifacts introduced by measurements. *Journal of Microbiology Methods* **3**, 321–30.

Nichols, P. D., Henson, J. M., Guckert, J. B., Nivens, D. E. and White, D. C. (1985). Fourier transform-infrared spectroscopic methods for microbial ecology: Analysis of bacteria, bacteria-polymer mixtures and biofilms. *Journal of Microbiology Methods* **4**, 79–94.

Nickels, J. S., King, J. D. and White, D. C. (1979). Poly-beta-hydroxybutyrate metabolism as a measure of unbalanced growth of the estuarine detrital microbiota. *Applied Environmental Microbiology* **37**, 459–65.

Odham, G., Tunlid, A., Westerdahl, G., Larsson, L., Guckert, J. B. and White, D. C. (1985). Determination of microbial fatty acid profiles at femtomolar levels in human urine and the initial marine microfouling community by capillary gas chromatography-chemical ionization mass spectrometry with negative ion detection. *Journal of Microbiology Methods* **3**, 331–44.

Parker, F. S. (1971). *Applications of Infrared Spectroscopy in Biochemistry, Biology and Medicine.* Plenum Press, New York.

Parkes, R. J. and Taylor, J. (1983). The relationship between fatty acid distribution and bacterial respiratory types in contemporary marine sediments. *Estuarine Coastal Marine Science* **16**, 173–89.

Ray, P. H., White, D. C. and Brock, T. D. (1971). Effect of growth temperatures on the lipid composition of *Thermus aquaticus*. *Bacteriology* **108**, 227–35.

Rizza, B., Tucker, A. N. and White, D. C. (1970). Lipids of *Bacteroides melanogenicus*. *Bacteriology* **101**, 84–91.

Russell, J. B. and Allen, M. S. (1984). Physiological basis for interactions among rumen bacteria: *Streptococcus bovis* and *Megasphaera elsdenii* as a model. In: *Current Perspectives in Microbiol Ecology*, Klug, M. J. and Reddy, C. A. (eds.), pp. 239–47. American Society of Microbiology, Washington, D.C.

Saddler, N. and Wardlaw, A. C. (1980). Extraction, distribution and biodegradation of bacterial lipopolysaccharides in estuarine sediments. *Antonie van Leeuwenhoek Journal of Microbiology* **46**, 27–39.

Sasser, M. (1985). Identification of bacteria by fatty acid composition. American Society of Microbiology Meeting, March 3–7, 1985.

Smith, R. L. and Klug, M. J. (1981). Reduction of sulfur compounds in the sediments of a hypereutrophic lake sediment. *Applied Environmental Microbiology* **47**, 5–11.

Taylor, J. and Parkes, R. J. (1983). The cellular fatty acids of the sulfate-reducing bacteria, *Desulfobacter* sp., *Desulfobulbus* sp. and *Desulfovibrio desulfuricans*. *General Microbiology* **129**, 3303–9.

Tunlid, A., Odham, G., Findlay, R. H. and White, D. C. (1985). Precision and sensitivity in the measurement of 15-N enrichment in D-alanine from bacterial cell walls using positive/negative ion mass spectrometry. *Microbiology Methods* **3**, 237–45.

Varel, V. H. (1984). Characteristics of some fermentative bacteria from a thermophilic methane-producing fermenter. *Microbiology Ecology* **10**, 15–24.

White, D. C. (1983). Analysis of microorganisms in terms of quantity and activity in natural environments. In: *Microbes in Their Natural Environments*, Slater, J. H., Whittenbury, R. and Wimpenny, J. W. T. (eds.). Society for General Microbiology Symposium, Vol. 34, pp. 37–66.

White, D. C. (1985). Quantitative physical-chemical characterization of bacterial habitats. In: *Bacteria in Nature, Vol. 2*, Poindexter, J. and Leadbetter, E. (eds.). Plenum Publishing Corporation, New York (in press).

White, D. C., Tucker, A. N. and Sweeley, C. C. (1969). Characterization of the iso-branched sphinganines from the ceramide phospholipids of *Bacteroides melaninogenicus*. *Biochima Biophysica Acta* **187**, 527–32.

White, D. C., Davis, W. M., Nickels, J. S., King, J. D. and Bobbie, R. J. (1979a). Determination of the sedimentary microbial biomass by extractible lipid phosphate. *Oecologia* **40**, 51–62.

White, D. C., Bobbie, R. J., Herron, J. S., King, J. D. and Morrison, S. J. (1979b). Biochemical measurements of microbial mass and activity from environmental samples. In: *Native Aquatic Bacteria: Enumeration, Activity and Ecology*, Costerton, J. W. and Colwell, R. R. (eds.), pp. 69–81. ASTM STP 695, American Society for Testing and Materials.

White, D. C., Bobbie, R. J., King, J. D., Nickels, J. S. and Amoe, P. (1979c). Lipid analysis of sediments for microbial biomass and community structure. In: *Methodology for Biomass Determinations and Microbial Activities in Sediments*, Litchfield, C. D. and Seyfried, P. L. (eds.), pp. 87–103. ASTM STP 673, American Society for Testing and Materials.

White, D. C., Bobbie, R. J., Nickels, J. S., Fazio, S. D. and Davis, W. M. (1980). Nonselective biochemical methods for the determination of fungal mass and community structure in estuarine detrital microflora. *Botanica Marina* **23**, 239–50.

White, D. C., Smith, G. A. and Stanton, G. R. (1984). Biomass, community structure and metabolic activity of the microbiota in Antarctic benthic marine sediments and sponge spicule mats. *Antarctic Journal of the United States* **29**, 125–6.

Wolin, N. R. (1979). The rumen fermentation: A model for microbial interactions in anaerobic ecosystems. *Advanced Microbiology Ecology* **3**, 49–77.

Zeikus, J. G. (1977). The biology of methanogenic bacteria. *Bacteriology Review* **41**, 514–41.

25
Fixed Film Reactors: Packing Media, Mathematical Models and Bulk Flow in Packed Beds

R. A. NORDSTEDT, M. V. THOMAS and C. Y. CHOU

Agricultural Engineering Department, University of Florida—IFAS, Gainesville, Florida, USA

INTRODUCTION

The fixed film reactor is a type of retained biomass reactor useful for conversion of feedstock streams which are low in solids content and/or contain materials that readily convert to methane. The biological organisms are retained on the surfaces of a packing medium in the reactor in 'fixed films' or 'biofilms'.

Although many materials have previously been evaluated for their suitability as packing media for fixed film reactors, wood has not been investigated for that use. The following work briefly reviews our efforts: (1) to develop wood media as packing material for fixed film anaerobic reactors and (2) to develop mathematical models of fixed film reactor operation. Although the porosity of wood is lower than other media, the materials cost is substantially less. Because the porosity of the wood chips is lower than that of some other types of medium, the reactor volume would probably have to be increased to achieve an equivalent media surface area for biofilm development. Wood chips cost 3% of the cost of some commercially available plastic media. Since 50% of the cost of a fixed film reactor with plastic media (not the complete system) may be in the packing media, the cost savings from using wood chip media could be significant. However, performance characteristics of wood media must be determined before the economics can be established. Wood's rough surface should promote rapid biofilm growth and reduce sloughing of the biofilm. Even though wood media may eventually degrade, they could be replaced. Other packing media may clog from excess biofilm development or from scale buildup and also require replacement. In either case, the reactor must be designed to permit replacement of the packing media and to minimize costs associated with downtime. In some cases, this may require multiple reactors so that individual reactors can be removed from operation for maintenance.

PREVIOUS WORK

The effectiveness of many materials as packing media has been evaluated in fixed film reactors. In one of the first major works on anaerobic filters, Young and McCarty (1969) used smooth quartzite stone, 2·5 – 3·8 cm diameter, as packing media. Lovan and Foree (1971) studied the treatment of brewery wastes using an anaerobic filter with crushed limestone media. Those media have the disadvantage of being very heavy and of having low porosity or void volume. Crushed rock was successfully used in a 1135 m^3 anaerobic filter at a starch processing plant (Taylor, 1972).

Jones et al. (1981) attempted to decrease the size of conventional dairy cattle manure digesters by using a nylon fabric as a bacterial support medium. The nylon fabric increased gas production but clogged in preliminary studies at two day retention times. The anaerobic fixed film reactor was applied as a second-stage reactor following an acid phase reactor using swine waste feedstock. The packing media were made by splitting 5 cm composition drain pipe lengthwise and then cutting it into 5 cm lengths (Smith et al., 1977). This resulted in a medium which was lighter than stone, having a porosity of 80% and a relatively high calculated specific surface area of 40·5 m^{2}m^{-3}.

Hudson et al. (1978) compared the performance of anaerobic filters containing whole oyster shells to filters containing 2·5–3·8 cm granitic stones as packing media. The porosities were 0·82 and 0·53 for the oyster shell and stones, respectively. Under similar organic loadings and hydraulic retention times, the reactors with oyster shell media performed better than the reactors with rock media. The better performance of the oyster shell media may have been due to: (a) an ability to augment buffering requirements, (b) a higher porosity, and/or (c) a higher specific surface area. They also noted that biofilm growth on oyster shells probably was better because of surface configuration and/or local contact characteristics. Van den Berg and Lentz (1981) found that reactors with baked clay media reached maximum performance in 1–3 months as compared to 7 months for polyvinylchloride plastic and 10–14 months for glass media. Similarly, Wilkie et al. (1983) compared plastic rings, coral, mussel shells and fired clay fragments as media in anaerobic filters. They found that maximum chemical oxygen demand (COD) conversion to CH_4 occurred within 20 days for clay fragments but that 39, 40 and 50 days were required for the coral, plastic and mussel shell media, respectively. Irrespective of the start-up time, the performance of all four filters at steady state was similar with COD removal efficiencies of 69–73% being attained. Start-up and steady-state performance did not correlate with either the unit surface area or the porosity of the support materials utilized.

Brumm (1980), building upon work by Nordberg (1977), treated dilute swine waste and found that the plastic media filters were more effective in both volatile solids (VS) removal and VS destruction than corncob media. However, gas production rates were higher for the corncob filters due to degradation of the corn cobs. Plugging was not a problem. Polypropylene plastic media to treat swine waste (Person, 1980) and plastic saddles to treat milk wastewater (Rittman et al., 1982) were used successfully in anaerobic fixed bed reactors. Limestone chips in an upflow anaerobic fixed bed reactor for digestion of pig slurry and later silage effluent (Barry and Colleran, 1982) were successful as was fired clay media for digestion of piggery waste (Kennedy and Van den Berg, 1982).

Work on the anaerobic filter initially included the development of an empirical equation predicting steady-state ultimate soluble BOD removal as a function of hydraulic retention time. A dynamic model of anaerobic fixed bed reactor kinetics, incorporating two-phase digestion, bacterial growth, pH, alkalinity, ammonia and gas production, was presented by Mueller and Mancini (1976). However, this model excluded considerations of biofilm activity and solids transport, and it relied heavily on empirically derived parameter values. While providing simulation of general trends in performance, this model was incapable of predicting effects due to media porosity (void volume), surface area and temperature, important considerations in design. A general model of single substrate utilization by bacterial films developed by Williamson and McCarty (1976) accounted for molecular diffusion and substrate uptake in the biofilm. This model was expanded by Rittman and McCarty (1980) to include biofilm growth and applied to steady-state biofilms.

Although the anaerobic reactor has been used for many years in treatment of wastes and wastewaters, the concept of commercial gas production from biomass and wastes in anaerobic reactors is relatively new. As design concepts improve, it may be possible to increase efficiency of the reactors by manipulating the microflora to favor the microbes that play crucial roles in biomass conversion to methane. Although several of the bacteria have been examined at the ultrastructural level in pure culture or in enriched mixed populations, the work reported here was the first on the examination of populations in anaerobic sludge or in biofilms (Robinson et al., 1984). They examined the biofilms from upflow fixed film reactors containing both wood block and plastic packing media. Their work was followed by that of Harvey et al. (1984) on biofilms on a needle-punched polyester support. Their reactor also received a swine waste influent, but it was operated in a downflow mode. Knowledge of the morphology and ultrastructure of the types of microbes present in biofilms is a prerequisite to modifying the reactor design or operation to enhance the growth of desired microbes.

MATHEMATICAL MODELING

A dynamic model was needed which would predict time variant behavior of the reactor under transient conditions, such as changes in the feedstock, temperature, loading rate, or other departures from steady state conditions. In this work, we extended the model of Rittman and McCarty (1980) to incorporate dynamic biofilm response (Bolte, 1983). A mathematical model was developed to include: (1) convective substrate transport in the bulk liquid through the reactor height, (2) one-dimensional diffusion through the biofilm and across the biofilm–liquid interface, and (3) two-culture bacterial substrate conversion in the biofilm. The bacterial processes controlling substrate transformation in the biofilm were modeled as a three-step process involving two classes of bacteria (Hill and Barth, 1977). The three substrates were complex organic matter, soluble organics and volatile organic acids. Bacterial growth is described using modified Monod kinetics with volatile acid inhibition. Parameters required as model inputs include temperature, packing media surface area and porosity, hydraulic retention time and influent composition. All of these parameters can be readily measured with the possible exception of surface area of wood chip media. Loading rates are calculated from influent composition and hydraulic retention time. The model predicts methane productivity, total methane production and COD reduction. Although only a generalized description of the model and the underlying assumptions follow, detailed descriptions are given by Bolte (1983) and Bolte et al. (1984).

Two distinct regions within the reactor are considered in the model: (a) bulk liquid flow through the interstitial spaces of the support medium, and (b) the biofilm, which contains active bacterial mass attached to the support medium. Substrate dynamics in the bulk liquid are governed primarily by convection due to hydraulic flow and flux into or out of the biofilm region resulting from substrate conversion. The reactor column is modeled assuming a plug flow hydraulic regime. The following assumptions were made: (a) no substrate transformation occurs in the bulk liquid, (b) longitudinal diffusion mass transport is negligible compared to convective transport, and (c) substrate concentration and biofilm flux are uniform throughout a differential volume. These assumptions are justified by the short hydraulic retention times. The assumption of no substrate transformation in the bulk liquid follows from the rapid washout of unattached bacteria. Although the model does not consider it, this could be negated by a high concentration of bacteria in the influent.

Substrate dynamics in the biofilm region are controlled primarily by diffusional mass transport into and across the biofilm and by biological substrate transformation within the biofilm. An idealized biofilm is conceptualized as two layers: (a) the biofilm, composed of bacterial mass and secreted gelatinous substances, and (b) a stagnant liquid zone between the biofilm and the bulk liquid. Bacterial population dynamics in the biofilm control the growth and decay of the biofilm. The stagnant liquid layer, termed diffusion layer and acting as a buffer zone between the biofilm and the bulk liquid, is assumed bacteria-free. Several assumptions are made in the biofilm model: (a) bacterial concentrations are uniform throughout the thickness of the biofilm, (b) bacterial concentration in the biofilm is relatively constant over time, (c) at a given height in the reactor column, the biofilm has a uniform thickness at any given time, and (d) losses of bacteria to shear forces are negligible.

The model was validated using the data of Young and McCarty (1969) for fixed bed anaerobic reactors operating on two synthetic soluble waste types: (a) a protein–carbohydrate mixture composed of nutrient broth and glucose, and (b) a volatile acid mixture composed of equal portions by COD of acetic and propionic acids. Loading rates ranged from 0·42 to 3·4 kg m^{-3} total reactor volume. Temperature was maintained at 25°C.

Under most operating conditions, the model was capable of predicting steady-state COD removal efficiencies within 10% of reported experimental values. Simulated removal efficiencies closely agreed with actual values for concentrated wastes (3·0–6·0 g COD liter^{-1}) at all retention times used. Greater discrepancy between compared values occurred under conditions of lower influent concentration (1·5 g COD liter^{-1}), particularly for protein–carbohydrate waste (Tables 1 and 2). Predicted removal efficiencies were generally lower than reported values, indicating greater bacterial activity in the experimental reactors. A possible source of this

TABLE 1
Chemical Oxygen Demand (COD) Removals for Protein–Carbohydrate Feedstock

Influent COD (g liter^{-1})	Hydraulic retention time (days)	Effluent COD (g liter^{-1})		COD removal (%)	
		Actual	Simulated	Actual	Simulated
1·5	1·5	0·112	0·243	92·1	84·2
1·5	0·75	0·122	0·287	91·5	82·4
1·5	0·375	0·312	0·508	79·3	66·1
1·5	0·187 5	0·950	0·864	36·7	42·1
3·0	3·0	0·204	0·144	93·4	95·0
3·0	1·5	0·347	0·240	88·4	92·1
3·0	0·375	1·105	1·146	63·0	62·6

From Bolte et al. (1984), reproduced by permission of the American Society of Agricultural Engineers.

TABLE 2
Chemical Oxygen Demand (COD) Removals for Volatile Acid Feedstock

Influent COD (g liter^{-1})	Hydraulic retention time (days)	Effluent COD (g liter^{-1})		COD removal (%)	
		Actual	Simulated	Actual	Simulated
1·5	1·5	0·024	0·211	99·4	86·1
1·5	0·75	0·136	0·239	90·5	86·0
1·5	0·375	0·314	0·266	79·0	82·0
1·5	0·187 5	0·476	0·350	68·4	77·1
3·0	3·0	0·042	0·183	98·6	94·3
3·0	1·5	0·240	0·206	92·0	92·8
6·0	1·5	0·139	0·291	97·7	95·5

From Bolte et al. (1984), reproduced by permission of the American Society of Agricultural Engineers.

discrepancy is the assumption of zero bacterial activity in the bulk liquid which reduces simulated removal efficiency, particularly under conditions of low flow. Inclusion of a bacterial reaction term in the bulk flow equation is currently under development.

Only limited data were available for validating simulated methane production. Unfortunately, laboratory results were reported primarily for dilute protein–carbohydrate waste (1·5 g COD liter^{-1}), for which the model was least capable of predicting state variable behavior (Table 3).

TABLE 3
Comparison of Actual and Simulated Steady-state Methane Production for Protein–Carbohydrate Feedstock

Influent COD concentration (g liter^{-1})	Hydraulic retention time (days)	Efficiency (liters CH$_4$ (g COD added)$^{-1}$)		Methane productivity (liters CH$_4$ (g COD removed)$^{-1}$)	
		Actual	Simulated	Actual	Simulated
3·0	3·0	0·31	0·36	0·33	0·39
1·5	1·5	0·31	0·38	0·33	0·30
1·5	0·75	0·35	0·25	0·38	0·32
1·5	0·375	0·29	0·20	0·39	0·30

From Bolte *et al.* (1984), reproduced by permission of the American Society of Agricultural Engineers.

Since simulated predictions were better for the volatile acid waste than the protein–carbohydrate waste, refinement of the kinetic description of the acetogenic reaction would possibly result in more accurate simulation of fixed bed reactors operating on complex biomass.

PACKING MEDIA

Actual surface area of the wood chips investigated as a potential packing medium for fixed-film reactors is difficult to determine, but the surface to volume ratio should be relatively high. Wood chips saturated with water form a mixture resembling a fluidized bed which may result in fewer clogging problems. In any case, the useful life and the performance characteristics of wood chip media versus several of the manufactured media in a fixed film reactor needed to be determined.

Thirteen bench scale anaerobic fixed-film reactors were placed in a temperature controlled cabinet at 31·1°C (Nordstedt and Thomas, 1985a). All reactors had a constant liquid or void volume of 5 liters, and they were operated in the upflow mode. Due to different media characteristics, the total reactor volumes and surface area to volume ratios were different (Table 4). Two reactors were set up for all types of media except for the plastic balls in a single reactor. The three types of wood media were chosen because of their commercial availability as wood chips. The wood was cut into rectangular blocks (oak, 4·5 × 2·1 × 2·6 cm; cypress, 4·7 × 2·5 × 2·5

TABLE 4
Characteristics of Fixed-film Reactors with 5 liter Liquid Volume

Media type	Media volume (liters)	Total volume (liters)	Media surface (m^2)	Column surface (m^2)	Media SV^{-1} ($m^2\,m^{-3}$)	Column SV^{-1} ($m^2\,m^{-3}$)	Porosity (%)
Plastic balls	5·6	10·6	1·32	0·448	125	167	47
Ribbed spheres	0·3	5·3	0·70	0·223	132	174	96
Plastic rings	0·5	5·5	0·60	0·252	109	157	91
Cypress wood	3·6	8·6	0·71	0·371	83	126	58
Pine wood	3·6	8·6	0·65	0·344	76	116	58
Oak wood	3·9	8·9	0·82	0·368	92	133	56
No media	—	5·0	—	0·205	—	41	—

From Nordstedt and Thomas (1985a), reproduced by permission of the American Society of Agricultural Engineers.

cm; pine, 4·9 × 2·6 × 2·6 cm) so that the surface area could be measured. Because of irregularities in the wood surface, it was recognized that linear measurements would give a minimum surface area for biofilm growth. Media were not washed or treated in any way to enhance biofilm growth or adhesion.

The feedstock for the reactors was supernatant from flushed and settled swine waste. The supernatant varied in solids content due to seasonal changes in water use and to changes in animal population and live weight. The average weekly solids content of the reactor influent ranged from 0·18 to 2·90% total solids (TS) and averaged 0·99% TS through week 52. Volatile solids (VS) averaged 62·3% of TS. The COD of the influent ranged from 2650 to 41 860 mg liter^{-1} and averaged 17 840 mg liter^{-1}. Initial loading of the reactors was at a mean hydraulic retention time of 35·6 days and gradually reduced to 14·5 days by the end of week 4, 8·0 days by the end of week 16, 3·0 days by the end of week 40, and 2·0 days by the end of week 48. After an average loading rate of 1380 mg VS liter^{-1}day^{-1} (based on void volume) during the first four weeks, the VS loading rate dropped to as low as 73 mg VS liter^{-1}day^{-1} in the summer when water use in the swine finishing house was high. (Cell mass was included in void volumes when VS loading rates were determined.) The VS loading rate gradually increased to 7430 mg VS liter^{-1}day^{-1} during weeks 49–52. The loading rates generally increased as the hydraulic retention times decreased. However, the higher solids content of the feedstock near the end of the study resulted in a dramatic increase in loading rates.

The oak wood media reactors produced the lowest effluent COD,

TABLE 5
Comparisons of Gas Production from Different Media on the Basis of Total and Void Volume, Organic Loading Rate and Surface Area

Media type	Gas production				
	liters day^{-1}	liters (liters void)$^{-1}$ day^{-1}	liters (liters total)$^{-1}$ day^{-1}	liters (g VS added)$^{-1}$	liters m^{-2} day^{-1}
Plastic balls	4·3	0·86	0·41	0·45	9·6
Ribbed spheres	3·7	0·74	0·70	0·36	16·6
Plastic rings	4·1	0·82	0·75	0·42	16·3
Cypress wood	4·4	0·88	0·51	0·48	11·8
Pine wood	4·4	0·88	0·51	0·50	12·8
Oak wood	4·3	0·86	0·48	0·38	11·7
No media	3·3	0·66	0·66	0·34	16·1

From Nordstedt and Thomas (1985a), reproduced by permission of the American Society of Agricultural Engineers.

followed by the pine, cypress, plastic rings, plastic balls, ribbed spheres, and no media reactors. On a void volume basis, the wood media had gas production rates of 0·86–0·88 liters gas liter^{-1} day^{-1} which was comparable to the 0·82 liters gas liter^{-1} day^{-1} from the plastic rings (Table 5). The gas production per gram VS added followed the same trend, although the oak media reactor was not as productive as the plastic ring media reactor. The gas production per unit of total reactor volume was approximately 50% higher for the plastic ring and ribbed sphere media than the wood media (Table 5). Average composition of the gas produced in the reactors ranged from 78·5 to 81·0% methane. The results indicate that a reactor with wood block media would have to be approximately 50% larger than a reactor with plastic ring media in order to produce the same quantity of gas. However, the total reactor cost could be as much as 25% less because of the lower cost of the wood media, assuming that future maintenance costs were similar. This comparison does not take into account the nonlinearities in reactor volume costs, type of reactor construction, or differences in costs of other reactor media types.

At the end of week 52, one each of the reactors with a given medium type was dismantled. Biofilm samples were collected for bacterial isolation and electron microscope studies (Robinson et al., 1984). Two morphologically distinct types of methanogenic bacteria were most prevalent in the biofilms. *Methanothrix* spp. were present in high numbers at the film surface and *Methanosarcina* spp. were commonly imbedded in the lower regions of the film. Inhabitants of the film were surrounded by an exopolysaccharide matrix that was very dense toward the base. An extensive network of channels was observed throughout the matrix that may facilitate gas and nutrient exchange to the lower regions of the film. This description of a biofilm is quite different from many of the assumptions which have been used in mathematical modeling of biofilms, i.e. uniform bacterial concentrations in the biofilm and uniform thickness of the biofilm at a given time.

Visual inspection of the media after 52 weeks of operation revealed no apparent decomposition of any of the wood or plastic media. It would have been extremely difficult to separate the biofilm from the porous wood structure and to measure decomposition rates of the wood. When eight of the reactors were opened after week 52, a grayish-brown crystalline deposit had begun to accumulate on both the plastic and wood media. It was found by X-ray analysis to contain calcium, phosphorus and a small amount of magnesium (Robinson *et al.*, 1984).

During the startup period for the reactors, the volatile fatty acid (VFA) levels in the effluents were much lower in the wood media reactors than in the plastic and no media reactors. This indicates that the wood media reactors might have shorter startup periods than reactors with plastic media.

To test effects of medium type on startup characteristics, a series of experiments was conducted using pine wood chips and plastic rings with a whey and cellulose feedstock (Nordstedt and Thomas, 1985b). The reactors were operated for periods of 30 days to monitor startup characteristics with no attempt to achieve steady-state operation. The reactors were inoculated with a mixture of 20% effluent from a stirred tank reactor operating on swine waste feedstock and 80% whey and cellulose feedstock. After an incubation period, regular feeding commenced at a hydraulic retention time of 2 days. The incubation periods which were examined were 5 days, 10 days and 10 days with pH control.

The reactors did not achieve stable operation or indicate that startup would occur within 30 days without the use of pH control with the whey and cellulose feedstock. With pH control, the reactors with both wood chip and plastic media were able to achieve gas production rates of 0·31 liter (g VS added)$^{-1}$ and 0·85 liter^{-1}day^{-1} within 30 days of startup. A 20% mixture of seed inoculum with the feedstock was adequate for a fixed bed reactor when a 10 day incubation period was provided prior to regular feeding at a 2 day hydraulic retention time.

BULK FLOW

Optimum performance from a fixed film reactor will not occur unless sufficient mixing and flow occurs in the bulk liquid to maintain good contact of substrate or feedstock with the microorganisms in the biofilm. This area has not been well defined in terms of nutrient uptake rates, biofilm surface area and thickness, populations of microorganisms, etc.

Clogging of the packing media may be due to buildup of biofilm, accumulation of undegraded feedstock, or a combination of both. Also, the mineral deposits on the media surfaces were sufficiently thick after 2 years to anticipate problems with bulk liquid flow within 5 years. If scale accumulations or other clogging problems were to occur in wood media, the media could be discarded and replaced. The removal of mineral scale from plastic media could pose serious removal problems or high replacement costs.

In startup studies comparing plastic ring and wood chip packing media, channelization was suspected in the wood chip media (Nordstedt and Thomas, 1985b) with the wood chips confined with retaining plates. In later studies, $14\,m^3$ of cypress wood chips were placed in a $20\,m^3$ reactor, and the wood chips were allowed to float freely in the bulk liquid to reduce the likelihood of channeling or short-circuiting of flow. The porosity of the wood chip portion ($14\,m^3$) of the reactor was 63%, and the hydraulic retention time was 9·2 days. The feedstock was supernatant from flushed and settled swine waste. During the first year of operation, the reactor was operated with no supplemental heating, and the average effluent temperature for the period January 1984 through September 1985 was 26·1°C. Since there was not sufficient feedstock available to reduce the hydraulic retention time or to increase the VS loading rate, the volumetric productivity of the reactor was only $0·43\,m^3\,m^{-3}day^{-1}$. However, the conversion rate for the period was $0·51\,m^3$ $(kg\,VS\,added)^{-1}$, essentially the same as the $0·48\,m^3$ $(kg\,VS\,added)^{-1}$ in earlier bench scale tests with wood block media.

SUMMARY

The fixed film reactor has many potential applications in methane production from biomass or waste feedstocks which are low in solids content and/or contain materials which are readily converted to methane. Wood blocks and wood chips were shown to be a feasible alternative to plastic as packing media in fixed film reactors in both bench scale and pilot scale reactors. Wood block media which were removed from reactors after 2 years of operation showed little indication of degradation of the wood. In studies of reactor startup characteristics, cypress wood chips performed equally as well as plastic rings as packing media. A dynamic mathematical model of a fixed film reactor was developed which included convective substrate transport in the bulk liquid through the reactor height, one-dimensional diffusion through the biofilm and across the biofilm–liquid interface, and two-culture bacterial substrate conversion in the biofilm. Under many operating conditions, the model was capable of predicting steady-state COD removal efficiencies within 10% of reported experimental values.

REFERENCES

Barry, M. and Colleran, E. (1982). Anaerobic digestion of silage effluent using an upflow fixed bed reactor. *Agricultural Wastes* **4**(3), 231–9.

Bolte, J. P. (1983). Dynamic modeling and simulation of fixed bed anaerobic reactors. M.S. thesis, Agricultural Engineering Department, University of Florida, Gainesville.

Bolte, J. P., Nordstedt, R. A. and Thomas, M. V. (1984). Mathematical simulation of upflow anaerobic fixed bed reactors. *Transactions of the ASAE* **27**(5), 1483–90.

Brumm, T. J. (1980). Dilute swine waste treatment in an anaerobic filter with an organic support media. M.S. thesis, Purdue University, West Lafayette, Indiana.

Harvey, M., Forsberg, C. W., Beveridge, T. J., Pos, J. and Ogilvie, J. R. (1984). Methanogenic activity and structural characteristics of the microbial biofilm on a needle-punched polyester support. *Applied and Environmental Microbiology* **48**(3), 633–8.

Hill, D. T. and Barth, C. L. (1977). A dynamic model for simulation of animal waste digestion. *Journal of the Water Pollution Control Federation* **49**, 2129–43.

Hudson, J. W., Pohland, F. G. and Pendergrass, R. P. (1978). Anaerobic packed column treatment of shellfish processing wastewaters. *Proceedings of the 33rd Purdue Industrial Waste Conference*, Ann Arbor Science Publishers, Ann Arbor, Michigan, pp. 560–74.

Jones, D. D., Dale, A. C., Nye, J. C. and Harrington, R. B. (1981). Fiber wall reactor digestion of dairy cattle manure. *Transactions of the ASAE* **24**(5), 1282–6, 1290.

Kennedy, K. J. and Van den Berg, L. (1982). Anaerobic digestion of piggery waste using stationary fixed film reactor. *Agricultural Wastes* **4**(2), 151–8.

Lovan, G. R. and Foree, E. G. (1971). The anaerobic filter for treatment of brewery press liquor waste. *Proceedings of the 26th Purdue Industrial Waste Conference*, Part Two, Purdue University, Lafayette, Indiana, pp. 1074–86.

Mueller, J. A. and Mancini, J. L. (1976). Anaerobic filter—kinetics and application. *Proceedings of the 30th Purdue Industrial Waste Conference*, Ann Arbor Science Publishers, Ann Arbor, Michigan, pp. 423–47.

Nordberg, W. V. (1977). Anaerobic methane fermentation of a solid cellulosic substrate in a closed-flow system. M.S. thesis, Purdue University, West Lafayette, Indiana.

Nordstedt, R. A. and Thomas, M. V. (1985a). Wood block media for anaerobic fixed bed reactors. *Transactions of the ASAE* **28**(6), 1990–6.

Nordstedt, R. A. and Thomas, M. V. (1985b). Startup characteristics of anaerobic fixed bed reactors. *Transactions of the ASAE* **28**(4), 1242–7, 1252.

Person, H. L. (1980). The anaerobic filter for treating liquid swine waste. Ph.D. thesis, University of Minnesota, St. Paul, Minnesota.

Rittman, B. E. and McCarty, P. L. (1980). Model of steady state biofilm kinetics. *Biotechnology and Bioengineering* **22**, 2343–57.

Rittman, B. E., Strubler, C. E. and Ruzicka, T. (1982). Anaerobic filter pretreatment kinetics. *Journal of the Environmental Engineering Division, American Society of Civil Engineers* **108**(EES), 900–12.

Robinson, R. W., Akin, D. E., Nordstedt, R. A., Thomas, M. V. and Aldrich, H. C. (1984). Light and electron microscopic examinations of methane-producing biofilms from anaerobic fixed-bed reactors. *Applied and Environmental Microbiology* **48**(1), 127–36.

Smith, R. E., Reed, M. J. and Kiker, J. T. (1977). Two-phase anaerobic digestion of swine waste. *Transactions of the ASAE* **20**(6), 1123–8.

Taylor, D. W. (1972). Full-scale anaerobic trickling filter evaluation. In: *Proceedings of the Third National Symposium on Food Processing Wastes*, New Orleans, Louisiana, March 28–30, 1972. EPA-R2-72-018, pp. 151–62.

Van den Berg, L. and Lentz, C. P. (1981). Effects of film area-to-volume ratio, film support, height and direction of flow on performance of methanogenic fixed film reactors. *Proceedings of the Workshop—Anaerobic Filters: An Energy Plus for Wastewater Treatment*. Argonne National Laboratory, Inc., Argonne, Illinois, pp. 1–10.

Wilkie, A., Faherty, G. and Colleran, E. (1983). The effect of varying the support matrix on the anaerobic digestion of pig slurry in the upflow anaerobic filter design. In: *Energy from Biomass, Second European Community Conference*, Elsevier Applied Science, New York, pp. 531–5.

Williamson, K. and McCarty, P. L. (1976). A model of substrate utilization by bacterial films. *Journal of the Water Pollution Control Federation* **48**, 9–23.

Young, J. C. and McCarty, P. L. (1969). The anaerobic filter for waste treatment. *Journal of the Water Pollution Control Federation* **41**(5), R160–73.

26
Anaerobic Digester Residues: Treatment and Utilization

D. A. GRAETZ

Soil Science Department, University of Florida—IFAS, Gainesville, Florida, USA

and

K. R. REDDY

Central Florida Research and Education Center, University of Florida—IFAS, Sanford, Florida, USA

INTRODUCTION

Anaerobic digestion of organic material for the production of methane produces a residue which may contain relatively large quantities of plant nutrients in a readily available form. Disposal or utilization of this residue becomes a significant consideration in an integrated 'methane from biomass' system. Because of the high nutrient content of this waste product, it may serve as a fertilizer material for either the production of terrestrial or aquatic crops which in turn could be cycled back into the methane digesters or be used for other purposes. Since use of biomass for production of methane is relatively new in the US, most of the research has dealt with optimizing methane yield and little research has been conducted on how to utilize the residue left after digestion. This chapter presents some characteristics of selected residues and provides examples of how these residues have been treated and/or utilized in both terrestrial and aquatic systems.

CHARACTERISTICS OF DIGESTER RESIDUE

Characteristics of digester residue are determined by digester operating conditions, residence time and feedstock material. Data for examples of several residues, including undigested material for comparative purposes, show large variations in the elemental properties (Table 1). Moorhead (1986) digested water hyacinth biomass of low and high N content in 55-liter batch digesters. After digestion for 4 months, the effluent, i.e. the liquid component, was separated from the solids by passing the contents of the digester through a 1·00 mm fiberglass screen. A major part of the plant K was solubilized during the digestion process. In contrast to K, the digested solids contained higher amounts of Ca, Na and Zn compared to the fresh plant tissue. Similar observations regarding distribution of elements between the liquid and solid phases have been observed for digested animal manures (Field et al., 1984). The increase in Na level in the digested material can be attributed to use of $NaHCO_3$ buffer in the digester. Only 20% of the initial organic N was found in the digested sludge for both high and low N plant materials. The C/N ratio of the plant material did not change significantly upon digestion for the high N plant tissue. However, for the low N plant tissue the C/N ratio decreased from 35 for the fresh material to 16 for the digested residue, indicating that for the low N material, the microbial population recycled much of the available N back into microbial biomass. Mineralization of organic N to NH_4 accounted for 72 and 35% of the added N for high and low N material respectively. Feldman (1986) analyzed digested water hyacinth material obtained from a digester which received 15 kg wet water hyacinth day^{-1}, three days $week^{-1}$, with a retention time of about 31 days. Composition of this material was similar to the digested residue of the previously mentioned high N residue (Table 1).

Terrestrial crops have also been used as digester feedstock. Atalay and Blanchar (1984) analyzed fresh and digested fescue hay as a soil amendment (Table 1). The N content of the fresh hay was relatively low, and as with the low N water hyacinth, the digested residue increased in N content. Phosphorus and Ca content increased in the sludge apparently due to addition of an inoculant with a high P content and limestone, respectively. More soluble elements, such as K and Na, showed decreases in concentration.

Animal manures were one of the first feedstocks used in anaerobic digesters. Residue obtained from digesters using swine manure feedstock showed much higher P levels (Table 1) than other residues reflecting high P input into the swine feed ration. The bovine waste residue represents composted solids remaining after liquid extraction. Thus, soluble elements such as K and Na are quite low in this material. Digestion of distillery residues (rum slops) as practiced by Bacardi Corporation (Szendry and

TABLE 1
Elemental Composition of Selected Fresh and Digested Biomass Materials, Animal Manures and Rum Slops

Material	TKN	NH_4-N	P	K	Ca	Mg	Na	Cu	Zn	Reference
Water hyacinth										
Fresh (g kg^{-1})[a]	34	NA[b]	NA	24	18	3	8	NA	0·52	Moorhead, 1986
Digested liquid (mg liter^{-1})	256	212	13	80	30	12	1 160	NA	NA	Moorhead, 1986
solid (g kg^{-1})	39	NA	NA	4	18	2	26	NA	0·90	Moorhead, 1986
Fresh (g kg^{-1})	11	NA	NA	22	21	7	11	NA	0·70	Moorhead, 1986
Digested liquid (mg liter^{-1})	72	48	5	245	53	53	1 240	NA	NA	Moorhead, 1986
solid (g kg^{-1})	27	NA	NA	3	24	2	17	NA	2·00	Moorhead, 1986
Digested solid (g kg^{-1})	29	NA	4	3	9	3	NA	5·00	32·00	Feldman, 1986
Fescue hay										
fresh (g kg^{-1})	11	NA	1	8	9	2	3	NA	0·13	Atalay and Blanchar, 1984
digested total (g kg^{-1})	20	NA	5	4	46	1	2	NA	0·15	Atalay and Blanchar, 1984
Swine manure Digested										
total (g kg^{-1})	1	NA	53	28	57	16	7	0·30	4·80	Sweeney et al., 1985
liquid (mg liter^{-1})	NA	1 170	17	610	37	4	91	0·04	0·01	Sweeney et al., 1985
Rum slops Digested										
total (g kg^{-1})	1	NA	4	120	32	18	14	0·09	0·10	Sweeney et al., 1985
liquid (mg liter^{-1})	NA	23	33	4 470	1 010	700	510	1·10	1·10	Sweeney et al., 1985
Bovine manure										
digested solids[c] (g kg^{-1})	14	NA	15	5	48	2	2	0·02	0·02	Sweeney et al., 1985

[a] All g kg^{-1} values are on a dry weight basis.
[b] NA = Not Available.
[c] Composted.

Dorion, 1986) produces an effluent high in K, Ca and Na, apparently due to additions of KOH and other materials during the distillation process.

DECOMPOSITION OF DIGESTER RESIDUE ADDED TO SOIL

One consequence of anaerobic digestion of organic material is a reduction in amount of readily decomposable C. This could have a significant effect on the rate at which residue decomposes in the soil and subsequently on the rate of nutrient release from the residue. Several of the above-mentioned residues have been evaluated with regard to decomposability when added to soil. Sweeney et al. (1985) evaluated three of the residues noted in Table 1 for their rate of decomposition in soil as measured by CO_2 evolution (Fig. 1). The bovine waste decomposed very slowly in the soil and CO_2 evolution was only slightly affected by application rate. In contrast, CO_2 evolution from the swine waste and molasses residues was relatively rapid for the first 40 days and continued at a slower rate for the duration of the study. Decomposition (CO_2 evolution) was initially faster for the molasses residue than for the swine residue, but by the end of the study there was no difference in total amount of CO_2 evolved. As the residue application rate increased, CO_2 evolution also increased but not at a rate proportional to the increase in residue application rate. Similar rate effects were found by Atalay and Blanchar (1984). These results indicate that anaerobic digester sludges, when added to the soil at reasonable rates, will decompose readily without adverse effects on the soil microbial population. The one exception, bovine sludge, had already gone through a composting process and thus the organic matter was relatively stable to decomposition.

Moorhead et al. (1987) compared fresh and digested water hyacinth biomass as soil amendments. After 90 days, 20% of the added C from the digested biomass had evolved as CO_2 compared to greater than 39% of the added C from the fresh biomass. Most likely, the readily decomposable C in the fresh biomass had already been released as CH_4 and CO_2 in the digestion process. The percentage C evolved from the anaerobically digested hyacinth sludge was similar to the results of Miller (1974) and Tester et al. (1977) for anaerobically digested and composted sewage sludge.

Moorhead et al. (1987) also evaluated N mineralization from fresh and digested water hyacinth biomass. Carbon/nitrogen ratio is commonly used as a guideline for predicting the relative decomposability or mineralization potential of organic materials added to soil (Reddy et al., 1980). However,

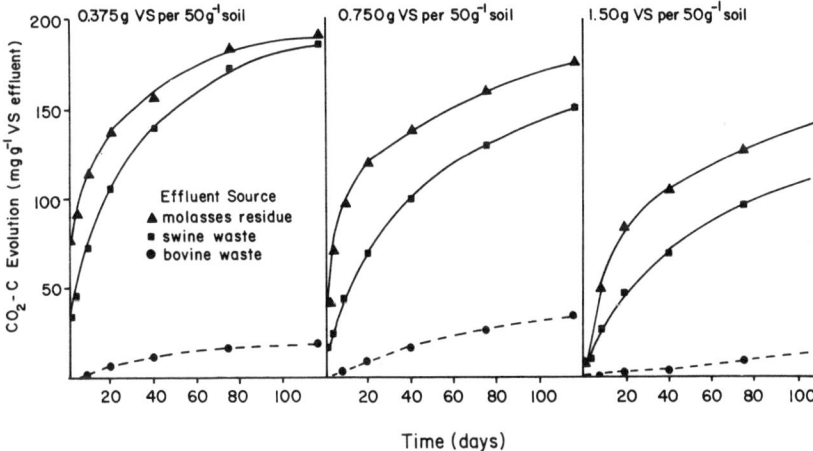

FIG. 1. Cumulative CO_2-C evolution (check corrected) as a function of time for three anaerobic digester effluents applied to soil at three application rates (Sweeney et al., 1985).

the use of C/N ratio of digested biomass sludges for predicting N mineralization may have limited applicability. For example, Moorhead et al. (1987) found that only 8% of the applied N was mineralized to NO_3 from digested biomass sludge compared to 33% from fresh hyacinth biomass even though both materials had a C/N ratio of 12. The amount of N mineralized was approximately one-third of the C lost as CO_2. This is in contrast to a 1:1 ratio observed between N mineralization and CO_2 evolution for sewage sludge added to soil (Gilmour et al., 1985). Thus, it appears that more research is needed to adequately define C and N mineralization of anaerobic digester residues, particularly with regard to plant feedstocks.

LAND APPLICATION OF DIGESTER RESIDUES

Land application is a traditional method of utilizing or disposing of crop residues and animal manures and this method appears to have potential for digester residues as well (Fischer et al., 1984). Organic materials added to soil can affect both soil properties and crop yields. In the case of fescue residue, one of the most beneficial effects of adding the residue to a silt loam soil was an increase in water-holding capacity of the soil (Atalay and

Blanchar, 1984). The residue was also a good source of P for wheat and sorghum but due to a high C/N ratio did not supply adequate N. Thus, inorganic N had to be added with this residue to make it a feasible fertilizer material. Rates of 10, 20, 40 and 80 g residue kg^{-1} had no adverse effect on seed germination.

Similar results were obtained when composted digester residue (bovine manure feedstock) was evaluated as a soil amendment on a sandy soil (Feldman, 1986). In this study, radish and corn were planted in succession in soil amended with up to 32 Mg ha^{-1} of residue. Radish yields were not significantly affected by addition of either residue but corn yields and N uptake were increased by both residues. Yields and N uptake were increased more with water hyacinth residue than with bovine residue. The high C/N ratio (86) of the bovine manure residue apparently limited N availability from the residue. The lack of effect on radish yield, with either residue, was likely due to insufficient time for the organic matter to mineralize and release nutrients in available form. Addition of water hyacinth residue increased soil pH and organic matter, NO_3-N, K, Ca, Mg and Zn levels. Soils data for the bovine residue-amended soils showed extreme variability and definite trends in the above parameters were not evident. Seed germination was not affected by either residue at any of the application rates.

In many cases, digester material must be disposed of in a somewhat continuous manner. Irrigation of a crop such as turfgrass would allow periodic (such as weekly) utilization of digester effluent in contrast to certain agricultural crops where only preplant applications would be feasible. Periodic application of effluent was evaluated in a study in which effluent from a digester using rum slop feedstock (Table 1) was applied weekly to St. Augustine grass for 222 days at rates of 0, 0·22, 0·44 and 0·88 cm week^{-1} (Sweeney et al., 1985). Growth rates for four approximately 60-day periods (Table 2) showed that during the first period all application rates resulted in similar growth rates, i.e., 14 to 18 kg dry matter ha^{-1} day^{-1}. Thus, there was no adverse effect of effluent at any application rate during this period. However, there was a definite decline in growth rate at the high application rate during the second period. A similar decline was noted for the intermediate application rate by the fourth period. At this point, there was still excellent growth at the low application rate.

Certain effluent constituents which adversely affected plant growth were accumulating in the soil. Results of soil analysis supported this hypothesis. For example, K levels in the soil were less than 50 ppm in the treatment

TABLE 2
Daily Growth Rate of St Augustine Grass as Affected by Effluent from a Digester Using Rum Slops (Sweeney et al., 1985)

Effluent application rate (cm/wk)	Growth rate ($kg\ ha^{-1}\ day^{-1}$)			
	Time period ($days^a$)			
	0–59	60–116	117–180	181–222
0	14	11	11	14
0^b	18	19	17	24
0·22	16	26	42	73
0·44	18	25	36	21
$0·88^c$	14	9	10	48

[a] Day 0 is the first day of effluent application.
[b] Nitrogen applied as NH_4NO_3.
[c] Effluent application stopped at 102 days.

receiving no effluent compared to 1260, 1980 and 3670 ppm for the 3 effluent application rates, respectively. P, Ca, Mg, and Na levels also increased with increasing effluent application rates. These data suggest that accumulation of soluble salts in the soil may have been responsible for the observed decline in growth rate. When the decline in growth rate was observed at the high effluent application rate, effluent was replaced with water at the same application rate and good recovery of growth rate was observed suggesting that the effect is reversible.

EFFLUENT TREATMENT WITH AQUATIC SYSTEMS

Integrated 'energy from biomass' systems which produce methane via anaerobic digestion may use aquatic plants such as water hyacinth both as a feedstock and as a means of 'cleaning up' digester effluent prior to disposal. Graetz et al. (1986) evaluated effluent from anaerobic digesters using swine manure feedstock as a nutrient source for production of water hyacinth. Water hyacinth were grown at an initial stocking rate of 10 kg wet weight m^{-2}. Effluent was added at rates to provide 0, 300 and 600 kg N ha^{-1} approximately every 20 days. Water hyacinth were harvested back to initial stocking rates at 20-day intervals. Yields averaged 7 Mg dry weight ha^{-1} $harvest^{-1}$ period for both the 300 and 600 kg N ha^{-1} application rates.

Plant N content was 2·64 and 3·92%, respectively. Excellent growth rates (32 g m^{-2} day^{-1}) were obtained and no adverse effects on plant growth were observed at either application rate. Inorganic N removal from the effluent was greater than 99% for both application rates during the summer, but during slower growth periods percentage N removal decreased, particularly at the high application rate.

Recent research has shown that water hyacinth should not be grown directly in anaerobic digester effluent. Placing water hyacinth in undiluted effluent resulted in reduced growth rate and, in some cases, death of the plants. Using effluent from a digester utilizing a mixture of water hyacinth and sewage sludge feedstock, DeBusk and Reddy (unpublished data, 1986) showed that water hyacinth yields were reduced until dilutions of at least 1:4 (effluent:water) were used (Table 3). Water hyacinth yields were lowest in 1:1 and 1:2 dilutions, with growth ceasing in both solutions after

TABLE 3
Growth of Water Hyacinth Cultured in Various Dilutions of Anaerobic Digester Effluent Obtained from Two Types of Biomass Feedstocks. Effluent Retention Period from Water Hyacinth/Sludge Feedstock was 24 Days and from Sorghum/ Raw Wastewater was 21 Days. Each Value is Average of 3 Replications. (EC = Electrical Conductivity or ionic strength, Eh = oxidation–reduction potential or redox potential (DeBusk, W. F. and Reddy, K. R., University of Florida, Gainesville, unpublished results))

Dilution	BOD_5 (mg liter^{-1})	EC	Eh (mV)	Biomass (g (dry wt) m^{-2} day^{-1})
Water hyacinth/sludge				
1:1	1 101	3·4	−205	1·2
1:2	694	2·5	−62	3·5
1:4	352	1·8	−14	8·8
1:8	241	1·1	64	11·1
1:16	152	0·7	195	9·9
1:32	76	0·5	194	10·6
1:64	38	0·4	270	8·0
Sorghum/raw wastewater				
1:1	1 740	6·5	−335	0·0
1:2	813	5·0	−362	0·0
1:4	271	3·5	−285	8·5
1:8	206	2·4	25	12·6
1:16	134	1·6	139	8·3
1:32	67	1·0	214	7·7
1:64	33	0·8	286	4·6

about 30 days. Highest yields were obtained with dilutions of 1:4–1:32. A slight yield reduction was noted with a dilution of 1:64 likely due to depletion of nutrients. Similar results were observed when hyacinths were used to treat anaerobic digester effluent obtained from sorghum feedstock (Table 3). At a 1:1 dilution, this effluent had electrical conductivity values twice those obtained for hyacinth/sludge feedstock. At 1:1 and 1:2 dilutions (effluent:water), plants did not survive, but growth was found to be normal at a 1:4 dilution.

Moorhead (1986) diluted a variety of effluents on the basis of electrical conductivity. Dilutions ranged from 1:3 to 1:8. Plant death resulted from use of several of the undiluted effluents. Although water hyacinth had difficulty growing in the undiluted effluents, algal activity was observed in all of them. Highest dry weight gains were obtained with plants growing in diluted effluents with electrical conductivity levels ranging from 0·7 to 2·3 dS m^{-1} and NH$_4$-N concentrations of 23–104 mg liter^{-1}. Plants survived, but showed very poor growth, in two undiluted effluents with electrical conductivity levels of 5·6 and 5·9 dS m^{-1} and NH$_4$-N levels of 24 and 49 mg liter^{-1}. All other conductivity and NH$_4$-N combinations resulted in plant death. Mortality was apparently due to salt injury and/or NH$_3$ toxicity.

Additional information is needed to establish optimum dilution of anaerobic digester effluents to promote maximum biomass yields. However, these studies demonstrate that at proper dilutions, anaerobic digester effluents can be a suitable growth medium for aquatic plant biomass production.

SUMMARY

The literature base for evaluating anaerobic digester residue characteristics is very limited because of the relatively recent advent of 'energy from biomass' programs on a large scale in the US. Available data suggest that residue characteristics will vary greatly depending on digester operating conditions and feedstock composition. Limited data available to date indicate that both terrestrial and aquatic systems can be used for treatment/utilization of digester residues. Practices which are commonly used in the application of animal wastes and sewage sludges to land provide some insight into how digester residues can be used as soil amendments. We found that continued application of high rates of rum slop residues led to salt build-ups. Some digester residues, when diluted, proved to be suitable for growing plants. However, additional data are needed on characterization of solid and liquid digester residues and their fate after application to soils or ponds containing aquatic plants. Additional data are also needed on optional loading rates of solid and liquid residues to achieve maximum treatment efficiency.

REFERENCES

Atalay, A. and Blanchar, R. W. (1984). Evaluation of methane generator sludge as a soil amendment. *Journal of Environmental Quality* **13**, 341–4.

Feldman, V. A. (1986). Effect of anaerobically digested water hyacinth and bovine waste on plant growth and soil characteristics. M.S. thesis, University of Florida, Gainesville, Florida.

Field, J. A., Caldwell, J. S., Jeyanayagam, S., Reneau, R. B., Jr, Kroontje, W. and Collins, E. R. Jr (1984). Fertilizer recovery from anaerobic digesters. *Transactions of the American Society of Agricultural Engineering* **27**, 1871–81.

Fischer, J. R., Paterson, J. A., Vandepopuliere, J. M. and Veum, T. L. (1984). Uses of effluent from a swine anaerobic digester. *Proceedings of the 4th Annual Solar and Biomass Energy Workshop*, Atlanta, Georgia, pp. 158–60.

Gilmour, J. T., Clark, M. D. and Sigua, G. C. (1985). Estimating net nitrogen mineralization from carbon dioxide evolution. *Soil Science Society of America Journal* **49**, 1398–402.

Graetz, D. A., Krottje, P. A. and Nordstedt, R. A. (1986). Treatment of anaerobic digester effluent by water hyacinth (*Eichhornia crassipes* (Mart) Solms). Program and Abstracts: Conference on Research Applications of Aquatic Plants for Water Treatment and Resource Recovery. Abstract APC-65, p. 126.

Miller, R. H. (1974). Factors affecting the decomposition of an anaerobically digested sewage sludge in soil. *Journal of Environmental Quality* **3**, 376–80.

Moorhead, K. K. (1986). Nitrogen cycling in an integrated 'Biomass from Energy' system. Ph.D. dissertation, University of Florida, Gainesville, Florida.

Moorhead, K. K., Graetz, D. A. and Reddy, K. R. (1987). Decomposition of fresh and anaerobically digested plant biomass in soil. *Journal of Environmental Quality* (in press).

Reddy, K. R., Khaleel, R. and Overcash, M. R. (1980). Carbon transformations in the land areas receiving organic wastes in relation to nonpoint source pollution: a conceptual model. *Journal of Environmental Quality* **9**, 434–42.

Sweeney, D. W., Graetz, D. A. and Sartain, J. B. (1985). Effect of anaerobic digester effluent on selected soil characteristics and growth of St. Augustine grass. *Agronomy Abstracts* 37.

Szendry, L. M. and Dorion, G. H. (1986). Methane production from anaerobic digestion of distillery residues. In: *Biomass Energy Development*, Smith, W. H. (ed.), pp. 517–31. Plenum Press, New York.

Tester, C. F., Sikora, L. J., Taylor, J. M. and Parr, J. F. (1977). Decomposition of sewage sludge in soil I. Carbon and nitrogen transformations. *Journal of Environmental Quality* **6**, 459–63.

27
Perspectives on Biomass Research

J. R. FRANK

Gas Research Institute, Chicago, Illinois, USA

and

W. H. SMITH

Center for Biomass Energy Systems, University of Florida—IFAS, Gainesville, Florida, USA

'Biotechnology begins with the identification of the gems provided by nature.'
(P. Abelson, 1986)

This chapter summarizes the current status of the Methane from Biomass Program, the directions our hypotheses are leading us, and some of the strategies for pursuing the opportunities with which we are now presented. The GRI/IFAS program has evolved considerably, both technically and programmatically, since its inception in July 1981. Initially, the program was a diverse collection of projects with a loose set of goals aimed primarily at screening biomass species for high biomass yields and developing suitable conversion systems. Today it has become a model of industry/ university cooperation with multidisciplinary teams focused on specific technical areas with a blend of applied and fundamental approaches. We have identified several 'gems nature provided us' and have made incremental improvements in some of these gems—organisms, germplasm, community structures and physical environments. Our current objective has become to substantially reduce the critical cost factors for crop production and conversion identified through systems analysis as the major contributors to the product gas cost. To achieve these goals requires that we focus on innovative application of 'state of the art' technologies.

The hypotheses and assumptions in our systems approach are explicitly defined in the 'BIOMET' models and the analyses performed by Reynold, Smith and Hills (see Chapters 3 and 4). Using these tools, the costs of

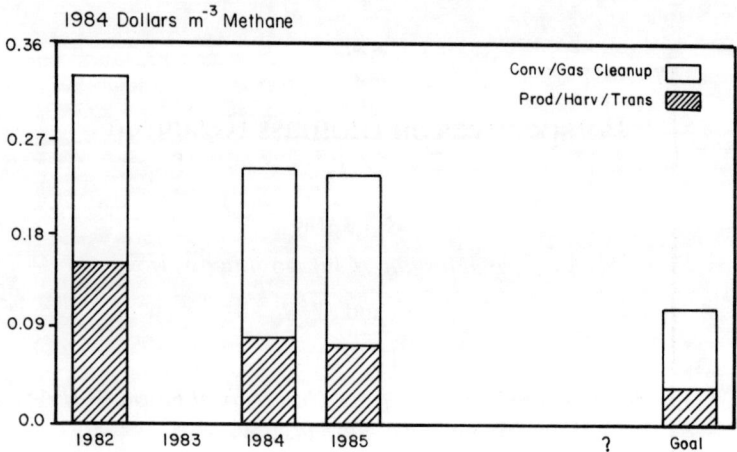

FIG. 1. Research progress and goals in the GRI/IFAS Biomass Program.

methane from biomass systems have been estimated (Fig. 1). These estimates project field and laboratory results to a full-scale Napiergrass system producing 2.5×10^5 m^3 day^{-1} of pipeline quality gas. Allowances have been made for the uncertainty of projecting bench and field-scale data to full-scale systems using the methodology developed by GRI (1983). These results indicate that through research in the past 5 years we have reduced estimated costs of pipeline quality methane from over \$9 GJ^{-1} to about \$6·50 GJ^{-1}. Most of these estimated reductions in cost were from incremental engineering and agronomic improvements including:

- Selection/development and management of new energy crops which has doubled biomass yields to ranges between 37 and 50+ dry Mg ha^{-1} yr^{-1} for a variety of grasses including sorghum, Napiergrass and hybrid sugarcanes and up to 112 dry Mg ha^{-1} yr^{-1} for aquatic plants such as water hyacinths (Fig. 2).
- Anaerobic digesters have been designed which reduced by one half the heat energy inputs by increasing solids loading and solids retention times.
- Methane yields have been increased from about 0·28 m^3 kg^{-1} VS to 0·34 m^3 kg^{-1} VS added by proper plant selection and increasing solids retention times in the anaerobic digesters.
- The use of systems analysis has defined facility sizes and land requirements and quantified cost components. This has allowed the program to

Perspectives on Biomass Research

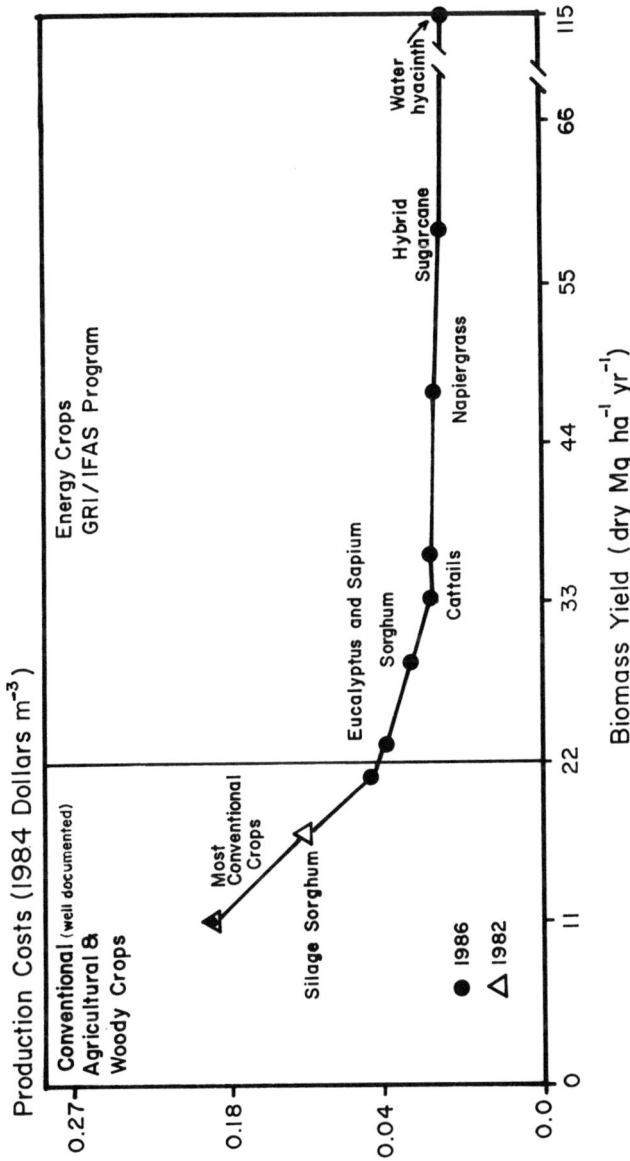

FIG. 2. Progress in improving biomass yields compared with estimated production costs.

focus on areas of research with the greatest cost reduction potential and avoid less economically important issues such as the utilization of marginal lands or avoidance of fertilization and pest management.

With these 'gems provided by nature' identified and initial incremental improvements made, it is more difficult to attain incremental cost improvements in the system through conventional approaches and short-term research. Also, the downward movement of gas prices and the possibility that this technology may not be needed until after the year 2000 has allowed the program to focus on sensitivities which require longer-term research involving more fundamental research focused on key biological regulators of production and conversion.

Analysis with BIOMET determined cost sensitivities of various production and conversion elements for identifying potential research areas. This analysis used a baseline Napiergrass system producing 2.5×10^5 m^3 day^{-1}, biomass yields of 44 dry Mg ha^{-1} yr^{-1}, and methane yields of 0·31 m^3 kg^{-1} VS. Vegetative propagation with at least six successive yearly crops regenerated annually from sprouting was assumed. The results (Fig. 3A, B) summarize the effects of:

- increasing methane yield to 0·43 m^3 kg^{-1} VS added;
- increasing methane generation rates to 10 reactor volumes day^{-1} from 1·5 volumes day^{-1};
- increasing methane content in the biogas to 95% before cleanup;
- increasing biomass yields to 67 dry Mg ha^{-1} yr^{-1};
- reducing planting costs by 90%; and
- reducing chemical inputs by 70%.

The research area of greatest cost sensitivity appears to be aspects that affect biomass convertibility to methane. This results because methane yield affects both conversion and production costs through reducing the size of the reactor and reducing the amount of biomass needed per unit output. In contrast, increasing the methane production rate tenfold, but with constant methane yield, merely decreases the capital cost of the reactor which represents only 10% of the overall capital cost (Warren et al., 1984). Not yet understood is what methane production rates may be attainable. An area with significant potential cost savings is the management of carbon dioxide in the reactor thereby increasing the methane concentration in the biogas. This may be achievable through proper digester design as described by Hayes and Isaacson (1986). Another area of potential cost reduction is improvements in biomass yield to 67 dry

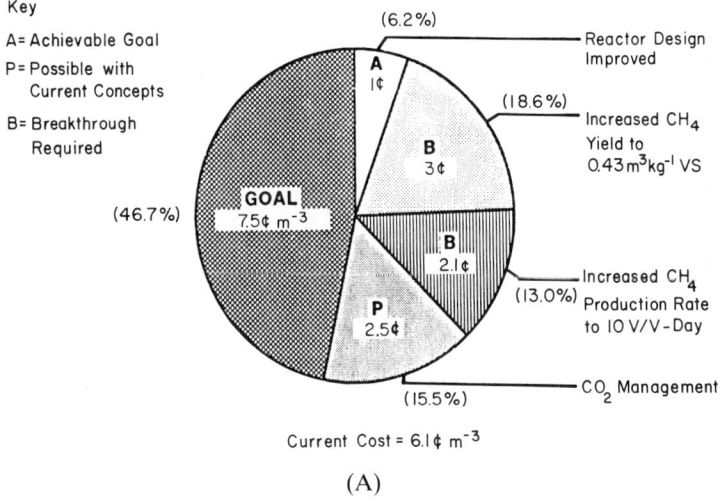

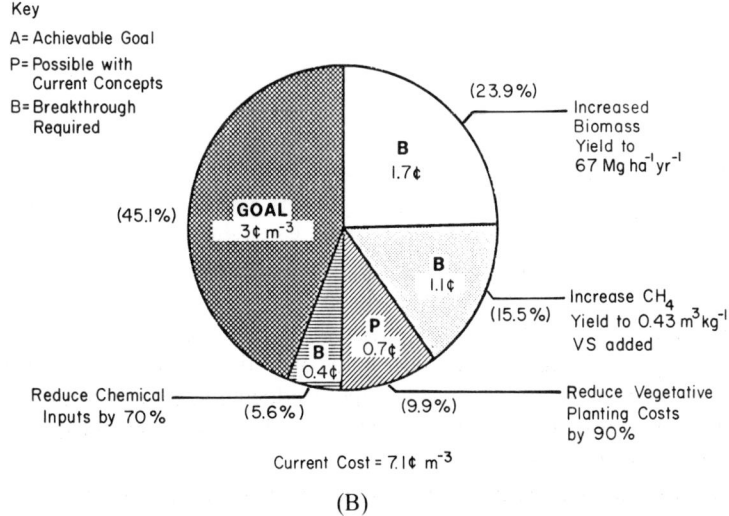

FIG. 3(A) Potential gas cost reductions from improvements in conversion and gas cleanup through research. (B) Potential cost reductions from improvements in production, harvesting and transportation through research. (A = Achievable goal; P = Possible with current concepts; B = Breakthrough required.)

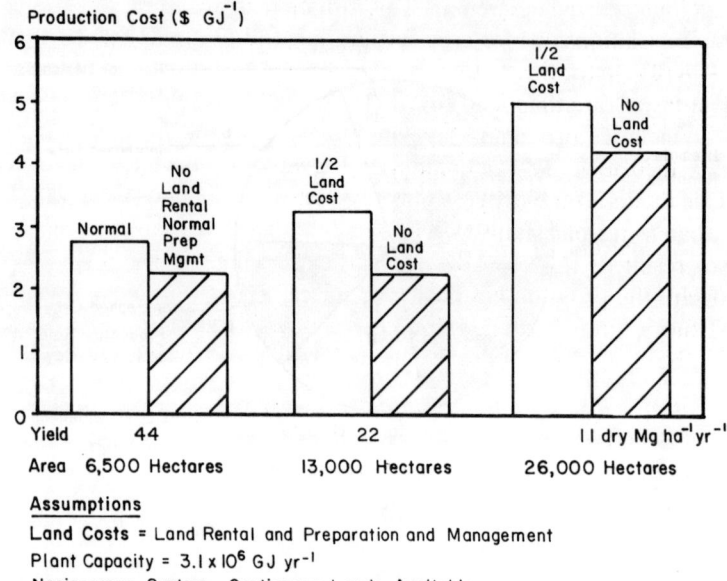

FIG. 4. Effects of biomass yield in comparison to land costs on production costs.

$Mg\,ha^{-1}\,yr^{-1}$. Yields of up to 60 dry $Mg\,ha^{-1}\,yr^{-1}$ have been observed for Napiergrass and over 67 dry $Mg\,ha^{-1}\,yr^{-1}$ for hybrid sugarcane and water hyacinths (see Chapters 6 and 16). However, biomass yields at this level are not yet sustainable on large scales. The production costs (Fig. 2) accelerate rapidly under 22 dry $Mg\,ha^{-1}\,yr^{-1}$ due to increasing transportation costs, and decrease gradually when yields are over 34 dry $Mg\,ha^{-1}\,yr^{-1}$. Therefore, biomass yields of 34 dry $Mg\,ha^{-1}\,yr^{-1}$ or greater appear to be suitable for energy crops. Reducing vegetative propagation costs (i.e. through tissue culture) also may have a significant impact on reducing cost (see Chapters 9 and 11).

Another similar analysis (Fig. 4) showed the cost sensitivity of land rental, land preparation and management on the cost of methane. At low biomass yields, land costs (rental, preparation and management) had a major impact, but at high biomass yields this sensitivity decreased. Furthermore, even if there were no land costs at low biomass yields, overall production costs were much higher than those at higher yields with normal inputs and land rental. This situation is analogous to the cost sensitivity relationships for crop fertilization and pest management (Fig. 3B). The reason for this result is because the costs of transporting biomass rise at a faster rate than land costs. The implications of these findings are that for facilities producing pipeline quality methane from a dedicated biomass feedstock, higher quality land used with normal fertilizer and pesticide inputs is far more cost-effective with high biomass yielding energy crops than using marginal land with low land rental costs and low inputs.

As a result of this type of interactive assessment using the 'systems approach', the program has coalesced into a few multidisciplinary teams with a more fundamental, advanced technology thrust addressing areas that can be significantly advanced by research. These teams include:

(1) *Integrated analysis* for developing, modifying and testing the validity of BIOMET and individual research models.
(2) *Microbial degradation* for defining microbial and biochemical processes in cell wall degradation (pectins, hemicellulose, cellulose) for developing methods of monitoring and regulation *in vivo*.
(3) *Microbiology and reactor systems* for developing cost-effective conversion systems by understanding and improving biotechnologies by use of the organisms that control processing mechanisms and their regulation, feedstock chemistry in relation to biodegradability, and by using microbial community structure analyses for assessing the health of the conversion system.
(4) *Advanced plant production* for developing methods of cellular and molecular biology for improving the yield and convertibility of energy crops through tissue culture, somatic hybridization, gene amplification and genetic engineering.
(5) *Production and materials handling* for increasing convertible biomass through developing new management strategies, using mycorrhizae and other stress-abatement techniques, improving materials handling/storage options, and field testing new plant genotypes.

These teams, by necessity, have to maintain effective communications with each other for their advancements to be properly used to positively improve the system. The growing thrusts of the research by these teams is to apply biotechnology to the energy crop and reactor systems identified as promising opportunities, and to replace classical chemical/physical controls with biosensors for biological regulation interfaced with computer controls. Considerable progress has encouraged the pursuit of these goals and the new thrusts for the program. Some examples of progress include:

- The tissue culture and genetic manipulation of the grasses (e.g. Napiergrass, sugarcane) including the first protoplast regeneration, the genetic transformation of grass by pioneering the use of electroporation technology, and the first grass somatic hybrids. In field tests, tissue culture plants have shown improved vigor and high genetic stability and included seed bearing hexaploid variants among crosses of Napiergrass/ pearl millet (see Chapters 9 and 10).
- Somatic embryo propagation was successful with sweet potato and these somatically derived plantlets are constituting a model system for exploring fluid or gel 'seeding' techniques (see Chapter 11).
- Techniques used to assess the microbial activity and biomass degradation in reactors have been developed including the use of monitoring phospholipids as chemical 'fingerprints' in concert with ultrastructural analyses, bacterial identifications and biochemical characterizations. The application of such techniques reveals the contrast between larger-scale and bench-scale reactors (see Chapters 19 and 22–24).
- Characterization of methanogens such as the acetate-utilizing *Methanosarcina mazei* has revealed an interesting life cycle as either a clumped cell mass (or 'cyst') or as disaggregated cells. A 'disaggretase' enzyme appears to be involved in these diverse morphologies (see Chapters 18 and 23).
- Cellulolytic organisms have been isolated from anaerobic digesters, one cellulase enzyme system has been characterized and the potential for cellulases to increase methane yield was demonstrated. Also the basis for using genetic engineering approaches for accelerating hydrolysis has been established (see Chapters 19 and 21).

Other topics now included or being expanded in the research program include: (1) applications of advances in genetic engineering to improvements in plant chemical composition and morphology in grasses, (2) use of gene amplification techniques to modify important metabolic pathways (e.g. nitrogen metabolism) in biomass species, (3) elucidation of the degradative steps using nuclear magnetic resonance spectroscopy and regulation of lignocellulose conversion to unit sugars using biochemical probes, (4) expression of modified genes, preferably those affecting hydrolysis in reactors and those affecting stress tolerance in biomass crops, (5) understanding the biochemistry of volatile acids, hydrogen and other important metabolites in the biomass to methane sequence, and (6) use of new detection/monitoring techniques applied to anaerobic digestion and plant growth such as advanced microbial analyses, biosensors and monoclonal antibody techniques.

Research thrusts in other GRI biomass programs and applied and fundamental research projects in IFAS will help with the problems listed above and support the program with scientific developments on important issues such as: (1) water stress effects on plant growth, (2) long-term environmental consideration, (3) improved production of short-rotation woody plants in north temperate regions, (4) biological pretreatment during storage, (5) gene identification, transformation and expression, (6) advanced understanding of important plant and microbial physiological processes, and (7) engineering advances that incorporate considerations of operation and design of solid-state anaerobic digesters.

SUMMARY

The strategy undertaken by the program was designed with the uncertainty concerning gas costs, supply and markets in mind. If a need for gas from this supplemental source does not arise in the next 15 years, major improvements will emerge that could reduce the costs and increase the feasibility of methane from biomass. Should a supply shortfall become an issue due to decreased exploration and production, a core team including engineers and scientists familiar with the technology will be in place allowing a more rapid response time for implementation. At the same time, this multidisciplinary team approach is leading to the development of new technology which can be applied to numerous industry needs including the gas and agricultural industries. For example, the phospholipid chemical fingerprinting methods are already seeing application to research areas of interest to the natural gas industry such as biocorrosion, environmental cleanup, and new potential markets for methane associated with methane use in groundwater treatment. The development of tissue culture propagation and other production technologies will be widely applicable to other agricultural enterprises. Therefore the synergistic efforts of the 5-year program in the use of biomass to produce methane are already significant. Expanding the application of biotechnologies and advanced computer control and analytical techniques should further strengthen the program.

At present, the energy industry appears to be in a period of significant uncertainty characterized by cycles in energy cost and supply while the agricultural industry is being challenged by the overproduction of the crops now available for farmers to grow. Biomass remains one of the few alternatives as a renewable source of gas and new cropping opportunities. The technology of methane from biomass appears amenable to cost reduction and improvement. With a sustained research effort the gas industry should be more prepared than in previous periods for the upcoming energy cycles to develop new supply technology when needed. Also, technology for producing a commercially important energy product (methane) could provide relief to the chronic overproduction in the agriculture industry by providing new crops and an alternative market for crops from farms and forests.

REFERENCES

GRI (1983). Guidelines for the evaluation of commercial fossil fuels gasification concepts (GRI/1983-0003). Gas Research Institute (GRI), Chicago, Illinois.

Hayes, T. D. and Isaacson, H. R. (1986). Advanced concepts for methane enrichment in anaerobic digestion. In: *AICHE Conference Proceedings of the 21st Intersociety Energy Conversion Engineering Conference (IECEC) Vol. 1.* San Diego, California, pp. 205–12. American Chemical Society, Washington, DC.

Warren, C. S., Bruderly, D. E., Angelieri, M., Bilello, L. J., Bucalo, S., Finger, G. W., Hart, R., Hinton, S. W., Newman, J. R. and Vincant, J. W. (1984). Evaluation of the Lake Apopka Natural Gas District. Gas Research Institute Topical Report No. 84-0015.1 (Document No. PB84-184969). National Technical Information Service, Springfield, Virginia.

Appendix A

List of GRI/IFAS Collaborative Program Publications, 1981–1986

Aldrich, H. C. (1986). Ultrastructure of *Methanosarcina mazei*. *Syst. Appl. Microbiology* **7**, 314–19.
Aldrich, H. C., Bleiweis, A. S., Smith, P. H., Robinson, B. W., Hurst, S. F., Mah, B. A. and Liu, I. (1985). Implications of the structure and metabolism of *Methanosarcina mazei* to anaerobic digestion. 1984 International Gas Research Conference, pp. 551–60. Government Institutes, Inc., Rockville, MD.
Arbelaez, J. A., Koopman, B. and Lincoln, E. P. (1983). Effects of dissolved oxygen and mixing on algal autoflotation. *Journal of the Water Pollution Control Federation* **55**, 1075–9.
Asokan, M. P., O'Hair, S. K. and Litz, R. E. (1983). *In vitro* plant development from bulbil explants of two *Dioscorea* species. *HortScience* **18**, 702–3.
Asokan, M. P., O'Hair, S. K. and Litz, R. E. (1984). Rapid multiplication of *Xanthosoma caracu* by *in vitro* shoot tip culture. *HortScience* **19**, 885–6.
Asokan, M. P., O'Hair, S. K. and Litz, R. E. (1984). *In vitro* plant regeneration from corn callus of *Amorphophallus rivieri* durieu. *Scientia Horticulturae* **24**, 251–6.
Asokan, M. P., O'Hair, S. K. and Litz, R. E. (1984). *In vitro* plant regeneration of Hausa Potato (*Coleus parviflorus*). *HortScience* **19**, 75–6.
Asokan, M. P., O'Hair, S. K. and Litz, R. E. (1985). *In vitro* responses of cocoyam *Xanthosoma caracu* lamina discs. TREC Research Report SB85-1. University of Florida, Homestead.
Bagnall, L. O. (1982). Bulk mechanical properties of waterhyacinth. *Aquatic Plant Management* **20**, 49–53.
Bagnall, L. O. (1983). Development of a water hyacinth biomass combine. ASAE Paper 83-5038.
Bagnall, L. O. (1984). Hydrodynamic characteristics of waterhyacinth plants. ASAE Paper 84-5030.
Bagnall, L. O. (1985). Harvesters: a new challenge for equipment designers. *Agricultural Engineering* **66**(9), 16.
Bagnall, L. O. (1986). Harvesting systems for aquatic biomass. In: *Biomass Energy Development*, Smith, W. H. (ed.), pp. 259–73. Plenum Publishing Corporation, New York.

Bagnall, L. O. and Lincoln, E. P. (1982). Feed and energy from swine waste aquaculture. ASAE Paper 82-5032.
Baker, C. M., Bryan, H. H., Cantliffe, D. J. and Litz, R. E. (1985). Influence of media on mature carrot (*Daucus carota* 'hicolor 9') somatic embryo development. Abstract. *HortScience* **20**(3), 576.
Bjorndal, K. A. and Moore, J. E. (1986). Prediction of fermentability of biomass feedstocks from chemicals. In: *Biomass Energy Development*, Smith, W. H. (ed.), pp. 447–54. Plenum Publishing Corporation, New York.
Bolte, J. P. (1983). Dynamic modeling and simulation of fixed bed anaerobic reactors. M.S. thesis, University of Florida, Gainesville, FL.
Bolte, J. P., Nordstedt, R. A. and Thomas, M. V. (1984). Mathematical simulation of upflow anaerobic fixed bed reactors. *Transactions of the ASAE* **27**(5), 1483–90.
Botti, C. and Vasil, I. K. (1983). Plant regeneration by somatic embryogenesis from parts of cultured mature embryos of *Pennisetum purpureum* (L.) K. Schum. *Zeitschrift fur Pflanzenphysiologie* **111**, 319–25.
Botti, C. and Vasil, I. K. (1984). Ontogeny of somatic embryos of *Pennisetum americanum*. II. In cultured immature inflorescences. *Canadian Journal of Botany* **62**, 1629–35.
Bowes, G. (1986). Aquatic plant photosynthesis: strategies to enhance carbon gain. *Journal Ecology* (in press).
Bowes, G. and Beer, S. (1987). Physiological plant process: photosynthesis. In: *Proc. Conference on Aquatic Plants for Water Treatment and Resource Recovery*, Magnolia Publishers, Orlando, FL. (in press).
Bowes, G. and Salvucci, M. E. (1984). *Hydrilla*: inducible C-4 type photosynthesis without Kranz anatomy. In: *Proc. the 6th International Congress on Photosynthesis*, Sybesma, C. (ed.), pp. 829–32. Martinus Nijhoff/Dr W. Junk Publishers, Boston, MA.
Brown, R. D. and Gritzali, M. (1984). Microbial enzymes and lignocellulose utilization. In: *Genetic Control of Environmental Pollutants*, Omenn, G. S. and Hollaender, A. (eds.), pp. 239–65. Plenum Publishing Corporation, New York.
Calhoun, D. S. and Prine, G. M. (1985). Response of elephantgrass to harvest interval and method of fertilization in colder subtropics. *Proceedings of the Soil and Crop Science Society of Florida* **44**, 111–15.
Campbell, M. S. F. (1983). Biomass yields of young, heavily stocked slash pine stands in north Florida. M.S. thesis, University of Florida, Gainesville, FL, 56 pp.
Campbell, M. S. F., Comer, C. W., Rockwood, D. L. and Henry, C. (1983). Biomass productivity of slash pine in young, heavily stocked stands. *Proc. 5th Southern Forest Biomass Workshop*, pp. 77–82. USDA Forest Service, Asheville, NC.
Cantliffe, D. J. (1984). Cloning biomass crops. *Sunspeak* **6**(2), 7.
Cantliffe, D. J. (1984). Creating 'artificial' seeds. *IFAS Impact* **1**, 1,2,11.
Chandler, S. F. and Vasil, I. K. (1984). Optimization of plant regeneration from long term embryogenic callus cultures of *Pennisetum purpureum* Schum. (Napiergrass). *Journal of Plant Physiology* **117**, 147–56.
Chandler, S. F. and Vasil, I. K. (1984). Selection and characterization of Mace tolerant cells from embryogenic cultures of *Pennisetum purpureum* Schum. (Napiergrass). *Plant Science Letters* **37**, 157–64.

Chandler, S. F., Rajasekaran, K. and Vasil, I. K. (1985). Large scale propagation of Napiergrass and Giant Napiergrass by tissue culture. 1984 International Gas Research Conference, pp. 561–6. Government Institutes, Inc., Rockville, MD.

Comer, C. W. and Rockwood, D. L. (1984). Screening of *Eucalyptus* species for coppice productivity. *Proc. 6th Southern Forest Biomass Workshop*, pp. 95–97. USDA Forest Service, Asheville, NC.

Comer, C. W., Rockwood, D. L. and Riekerk, H. (1984). Frost tolerant *Eucalyptus* for North Florida. *Proc. IUFRO, CSIRO, and AFOCEL Joint Conf. on Frost Resistant Eucalyptus*, pp. 377–382. AFOCEL, Nantes, France.

Comer, C. W., Conde, L. F., Rockwood, D. L. and Geary, T. F. (1986). *Casuarina* cultural improvement performance in Florida. *Nitrogen Fixing Tree Research Reports* 4, 53–6.

Csizinszky, A. A. and Gilreath, J. P. (1985). The utilization of residual nutrients in fallow vegetable fields for the production of herbaceous phytomass in West Central Florida. *Proceedings of the Soil and Crop Science Society of Florida* 44, 223–27.

Dangler, J. M. (1982). Root and top yields of cassava and sweet potato as effected by cultivar, fertilizer rate, cowpea intercrop or time of harvest. M.S. thesis, University of Florida, Gainesville, FL.

Dangler, J. M. and Locascio, S. J. (1984). Sweet potato for biomass. *Biomass* 4, 253–61.

Dangler, J. M., Locascio, S. J. and O'Hair, S. K. (1985). Biomass production by cassava as affected by fertilizer and cowpea intercrop. *Proceedings of the Soil and Crop Science Society Florida* 44, 116–18.

DeBusk, W. F., Reddy, K. R. and Tucker, J. C. (1986). Management strategies for water hyacinth production in a nutrient-limited system. In: *Biomass Energy Development*, Smith, W. H. (ed.), pp. 275–86. Plenum Publishing Corporation, New York.

Dehgan, B. and Wang, S. C. (1983). Evaluation of hydrocarbon plants suitable for cultivation in Florida. *Proceedings of the Soil and Crop Science Society of Florida* 42, 17–19.

Dippon, D. R., Rockwood, D. L. and Comer, C. W. (1985). Slash and sand pine intensive short rotation culture, economic energy feedstock? *Proc. 7th Southern Forest Biomass Workshop*, pp. 52–7. Institute of Food and Agricultural Sciences, University of Florida, Gainesville, FL.

Dippon, D. R., Rockwood, D. L. and Comer, C. W. (1986). Cost sensitivity analysis of *Eucalyptus grandis* woody biomass systems. In: *Biomass Energy Development*, Smith, W. H. (ed.), pp. 143–56. Plenum Publishing Corporation, New York.

Dowling, N. J. E., Widdel, F. and White, D. C. (1986). Analysis of phospholipid ester-linked fatty acid biomarkers of acetate-oxidizing sulfate reducers and other sulfide forming bacteria. *General Microbiology* (in press).

Dunavin, L. S. (1983). Cool season forage trials, 1982–83. Jay, Florida. AREC Res. Rept. WF83-3. IFAS, University of Florida, Gainesville, FL.

Dunavin, L. S. (1985). Cool-season forage trials, 1984–85. Jay, Florida. AREC Res. Rept. WF85-4. IFAS, University of Florida, Gainesville, FL.

Earle, J. F. K., Koopman, B. and Lincoln, E. P. (1984). Role of purple sulfur bacteria in swine waste reclamation. *Agricultural Wastes* 10, 297–312.

Edlund, A., Nichols, P. D., Roffey, R. and White, D. C. (1985). Extractable and lipopolysaccharide fatty acid and hydroxy acid profiles from *Desulfovibrio* species. *Lipid Research* **26**, 982–8.

Ferguson, T. J. and Mah, R. A. (1983). Effect of H_2-CO_2 on methanogenesis from acetate or methanol in *Methanosarcina* spp. *Applied and Environmental Microbiology* **46**, 348–55.

Findlay, R. H. and White, D. C. (1983). Polymeric beta-hydroxy-alkanoates from environmental samples and *Bacillus megaterium*. *Applied and Environmental Microbiology* **45**, 71–8.

Fisher, R. F. and Mollitor, A. V. (1982). Eucalyptus responds dramatically to fertilization. Agronomy Abstracts, p. 264. American Society of Agronomy, Madison, WI.

Frampton, L. J., Jr and Rockwood, D. L. (1983). Genetic variation in biomass traits of sand and slash pines. *Silvae Genetica* **319**(2–3), 18–23.

Frank, J. R., Hayes, T. D. and Smith, P. H. (1985). The production of methane from biomass in the United States: economics, tradeoffs and prospects. In: *Energy from Biomass*, Palz, W., Coombs, J. and Hall, D. O. (eds.), pp. 484–8. Elsevier Applied Science, London.

Freedman, D., Koopman, B. and Lincoln, E. P. (1983). Chemical and biological flocculation of purple sulfur bacteria in anaerobic lagoon effluent. *Journal of Agricultural Engineering Research* **28**, 115–25.

Gehron, M. J. and White, D. C. (1983). Sensitive assay of phospholipid glycerol in environmental samples. *Journal of Microbiology Methods* **1**, 23–32.

Gilreath, J. P. (1985). Description of a BASIC program for data collection using a portable microcomputer. *HortScience* **20**(2), 301.

Gilreath, J. P. (1985). Formatplus, a microcomputer program for conversion of data files created using GATHER to files structured for SAS. *HortScience* **20** (4), 773.

Gilreath, J. P. (1986). Effect of plant population on biomass production by six weed species. In: *Biomass Energy Development*, Smith, W. H. (ed.), pp. 207–16. Plenum Publishing Corporation, New York.

Gilreath, J. P. (1986). Development of a production system for weeds as biomass crops. *Biomass* **9**(2), 135–44.

Gilreath, J. P. and Jones, J. B. (1985). White rust epiphytotic on *Amaranthus hybridus* in south Florida incited by *Albugo bliti*. *Plant Disease* **69**, 542.

Gilreath, J. P., Pitman, W. D. and Rockwood, D. L. (1983). Production of nonconventional crops for methane feedstocks. *Proc. of 1983 International Gas Research Conference*, pp. 340–50. Government Institutes, Inc., Rockville, MD.

Goddard, R. E. and Rockwood, D. L. (1983). Cooperative forest genetic research program 25th progress report. Research Report No. 34. School of Forest Resources and Conservation, IFAS, University of Florida. 18 pp.

Gonzalez, A., Miller, D. and Lee, H. (1983). Selection of optimum biomass gasifier operating conditions to achieve intrinsic kinetics. *Energy from Biomass and Wastes VII*, pp. 455–72. Institute of Gas Technology, Chicago. IL.

Gritzali, M., Chirico, W. J. and Brown, R. D., Jr (1982). Cellulose: utilization via enzymatic processes. XIth International Carbohydrate Symposium, p. V-12. International Union of Biochemistry, Vancouver, Canada.

Guckert, J. B., Antworth, C. P., Nichols, P. D. and White, D. C. (1985). Phospholipid, ester-linked fatty acid profiles as reproducible assays for changes

in prokaryotic community structure of estuarine sediments. *FEMS Microbiology Ecology* **31**, 147–58.
Habig, C. and Ryther, J. H. (1983). Methane production from the anaerobic digestion of some marine macrophytes. *Resources and Conservation* **8**, 271–9.
Habig, C. and Ryther, J. H. (1984). Some correlations between substrate composition and biogas yields. *Energy from Biomass and Wastes VIII*, pp. 817–32. Institute of Gas Technology, Chicago, IL.
Habig, C., Andrews, D. A. and Ryther, J. H. (1984). Nitrogen recycling and methane production using *Gracilaria tikvahiae*: a closed system approach. *Resources and Conservation* **10**, 303–13.
Habig, C., DeBusk, T. A. and Ryther, J. H. (1984). The effect of nitrogen content on methane production by the marine algae *Gracilaria tikvahiae* and *Ulva* sp. *Biomass* **4**, 239–51.
Hanisak, M. D. and Ryther, J. H. (1984). Cultivation biology of *Gracilaria tikvahiae* in the United States. *Hydrobiologia* **116/117**, 295–8.
Hanisak, M. D. and Ryther, J. H. (1986). The experimental cultivation of the red seaweed *Gracilaria tikvahiae* as an 'energy crop': an overview. Barclay, W. R. and McIntosh, R. P. (eds.). In: *Algal Biomass Technologies, Beihefte Zur Nova Hedwigia* **83**, 212–17.
Haydu, A. and Vasil, I. K. (1981). Somatic embryogenesis and plant regeneration from leaf tissues and anthers of *Pennisetum purpureum* Schum. *Theoretical Applied Genetics* **59**, 267–73.
Hayes, T. D., Chynoweth, D. P., Reddy, K. R. and Schwegler, B. (1984). The integration of biogas production with wastewater treatment. In: *Energy Applications of Biomass*, Lowenstein, M. Z. (ed.), pp. 189–99. Elsevier Applied Science, London.
Hayes, T. D., Biljetina, R., Reddy, K. and Chynoweth, D. P. (1986). An integrated wastewater energy production system. *Energy from Biomass and Wastes X*. Institute of Gas Technology, Washington, D.C. (in press).
Henson, J. M. (1983). Isolation of butyrate-utilizing bacteria from thermophilic and mesophilic methane-producing ecosystems. Ph.D. dissertation, University of Florida, Gainesville, FL.
Henson, J. M. and Smith, P. H. (1982). Utilization of propionate and butyrate in methanogenic ecosystems. *Proc. Southeastern and South Carolina Branches ASM*, p. 25. American Society for Microbiology, Washington, DC.
Henson, J. M. and Smith, P. H. (1983). Isolation of an anaerobic thermophilic butyrate-utilizing bacterium in coculture with *Methanobacterium thermoautotrophicum*. Abstracts of Annual Meeting of American Society of Microbiology, p. 147. Washington, DC.
Henson, J. M. and Smith, P. H. (1984). Isolation of a thermophilic butyrate-utilizing bacterium in coculture with methanogenic bacteria. Abstracts of Annual Meeting of American Society for Microbiology, p. 126. Washington, DC.
Henson, J. M. and Smith, P. H. (1985). Isolation of a butyrate-utilizing bacterium in coculture with *Methanobacterium thermoautotrophicum* from a thermophilic digester. *Applied and Environmental Microbiology* **49**, 1461–6.
Henson, J. M., Smith, P. H. and White, D. C. (1985). Examination of thermophilic methane-producing digesters by analysis of bacterial lipids. *Applied and*

Environmental Microbiology **50**, 1428–33.

Henson, J. M., Bordeaux, F. M. and Smith, P. H. (1985). Effects of the addition of propionate and butyrate to thermophilic methane-producing digesters. Biotechnological Advances in Processing Municipal Wastes for Fuels and Chemicals, *Proc. of the First Symposium Argonne National Laboratory*, pp. 235–46.

Henson, J. M., Bordeaux, F. M., Rivard, C. R. and Smith, P. H. (1986). Quantitative influence of butyrate and propionate on thermophilic production of methane from biomass. *Applied and Environmental Microbiology* **51**, 288–92.

Hetrick, V. R. and Portier, K. M. (1982). CSIMS—Comprehensive Scientific Information Management System: a tool for managing information in biomass programs. In: *Energy from Biomass*, Strub, A., Chartier, P. and Schleser, G. (eds.), pp. 666–70. Elsevier Applied Science, London.

Hodgson, L. M. (1984). Tolerance of desiccation by the subtidal red alga *Gracilaria tikvahiae* rhodophyta. *Journal of Phycology* **20**, 444–6.

Holoday, A. S., Salvucci, M. E. and Bowes, G. (1983). Variable photosynthesis/photorespiration ratios in *Hydrilla* and other submersed aquatic macrophyte species. *Canadian Journal of Botany* **61**, 229–36.

Hurst, S. F., Robinson, R. W. and Bleiweis, A. S. (1982). Cell wall composition of *Methanococcus mazei*. *Abstracts of American Society of Microbiology* **82**, J10.

Kalmbacher, R. S., Martin, F. G. and Mislevy, P. (1985). Fermentation substrate and forage from south Florida cropping sequences. *Biomass* **7**(1), 1–11.

Karlsson, S. B. and Vasil, I. K. (1986). Morphology and ultrastructure of embryogenic cell suspension cultures of *Panicum maximum* Jacq. (Guinea grass) and *Pennisetum purpureum* Schum. (Napiergrass). *American Journal of Botany* **73**, 894–901.

Karlsson, S. B. and Vasil, I. K. (1986). Growth, cytology and flow cytometry of embryogenic cell suspension cultures of *Panicum maximum* Jacq. (Guinea grass) and *Pennisetum purpureum* Schum. (Napiergrass). *Journal of Plant Physiology* **123**, 211–27.

Koopman, B. and Lincoln, E. P. (1983). Autoflotation harvesting of algae from high-rate pond effluents. *Agricultural Wastes* **5**, 231–46.

Koopman, B., Lincoln, E. P. and Nordstedt, R. A. (1981). Anaerobic-photosynthetic reclamation of swine waste. *Proc. Water Reuse Symposium II*, pp. 924–34. AWWA Research Foundation, Washington, D.C.

Kossuth, S. V., Roberts, D. R., Huffman, J. B. and Wang, S. (1982). Resin acid, turpentine, and caloric content of paraquat-treated slash pine. *Canadian Journal of Forest Research* **12** (3), 489–92.

Kossuth, S. V., Roberts, D. R., Huffman, J. B. and Wang, S. (1984). Energy value of paraquat treated and resin-soaked loblolly pine. *Wood and Fiber Science* **16**(3), 398–402.

Lapointe, B. E. (1986). Strategies for pulse nutrient supply to *Gracilaria* cultures in the Florida Keys: interactions between concentration and frequency of nutrient pulses. *Journal of Experimental Marine Biology & Ecology* **92**, 211–22.

Lapointe, B. E. (1986). Phosphorus-limited photosynthesis and growth of *Sargassum natans* and *Sargassum fluitans* (Phaeophyceae) in the Western North Atlantic. *Deep-Sea Research* **33**, 391–9.

Lapointe, B. E. and Miller, P. K. (1986). *In situ* nutrient limited photosynthesis and

growth of a red seaweed in the Florida Keys: nitrogen and phosphorus. *Limnology and Oceanography* (in press).
Latorre, C., Lee, J. H., Spiller, H. and Shanmugam, K. T. (1986). Ammonium ion-excreting cyanobacterial mutant as a source of nitrogen for growth of rice: a feasibility study. *Biotechnology Letters* **8**, 507–12.
Lee, J. H., Patel, P., Sankar, P. and Shanmugam, K. T. (1985). Isolation and characterization of mutant strains of *Escherichia coli* altered in H_2 metabolism. *Journal of Bacteriology* **162**, 344–52.
Le Grand, F. (1983). Pre-treatment method for specific crops to increase reducing sugars content for potential methane production. *Proc. 3rd Annual Solar and Biomass Workshop*, Atlanta, GA, pp. 310–13. USDA, Washington, DC.
Lincoln, E. P. (1985). Resource recovery with microalgae. *Arch. Hydrobiol. Beih. Ergebn. Limnology* **20**, 25–34.
Lincoln, E. P. and Koopman, B. (1982). Algal culture as an adjunct to livestock production. *Proc. Seminario Econtelnologico Para el Desarrollo de Mexico*, pp. 38–43. Programa MAB-UNESCO, Mexico DF.
Lincoln, E. P., Hall, T. W. and Koopman, B. (1983). Zooplankton control in mass algal cultures. *Aquaculture* **32**, 331–7.
Lincoln, E. P., Koopman, B., Bagnall, L. O. and Nordstedt, R. A. (1986). Aquatic system for fuel and food production from livestock wastes. *Journal of Agricultural Engineering Research* **33**, 159–69.
Little, M. C. and Preston, J. F. III (1984). The fluorimetric of amatoxins, phallotoxins, and other peptides in *Amanita suballiacea*. *Journal of Natural Products* **47**(1) 93–9.
Little, M. C. and Preston, J. F. III (1985). Plant physiology, sensitivity of carrot cultures and RNA polymerase II to amatoxins: evidence for the inactivation of 5'-hydroxyamatoxins. *Plant Physiology* **77**, 443–9.
Little, M. C. and Preston, J. F. III (1986). The improved synthesis and purification of methylated amanitins using Diazomethane. *International Journal of Peptide Protein Research* (in press).
Liu, J. R. and Cantliffe, D. J. (1983). Somatic embryogenesis and plant regeneration in tissue cultures of sweet potato (*Ipomoea batatas* Poir.). *HortScience* **18**, 618.
Liu, J. R. and Cantliffe, D. J. (1984). Improved efficiency of somatic embryogenesis and plant regeneration in tissue cultures of sweet potato (*Ipomoea batatas* Poir.). *HortScience* **19**(3), 589.
Liu, J. R. and Cantliffe, D. J. (1984). Tissue culture propagation development and its application to energy crops. 1984 International Gas Research Conference, pp. 622–9. Government Institutes, Inc., Rockville, MD.
Liu, J. R. and Cantliffe, D. J. (1984). Somatic embryogenesis and regeneration in tissue cultures of sweet potato (*Ipomoea batatas* Poir.). *Plant Cell Reports* **3**, 112–15.
Liu, Y., Boone, D. R., Sleat, R. and Mah, R. A. (1985). *Methanosarcina mazei* LYC, a new methanogenic isolate which produces a disaggregating enzyme. *Applied and Environmental Microbiology* **49**, 608–13.
Locascio, S. J. and Dangler, J. M. (1986). Starch and mineral nutrient accumulation by sweet potato cultivars. In: *Biomass Energy Development*, Smith, W. H. (ed.), pp. 197–205. Plenum Publishing Corporation, New York.

Lorber, M. N., Fluck, R. and Mishoe, J. W. (1982). Analysis of sugarcane biomass production systems. *Transactions of the ASAE*. ASAE Paper No. 82-3088.

Lorber, M. N., Mishoe, J. W. and Reddy, K. R. (1983). Modeling and analysis of water hyacinth. ASAE Paper No. 83-3023.

Lorber, M. N., Fluck, R. C. and Mishoe, J. W. (1984). A method for analysis of sugarcane (*Saccharum* sp.) biomass production systems. *Transactions of the ASAE* **27**(1), 146–152, 158.

Lorber, M. N., Mishoe, J. W. and Reddy, K. R. (1984). Modeling and analysis of water hyacinth biomass. *Ecological Modeling* **24**, 61–77.

Mah, R. A. and Kuhn, D. A. (1984). Transfer of the type species of the genus *Methanococcus* to the genus *Methanosarcina*, naming it *Methanosarcina mazei* (Barker 1936) comb. nob. et emend. and conservation of the genus *Methanococcus* (Approved Lists 1980) with *Methanococcus vannielii* (Approved Lists 1980) as the type species. *International Journal of Systematic Bacteriology* **34**, 263–5.

Mancuso, C. A., Nichols, P. D. and White, D. C. (1986). A method for the separation and characterization of Archaebacterial signature ether lipids. *Journal of Lipid Research* **27**, 49–56.

Martz, R. F., Sebacher, D. L. and White, D. C. (1983). Biomass measurement of methane-forming bacteria in environmental samples. *Journal of Microbiology Methods* **1**, 53–61.

McCluskey, D. N. (1983). The influence of inoculum source and fertilization on root nodule formation in *Casuarina glauca*. M.S. thesis, University of Florida, Gainesville, FL.

McCluskey, D. N. and Fisher, R. F. (1983). The effect of inoculum source on nodulation in *Casuarina glauca*. *Commonwealth Forest Review* **62**(2), 117–24.

McSorley, R., O'Hair, S. K. and Parrado, J. L. (1983). Nematodes associated with the edible aroid genera *Xanthosoma* and *Colocasia*. *Nematropica* **13**, 165–80.

McSorley, R., O'Hair, S. K. and Parrado, J. L. (1983). Nematodes of cassava, *Manihot esculenta* crantz. *Nematropica* **13**, 261–87.

Mishoe, J. W. (1985). Integration and assessment of biomass research information by use of system simulation. In: *Energy From Biomass*, Palz, W., Coombs, J. and Hall, D. O. (eds.), pp. 1020-4. Elsevier Applied Science, London.

Mishoe, J. W., Lorber, M. N., Peart, R. M., Fluck, R. C. and Jones, J. W. (1983). Analysis of biomass production systems. ASAE Paper No. 83-3019.

Mishoe, J. W., Lorber, M. N., Peart, R. M., Fluck, R. C. and Jones, J. W. (1983). Modeling and analysis of biomass production systems. Summary of Southern Biomass Energy Research Conference. Tuscaloosa, AL.

Mishoe, J. W., Lorber, M. N., Fluck, R. C., Kirmse, D. and Jones, J. W. (1983). BIOMET: Biomass to methane system simulation. Model Description and Users' Guide. Agricultural Engineering Department, IFAS, University of Florida, Gainesville, FL.

Mishoe, J. W., Lorber, M. N., Peart, R. M., Fluck, R. C. and Jones, J. W. (1984). Modeling and analysis of biomass production systems. *Biomass* **6**(1&2), 119–30.

Mislevy, P., Dantzman, C. L., Prevatt, J. W., Overman, A. J., Horton, G. M. J. and Johnson, F. A. (1982). Forage production and utilization from a south Florida multicropping system. Bulletin No. 830. IFAS, University of Florida, Gainesville, FL. 37 pp.

Mislevy, P., Overman, A. J. and Dantzman, C. L. (1983). Seasonal forage yield and quality as affected by crop sequence. *Proceedings of the Soil and Crop Science Society of Florida* **42**, 99–103.

Moore, J. E. and Bjorndal, K. A. (1983). Evaluation of near infra-red reflectance spectroscopy for estimating the composition and quality of tropical grass hays. 1983 Beef Cattle Research Report. Animal Science Department, IFAS, University of Florida, Gainesville, FL.

Moore, J. E., Kunkle, W. E., Bjorndal, K. A., Sand, R. S., Chambliss, C. G. and Mislevy, P. (1984). Extension forage testing program utilizing near infra-red reflectance spectroscopy. *Proc. Forage and Grassland Conference*, pp. 41–52. American Forage and Grassland Council, Houston, TX.

Nassar, N. M. A. and O'Hair, S. K. (1986). Variation among cassava clones in relation to seed germination. *Industrial Journal of Genetics and Plant Breeding* (in press).

Nichols, P. D., Henson, J. M., Guckert, J. B., Nivens, D. E. and White, D. C. (1985). Fourier transform-infrared spectroscopic methods for microbial ecology: analysis of bacteria, bacteria-polymer mixtures and biofilms. *Journal of Microbiology Methods* **4**, 79–94.

Nordstedt, R. A. (1981). Anaerobic digestion of dairy farm wastes for methane generation. *Southeastern Dairy Review* **18**(12), 12.

Nordstedt, R. A. (1986). Experience with wood media for anaerobic fixed bed reactors. Abstract. In: *Biomass Energy Development*, Smith, W. H. (ed.), p. 647. Plenum Publishing Corporation, New York.

Nordstedt, R. A. and Young, D. T. (1981). Anaerobic digestion system for flushed swine waste. ASAE Paper No. SER-91-005.

Nordstedt, R. A. and Thomas, M. V. (1983). Wood media for anaerobic fixed bed reactors. ASAE Paper No. 83-4051.

Nordstedt, R. A. and Thomas, M. V. (1984). Inoculum requirements for startup of anaerobic fixed bed reactors. ASAE Paper No. 84-4090. 31 pp.

Nordstedt, R. A. and Thomas, M. V. (1985). Startup characteristics of anaerobic fixed bed reactors. *Transactions of the ASAE* **28**(4), 1242–7, 1252.

Nordestedt, R. A. and Thomas, M. V. (1985). Wood block media for anaerobic fixed bed reactors. *Transactions of the ASAE* **28**(6), 1990–6.

Nordstedt, R. A., Thomas, M. V. and Chou, C. Y. (1987). Fixed film reactors: packing media, mathematical models and bulk flow in packed beds. In: *Methane from Biomass—A Systems Approach*, Smith, W. H. and Frank, J. R. (eds.). Elsevier Applied Science, London, pp. 431–43.

Ogwada, R. A., Reddy, K. R. and Graetz, D. A. (1984). Effects of aeration and temperature on nutrient regeneration from selected aquatic macrophytes. *Journal of Environmental Quality* **13**, 239–43.

O'Hair, S. K. (1982). Root crop evaluation, selection and improvement in Florida for energy applications. *Energy from Biomass and Wastes VI*, pp. 135–65. Institute of Gas Technology, Chicago, IL.

O'Hair, S. K. (1983). Breeding goals. In: Breeding New Sweet Potatoes for the Tropics, *Proceedings of the American Society of Horticultural Science, Tropical Region*, Vol, 27B, Martin, F. W. (ed.), pp. 139–40.

O'Hair, S. K. (1984). Farinaceous crops. In: *CRC Handbook of Tropical Food Crops*, Martin, F. W. (ed.), p. 304. CRC Press, Boca Raton, FL.

O'Hair, S. K. (1984). Cassava root qualities and yield in a subtropical environment. In: *Root Crops in the Caribbean*, Proc. *Caribbean Regional Workshop on Tropical Root Crops*, Dolly, D. (ed.), pp. 161–6. Faculty of Agriculture, University of West Indies, St. Augustine, Trinidad.

O'Hair, S. K. (1985). Potato growth in the subtropics of Florida. *American Potato Journal* **62**, 391–401.

O'Hair, S. K. (1986). Edible aroids: taxonomy and horticulture. *Horticultural Reviews* **6**, 43–99.

O'Hair, S. K. and Snyder, G. H. (1983). Taro and cocoyam production in Florida. *Proc. Taro and Other Aroids for Food, Feed and Fuel Conference*, IFAS, University of Florida, Center for Tropical Agriculture, 35 pp.

O'Hair, S. K., Dangler, J. M., Everett, P. H., Forbes, R. B., Halsey, L. H., Locascio, S. J., Ozaki, H., Rich, J. R., Stanley, R. L., Trafford, H. J. and White, J. M. (1981). Location, growing season and soil type effects on Florida cassava yields. *HortScience* **16**(3), 98.

O'Hair, S. K., Snyder, G. H. and Morton, J. F. (1982). Wetland taro: a neglected crop for food, feed and fuel. *Proceedings of the Florida State Horticultural Society* **96**, 367–74.

O'Hair, S. K., Forbes, R. B., Locascio, S. J., Rich, J. R. and Stanley, R. L. (1983). Starch and glucose distribution within cassava roots as affected by cultivar and location. *HortScience* **18**, 735–7.

O'Hair, S. K., Asokan, M. P., Litz, R. E. and Bryan, H. H. (1983). Fluid drilling of somatic potato embryos as a means of crop establishment. In: *Proc. Research for the Potato in The Year 2000 Congress*, Hooker, W. J. (ed.), pp. 144–6. International Potato Center, Lima, Peru.

O'Hair, S. K., Locascio, S. J., Forbes, R. B., White, J. M., Hensel, D. R., Shumaker, J. R. and Dangler, J. M. (1983). Root crops and their biomass potential in Florida. *Proceedings of the Soil and Crop Science Society of Florida* **42**, 13–17.

O'Hair, S. K., McSorley, R., Parrado, J. and Matthews, R. F. (1983). The production and qualities of Cuban sweetpotato cultivars in Florida. In: Breeding New Sweet Potatoes for the Tropics, *Proceedings of the American Society of Horticultural Science, Tropical Region*, Vol. 27B, Martin, F. W. (ed.), pp. 35–41.

O'Hair, S. K., Volin, R. B. and Asokan, M. P. (1984). Starch distribution in cocoyam *Xanthosoma spp.* corms and cormels. *Proceedings of the International Society for Tropical Root Crops* **6**, 161–4.

O'Hair, S. K., Dangler, J. M., Everett, P., Forbes, R. B., Locascio, S. J., Olson, S. M., Shumaker, J. R. and White, J. M. (1986). Cruciferous and root crops for year-round biomass production. In: *Biomass Energy Development*, Smith, W. H. (ed.), pp. 173–84. Plenum Publishing Corporation, New York.

Oki, Y., Nakagawa, K. and Reddy, K. R. (1985). Uptake and translocation to 15N in water hyacinth. In: *Proc. 10th Asian-Pacific Weed Sci. Soc. Conf.*, pp. 317–24.

Ottman, A. B. (1984). Evaluation of leucaena for biomass and forage. M.S. thesis, University of Florida, Gainesville, FL, 100 pp.

Ottman, A. B. and Prine, G. M. (1984). Leucaena for biomass in humid subtropics. Agronomy Abstracts. American Society of Agronomy, p. 134.

Ozias-Akins, P., Tabaeizadeh, Z. and Vasil, I. K. (1985). Somatic hybridization in

the Gramineae. Abstract in the First International Congress Plant Molecular Biology, p. 44. University of Georgia, Athens, GA.
Paterek, R. (1983). Ecology of methanogenesis in two hypersaline biocoenoses: Great Salt Lake and a San Francisco Bay saltern. Ph.D. dissertation, University of Florida, Gainesville, FL.
Paterek, R. and Smith, P. H. (1982). Methanogenic activity of sediment and water column samples from Great Salt Lake and a San Francisco Bay saltern. *Proc. Southeastern and South Carolina Branches ASM*, p. 24.
Paterek, R. and Smith, P. H. (1983). Isolation of a halophilic methanogenic bacterium from the sediment of Great Salt Lake and a San Francisco Bay saltern. Abstracts of Annual Meeting of American Society of Microbiology, p. 140.
Paterek, R. and Smith, P. H. (1984). Methanogenesis in two hypersaline ecosystems: Great Salt Lake and a San Francisco Bay saltern. Abstracts of Annual Meeting of American Society for Microbiology, p. 134.
Paterek, R. and Smith, P. H. (1985). Isolation of a halophilic methanogen from Great Salt Lake. *Applied and Environmental Microbiology* **50**, 877–81.
Pena, J. E., Waddill, V. H. and O'Hair, S. K. (1984). Mites attacking cassava in southern Florida: damage descriptions and density estimate methods. *Entomologist* **67**, 141–6.
Phlips, E. J., Willis, M. and Verchick A. (1986). Aspects of nitrogen fixation in *Sargassum* communities off the coast of Florida. *Experimental Marine Biology and Ecology* (submitted for publication).
Polne-Fuller, M. and Gibor, A. (1987). Tissue culture of seaweeds. In: *Seaweed Cultivation for Renewable Resources*, Bird, K. and Benson, P. H. (eds.). Gas Research Institute, Chicago, IL (in press).
Polne-Fuller, M. and Gibor, A. (1987). Callus growth in seaweeds, induction and culture. *Journal of Hydrobiologia* (in press).
Polne-Fuller, M. and Gibor, A. (1987). Calluses, cells, and protoplasts in studies towards genetic improvement of seaweeds. 2nd Int. Symp. Genetic in Aquaculture, Davis, CA (in press).
Polne-Fuller, M., Saga, N. and Gibor, A. (1986). Algal cell, callus, and tissue cultures, and selection of algal strains. *Proc. of the Algal Biomass Symposium*, Golden, CO. *Nova Hedwigia Beiheft*, **83**, 30–6.
Portier, K. M. and Hetrick, V. R. (1982). CSIMS: Comprehensive Scientific Information Management System. *Proc. COMSTAT*. International Association of Statistical Computing, Physicia-Verlag, Vienna.
Preston, J. F. and Jiminez, H. (1986). Quantification of mannitol in *Sargassum* species. *Journal of Economic Botany* (in press).
Preston, J. F., Romeo, T., Bromley, J. C., Robinson, R. W. and Aldrich, H. C. (1985). Alginate lyase secreting bacteria associated with the algal genus, *Sargassum*. *Developments in Industrial Microbiology* **26**, 727-40.
Preston, J. F., Romeo, T., Gibor, A. and Polne-Fuller, M. (1985). Investigations of *Sargassum* species for bioconversion to methane: mannitol levels, temperature requirements, and protoplast formation. 1984 International Gas Research Conference, pp. 567–579. Government Institutes, Rockville, MD.
Preston, J. F., Romeo, T. and Bromley, J. C. (1985). Agarase, alginase, and cellulase activities secreted by marine bacteria associated with the brown algal

genus *Sargassum*. *Abstracts of Society for Industrial Microbiology News* **30**(4), 47.

Preston, J. F., Romeo, T. and Bromley, J. C. (1986). Selective alginate degradation by marine bacteria associated with the algal genus *Sargassum*. *Journal of Industrial Microbiology* (in press).

Prine, G. M. (1985). Leucaena and elephantgrass as energy crops in colder subtropics. *Proc. of the 5th Annual Solar and Biomass Energy Workshop* (Including Wind), pp. 149–52. USDA, Washington, DC.

Prine, G. M. and Mislevy, P. (1983). Grass and herbaceous plants for biomass. *Proceedings of the Soil and Crop Science Society of Florida* **42**, 8–12.

Prine, G. M., Mislevy, P., Shiralipour, A. and Smith, P. H. (1984). Elephantgrass, energy crop for humid subtropics. *Proc. of the 4th Annual Solar and Biomass Energy Workshop*, pp. 124–7. USDA, Washington, DC.

Rajasekaran, K., Hein, M. B., Davis, G. C., Carnes, M. G. and Vasil, I. K. (1985). Endogenous plant growth regulators in leaves and tissue cultures of *Pennisetum purpureum* (Napiergrass). *American Journal of Botany* **72**, 910.

Rajasekaran, K., Schank, S. C. and Vasil, I. K. (1986). Characterization of biomass production, cytology and phenotypes of plants regenerated from embryogenic callus cultures of a *Pennisetum americanum* × *P. purpureum* (hybrid triploid Napiergrass). *Theoretical and Applied Genetics* (in press).

Reddy, K. R. (1983). Fate of nitrogen and phosphorus in a wastewater retention reservoir containing aquatic macrophytes. *Journal of Environmental Quality* **12**, 137–41.

Reddy, K. R. (1984). Nutrient removal potential by aquatic plants. *Aquatics* **6**, 15–16.

Reddy, K. R. (1984). Nutrient transformations in aquatic macrophyte filters used for water purification. *Proc. Water Reuse Symposium III*, Vol. 2, pp. 660–678. AWWA Research Foundation, San Diego, CA.

Reddy, K. R. (1984). Water hyacinth biomass production in Florida. *Biomass* **6**, 167–81.

Reddy, K. R. (1986). Nitrogen utilization by *Typha latifolia* L. as affected by temperature and rate of labeled nitrogen application. *Aquatic Botany* (in press).

Reddy, K. R. and Tucker, J. C. (1983). Productivity and nutrient uptake of water hyacinth (*Eichhornia crassipes* [Mart] Solms): I. Effect of nitrogen source. *Journal of Economic Botany* **37**, 237–47.

Reddy, K. R. and DeBusk, W. F. (1984). Phosphorus removal by *Azolla caroliniana* cultured in nutrient enriched waters. Symp Proc. Practical Application of Azolla for Rice Production. Mayaguez, Puerto Rico. *Developments in Plant Science* **13**, 151–62.

Reddy, K. R. and DeBusk, W. F. (1984). Growth characteristics of aquatic macrophytes cultured in nutrient enriched water. I. Water hyacinth, water lettuce, and pennywort. *Journal of Economic Botany* **38**, 229–39.

Reddy, K. R. and DeBusk, W. F. (1985). Growth characteristics of aquatic macrophytes cultured in nutrient enriched water. II. Azolla, duckweed, and salvinia. *Journal of Economic Botany* **39**, 200–8.

Reddy, K. R. and DeBusk, W. F. (1985). Nutrient removal potential of selected aquatic macrophytes. *Journal of Environmental Quality* **14**, 459–62.

Reddy, K. R. and Sutton, D. L. (1984). Water hyacinths for water quality improvement and biomass production. *Journal of Environmental Quality* **13**, 1–8.

Reddy, K. R. and Tucker, J. C. (1985). Growth and nutrient uptake of pennywort (*Hydrocotyle umbellata* L.) as influenced by the nitrogen concentration of water. *Journal of Aquatic Plant Management* **23**, 35–40.

Reddy, K. R., Sacco, P. D., Graetz, D. A., Campbell, K. L. and Portier, K. M. (1983). Effect of aquatic macrophytes on physico-chemical parameters of agricultural drainage water. *Journal of Aquatic Plant Management* **21**, 1–7.

Reddy, K. R., Hueston, F. M. and McKim, T. (1983). Water hyacinth biomass production in sewage effluent. *Energy from Biomass and Wastes VII*, pp. 101–16. Institute of Gas Technology, Chicago, IL.

Reddy, K. R., Sutton, D. L. and Bowes, G. E. (1983). Freshwater aquatic plant biomass production in Florida. *Proceedings of the Soil and Crop Science Society of Florida* **42**, 28–40.

Reddy, K. R., DeBusk, W. F. and Bagnall, L. O. (1984). Water hyacinth biomass production in eutrophic lake water. 1984 International Gas Research Conference, pp. 336–50. Institute of Gas Technology, Chicago, IL.

Reddy, K. R., Hueston, F. M. and McKim, T. (1985). Biomass production and nutrient removal potential of water hyacinth cultured in sewage effluent. *Journal of Solar Engineering* **107**, 128–35.

Reddy, K. V., Rockwood, D. L., Comer, C. W. and Meskimen, G. F. (1985). Predicted genetic gains adjusted for inbreeding for an *Eucalyptus grandis* seed orchard. *Proc. 18th Southern Forest Tree Improvement Committee*, pp. 283–9. Long Beach, MS.

Reddy, K. V., Rockwood, D. L., Comer, C. W. and Meskimen, G. F. (1986). Genetic improvement of *Eucalyptus grandis* for biomass production in Florida. In: *Biomass Energy Development*, Smith, W. H. (ed.), pp. 103–10. Plenum Publishing Corporation, New York.

Reddy, K. C., Soffes, A. R. and Prine, G. M. (1986). Tropical legumes for green manure. I. Nitrogen production and the effects on succeeding crop yields. *Agronomy* **78**(1), 1–4.

Reighard, G. L. and Rockwood, D. L. (1985). Plot subsampling in intensively cultured slash and sand pine tests. *Proc. 7th Southern Forest Biomass Workshop*, pp. 27–30. Institute of Food and Agricultural Sciences, University of Florida, Gainesville, FL.

Reighard, G. L., Rockwood, D. L. and Comer, C. W. (1985). Genetic and cultural factors affecting growth performance of slash pine. *18th Southern Forest Tree Improvement Committee*, pp. 100–106. Long Beach, MS.

Rivard, C. J. (1983). Effects of continuous addition of nitrate to a thermophilic anaerobic digestion system. Ph.D. dissertation, University of Florida, Gainesville, FL.

Rivard, C. J. and Smith, P. H. (1981). Isolation and characterization of a thermophilic marine methanogenic bacterial species. *Bacteriol. Proc.*

Rivard, C. J. and Smith, P. H. (1982). Isolation and characterization of a thermophilic marine methanogenic bacterium, *Methanogenium thermophilicum* sp. nov. *International Journal of Systematic Bacteriology* **32**, 430–6.

Rivard, C. J. and Smith, P. H. (1984). Effects of continuous addition of nitrate to a thermophilic anaerobic digestion system. Abstracts of Annual Meeting of

American Society for Microbiology, p. 190.
Rivard, C. J., Henson, J. M., Thomas, M. V. and Smith, P. H. (1983). Isolation and characterization of a mesophilic marine methanogen, *Methanomicrobium paynteri*. Abstracts of Annual meeting of American Society of Microbiology, p. 140.
Rivard, C. J., Henson, J. M., Thomas, M. V. and Smith, P. H. (1983). Isolation and characterization of *Methanobacterium paynteri* sp. nov. a mesophylic isolate from marine sediments. *Applied and Environmental Microbiology* **64**, 484–90.
Robinson, R. W. and Erdos, G. W. (1985). Immuno-electron microscopic identification of Methanosarcina spp. in anaerobic digester fluid. *Canadian Journal of Microbiology* **31**, 839–44.
Robinson, R. W., Akin, D. E., Nordstedt, R. A., Thomas, M. V. Aldrich, H. C. (1984). Light and electron microscopic examinations of methane-producing biofilms from anaerobic fixed bed reactors. *Applied and Environmental Microbiology* **48**(1), 127–36.
Robinson, R. W., Aldrich, H. C., Hurst, S. F. and Bleiweis, A. S. (1985). Role of the cell surface of *Methanosarcina mazei* in cell aggregation. *Applied and Environmental Microbiology* **49**, 321–7.
Rockwood, D. L. (1983). Alternative designs for progeny testing slash pine. *Proc. 17th Southern Forest Tree Improvement Conference*, pp. 179–185. Athens, GA.
Rockwood, D. L. (1983) Biomass feedstocks: improvement in quality and quantity through genetic selection and breeding. Summary Southeastern Biomass Energy Research Conference, pp. 72–6. Athens, GA.
Rockwood, D. L. (1984). Genetic improvement in biomass quantity and quality. *Biomass* **6** (1 & 2), 37–45.
Rockwood, D. L. (1986). Development of woody biomass cultural systems for Florida. In: *Biomass Energy Development*, Smith, W. H. (ed.), pp. 85–94. Plenum Publishing Corporation, New York.
Rockwood, D. L. and Geary, T. F. (1982). Genetic variation in biomass productivity and coppicing of intensively grown *Eucalyptus Grandis* in southern Florida. *Proc. 7th North American Forest Biology Workshop*, pp. 400–5. University of Kentucky, Lexington, KY.
Rockwood, D. L. and Devalerio, J. T. (1986). Promising species for woody biomass production in warm-humid environments. *Biomass* (in press).
Rockwood, D. L., Geary, T. F. and Bourgeron, P. S. (1982). Planting density and genetic influences on seedling growth and coppicing of Eucalyptus in southern Florida. *Proc. 1982 Southern Forest Biomass Working Group Workshop*, pp. 95–102. Southern Forest Experiment Station, Alexandria, LA.
Rockwood, D. L., Comer, C. W., Dippon, D. R., Huffman, J. B., Riekerk, H. and Wang, S. (1983). Current status of woody biomass production research in Florida. *Proceedings of the Soil and Crop Science Society of Florida* **42**, 19–27.
Rockwood, D. L., Reddy, K. V., Comer, C. W., Webley, O. J., Meskimen, G. F. and Geary, T. F. (1984). Variation in coppicing of eucalypts in southern Florida. Abstract. *Proc. 8th North American Forest Biology Workshop*, p. 147. Society of American Foresters, Loggin, UT.
Rockwood, D. L., Dippon, D. R. and Comer, C. W. (1984). Potential of *Eucalyptus grandis* for biomass production in Florida. *Proc. BioEnergy 84*, **2**, 86–93.

Rockwood, D. L., Comer, C. W., Dippon, D. R. and Huffman, J. B. (1985). Woody biomass production options for Florida. Florida Agriculture Experiment Station Technical Bulletin 856, 29 pp.

Rockwood, D. L., Kellison, R. C., Franklin, E. C. and Meskimen, G. F. (1986). Operational advanced generation improvement programs for minor species in the South. Southern Coop. Series Bulletin, Vol. 309, pp. 27–37.

Roeder, K. R. and Rockwood, D. L. (1982). Potential stem biomass and energy content yields of *Eucalyptus grandis* and *Eucalyptus robusta* in south Florida. Florida Agricultural Experiment Station Technical Bulletin No. 831, 6 pp.

Romeo, T. (1986). Catalytic and structural properties of alginate lyases from bacterial epiphytes of *Sargassum*. Ph.D. Dissertation, University of Florida, Gainesville, FL.

Romeo, T. and Preston, J. F. (1983). Partial purification and substrate specifications of alginase enzymes. Abstract. Southeastern Regional Meetings, American Chemical Society, Vol. 35, p. 85.

Romeo, T. and Preston, J. F. (1984). Metabolic inactivation of amatoxins by non-toxic *Amanita* species. *Experimental Mycology* **8**, 25–36.

Romeo, T. and Preston, J. F. (1986). L.C. analysis of the depolymerization of (1→4)-β-D-manmuronan by an extracellular alginate lyase from a marine bacterium. *Carbohydrate Chemistry* (in press).

Romeo, T. and Preston, J. F. (1986). Purification and structural properties of an extracellular (1→4)-β-D-manmuronan specific alginate lyase from a marine bacterium. *Biochemistry* (in press).

Romeo, T. and Preston, J. F. (1986). Depolymerization of alginate by an extracellular alginate lyase from a marine bacterium: substrate specificity and accumulation of reaction production. *Biochemistry* (in press).

Romeo, T., Bromley, C. and Preston, J. F. (1984). Alginate lyase activities from bacteria of the marine algal genus, *Sargassum*. 84th Annual Meetings of the American Society for Microbiology.

Romeo, T., Bromley, J. C. and Preston, J. F. (1985). Mechanisms of alginate depolymerization by lyases from marine bacteria. Abstracts of the Annual Meeting of the American Society for Microbiology, p. 179.

Romeo, T., Bromley, J. C. and Preston, J. F. (1985). Alginate lyases of varying substrate specificities from marine bacteria. In: *Biomass Energy Development*, Smith, W. H. (ed.), pp. 303–20. Plenum Publishing Corporation, New York.

Ryther, J. H. (1983). Marine biomass research in Florida. *Proceedings of the Soil and Crop Science Society of Florida* **42**, 40–8.

Ryther, J. H. (1986). Marine biomass production. In: *Biomass Energy Development*, Smith, W. H. (ed.), pp. 241–57. Plenum Publishing Corporation, New York.

Ryther, J. H., DeBusk, T. A., Andrews, D. A. and Habig, C. (1983). Cultivation of *Graciliaria* as a biomass source for energy. *Proc. World Mariculture Society Annual Meeting*, Washington, D.C.

Saga, N., Polne-Fuller, M. and Gibor, A. (1986). Protoplasts from seaweeds; production and fusion. In: *Proc. Biotech. of Algal Biomass Production*. University of Colorado Press, Boulder, CO (in press).

Sankar, P., Lee, J. H. and Shanmugam, K. T. (1985). Cloning of hydrogenase genes and fine structure analysis of an operon essential for H_2 metabolism in

Escherichia coli. Journal of Bacteriology **162**, 355–60.
Shih, S. F. and Snyder, G. H. (1984). Leaf area index and dry biomass of taro. *Agronomy Journal* **76**, 750–3.
Shih, S. F. and Snyder, G. H. (1985). Leaf area index and evapotranspiration of taro. *Agronomy Journal* **77**, 554–6.
Shih, S. F., Rahi, G. S. and Snyder, G. H. (1982). Evapotransporation studies on taro in the Everglades. ASAE Paper No. 82-2595.
Shih, S. F., O'Hair, S. K. and Snyder, G. H. (1986). Using leaf dry biomass to improve taro (*Colocasia esculenta*) production systems. *Proc. Inter. Soc. Trop. Root Crops.* (in press).
Shiralipour, A. and Smith, P. H. (1984). Conversion of biomass resources into methane gas. *Biomass* **6**, 85–92.
Shiralipour, A. and Smith, P. H. (1985). Variation in methane production associated with plant resource groups, plant parts and growth conditions. *Proc. of the 5th Annual Solar and Biomass Energy Workshop* (Including Wind), Atlanta, GA, pp. 59–62.
Shiralipour, A., Garrard, L. A. and Haller, W. T. (1981). Nitrogen source, biomass production and phosphorus uptake in water hyacinth. *Journal of Aquatic Plant Management* **19**, 40–3.
Shiralipour, A., Haller, W. T. and Garrard, L. A. (1981). Effects of nitrogen containing sprays on biomass production and phosphorus uptake in water hyacinth. *Journal of Aquatic Plant Management* **19**, 44–7.
Shiraliour, A., Smith, P. H., Bjorndal, K. A. and Moore, J. E. (1983). Screening for conversion of biomass resources into methane gas. In: *Proc. 2nd Southern Biomass Energy Research Conference*, Tuskaloosa, AL pp. 108–11.
Shiralipour, A., Smith, P. H. and Portier, K. M. (1986). Prediction of methane yields from biomass. In: *Biomass Energy Development*, Smith, W. H. (ed.), pp. 439–46. Plenum Publishing Corporation, New York.
Sivakumaran, K. and Bagnall, L. O. (1984). In-situ mechanical properties of water hyacinth. ASAE Paper 84-5029.
Sleat, R., Robinson, R. and Mah, R. A. (1984). Isolation and characterization of an anaerobic, cellulolytic bacterium, *Clostridium cellulovorans*. sp. nov. *Applied and Environmental Microbiology* **48**, 88–93.
Smith, R. L., Schank, S. C., Milam, J. R. and Baltensperger, A. A. (1984). Responses of *Sorghum* and *Pennisetum* species to the N_2-fixing bacterium *Azospirillum brasilense*. *Applied and Environmental Microbiology* **47**, 1331–6.
Smith, W. H. (1982). Biomass: renewable energy for Florida's future. *Sunspeak* **4**, 4.
Smith, W. H. (1983). Energy from biomass: a new commodity. In: *Agriculture in the 21st Century*, Rosenblum, J. W. (ed.), pp. 61–9. John Wiley, New York.
Smith, W. H. (1984). University of Florida's biomass program. *Proc. 4th Annual Solar and Biomass Energy Workshop*, Atlanta, GA, pp. 17–19. USDA, Washington, DC.
Smith, W. H. (1984). Biomass energy programs funded by the state of Florida. *Energy from Biomass and Wastes VIII*, pp. 1495–504. Institute of Gas Technology, Chicago, IL.
Smith, W. H. (1985). The southern United States biomass energy programs with emphasis on Florida. In: *Energy From Biomass*, Palz, W., Coombs, J. and Hall, D. O. (eds.), pp. 484–8. Elsevier Applied Science, London.

Smith, W. H. (1986). Biomass energy crops—what, where, when, and who. In: *Proc. of U.S. Department of Agriculture Solar and Biomass Workshop*, Tifton, GA, pp. 51–6.
Smith, W. H. and Dowd, M. L. (1981). Biomass production in Florida. *Journal of Forestry* **79**, 508–11.
Smith, W. H. and Eoff, K. M. (1982). Developing biomass for biofuels. *Proc. the First Congress of Mechanical, Electrical and Chemical Engineers of Honduras*, Tegucigalpa, Honduras, pp. 100–6.
Smith, W. H. and Frank, J. R. (1982). Methane from biomass and waste: a comprehensive research program. *Proc. the 9th Energy Technology Conference*, pp. 1429–34. Government Institutes, Inc., Rockville, MD.
Smith, W. H. and Frank, J. R. (1985). Comparative biomass yields of energy crops. In: *Energy From Biomass*, Palz, W., Coombs, J. and Hall, D. O. (eds.), pp. 323–9. Elsevier Applied Science, London.
Smith, W. H., Smith, P. H. and Frank, J. R. (1983). Biomass feedstocks for methane production. In: *Energy from Biomass*, Strub, A., Chartier, P. and Schleser, G. (eds.), pp. 122–6. Elsevier Applied Science, London.
Snyder, G. H. and O'Hair, S. K. (1986). Biomass production from taro (*Colocasia esculenta*) in subtropical wetlands. In: *Biomass Energy Development*, Smith, W. H. (ed.), pp. 185–96. Plenum Publishing Corporation, New York.
Spencer, W. E. and Bowes, G. (1985). Growth and photosynthesis of waterhyacinth at elevated CO_2. *Plant Physiology* **77**, 84.
Spencer, W. E. and Bowes, G. (1986). Photosynthesis and growth of waterhyacinth at elevated CO_2 levels. *Plant Physiology* (submitted for publication).
Spiller, H., Latorre, C., Hassan, M. E. and Shanmugam, K. T. (1986). Isolation and characterization of nitrogenase-derepressed mutant strains of cyanobacterium *Anabaena variabilis*. *Journal of Bacteriology* **165**, 412–19.
Srinivasan, C. and Vasil, I. K. (1986). Regeneration of plants from sugarcane protoplasts. *Plant Physiology* (in press).
Stanley, R. L., Jr and Dunavin, L. S. (1986). Potential sorghum biomass production in north Florida. In: *Biomass Energy Development*, Smith, W. H. (ed.), pp. 217–26. Plenum Publishing Corporation, New York.
Swedlund, B. and Vasil, I. K. (1985). Cytogenetic characterization of embryogenic callus and regenerated plants of *Pennisetum americanum* (L.) K. Schum. *Theoretical and Applied Genetics* **69**, 575–81.
Tabaeizadeh, Z., Ferl, R. J. and Vasil, I. K. (1986). Somatic hybridization of sugarcane (*Saccharum officinarum* L.) and pearl millet (*Pennisetum americanum* [L.]K. Schum.). *Proc. National Academy of Science USA* (in press).
Vasil, I. K. (1983). Regeneration of plants from single cells of cereals and grasses. In: *Genetic Engineering in Eukaryotes*, Lurquin, P. and Kleinhofs, A. (eds.), pp. 233–52. Plenum Publishing Corporation, New York.
Vasil, I. K. (ed.). (1984). *Cell Culture and Somatic Cell Genetics of Plants. Vol. 1, Laboratory Procedures and Their Applications*. Academic Press, Orlando, FL, 825 pp.
Vasil, I. K. (1985). Somatic embryogenesis and its consequences in the Gramineae. In: *Tissue Culture in Forestry and Agriculture*, Henke, R. R., Hughes, K. W., Constantin, M. P. and Hollaender, A. (eds.), pp. 31–47. Plenum Publishing Corporation, New York.

Napiergrass
  biomass yield of, xii, 55, 56, 92, 93, 95, 98–9, 457
  BIOMET model studies of
    harvest schedules, 46
    transportation costs, 44–6
  cellulase enzymic action on, 373
  cold tolerance of, 94, 95, 172, 173
  composition of, 358, 359, 362
  conceptual design BTM system study for, 54–60
  conversion of, xiv, 57–60
    effect of micronutrients on, 305–9
    fermentability studied, 358, 359, 360, 362
  crop growth models used in BIOMET, 31, 33–5
  fertilization of, 96, 98, 99
  flow diagrams for BTM systems, 64
  genotype selection of, 92–3
  growing conditions for, xiii, 91
  growth analyses of, 96, 97
  harvesting of, 55, 56, 93, 95, 100
  hybridization with pearl millet, 156, 159, 163, 170–5
  lodging resistance of, 94
  methane yield of, 299
  planting of, 54, 57
  pretreatment of, 57–8
  propagation of, xii, 96, 98–100, 170
  storage of biomass, 100
  tissue culture of, 156, 158–9, 160
    salt-tolerance selection procedures, 161–2
  transportation of, 56, 71–2, 461
  winter survival of, 94, 95, 100, 172, 173
Natrosol gel, 192
Nematocides, grasses affected by, 88–9, 267
Nitrogen
  fixation
    Leucaena leucocephela, 282
    Sargassum spp., 198–208
    terrestrial bacteria, 197–8
    water hyacinth use of, 123
  nutrients
    anaerobic digestion affected by, 300–2, 302–3, 422

Nitrogen—*contd.*
  nutrients—*contd.*
    arrowleaf sida affected by, 274
    Gracilaria requirements, 251–2
    reduction reactions for, 300, 303
    Sargassum requirements, 251–2
    sorghum affected by, 87–8
    types in water, 107–8
    water hyacinth affected by, 28, 29, 30–1, 107–10, 135
Nonconventional crops, 261–88
  advantages of, 262
  compared to high-input crops, 261
Nutrient media, water hyacinths grown in, 119, 130

Office of Technology Assessment (OTA), biomass/energy study, 9
Organic acids
  accumulation of, prehistoric plant residues, 291
  anaerobic oxidation of, 293
  digestion processes affected by, 303–5, 306, 344–6, 422
  see also Volatile fatty acids

Packed bed digesters
  combined with leachbeds, 59, 309–11, 327, 330
  mathematical modeling of, 323, 325–30
    assumptions made, 326–7
    material balances, 327, 328–9
    reactor characterization, 323, 325–6
    systems model equations for, 37–9
  see also Fixed film reactors
Palmetto prairie, eucalyptus trees on, 282, 294
Pearl millet
  growing conditions for, 171
  hybridization with Napiergrass, 156, 159, 163, 170–5
  tissue culture of, 156, 158
    salt-tolerance selection procedures, 161
Pelvetia spp. (seaweeds), 213, 214, 215
Pennisetum spp.
  dwarf vs. tall plants, 172–4

Vasil, I. K. (1986). Advantages of embryogenic cell cultures of Gramineae. In: *Nuclear Techniques and In Vitro Culture for Plant Improvement*, pp. 71–5. International Atomic Energy Agency, Vienna.
Vasil, V. and Vasil, I. K. (1981). Somatic embryogenesis and plant regeneration from tissue cultures of *Pennisetum americanum* and *P. americanum* × *P. purpureum* hybrid. *American Journal of Botany* **68**, 864–72.
Vasil, V. and Vasil, I. K. (1981). Somatic embryogenesis and plant regeneration from suspension cultures of pearl millet (*Pennisetum americanum*). *Annals of Botany* **47**, 669–78.
Vasil, V. and Vasil, I. K. (1984). Induction and maintenance of embryogenic callus cultures of Gramineae. In: *Cell Culture and Somatic Cell Genetics of Plants*, Vol. 1, Vasil, I. K. (ed.), pp. 36–42. Academic Press, Orlando, FL.
Vasil, V., Wang, D. Y. and Vasil, I. K. (1983). Plant regeneration from protoplasts of *Pennisetum purpureum* Schum. (Napiergrass). *Zeitschrift fur Pflanzenphysiologie* **111**, 233–9.
Volin, R. B., O'Hair, S. K. and Beale, A. J. (1986). Genetic variation and production potential in cocoyam and taro. *Proc. Improved Exploitation and Development of Plant Resources Symposium*, Taipei, Taiwan (in press).
Wang, D. Y. and Vasil, I. K. (1982). Somatic embryogenesis and plant regeneration from inflorescence segments of *Pennisetum purpureum* Schum. (Napiergrass). *Plant Science Letters* **25**, 147–54.
Wang, S. (1983). Screening biomass species for methane production in Florida. *Forestry Products Journal* **33**(10), 66–8.
Wang, S. and Huffman, J. B. (1981). Botanochemicals: supplements to petrochemicals. *Journal of Economic Botany* **35**(4), 369–82.
Wang, S. and Huffman, J. B. (1982). Effect of extractives on heat content of *Melaleuca* and *Eucalyptus*. *Wood Science* **15**(1), 33–9.
Wang, S. and Littell, R. C. (1983). Phenotypic variation in calorific value of melaleuca material from south Florida. *Journal of Economic Botany* **37**(3), 292–8.
Wang, S., Huffman, J. B. and Littell, R. C. (1981). Characterization of *Melaleuca* biomass as a fuel for direct combustion. *Wood Science* **13**(4), 216–19.
Wang, S., Huffman, J. B. and Rockwood, D. L. (1982). Qualitative evaluation of fuelwood in Florida—a summary report. *Journal of Economic Botany* **36**(4), 381–8.
Wang, S., Littell, R. C. and Rockwood, D. L. (1984). Variation in density and moisture content of wood and bark among 20 *Eucalyptus grandis* progenies. *Wood Science Technology* **18**, 97–100.
Warren, C. S., Bruderly, D. E., Angelieri, M., Bilello, L. J., Bucalo, S., Finger, G. W., Hart, R., Hinton, S. W., Newman, J. R. and Vinzant, J. W. (1984). Evaluation of biomass to methane systems in the Lake Apopka natural gas district. Reynolds, Smith and Hills. Jacksonville, FL. (GRI84/0015.1).
Webley, O. J., Geary, T. F., Rockwood, D. L., Comer, C. W. and Meskimen, G. F. (1986). Seasonal coppicing variation for three eucalyptus in Southern Florida. *Australian Journal of Forest Research* (accepted for publication).
White, D. C. (1983). Analysis of microorganisms in terms of quantity and activity in natural environments. In: *Microbes in their Natural Environments*, Slater, J. H., Whittenbury, R. and Wimpenny, J. W. T. (eds.). Society for General Microbiology Symposium, Vol. 34, pp. 37–66.

White, D. C. (1986). Quantitative physical-chemical characterization of bacterial habitats. In: *Bacteria in Nature, Vol. 2*, Poindexter, J. and Leadbetter, E. (eds.). Plenum Publishing Corporation, New York (in press).

White, J. M. and Forbes, R. B. (1982). Biomass potential of five root crops grown on two central Florida soils. *Proceedings of Florida State Horticultural Society* **95**, 365–7.

White, J. M., Forbes, R. B. and Reddy, K. R. (1986). Residues of selected vegetable crops and their potential as biofuel. *Proceedings of the Soil and Crop Science Society of Florida* **45**, 105–7.

Wi, J. R. and Cantliffe, D. J. (1984). Somatic embryogenesis and plant regeneration in tissue cultures of sweet potato (*Ipomoea batatas* Poir.). *Plant Cell Reports* **3**, 112–15.

Wilkie, A., Goto, M., Bordeaux, F. M. and Smith, P. H. (1986). Enhancement of anaerobic methanogenesis from Napiergrass by addition of micronutrients. *Biomass* (in press).

Woodard, K. R., Prine, G. M. and Ocumpaugh, W. R. (1985). Techniques in the establishment of elephantgrass (*Pennisetum purpureum* Schum). *Proceedings of the Soil and Crop Science Society of Florida* **44**, 216–21.

Yang, J. and Smith, P. H. (1985). Isolation of methanogenic bacteria from termites. Abstracts of Annual Meeting of American Society of Microbiology, p. 160.

Appendix B

Conversion Table: SI to Imperial Units

SI units	Multiply by	Imperial units
Hectares (ha)	2·471	acres (A)
Square centimeter (cm^2)	0·155	inches square (in^2)
Square meter (m^2)	10·764	feet square (ft^2)
Liter	0·264	gallon (gal)
Cubic meter (m^3)	35·7	cubic feet (ft^3)
Meter (m)	3·28	feet (ft)
Centimeter (cm)	0·394	inch (in)
Kilometer (km)	0·621	mile (mi)
Grams (g)	0·0353	ounce (oz)
Kilogram (kg)[a]	2·202	pound (lb)
Megagram (Mg)[a]	1·102	short ton (t)
Kilogram per hectare (kg ha^{-1})	0·892	pounds per acre (lb/A)
Megagram per hectare (Mg ha^{-1})	0·446	tons per acre (t/A)
Grams per square meter (g m^{-2})	$4·46 \times 10^{-3}$	tons per acre (t/A)
Grams per square meter (g m^{-2})	8·93	pounds per acre (lb/A)
Liter per hectare (liter ha^{-1})	0·107	gallons per acre (gal/A)
Megapascal (m Pa)	9·90	atmosphere
Joule (J)	$0·948 \times 10^{-3}$	British thermal unit
Joule (J)	0·239	calorie
Megajoule per kilogram (MJ kg^{-1})	430	British thermal unit per pound (BTU/lb)
Joule per cubic meter (J m^{-3})	$2·685 \times 10^{-5}$	British thermal unit per cubic foot (BTU/ft^3)
Million Gigajoule[a] (GJ $\times 10^6$)	$2·778 \times 10^5$	megawatt-hours
Billion Gigajoule (GJ $\times 10^9$)	0·949	quad or (Q) (10^{15} British thermal unit; BTU)

[a] 10^9 = giga; 10^6 = mega; 10^3 = kilo.

Index

Note: 'BTM' means 'biomass-to-methane'

Aceticlastic methanogens, 347–8, 462
  see also Methanosarcina barkeri; M. mazei
Agricultural drainage water, 117–18
  composition of, 117
  water hyacinths grown in, 119, 130, 131
Agricultural land availability, 7–8
Algae. *See* Brown . . .; Marine algae; Red . . .
Alginates
  marine algae, 213–14, 215
  cleavage by bacteria, 218–22, 223
  utilization by bacteria, 218–22, 223
  Sargassum benthic species, 255
Anaerobic digestion processes
  chemical reactions in, 336
  effectiveness of, 292
  experimental data for, 67
  future research necessary, 458
  inocula used, 336–7
  leachbeds used, 35–7, 59, 312–23, 324
  limitations of, 367
  mathematical modeling of, 35–9, 310–30
    fixed film reactors, 433, 434–6
    leachbed model, 312–23
      chemical reactions modelled, 318–21
      material balance equations, 322–3, 324
      reaction kinetics for, 314–18
      reaction network used, 312–13

Anaerobic digestion processes—*contd.*
  mathematical modeling of—*contd.*
    leachbed model—*contd.*
      reactor characterization, 321–2
      stoichiometric relationships used, 313–14
    leachbed/packed bed combined system, 327, 330
    packed bed model, 323–7
      assumptions made, 326–7
      material balance equations, 327, 328–9
      reactor characterization, 323, 325–6
  microbial populations involved, xiii, 335–49, 397, 406, 413–14
    layering effect in, 423–5
    microbial biomass analysed, 421–2
    on-line monitoring of. 419–21
    phospholipid analysis of, 415–19
  multiphase digesters used, 295–310
    effect of
      micronutrients, 305–10
      nitrate, 300–2, 302–3
      oxidants, 300–3
      sulfate, 302, 303
    evaluation of feedstock, 296–300
    heat transfer considerations, 295–6
    mass transfer considerations, 295
    organic acids in, 303–5
    rate control considerations, 296
  natural use of, 292
  packed beds used, 37–9, 59, 323–7

485

Anaerobic digestion processes—*contd.*
　packed beds used—*contd.*
　　mathematical modeling of, 323–7
　　packing media used, 431, 432–3, 437–40
　　pretreatment used, 57–9
　　reactions involved, 293–4
　　residues from
　　　characteristics of, 445–8
　　　decomposition in soil, 448–9
　　　disposal on land, 449–51
　　retention times for, 58–9
　　single-stage digesters used, 294
　　solids retention time for, 337
Aquatic plants
　biomass yields of, 104
　composition of, 358, 362
　fermentability of, 358, 360, 362
　methane yield of, 299
　selection of, 104–5
　see also Azolla; Bulrush; Cattails; Duckweed; Elodea; Giant duckweed; Pennywort; Salvinia; Water hyacinth; Water lettuce
Aroids
　biomass produced by, 238–42
　see also Taro
Arrowleaf sida
　biomass potential of, 265, 271, 274
Attenuated total reflectance (ATR) spectroscopy, microbial ecology examined by, 420
Avicel, 372
　hydrolysis of, 377–8
Azolla
　biomass yield of, 104
　fermentability of, 360

Battelle study (on biomass/energy), 7, 9
Beets, 238, 244–5
　biomass yield of, 244, 245
　fermentability of, 360
Bibliographic information base, 18
Biogasification
　advantages of, xiii, 2
　enzyme pretreatment effects on, 372–80

BIOINQ computer module, 18, 19
BIOMAIL computer module, 19
Biomass
　decentralized processing of, 5
　definition of, 2
　energy
　　crop yield considerations, 8–9
　　land availability constraint on, 7–8
　　opportunities for, 5–11
　　present capacity figures, 6
　　size of facilities, 5, 6
　　systems analysis approach, 9–11
　quality of
　　sorghum, 88
　　water hyacinth, 129–31
　research. *See* Research program (GRI/IFAS)
Biomass-to-methane (BTM) systems
　analytical assumptions for, 55
　computer model of, 21–47
　economics of, 69–74
　economies of scale for, 74–5
　elements of, 52
　flow diagrams for
　　Napiergrass, 64
　　water hyacinth, 64
　improvement of, 292–3, 456, 458
BIOMET systems model, 10, 21–2
　block diagram for, 22–3
　comparison with other analysis procedures, 24
　conversion models used, 35–40
　crop growth models used, 25–35
　　Napiergrass, 31, 33–5
　　water hyacinth, 25–31, 32
　economic models used, 42–4
　energetic analyses used, 41
　future research indicated by, 458
　system simulation by, 44–6
　transportation costs examined by, 44–6
BIONEWS computer module, 19
BIOTEXT computer module, 17
Boom–winch (water hyacinth) harvester, 43, 62, 70–1, 145–9
Bovine manure
　composition of digested solids, 447
　digestion of, 432
　effluent

Bovine manure—*contd.*
    effluent—*contd.*
        application to soil, 448, 449, 450
        composition of, 447
Brassica crops, 244
    biomass yields of, 238, 244
    composition of, 358, 362
    cultivation details for, 238, 244
    fermentability of, 358, 360, 362
Brewery waste, treatment of, 432
Brown marine algae
    chemical composition of, 211, 213–15
    composition of, 358
    fermentability of, 358
    see also Fucus; Laminaria; Macrocystis; Pelvetia; Sargassum
Bulrush, biomass yield of, 104
Butyric acid, anaerobic digestion affected by, 304, 305, 344–6, 422

Calcium nutrients, water hyacinth requirements for, 112–13
CALENDAR computer module, 19
Callus culture
    Napiergrass plants, 156–60
    Sargassum spp., 226–9
    sweet potato, 185–90
Carbon dioxide, water hyacinth requirements for, 106–7, 135
Carrots
    biomass potential of, 245
    fermentability of, 360
Cassava
    advantages as biomass crop, 243
    biomass yields of, 238
    disadvantages of, 243
    fermentability of, 360
    fertilization of, 243
    growing conditions for, 242
    propagation of, 183, 242
Castorbean
    biomass potential of, 264, 271
    fermentability of, 360
    growing conditions for, 273

Cattails, biomass yield of, 9, 104, 457
Cell wall constituents (CWC)
    analytical procedure for, 356
    biodegradability model affected by, 361–4
    fermentability affected by, 355, 357
    ranking of species by, 357, 360
    values quoted, 358–9
Cellobiohydrolase (CBH) enzymes, 368, 369–70
    molecular biological considerations, 378, 380
    synergism of fungal with bacterial cellulases, 378, 379
Cellulase enzymes
    bioassay method for, 370–2
    biogasification affected by, 372–80
    kinetics of disappearance, 376–8
    nature of, 368–70, 462
    persistence of activity, 374
    synergistic action of, 367–8, 377, 378
Cellulolytic bacteria, 339–40
Cellulolytic enzymes, 367, 462
Cellulose
    hydrolysis of, xiii, 293, 336, 339–40, 367
    effect of cellulase enzyme on, 373, 377–8
    marine algae, 214–15
Chemical oxygen demand equivalents (CODe), 314
Chicory
    advantage as biomass crop, 243
    biomass yield of, 238, 243
    fermentability of, 360
Clostridium cellulovorans, 340
Clostridium populeti, 340
Clostridium thermocellum, cellulase enzymes from, 369
Coal forests, reason for plant residues in, 291
Cold tolerance
    cassava, 243
    cruciferous crops, 244
    energy cane, 268
    Napiergrass, 94, 95, 172, 173
    pennywort, 105, 116, 125
    sweet potato, 246
    water hyacinth, 105, 113, 116, 125

Combine harvesting system (for water hyacinths), 149–50
Comprehensive Scientific Information Management System (CSIMS), 16, 17, 19
Computer
  image enhancing software, 414
  information management system, 10–11, 15–20
  models used, 10, 21–46
Conceptual design study, 49–75
  approach adopted, 50–67
  cost analysis used, 63–7
  economic analytical results for, 69–74
  economies of scale studied, 74–5
  Napiergrass systems studied, 54–60
  questions to be answered by, 49
  results of, 67–75
  water hyacinth systems studied, 55, 60–3
Continuous stirred tank reactors (CSTRs), 58
Conversion processes
  Napiergrass, 57–60
  systems model equations for, 35–9
  water hyacinth, 61, 63
  see also Anaerobic digestion processes
Conversion table (SI/Imperial units), 485
Corn, fermentability of, 360
Cost analysis, xi–xii
  assumptions made for, 55, 65, 66
  BIOMET model, 42–4
  conceptual design, 63–7, 69–74
  future research indicated by, 458–60
  see also BIOMET systems model
Cruciferous crops, 244
  biomass yields of, 238
  fermentability of, 360
  see also Brassica crops
Cyanobacteria
  isolation in pure culture, 203–4
  nitrogen fixation by, 198, 201
  nitrogenase activity of, 204–5
    effect of organic compounds on, 205–6

Cyanobacteria—contd.
  nitrogenase activity of—contd.
    effect of physical environment on, 206–8
    occurrence on *Sargassum*, 201–3

Denitrification, water hyacinths involved in, 129
Detritus
  BIOMET model of, 28–9
  water hyacinth use of, 122–3, 128
2,4-Dichlorophenoxyacetic acid (2,4-D), 156, 157, 185
Diffuse reflectance infrared Fourier transform (DRIFT) spectroscopy, microbial ecology examined by, 420, 421
Digester residues
  characteristics of, 445–8
  composition of, 447
  disposal of, 56, 60, 449–51
  water hyacinths grown in effluent, 120–1, 126–7, 130, 131, 133–4, 451–3
Disaggregatase, 404, 406, 462
  pH effects on, 404, 409
Disney World. *See* Walt Disney World Resort Complex
Dissimilatory nitrate reduction, 300
Distillery waste, composition of digested material, 446, 447, 448
Dogfennel
  biomass potential of, 265, 271, 273
  fermentability of, 360
  growing conditions for, 273
  winter growth of, 273
Duckweed, biomass yield of, 104

Eastern gamagrass
  biomass potential of, 264, 267, 269, 271
  fermentability of, 360
  growing conditions for, 269
  winter growth of, 269

Economic models
  generalized equation for, 42–3
  Napiergrass production/conversion costs, 43–4
  water hyacinth production/conversion costs, 43
Effluents
  composition of, 117, 118, 447
  disposal of, 56, 60, 449–51
  water hyacinths grown in, 118, 120–1, 126–7, 130, 131, 133–4, 451–3
Eichhornia crassipes. See Water hyacinth
Elephantgrass. See Napiergrass
Elodea, biomass yield of, 104
Endoglucanases, 369
  chemical reactions of, 369
Energy
  analyses by BIOMET model, 41
  cane
    biomass potential of, 264
    biomass yield of, 84, 179, 267, 268, 457
    cold tolerance of, 268
    fertilization of, 267
    growing conditions for, 263
    herbicides used, 263
    propagation of, 263
    see also Sugar cane
  crops
    development of, 79–81
    yield of, 8–9
    opportunities for biomass, 5–11
    prices, xi, 4, 66–7
    Research Advisory Board (ERAB), biomass/energy study, 9
    supply situation, 4–5
    Yield Ratios, 41
Erianthus, 84, 262–3, 264
  biomass potential of, 264
  biomass yield of, 267
  fermentability of, 360
  growing conditions for, 263, 266
  growth habit of, 262–3
Estuarine conditions, *Sargassum* communities in, 256

Eucalypts
  biomass yield of, 281, 284, 287, 457
  crop management effects on, 282–3
  cultivation effects, 282, 284
  fermentability of, 339, 360
  planting zones for, 279
  production system effects, 284–6
  selection for biomass production, 277–8, 279–81
  winter tolerance of, 280
Eutrophic lake waters, 103
  composition of, 117
  water hyacinths grown in, 119–20, 130, 131, 132–3
  see also Lake Apopka

Farmers, alternative crops for, 2, 464
Fatty acids, anaerobic digestion affected by, 303–5, 306, 344–6
Fermentability
  prediction of, 361–4
  ranking by cell wall constituents percentage, 360
Fertilizer effects
  arrowleaf sida, 274
  cassava, 243
  cost assumptions for, 65
  economics of, 197
  grasses, 267
  Leucaena, 282
  Napiergrass, 54, 96, 98, 99
  sorghum, 86–8
  taro, 241–2
  water hyacinth, 44, 122, 130, 131
Fescue hay
  composition of, 447
  digester residues
    application to soil, 449–50
    composition of, 447
Fixed film reactors, 323, 325
  bulk flow characteristics of, 441
  function of, 327
  mathematical modeling of, 433, 434–7
    chemical oxygen demand forecast, 435, 436

Fixed film reactors—*conta*.
  mathematical modeling of—*contd*.
    methane production forecast,
      436–7
    packing media for, 431–4, 437–40
      clay pipe used, 432
      coral used, 432–3
      corncobs used, 433
      experimental studies for, 437–40
      mussel shells used, 432–3
      oyster shells used, 432
      plastic media used, 432, 433, 434,
        437–40
      rock used, 432
      wood chips used, 431, 437–40
    startup characteristics of, 440
  see also Packed bed digesters
Flooding tolerance
  arrowleaf sida, 271, 274
  dogfennel, 271
  herbaceous crops, 271
  sweet potato, 245, 246
  taro, 239, 240, 241
Florida
  biomass crop opportunities in, 2
  renewable energy sources in, 6
Flowers, Dr Ab, v–vii
Fluid (seed-)drilling techniques, 184,
    462
  gels used, 192
  Napiergrass, 57
Fluidized transport, water hyacinth
    feedstock, 152–3
Fodder beet, 238, 244–5
  fermentability of, 360
Fodder carrot, 245
Foliar fertilizers
  water hyacinth, 122, 130, 131
Fourier transform infrared (FT/IR)
    spectrometry, 419–20
  living biofilms examined by, 420
  microbial nutritional status markers
    analysed by, 420–1
Foxtail. *See* Giant foxtail
Freshwater aquatics. *See* Aquatic plants
Fucoidin, 215
Fucus spp. (seaweeds), 214, 215

Fungal cellulase, 367–8
  bioassay of, 371–2
  methane yield affected by, 372–3,
    374–8
  molecular biological considerations,
    378, 380
  persistence of activity, 374
  synergism with bacterial cellulases,
    378, 379

Gamagrass. *See* Eastern gamagrass
Gas Research Institute (GRI)
  biomass project, xi, 1–2
  formation of, vi
  see also Research program (GRI/
    IFAS)
Genetic engineering
  grass species, 161
  halophilic bacteria, 343
  marine algae, 212, 216, 258
Giant duckweed, 104
Giant foxtail
  biomass potential of, 264
  crop rotation for, 269–70
  growing conditions for, 269
Giant pigweed, 265, 270, 271
Gibberellic acid, water hyacinth
    affected by, 114, 115
β-Glucosidase, 368, 369, 377
Gracilaria spp.
  biomass yields of, 249–50, 252, 253
  cultivation of, 250
  nutrient requirements of, 251–2
  salinity tolerance of, 251
Gramineae
  callus culture of, 156–60
  protoplast culture of, 160–1
  selection for salt tolerance in, 161–2
  tissue culture of, 156–61
    biomass yield of plants, 164, 165
    compared with vegetative
      propagation, 163, 165
    uniformity of plants, 162–4
  see also Corn; Grasses; Guinea grass;
    Maize; Napiergrass; Pearl
    millet; Sorghum; Sugarcane

Grasses, 262–70
  biomass potential of, xii, 264
  BTM performance data for, 67
  composition of, 358, 362
  disadvantages of, 83–4
  fermentability of, 358, 360, 362
  methane yield of, 299
  uses of, 169
  see also Corn; Guinea grass; Maize; Napiergrass; Pearl millet; Sorghum; Sugarcane
Great Salt Lake, bacteria from, 342
Growth regulators. *See* Plant growth regulators
Guinea grass, tissue culture of, 191

Halomethanococcus mahi, 343
Halophilic microflora, 341–4
Harvesting
  frequency
    effects on costs, 46
    Napiergrass biomass affected by, 93, 95
    water hyacinth affected by, 124–5
  Sargassum spp., 250
  sweet potato, 246
  taro, 240, 242
Heat loss calculations, systems model equations for, 39–41
Henry's law, 319
Herbaceous crops, 264–5, 270–5
  biomass potential of, 239, 243, 245, 264–5
  biomass yields of, 238
  cultivars examined in Florida, 234–7
  fermentability of, 360
  methane yield of, 299
Herbicides
  energy canes, 263
  sorghum affected by, 90, 91
High pressure liquid chromatography (HPLC)
  phytanyl glycerol diethers assayed by, 417–18
  water hyacinth digester analysis, 426

Hybridization possibilities, *Sargassum* spp., 257–8
Hydrogen-producing organic acid oxidizer (HPOAO) bacteria, 336, 344–6

IBM 3090 mainframe computer, use by researchers, 16
Indiangrass, fermentability of, 360
Information management system used, 10–11, 15–20
  acquisition of information, 16–17
  retrieval of information, 17–18
  supportive facilities, 19
Institute of Food and Agricultural Sciences (IFAS)
  biomass project, xi, 1–2
  see also Research program (GRI/IFAS)
Iron nutrients, water hyacinth requirements for, 113–14, 135

Jerusalem artichoke, 238, 245

Kale, 238, 244
Kelp farming, 212
  economics analyses of, 250
Kryosol process, 56

Lake Apopka (Florida, USA)
  composition of water, 117
  Natural Gas District (LANGD)
    agricultural crops in, 51
    feasibility study, 50–75
    lake areas in, 53, 60
    maps of, 53
    reasons for choice as study area, 51
    water hyacinth trials, 31, 32, 115, 119–20, 130, 131, 132–3
Lake Okeechobee (Florida, USA), 118
Laminaria spp. (seaweeds), 212, 214
Land
  application of digester residue on, 56, 60, 449–51

Land—contd.
  availability, 7–8
  costs, methane production costs
    affected by, 460–1
  LANGD study, 67–9
Leachbeds
  combined with packed bed reactors,
    59, 309–11, 327, 330
  mathematical modeling of, 312–23
    chemical reaction considerations,
      318–21
    material balances calculated, 322–
      3, 324
    reaction kinetics, 314–18
    reaction network, 312–13
    reactor characterization, 321–2
    stoichiometric relationships, 313–
      14
  systems model equations for, 35–7
Leucaena sp.,
  biomass yield of, 281, 287
  fermentability of, 360
  fertilization of, 282
  planting zone for, 279
  selection for biomass production,
    278, 280–1
  winter tolerance of, 281
Levelized costs, meaning of term, 65
Lignin
  analytical procedure for, 356
  biodegradability model affected by,
    361–3
  digestibility and, 293, 355
  hydrolysis of, 293
  values quoted, 358–9

Macrocystis spp. (seaweeds), 211–12,
    214, 215
Magnesium nutrients, water hyacinth
    requirements for, 112–13
Maize, tissue culture of, 161, 190
Marine algae
  alginate content of, 214
  chemical composition of, 213–15
  composition of, 358
  fermentability of, 358, 360
  mannitol content of, 214

Marine algae—contd.
  see also *Fucus*; *Laminaria*;
    Macrocystis; *Pelvetia*;
    Sargassum; Seaweeds
Marine crops
  Florida coast study, 249
  methane yield of, 299
Marine microflora, 346–7
  mesophilic isolate, 347
  thermophilic isolate, 346
Mathematical models
  anaerobic digestion process models,
    35–41, 310–30
    development of model, 311
    leachbed model, 312–23
      chemical reactions for, 318–21
      material balances for, 322–3, 324
      reaction kinetics for, 314–18
      reaction network for, 312–13
      reactor characterization for,
        321–2
      stoichiometry relationships for,
        313–14
    packed-bed model, 323–30
      assumptions made, 326–7
      material balance equations for,
        327, 328–9
      reactor characteristics, 323, 325–
        6
  BIOMET systems model, 21–46
    conversion process model, 35–41
    crop growth models, 25–35
    economic models, 42–4
    energetics analyses, 41
    system stimulation for, 44–6
  digestibility, 361–4
  fixed film reactors, 433, 434–7
    chemical oxygen demand forecast
      by, 435, 436
    methane production forecast by,
      436–7
Mercerizing process, 57
Methane
  digester effluents
    composition of, 117, 118, 447
    disposal of, 56, 60
    water hyacinths grown in, 121, 130,
      451–3

Index   493

Methane—contd.
  price of, xi, 4, 66–7
  yield
    aquatic plants, 299
    assay method used, 297–8
    evaluation of plant species for, 296–300
    grasses, 299
    herbaceous crops, 299
    marine crops, 299
    Napiergrass, 299
    research target for, 458, 459
    root crops, 299
    sweet potato, 298
    water hyacinth, 298, 299
Methane from Biomass Program. *See* Research program (GRI/IFAS)
Methanobacterium bryantii, 341, 388, 391
Methanobacterium formicicum, 341, 388, 391
Methanobacterium thermoautotrophicum, 344, 345, 388, 391
Methanobrevibacter arboriphilus, 341, 388, 390
Methanobrevibacter smithii, 388, 390
Methanococcus mazei. See Methanosarcina mazei
Methanococcus vannielii, 393
Methanococcus voltae, 393
Methanogenesis, rates in water hyacinth digester, 425
Methanogenic bacteria
  aggregation/dispersion of, 398–410
  antiquity of, 291
  ultrastructural analyses of, 385–96
Methanogenium marisnigri, 392, 395
Methanogenium thermophilicum, 346, 392, 395
Methanomicrobium paynteri, 347
Methanosarcina barkeri, 347–8, 388, 389, 406
Methanosarcina mazei, 348, 386–8, 389
  chemical analysis of matrix material from, 402, 405
  cytochemistry of, 400, 402

Methanosarcina mazei—contd.
  disaggregating variant of, 402, 404, 406, 407, 408, 409
  morphotypes of, 397–8, 462
  ultrastructure of, 386–8, 389, 398–400, 401, 403
Methanosarcina spp., 344, 345, 347–8
Methanospirillum hungatei, 345, 392, 394
Microbial consortia, 413–14
  anaerobic digesters, 413–14
  metabolic activity of components, 418–19
  nutritional status of, 418
  on-line monitoring of, 419–21
  phospholipid analysis of, 415–19
Microcomputers, use by researchers, 17, 19–20
Micronutrients
  marine plants affected by, 251–2
  Napiergrass fermentation affected by, xiv, 306, 308–9
  water hyacinth growth affected by, 113–14
Microscopy, direct, bacteria underestimated by, 414
Millet
  fermentability of, 360
  growing conditions for, 171
  hybridization with Napiergrass, 156, 159, 163, 170–5
  tissue culture with, 156, 158, 161
Molasses waste
  effluent applied to soil, 448, 449
  see also Rum slops
Monod functions, 315, 316
Multidisciplinary approach, 3, 10–11, 49, 455, 461
Multiphase digesters
  affected by
    nitrate, 300–2, 302–3
    oxidants, 300–3
    sulfate, 302, 303
  evaluation of feedstock for, 296–300
  heat transfer in, 295–6
  mass transfer in, 295
  rate control of, 296

Pennisetum spp.—*contd.*
  hybrid breeding of, 156, 159, 163, 170–5
  selection and breeding of, 170–5
  tissue culture of, 156–64, 189–90
  see also Napiergrass; Pearl millet
Pennywort
  biomass yield of, 104, 125
  co-production with water hyacinth, 60, 125–6
    see also Water hyacinth
  cold tolerance of, 105, 116, 125
  growth rates of, 125
  harvesting of, 126
  winter growth of, 60, 105, 125
Phaeophyta
  chemical composition of, 213–15
  see also *Fucus*; *Laminaria*; *Macrocystis*; *Pelvetia*; *Sargassum*
Phospholipids
  biomass estimation by analysis of, 415–16
  ester-linked fatty acids (PLFA), 413, 416
    anaerobic digesters analysed by, 421–2, 423
    bacterial community structure analysed by, 416–18
    water hyacinth digester analysed by, 422–7
Phosphorus nutrients
  Gracilaria affected by, 251–2
  nitrogen fixation on *Sargassum* affected by, 199–200
  Sargassum affected by, 251–2
  sorghum affected by, 87–8
  water hyacinth affected by, 28, 29–30, 110, 111, 135
Photosynthetic organisms
  methanogenic bacteria, 426
  Sargassum spp., 256, 257
Phytanyl glycerol diethers, assay of, 417
Pigweed
  biomass potential of, 264, 270–2
  fermentability of, 360
  growing conditions for, 270, 272
Pine trees
  biomass yield of, 281, 283, 287

Pine trees—*contd.*
  cultivation effects, 282–3
  fermentability of, 360
  production system effects, 285, 286
  selection for biomass production, 281
Plant-growth regulators, effect on
  sorghum, 90–1
  water hyacinth, 114, 115
Planting density effects
  arrowleaf sida, 274
  dogfennel, 273
  eucalyptus trees, 282, 284
  giant foxtail, 269
  herbaceous crops, 269, 272, 273, 274
  pigweed, 272
  pine trees, 282, 283
  sorghum, 89
  water hyacinth, 124
Poly β-hydroxy butyrate, 418
Polymers, hydrolysis of, 293, 336
Poplar trees/wood
  biomass yield of, 287
  fermentability of, 337, 339, 360
Potassium nutrients
  water hyacinth requirements for, 110–11, 112, 135
Price comparison (for methane), xi, 4, 66–7
Process-related bacterial species
  aceticlastic methanogens, 347–8
  cellulolytic bacteria, 339–40
  halophilic microflora, 341–4
  hydrogen-producing organic acid oxidizer bacteria, 336, 344–6
  marine microflora, 346–7
  mesophilic bacteria, 347
  thermophilic bacteria, 346
  termite-gut methanogenic bacteria, 340–1
Product-oriented research approach, 11
Propagation techniques
  cassava, 183, 242
  energy cane, 263
  Napiergrass, 96, 98–100, 155–65, 170
  Sargassum spp., 222–9
  sweet potato, 185–93
  tissue culture, 155–65
  water hyacinth, 105–6

Propionic acid, anaerobic digestion
    affected by, 304, 306, 422
Protoplast culture techniques
    Napiergrass plants, 160–1, 462
    Sargassum spp., 222–6
Public opinion effect, xii
Pumps, water hyacinth feedstock, 152–3
Purification processes, xiv

Radish, 238, 244
Ragweed
    biomass potential of, 264
    growing conditions for, 272
Red marine algae, 211, 249
    composition of, 358
    fermentability of, 358
    see also *Gracilaria*
Research program (GRI/IFAS)
    areas of research in, 4
    conceptual design study, 49–75
        approach adopted, 50–67
        cost analysis used, 63–7
        economic analytical results for, 69–74
        economies of scale studied, 74–5
        Napiergrass systems studied, 54–60
        questions to be answered by, 49
        results of, 67–75
        water hyacinth systems studied, 55, 60–3
    cost sensitivity analysis used, 458, 459
    future research necessary, 461–3
    improvements achieved, 456–8
    information management system for, 10–11, 15–20
    management structure for, 3–4
    operating premises for, 2–3
    publications listed, 465–83
    size of, 15, 16
RIM database management system, 18
Root crops
    biomass yields of, 238
    cultivars examined in Florida, 234–7
    methane yield of, 299
Rum slops
    composition of digested material, 446, 447, 448

Rum slops—*contd.*
    effluent applied to soil, 448, 449, 450–1
Rumen fermentation
    biodegradability measured by, 355, 356
    multiphase nature, of, 294
Ruminant animals
    digestion by, 292
    microflora in, 413
Rutabaga, 238, 244

Salinity tolerance
    Gracilaria spp., 251
    methanogenic bacteria, 342–4, 346–7
    Napiergrass and hybrids, 161–2
    nitrogen-fixing bacteria, 208
    Sargassum spp., 251
    taro, 242
Salt domes, digesters constructed from, 341
Salvinia, biomass yield of, 104
Sanford (Florida, USA), water hyacinth trial, 31, 32
Sapium sp.
    biomass yield of, 281, 287, 457
    fermentability of, 360
    planting zone for, 279
Sargasso Sea, biomass yield from, 216
Sargassum spp.
    alginates cleaved by bacteria, 218–22, 223
    bacterial flora on, 201–3, 216–22
    benthic species, 199
        alginate content of, 255
        compared to pelagic species, 254
        growth rate of, 254, 258
    bluewater vs. greenwater yields, 252, 253
    callus culture of, 226–9
    containment of, 250
    floating nature of, 250
    fungus infection of, 223–4
    harvesting of, 250
    nitrogen fixation on, 198–208
        conditions for nitrogenase activity, 204–5
        cyanobacterial epiphyte isolated, 203–4

Sargassum spp.—*contd.*
  nitrogen fixation on—*contd.*
    effect of organic compounds on nitrogenase activity, 205–6
    effect of physical environment on nitrogenase activity, 206–8
    estuarine communities, 200
    oceanic communities, 200
    organisms responsible, 201–3
    quantity of nitrogen fixed, 198–200
  nutrient requirements of, 251–2
  pelagic species, 200
    bacteria on, 201–2, 203, 216–22
    genetic improvement of, 216
    nitrogen fixation associated with, 200
    salinity tolerance of, 251
    utilization of light by, 251
  photosynthesis/light intensity studies, 256, 257
  population off Florida coast, 254–6
  protoplasts prepared from, 222–3
    enzymes used, 224–6
    micrographs of, 225
    wall regeneration in, 226
  reproduction of, 225, 256
  salinity tolerance of, 251
  solar radiation effects on, 251
Scale, economies of, 74–5
Seaweeds
  chemical composition of, 213–15
  see also Fucus; *Laminaria*; *Macrocystis*; *Pelvetia*; *Sargassum*
Sediments, water hyacinth use of, 122, 128
Sewage effluents, 118
  composition of, 117, 425
  water hyacinths grown in, 120–1, 126–7, 130, 131, 133–4, 451
Sida
  biomass potential of, 265, 271, 274
  fermentability of, 360
Slurry reactors, 323, 325
  function of, 327
Smooth pigweed, 265, 270, 271, 272
Sodium nutrients, water hyacinth requirements for, 111–12
Solka floc, effect of cellulase enzyme on, 373, 374

Somatic embryogenesis
  Napiergrass propagated by, 156–62
  sweet potato propagated by, 185–93, 462
Somatic hybridization techniques, Napiergrass plants, 161
Sorghum
  biomass yield of 85–6, 176, 457
  breeding of, 175–8
  cellulase enzymic action on, 373, 374
  composition of, 358, 359, 361, 362, 363
  crop management effects, 88–91
  cultivars yields for, 85–6
  digester waste, water hyacinths grown on, 452, 453
  fermentability of, 358, 359, 360, 362
  fertilization of, 86–8
  growing conditions for, xiii, 84
  herbicides, effect on, 90–1
  legumes grown as prior crop, 89–90
  plant-growth regulators, effect on, 90–1
  quality of biomass from, 88, 89
  row spacing effects, 89
  seeding rate effects, 89
  selection for biomass use, 175–8
Storage (of biomass), 100
Sugarbeet, 244, 245
Sugarcane
  biomass yields of, 84, 91, 178, 179, 457
  compared with Napiergrass, 84, 91
  fermentability of, 360
  growing conditions for, xiii, 178
  hybrid breeding of, 178
  selection and breeding of, 178–80
  see also Energy cane
Sulfate, anaerobic digestion affected by, 302, 303, 422, 423
Sulfate-reducing bacteria, analysis of, 418, 422
Sweet potato, 245–6
  biomass yields of, 238, 246
  composition of, 358, 359, 362
  fermentability of, 358, 359, 360, 362
  harvesting of, 246
  methane yield of, 298
  plantlet handling, 192–3

Sweet potato—*contd.*
somatic embryogenesis for, 185–91, 462
Swine manure
composition of digested material, 447
digestion of, 433
effluent
application to soil, 448, 449
composition of, 447
Synergism, cellulase enzymes, 367–8, 377, 378
Systems approach, 3, 9–11, 461

Tannins, 355–6
analytical procedure for, 357
biodegradability model affected by, 363, 364
values quoted, 361
Taro
advantages as biomass crop, 242
biomass yields of, 238, 239
fermentability of, 360
fertilization of, 241–2
flooding tolerated by, 239, 240, 241
growing conditions for, 239
harvesting of, 240, 242
planting density effects, 239, 241
salinity tolerated by, 242
Temperature effects
systems model equations for, 39–41
water hyacinth growth, 28, 44, 116
Termites
cellulolytic bacteria from, 340–1
methanogenic bacteria from, 341
Thesaurus, information management, 18
Tissue culture
Napiergrass, xii, 80, 462
stability of resulting plants, 162–4, 190–1
sweet potato, 80, 185–93, 462
Trace elements
Gracilaria requirements, 251–2
Napiergrass fermentation affected by, xiv, 306, 308–9
Sargassum requirements, 251–2
water hyacinth growth affected by, 113–14

Transportation
BIOMET model study of costs, 44–6
Napiergrass, 71–2
cost analysis of, 44–6, 461
water hyacinth, 71, 72–3, 152–3
Trichoderma reesei
cellulolytic enzymes from, 367–8
bioassay of, 370–2
methane production affected by, 374–5
polysomes from, 378, 379
Turnip, 238, 244

Ultrastructural analyses
methanogenic bacteria, 385–96, 398–400, 401, 403
techniques used, 386

VAX 750 minicomputers, use by researchers, 16, 17–18, 19
Volatile fatty acids (VFA), 303
anaerobic digestion inhibited by, 303, 304, 305
chemical reactions of, 318–19, 320
dynamics involved, 303–5

Walt Disney World Resort Complex
biomass project first started, vi
sorghum digester trial, 67
water hyacinth trials, 110–11, 120, 131, 423, 425, 426
Wastewater disposal, digestion process, 56, 60, 449–51
Water
hyacinth
aerodynamic properties of, 142–3, 144
agricultural drainage water used for growth of, 119
biochemical transformations caused by, 127–9
biomass yields of, xii, 55, 61, 104, 119–21, 457
BIOMET model study of growth of, 44
bulk specific volume of, 150

Water—contd.
  hyacinth—contd.
    calcium requirements for, 112–13
    carbon requirements of, 106–7
    cellulase enzymic action on, 373
    chopping of, 151
    composition of biomass, 358, 362, 425, 447
    composition of digested material, 447
    conceptual design BTM system study for, 55, 60–3
    containment of, 60
    conversion of, 61, 63
    co-production with pennywort, 60, 125–6
    crop growth models used in BIOMET, 25–31, 32
    culture media used, 116–18
    detrital supply of nutrients for, 122–3
    digester effluents used for growth of, 121, 130, 451–3
    distribution of, 103
    effluents treated with, 120–1, 126–7, 130, 131, 133–4, 451–3
      dilution effects, 452–3
    environmental effects on, 114–16
    eutrophic lake water used for growth of, 119–20
    extrusion of, 151–2
    fermentability of, 358, 360, 362
    flow diagrams for BTM systems, 64
    foliar fertilization of, 122
    growing conditions for, xiii, 103
    handling system for, 150
    harvesting frequency, effect on, 124–5
    harvesting methods for, 61, 62, 145–50
      combine system, 149–50
      Zellwood system, 145–9
    hydrodynamic properties of, 143–5
    in-situ properties of, 141–5
    magnesium requirements for, 112–13
    management practices/system for, 123–7

Water—contd.
  hyacinth—contd.
    methane yield of, 298, 299
    microbial components in digester, 422–7
    nutritional status monitored, 426
    monoculture vs. polyculture of, 125–6
    morphology of, 105–6
    nutrient effect on, 109–10
    nitrogen fixation as nutrient supply for, 123
    nitrogen requirements for, 28, 29, 30–1, 107–10
    nutrient medium used for growth of, 119
    nutrient requirements of, 106–14
    particle size reduction of, 151–2
    phosphorus requirements for, 28, 29–30, 110, 111
    physical-mechanical properties of, 142
    plant density effects on, 124
    potassium requirements for, 110–11, 112
    production in
      lakes/streams, 126
      sewage treatment systems, 126–7, 451
    propagation of, 105–6
    sedimental supply of nutrients for, 122
    sodium requirements for, 111–12
    solar radiation effects on, 114–15
    supply of nutrients for, 122–3
    temperature effects on, 116
    trace elements required for, 113–14
    transportation of, 61, 72–3, 152–3
    uses of, 103
    volume reduction of, 150–2
    winter growth of, 105, 113, 116, 125
    yields quoted, xii, 55, 61, 104, 119–21
  lettuce, biomass yield of, 104
  treatment by water hyacinths, 21, 103, 131–4, 135

Winter growth/survival
  cassava, 243
  dogfennel, 273
  eastern gamagrass, 269
  energy cane, 268
  eucalyptus trees, 280
  Leucaena leucocephala, 281
  Napiergrass, 94, 95, 100, 172, 173
  pennywort, 60, 105, 125
  water hyacinth, 105, 113, 125
Wood, Dr F. Aloysius, vii–viii
Wood chips
  cost compared with plastic media, 431
  fixed film reactor packing use, 431, 437–40
Woody crops
  biomass yields for, 287, 457
  culture effects on, 282–4
  digestion of, 337–9
    biochemical methane potential studies, 338–9
    long-term studies, 338
  fermentability of, 337–9, 360
  production systems evaluated for, 278–9, 284–6
  research required, 288, 463
  selection for biomass production, 277–8

Yields
  biomass
    agricultural waste biomass, 8
    aquatic plants, 104, 456
    cattails, 9, 104, 457
    climatic effects on, 9
    energy cane, 84, 179, 267, 268, 457
    energy crops, 8–9
    Erianthus, 267

Yields—*contd.*
  biomass—*contd.*
    gamagrass, 267
    grasses, xii, 267, 456; *see also* individual crops
    herbaceous crops, 238
    improvements in, 456, 457
    marine algae, 249, 252–4
    minimum required, 80, 460
    Napiergrass, xii, 55, 56, 92, 93, 95, 98–9, 172, 457
      hybrid tissue-culture plants, 164, 165
    production costs affected by, 81, 457, 459, 460
    research target for, 458, 459
    root crops, 238
    sorghum, 85–6, 176, 457
    sugarcane, 84, 91, 178, 179, 457
    water hyacinth, xii, 55, 61, 104, 119–21, 457
    woody crops, 281, 283, 287, 457
  methane from biomass
    aquatic plants, 298, 299
    grasses, 299
    herbaceous crops, 299
    improvements in, 456
    marine crops, 299
    Napiergrass, 299
    research target for, 458, 459
    root crops, 298, 299
    sweet potato, 298
    water hyacinth, 298, 299

Zellwood harvesting system (for water hyacinths), 145–9
  conveyor used, 147
  rake used, 147
  winch system used, 146

Winter growth/survival
  cassava, 243
  dogfennel, 273
  eastern gamagrass, 269
  energy cane, 268
  eucalyptus trees, 280
  Leucaena leucocephala, 281
  Napiergrass, 94, 95, 100, 172, 173
  pennywort, 60, 105, 125
  water hyacinth, 105, 113, 125
Wood, Dr F. Aloysius, vii–viii
Wood chips
  cost compared with plastic media, 431
  fixed film reactor packing use, 431, 437–40
Woody crops
  biomass yields for, 287, 457
  culture effects on, 282–4
  digestion of, 337–9
    biochemical methane potential studies, 338–9
    long-term studies, 338
  fermentability of, 337–9, 360
  production systems evaluated for, 278–9, 284–6
  research required, 288, 463
  selection for biomass production, 277–8

Yields
  biomass
    agricultural waste biomass, 8
    aquatic plants, 104, 456
    cattails, 9, 104, 457
    climatic effects on, 9
    energy cane, 84, 179, 267, 268, 457
    energy crops, 8–9
    Erianthus, 267

Yields—*contd.*
  biomass—*contd.*
    gamagrass, 267
    grasses, xii, 267, 456; *see also* individual crops
    herbaceous crops, 238
    improvements in, 456, 457
    marine algae, 249, 252–4
    minimum required, 80, 460
    Napiergrass, xii, 55, 56, 92, 93, 95, 98–9, 172, 457
      hybrid tissue-culture plants, 164, 165
    production costs affected by, 81, 457, 459, 460
    research target for, 458, 459
    root crops, 238
    sorghum, 85–6, 176, 457
    sugarcane, 84, 91, 178, 179, 457
    water hyacinth, xii, 55, 61, 104, 119–21, 457
    woody crops, 281, 283, 287, 457
  methane from biomass
    aquatic plants, 298, 299
    grasses, 299
    herbaceous crops, 299
    improvements in, 456
    marine crops, 299
    Napiergrass, 299
    research target for, 458, 459
    root crops, 298, 299
    sweet potato, 298
    water hyacinth, 298, 299

Zellwood harvesting system (for water hyacinths), 145–9
  conveyor used, 147
  rake used, 147
  winch system used, 146

Water—*contd.*
  hyacinth—*contd.*
    calcium requirements for, 112–13
    carbon requirements of, 106–7
    cellulase enzymic action on, 373
    chopping of, 151
    composition of biomass, 358, 362, 425, 447
    composition of digested material, 447
    conceptual design BTM system study for, 55, 60–3
    containment of, 60
    conversion of, 61, 63
    co-production with pennywort, 60, 125–6
    crop growth models used in BIOMET, 25–31, 32
    culture media used, 116–18
    detrital supply of nutrients for, 122–3
    digester effluents used for growth of, 121, 130, 451–3
    distribution of, 103
    effluents treated with, 120–1, 126–7, 130, 131, 133–4, 451–3
    dilution effects, 452–3
    environmental effects on, 114–16
    eutrophic lake water used for growth of, 119–20
    extrusion of, 151–2
    fermentability of, 358, 360, 362
    flow diagrams for BTM systems, 64
    foliar fertilization of, 122
    growing conditions for, xiii, 103
    handling system for, 150
    harvesting frequency, effect on, 124–5
    harvesting methods for, 61, 62, 145–50
      combine system, 149–50
      Zellwood system, 145–9
    hydrodynamic properties of, 143–5
    in-situ properties of, 141–5
    magnesium requirements for, 112–13
    management practices/system for, 123–7

Water—*contd.*
  hyacinth—*contd.*
    methane yield of, 298, 299
    microbial components in digester, 422–7
    nutritional status monitored, 426
    monoculture vs. polyculture of, 125–6
    morphology of, 105–6
    nutrient effect on, 109–10
    nitrogen fixation as nutrient supply for, 123
    nitrogen requirements for, 28, 29, 30–1, 107–10
    nutrient medium used for growth of, 119
    nutrient requirements of, 106–14
    particle size reduction of, 151–2
    phosphorus requirements for, 28, 29–30, 110, 111
    physical-mechanical properties of, 142
    plant density effects on, 124
    potassium requirements for, 110–11, 112
    production in
      lakes/streams, 126
      sewage treatment systems, 126–7, 451
    propagation of, 105–6
    sedimental supply of nutrients for, 122
    sodium requirements for, 111–12
    solar radiation effects on, 114–15
    supply of nutrients for, 122–3
    temperature effects on, 116
    trace elements required for, 113–14
    transportation of, 61, 72–3, 152–3
    uses of, 103
    volume reduction of, 150–2
    winter growth of, 105, 113, 116, 125
    yields quoted, xii, 55, 61, 104, 119–21
  lettuce, biomass yield of, 104
  treatment by water hyacinths, 21, 103, 131–4, 135